Enterprise Architecture with SAP®

PRESS

Anup Das, Peter Klee, Johannes Reichel

With forewords by Thomas Saueressig and André Christ

Enterprise Architecture with SAP®

Planning, Management, and Transformation

Editor Megan Fuerst
Acquisitions Editor Hareem Shafi
Copyeditor Doug McNair
Cover Design Graham Geary
Photo Credit Shutterstock: 532961026/© fanjianhua
Layout Design Vera Brauner
Production Graham Geary
Typesetting SatzPro, Germany
Printed and bound in Canada, on paper from sustainable sources

ISBN 978-1-4932-2573-6

1st edition 2025

Library of Congress Cataloging-in-Publication Control Number: 2024044380

Contents at a Glance

Contents

Part II SAP Enterprise Architecture Use Cases and Patterns

Foreword by Thomas Saueressig

Imagine a colossal ship navigating waters without a map or compass. This is the reality for many businesses without a solid enterprise architecture practice. Just as sailors rely on their navigational tools to reach their destination, modern organizations depend on enterprise architecture to steer through the complexities of today's digital landscapes.

In a fast-changing world where new business models emerge rapidly and the boundaries of industries blur, businesses face unprecedented challenges. New technologies, such as business AI, are disrupting how business is done. Sustainability has become a key priority for our future, and geopolitical challenges are impacting supply chains. Engaging in mergers, acquisitions, and divestitures (MAD) has become a growth strategy, and consumer behavior and expectations are constantly changing. This ever-evolving environment requires businesses to continuously transform themselves to stay ahead.

The relevance of enterprise architecture has never been greater for our customers and partners, as well as for SAP. With the introduction of RISE with SAP, which is more than a technical migration and rather is a comprehensive solution for business transformation, enterprise architecture becomes even more critical. Enterprise architecture is key to providing direction and clarity, bridging business and IT, and translating strategy into architectural transformation. Consequently, the enterprise architect has gained paramount importance as a key player in any transformation.

Enterprise Architecture with SAP: Planning, Management, and Transformation is a comprehensive guide for organizations navigating the complexities of digital transformation. The book introduces SAP Enterprise Architecture Framework and offers detailed insights into its methodology, reference content, tools, and practices. The book covers a wide range of architectural use cases based on experience, and it explores the application of SAP LeanIX. By combining theoretical foundations with real-world examples, this book equips organizations to effectively implement and manage their own enterprise architecture practice.

Let's set sail on this exciting adventure!

Thomas Saueressig
Member of the Executive Board of SAP SE, Customer Services & Delivery

Foreword by André Christ

The success and future viability of companies today depends more than ever on their ability to anticipate future disruption, innovate, and, above all, adapt to change. Because these abilities are rooted in technology, IT leaders face increasing pressure to find the perfect balance between promoting agility and reducing costs and minimizing risks in various technology landscapes. Unfortunately, the incredible complexities of these landscapes get in the way. The visible part of IT—the applications people use on a daily basis—is just the tip of the iceberg. The real complexity lies below the surface.

Technology landscapes, especially in mature organizations, are characterized by a high level of functional complexity, a huge number of interfaces with mutual dependencies, and a continuously expanding volume of data. Driving transformation in these landscapes requires transparency. Without insight into the technology you own and how it all interconnects, transformation becomes either impossible or extremely painful. At the end of the day, you can only change what you understand.

This is where enterprise architects come in. We have been fortunate over the last few years to witness the remarkable rise of enterprise architects from niche specialists in IT to strategic partners for the entire business. Modern enterprise architecture tools—such as SAP LeanIX, with its data-driven, automated approach that is now enhanced with AI capabilities—have played a crucial role in this transformation of the discipline. Today, enterprise architects can provide business leaders and other stakeholders a comprehensive overview of the technology landscape that takes account of its full complexity. At the same time, they can present this information in the most relevant business context, enabling better collaboration across the company by creating a common language for business and IT.

Collaboration is critical for transformation. As it turns out, collaboration is a hallmark of the enterprise architect community, which is characterized by the active and open sharing of tools, approaches, and experiences. This book, with all its valuable insights, illuminating use cases, and practical frameworks, is further evidence of the remarkable willingness of enterprise architects to grow and do great work together.

André Christ
Cofounder and General Manager of SAP LeanIX

Preface

You may look at this book and wonder, "Why another book on enterprise architecture—and why in connection with SAP?"

We agree that the story of enterprise architecture has been told many times, and sometimes, the practice still has an ivory tower connotation, even though it has become widely accepted. But we've set out to tell a different story, one of active enterprise architecture transformation, where the practice is not isolated but integrated and stringently laid out.

And why the SAP perspective? Customers often ask their primary software partner, SAP, for guidance in their transformations. We believe that enterprise architecture with SAP can make a difference if and when the transition team uses a stringent framework and methodology with a strong and comprehensive SAP reference architecture, as well as an end-to-end toolchain around SAP LeanIX.

At SAP, we've realized that we need to better support our customers' transformations and provide the relevant tooling, and that has led SAP to acquire Signavio and LeanIX. In this book, we'll cover the tooling perspective with a special focus on SAP LeanIX, but before that, we'll show the relevance of enterprise architecture with SAP by investigating typical SAP-related transformation patterns.

The subtitle of this book, *Planning, Management, and Transformation*, stands for the evolution of enterprise architecture and the corresponding mindset. The details of these three components of enterprise architecture evolution are as follows:

- **Planning**
  This is the traditional approach: put a plan on the wall and be done with it. We've seen enterprise architecture practices proudly point to a poster on their wall when asked if they'd modeled their business capabilities, but we rarely hear that these organizations then use these models to align their business with their quarterly priorities.
- **Management**
  With management, we mean further developing the plan with all domain perspectives considered. It's the attempt to make an enterprise architecture practice big, comprehensive, and self-contained, to incorporate all domains and drive decisions. In our view, efforts can get too complex, and agility can possibly be lost by adopting frameworks like The Open Group Architecture Framework (TOGAF) in an overly academic way. Internally, organizational complexities can be other root causes of why these architecture practices tend to become less relevant to other parts of the organization.

- **Transformation**
  Enterprise architecture needs to be understood as an integrated discipline, rather than an ivory tower where a chief architect drives roadmaps from the top down. Its practice requires collaboration with change management functions, process management, data offices, and platform teams, and it also requires an orchestrated implementation. Architects, in this function, don't claim mandates, but rather, they provide clarity and help lead transformation in the right direction.

This is not to say that core enterprise architecture disciplines and techniques like business capability modeling or application rationalization are no longer relevant. On the contrary! Their impact is even greater if they are executed in concert with others and thus deliver on the promise to align all key stakeholders.

Forward-Looking Statements

This book may include predictions and projections about future events. These statements are based on current expectations and are subject to uncertainty. While we will discuss potential future trends, we do not make any claims regarding planned functionalities from SAP, as these are subject to change. No promises are made on behalf of SAP.

Who This Book Is For

This book is a plea for change. It's designed for enterprise architects, SAP professionals, and C-level executives who are navigating the complexities of modern IT landscapes. It is designed to help these three groups in the following ways:

- **Enterprise architects**
  For those new to enterprise architecture, this book provides a comprehensive guide to key methodologies, frameworks, and best practices, making it an invaluable resource to help them build a solid foundation in the discipline. Seasoned enterprise architects—both within and outside of the SAP ecosystem—will find this book an insightful update on SAP's latest developments in the field. We dive into SAP's approach to enterprise architecture and how to apply it through real-world examples, showcasing best practices that ensure success in transformation initiatives.
- **SAP professionals**
  This book serves SAP professionals who are either looking to transition into an enterprise architecture role or seeking a deeper understanding of how their work fits into broader enterprise architecture strategies. If you're an SAP professional, this book will help you see how your expertise aligns with enterprise-wide planning and decision-making, offering you the tools to evolve your role and collaborate more effectively with enterprise architects.

- **C-level executives**
  This book also speaks directly to today's forward-thinking C-level executives, especially CIOs who have risen through the ranks and possess a hands-on understanding of required strategic interventions. Modern CIOs recognize the importance of enterprise architecture, especially when integrated with SAP, as a key enabler of business transformation. We aim to equip them with the insights to champion these ideas within their organization so that they can lead with innovation.

As we mentioned before, many organizations still have IT departments that grew organically around specific platforms and solutions and often operate with competing agendas. This book advocates for change. By fostering a unified approach—one that brings together SAP and non-SAP teams under a cohesive, collaborative practice—companies can unlock new opportunities for innovation and growth.

How to Read This Book: The Reading Metro Map

We hope to win over readers of all backgrounds and different motivations. We'd like to give you the best reading experience as you navigate this book. We strongly recommend that everyone start out with Chapter 1. This will give you an initial sense of whether you would like to follow the canonical flow. Alternatively, we suggest the following alternative reading sequences outlined in our *reading metro map* in Figure 1.

Let's explore this metro map so you can prepare for your alternative journeys through the book:

- Blue line: Enterprise architecture use cases and patterns
  In Part II, Chapter 5 to Chapter 10 will describe a wide range of practical, SAP-centric use cases. This can be the right part to start reading if some of these use cases match current challenges in your company.

  As the use cases themselves refer to SAP Enterprise Architecture Framework (SAP EA Framework), you'll get the most from these chapters if you have a decent understanding of enterprise architecture. Depending on your depth of enterprise architecture knowledge, we recommend returning to Part I when you're done with the use cases to strengthen your foundation in the specifics of SAP EA Framework.
- Gold line: Enterprise architecture practice (including the SAP perspective)
  Chapter 15 is the starting point of your tour and goes hand in hand with Chapter 3, where we highlight relationships to other functions. We recommend this approach to experienced enterprise architects who have sufficient exposure to the SAP world.

  If you wish to deepen your knowledge of SAP EA Framework, Chapter 2 is your next best chapter.

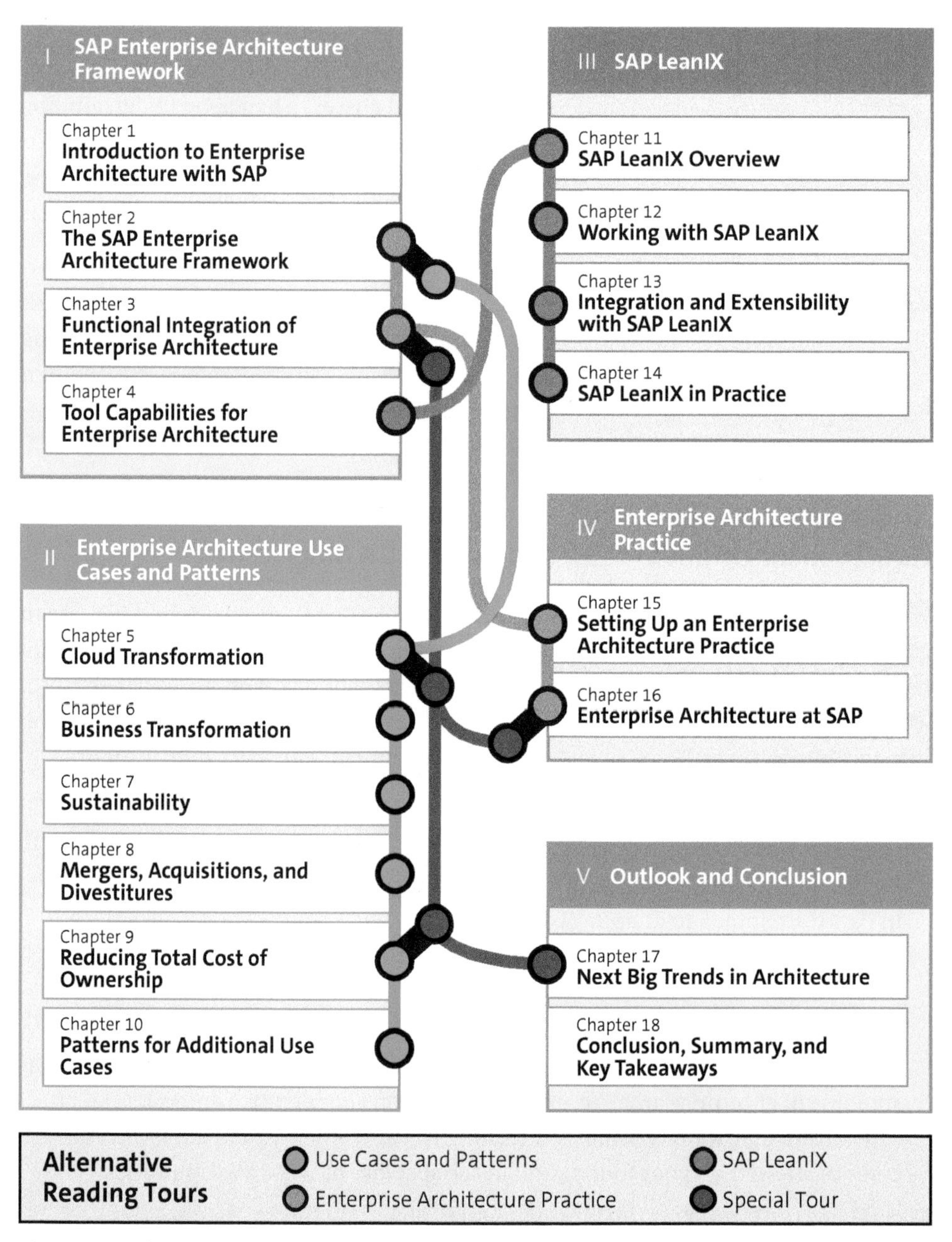

Figure 1 Reading Metro Map

- Orange line: SAP LeanIX
  If you'd like a deep dive into SAP LeanIX, the Part III stops on the Orange line are yours. If you're about to select an enterprise architecture tool or would first like to get a generic understanding of the importance of enterprise architecture tools, hop on at Chapter 4. But as a word of caution, a tool will stay just that as long as you have

a tool-only focus, rather than a targeted approach and methodology and/or a well-organized practice. So, don't forget to take a ride later on the Gold line.

- **Purple line: Special tour**
  These special stops might satisfy your curiosity or meet your current needs:
  - Chapter 3 is worth a read, as it looks at the boundaries of enterprise architecture. This can be of special interest from a CIO perspective, where there may be a need to reposition internal IT functions in the light of an upcoming transformation.
  - Chapter 5 covers SAP's cloud architecture and the clean core principle. *Clean core* initially originated in the context of new extensibility approaches, but today, the term is also used for data, processes, integrations, and operations.
  - Chapter 16 sheds light on different architecture practices at SAP that bring different perspectives: product development, internal IT, services, presales, and tool development.
  - Chapter 17 describes current trends and outlooks. As generative AI in enterprise architecture has already arrived, we'll not just focus on AI but also consider other perspectives. To see how AI is already used in SAP LeanIX, visit Chapter 12 on the Orange line.
- **Free ride: Hop on and hop off**
  Travel through the network by switching lines and hop on and hop off as you wish.

Acknowledgments

The thoughts and ideas that make up this book come from various sources, and we need to thank many people. Without our conversations with them, this project would not have been possible.

In the global SAP enterprise architecture community, we thank Andreas Poth (co-owner of SAP Enterprise Architecture Methodology) and Ivana Trickovic. They are the ones who've been driving the conversation from the product engineering perspective for many years.

From our own SAP Transformation Hub team, we'd like to thank Darius Golshani, who not only helped build the team from day one but, with his continuous focus on scaling and modularization, had a large part in bringing us to where we are today.

Working as a global SAP Transformation Hub team is a daily inspiration, and the way this team reinvents former architecture approaches is unmatched! This is why we'd like to thank the entire team for building such great content that, ultimately, was a stimulus for this book.

Thank you also to Christian Langpape. Your leadership and guidance to the team were way beyond anywhere near the norm!

We'd also like to thank Thomas Walther for his continuous support and sponsorship of our writing this book.

A special mention, too, for Joern Bartelheimer, Rene de Daniel, Martin Effenberger, Stefan Fassmann, Andrey Hoursanov, Marko Korhonen, Marco Michel, Sven York Pohl, Peter Schmidt, and Kai Wussow.

With LeanIX joining the SAP family, we can acknowledge the entire team and thank them for their willingness to collaborate and innovate the tooling in an agile way. A big shout-out to André Christ and his team!

Another relentless enterprise architecture activist is Paul Kurchina. He brings the community together and has cheered us on in our work on this book from the first minute.

We can't number or name the multitude of interactions we've had with so many great customers, who always challenge us, make us rethink our approach, and help evolve it.

We'd also like to thank the team at SAP PRESS. They were always there to answer whatever questions we had and were part of our journey from the beginning.

Finally, a big thank-you goes to our families, who supported us and made do without us for long periods of time.

I (Peter) am deeply grateful to my wife, Constanze, for her steady moral support and for dedicating considerable time to reviewing and improving my writing.

My (Johannes') heartfelt thanks go to my wife, Barbara, who kept our kids so entertained that they didn't even notice me hiding with my laptop on weekends and holidays. Your persistent encouragement made it all possible!

From my (Anup's) side, I want to express my gratitude to my wife, Suhita, for her constant encouragement of my dream of writing a book one day. She kept me grounded so I could get this to the finish line.

– **Anup**, **Peter**, and **Johannes**

Tell Us What You Think!

Our aim in this book is to enable change, and there surely is no better way to do so than by fostering exchange on many different levels:

- In the growing external community of SAP enterprise architects.
- In direct dialog with our customers.
- Internally at SAP, where we've come a long way but are far from being finished!
- With you as readers. We treasure your feedback, both on the book and on your approach to maturing your enterprise architecture practice.
- At SAP events or other transformation and architecture-related conventions of different sizes.

Now hop on board, and let's begin the journey!

PART I

SAP Enterprise Architecture Framework

This part introduces enterprise architecture as it relates to SAP-centric landscapes. You'll learn about the importance of enterprise architecture transformation and its challenges. After an introduction to the SAP-developed framework for architecture design, you'll explore the tools that can be used, and how specific business functions are integrated with architecture planning and design.

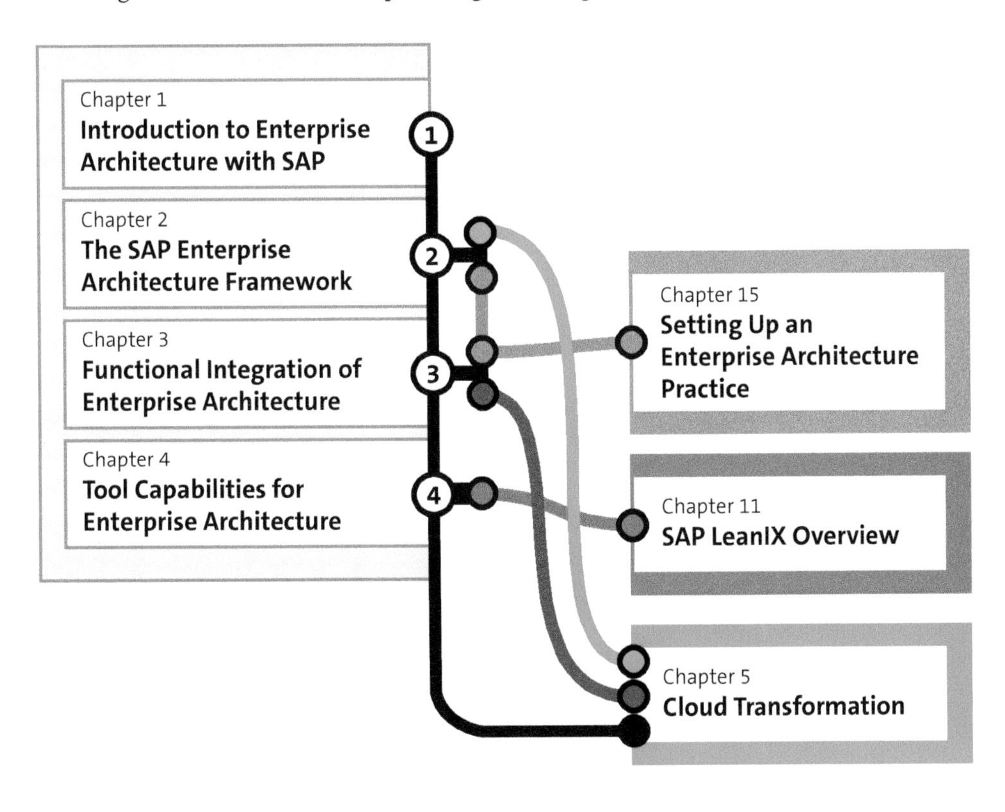

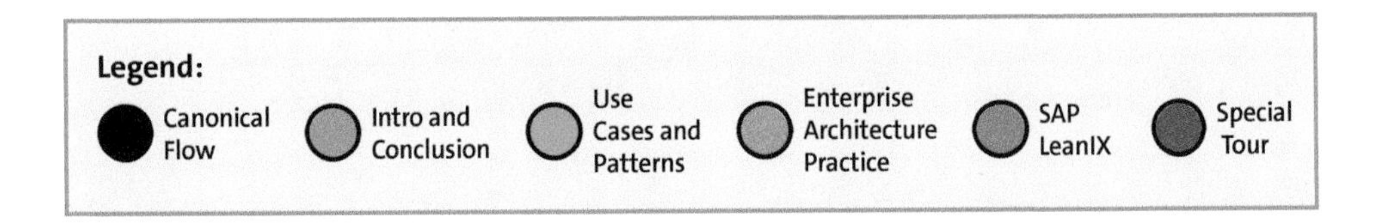

Chapter 1
Introduction to Enterprise Architecture with SAP

Is your organization's architecture foundation as solid as you think, or are there cracks hidden beneath the surface? In this chapter, we delve into enterprise architecture, revealing its essential role in not only supporting but also driving modern business strategies. Join us as we explore how this critical discipline is reshaping itself to meet the demands of a rapidly evolving SAP landscape.

In the late twelfth century, a majestic bell tower began to rise as part of a cathedral complex in a prosperous Italian city, aimed at showcasing the region's wealth and engineering expertise. The project became famous not just for its striking Romanesque design but also for its notorious tilt. The foundation of the tower was laid in soft soil, which could not properly support its weight. Over the decades, as more levels were added, the tower began to lean dangerously, turning what was meant to be a magnificent bell tower into a symbol of flawed planning.

The underlying problem of today's famous Leaning Tower of Pisa was a failure to adequately assess and prepare the foundation—the soil itself—before proceeding with construction. Despite subsequent stabilization efforts, the tower remains a classic example of how foundational oversights can lead to ongoing problems and require costly corrections.

In the realm of enterprise architecture, this mirrors the risk businesses face when they rush to deploy new IT systems or expand their IT footprint without first ensuring a solid foundation and considering whether there is a strong strategic business and IT alignment. If the basic architecture isn't suited to the scale or type of business operations, or if it can't support emerging technologies, then the entire IT structure may become unstable, like a tower built on weak soil. Such oversight can lead to inefficiencies, increased costs, and lost opportunities, continually requiring adjustments and fixes to align with business needs and capabilities.

This chapter will explore the fundamental aspects of enterprise architecture, tracing such architecture's evolution from a purely technical framework to a strategic imperative for businesses today. We'll give special attention to the advancing role of the enterprise architect and the ever-growing relevance of enterprise architecture within evolving SAP landscapes. As we navigate through these topics, we'll also address the common challenges and obstacles that organizations face in implementing effective enterprise architectures.

1.1 What Is Enterprise Architecture?

It's challenging to find a standard definition for the term *enterprise architecture*, not only due to the inherent complexity of the field it describes but also due to the diversity of perspectives on that field. There are myriad definitions that are shaped by different authors' academic and professional backgrounds. This variability often results in descriptions that are either highly theoretical or rich in analogies, which, while helpful for conceptual understanding, can obscure the practical elements of the discipline. Despite being an established field, enterprise architecture elicits widely varying responses from professionals who are asked to define it, reflecting its adaptation to different business needs and technological environments over time.

Furthermore, enterprise architecture is sometimes viewed skeptically as an "ivory tower"—a kingdom of high-level strategic decision-making disconnected from the day-to-day operational realities of a business. This perception further complicates enterprise architecture's definition, as it must bridge the gap between strategic oversight and practical, actionable architecture. Furthermore, enterprise architecture is frequently associated with common misconceptions like those illustrated in Figure 1.1, leading to some considering it to be a complex endeavor with unclear practical relevance.

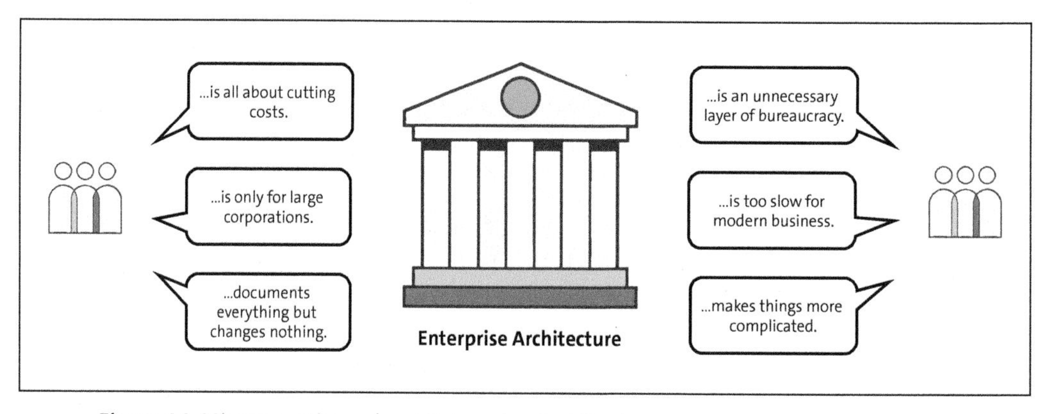

Figure 1.1 Misconceptions about Enterprise Architecture

Let's follow the common approach to defining enterprise architecture and begin with an analogy. Think of city planning, in which how streets are laid out, areas are zoned, and public services are placed shapes the whole city's functioning and flow. Similarly, enterprise architecture organizes a company's digital and operational framework. Just like good city planning can improve traffic, make services easier to access, and boost economic growth, well-designed enterprise architecture helps information move smoothly, aligns technology with business goals, and prepares the organization for future changes and growth.

Enterprise architecture and city planning are frequently compared because both disciplines involve designing complex systems with multiple interconnected components that must function together in harmony. The following are general points of comparison:

- **Strategic planning**
  Both city planning and enterprise architecture involve detailed planning and design to align various components (like roads and IT systems) with broader goals, ensuring efficiency and scalability.
- **Integration and interoperability**
  Just as city systems (e.g., transportation, utilities) must function cohesively, enterprise architecture requires diverse IT systems and business processes to work seamlessly together.
- **Adaptability to change**
  City planning and enterprise architecture must both be flexible, adapting to changes such as population growth or shifting business strategies to remain effective and relevant without rebuilding everything from scratch.

The analogy between city planning and enterprise architecture originates from earlier days, when the focus and pace of change in IT and business environments were different and large enterprise systems were indeed planned like cities, with the emphasis on operational stability over speed of innovation. Those systems have tended to run over decades, and while they've provided the necessary stability to support growing businesses, they've increasingly inhibited rapid responses to the ever-changing business environments and technology advancements that we've observed over time. Therefore, this analogy no longer holds as effectively as before, due to evolving demands on enterprise architecture, such as the following:

- **Acceleration of technological change**
  Modern enterprise architecture has to adapt to the rapid pace of technological innovation, where change occurs almost continuously. This contrasts sharply with city planning, where developments are planned over years or even decades.
- **Agility and scalability**
  Today's enterprise architecture must be inherently agile and scalable to respond quickly to business needs and opportunities. This need for agility is at odds with the relatively static and long-term nature of city planning, where adjustments and scalability are constrained by many factors.
- **Ecosystem dynamics**
  While city planning traditionally centers on the development of a single entity (the city), enterprise architecture now needs to consider the interconnectedness of various and steadily changing ecosystems (partners, suppliers, competitors, etc.).

- **Dynamic governance models**
  While city planning typically relies on hierarchical governance structures with centralized decision-making, enterprise architecture increasingly adopts more dynamic and decentralized governance models.

In our experience, the enterprise architecture process needs to consist of the following three main building blocks: planning, management, and transformation (see Figure 1.2).

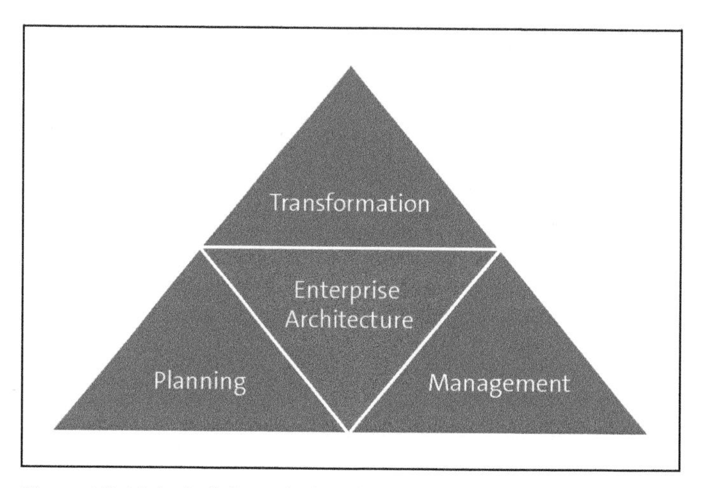

Figure 1.2 Main Building Blocks of Enterprise Architecture

Let's take a closer look at each building block:

- **Enterprise architecture planning**
  Effective enterprise architecture begins with robust planning. This involves creating comprehensive models and roadmaps that align the organization's IT capabilities with its strategic business goals. However, planning goes beyond merely developing these models; it requires active engagement with business stakeholders to ensure these plans are not only visually represented but also practically implemented and updated regularly. Successful enterprise architecture planning involves continuous collaboration and alignment with business units to prioritize initiatives and adapt to changing market conditions, ensuring that the architectural vision is both relevant and actionable.
- **Enterprise architecture management**
  Managing enterprise architecture entails overseeing the execution and evolution of the established plans in all domains of the organization. This aspect requires a holistic approach, taking into account the interdependencies among different systems, processes, and business units. While comprehensive frameworks like The Open Group Architecture Framework (TOGAF) can provide structure, it's crucial for enterprise architecture management to avoid becoming overly bureaucratic or isolated from the rest of the organization. Effective enterprise architecture management

focuses on maintaining relevance and agility, ensuring that architectural decisions are informed by and integrated with the broader business context, thereby driving meaningful progress and value.

- **Enterprise architecture transformation**
  Transformation in enterprise architecture is about dynamically adapting to changing business needs and technological advancements. This requires a shift from traditional, siloed approaches to a more integrated and collaborative model. Enterprise architects must work closely with various functions such as change management, process management, data governance, and other teams to orchestrate comprehensive solutions. By fostering strong interdepartmental collaboration and maintaining flexibility, the enterprise architecture practice can effectively guide the organization through continuous innovation and transformation, ensuring that the architecture evolves in tandem with business objectives and market demands.

In the following sections, we'll explore the increasing importance of enterprise architecture in SAP landscapes and the evolution of the discipline in response to the unprecedented revolution brought by artificial intelligence (AI).

Is Enterprise Architecture an Art?

The question of whether enterprise architecture is an art or a teachable skill has sparked many debates. While some argue that it requires inborn creativity, many years of experience, and intuition similar to artistry, we, as authors of this book, firmly believe that enterprise architecture can be learned. Our conviction stems from the belief that with a robust methodology, comprehensive reference content, and adequate tool support, individuals can acquire the necessary skills to excel in enterprise architecture. In this book, we aim to demonstrate how these elements combine to empower practitioners to master enterprise architecture and effectively navigate complex organizational landscapes.

1.2 How Is Enterprise Architecture Relevant in SAP Landscapes?

Enterprise architecture is essential for navigating the complexities of transformation, particularly in shifting from legacy systems to modern, streamlined, and agile environments. Traditionally, customer landscapes are burdened with heterogeneous, often siloed systems accumulated over years, creating a convoluted as-is state that hampers flexibility and speed. These legacy systems typically suffer from high maintenance costs and are resistant to change, which starkly contrasts with the dynamic needs of modern business. Enterprise architecture addresses this by introducing a structured methodology for transitioning from a complex as-is state to a future to-be state that emphasizes a clean-core approach. This transformation not only simplifies the IT landscape but also aligns it more closely with business goals, enabling companies to respond swiftly to market changes and technological advancements with greater efficiency and less disruption.

We discuss some drivers of these transformations in the following list, and we'll revisit them in the use case chapters:

- **Business transformation**
  Enterprise architecture is essential to aligning business strategy with IT infrastructure, which is particularly crucial for SAP customers undergoing business transformation. Enterprise architecture helps ensure that every technological investment and decision supports the overarching business goals, facilitating transformations that are both scalable and sustainable. For companies utilizing SAP, this alignment is key to leveraging the full potential of SAP solutions, ensuring that these tools drive business objectives forward effectively.
- **Cloud transformation**
  With the shift toward cloud environments, SAP customers find themselves needing to adapt to cloud architectures that offer flexibility, scalability, and cost effectiveness. Enterprise architecture plays a crucial role in this transition through the design of frameworks that not only integrate seamlessly with SAP's cloud solutions but also align with the company's long-term cloud strategy. This integration enables businesses to achieve enhanced performance and agility, preparing them for a future where cloud computing holds a central role.
- **Maintaining a clean core**
  The concept of a *clean core*—keeping the core SAP system lean and free of modifications—ensures that updates and upgrades are simpler and less disruptive. Enterprise architecture is integral to maintaining a clean core because it facilitates the oversight of the architecture and ensures that customizations and extensions are built outside the core system, on platforms like *SAP Business Technology Platform* (SAP BTP). This approach reduces complexity and streamlines processes, thereby enhancing system stability and agility.
- **Leveraging AI**
  As AI becomes more embedded in business processes, SAP customers stand to gain significantly from AI-driven insights and automation. Enterprise architects are at the forefront of integrating AI capabilities with existing SAP systems, facilitating smarter, data-driven decisions that propel businesses forward. By designing architectures that support AI integration, enterprise architects help SAP customers harness the power of AI for everything from predictive analytics to automated business processes.
- **Driving sustainability**
  Sustainability is becoming a mandate rather than a choice, and SAP customers are looking to incorporate green practices into their core operations. Enterprise architecture supports these initiatives by designing systems that optimize resource use and reduce waste. This could involve implementing SAP's sustainability solutions,

which track carbon emissions and other sustainability metrics, thereby enabling companies to achieve their environmental goals effectively.

- **Total cost of ownership**
  The *total cost of ownership* (TCO) is a critical factor in any SAP implementation. Enterprise architects help minimize TCO by designing efficient, scalable architectures that reduce the need for expensive customizations and ensure that future upgrades are less costly and disruptive. By focusing on a clean core and leveraging cloud solutions, enterprise architects contribute to a reduction in both direct and indirect costs associated with maintaining and upgrading SAP systems.
- **Adapting to future trends**
  As the digital landscape evolves, new trends will inevitably emerge. Enterprise architects need to stay ahead of these trends, whether they be future waves of digital innovation, evolving cybersecurity threats, or new regulatory requirements. For SAP customers, enterprise architects are essential in ensuring that the architecture can adapt and evolve, thus enabling the business to stay competitive and relevant.

In conclusion, enterprise architecture is not just relevant; it is a strategic imperative for SAP customers. By ensuring alignment between business goals and IT infrastructure, facilitating cloud and AI integrations, and supporting sustainability and cost efficiency, enterprise architects help shape businesses that are not only prepared for the future but also capable of leading the charge in innovation.

Does SAP Mandate a Dedicated Enterprise Architecture Framework?

SAP doesn't mandate the use of a specific enterprise architecture framework; many of our customers opt for common frameworks such as TOGAF. However, for landscapes primarily centered on SAP solutions, we recommend *SAP Enterprise Architecture Framework* (SAP EA Framework). This framework is specifically designed for practitioners, offering a structured methodology, comprehensive reference content, and robust tool support tailored to SAP landscapes. In Chapter 2, we'll define in detail what constitutes an enterprise architecture framework, compare common frameworks, and provide an in-depth introduction to SAP EA Framework, illustrating how it aligns business with IT across all relevant enterprise architecture domains.

1.3 Evolution of Enterprise Architecture and the Enterprise Architect Role

Enterprise architecture has continually evolved to address the expanding complexities of technology in business contexts. This evolution reflects the changing priorities and strategies of organizations as they adapt to new technologies and business demands. Figure 1.3 represents a typical trajectory in the evolution of enterprise architecture, spanning from a narrow IT focus to a broader perspective that includes digital transformation and the current and future AI revolutions in enterprise architecture.

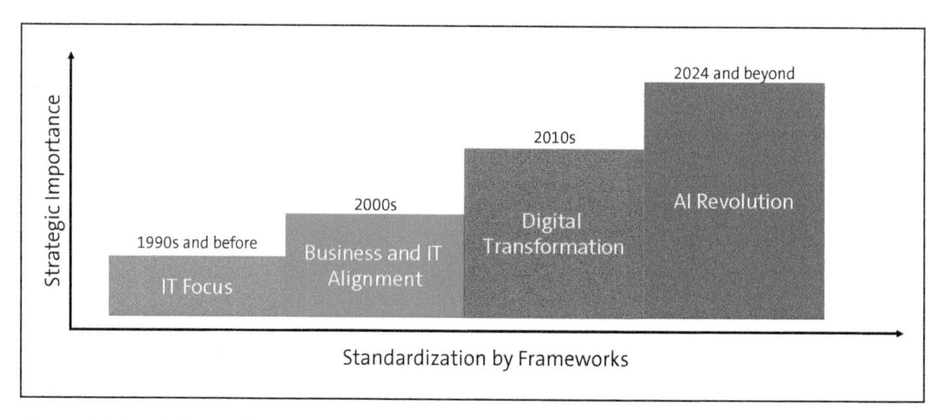

Figure 1.3 Evolution of Enterprise Architecture

The depicted aspects of enterprise architecture are not strictly defined, yet it is evident that the focus of enterprise architecture has shifted over time, gaining strategic significance. Concurrently, the standardization facilitated by enterprise architecture frameworks has also progressed (refer to Chapter 2 for details).

Let's break down the columns shown in Figure 1.3 in more detail:

- **IT focus: The foundation of enterprise architecture**
  The initial phase of enterprise architecture is characterized by a focus on IT. During this stage, the primary goal of enterprise architecture is to streamline and manage the technical infrastructure of an organization. The emphasis is on optimizing the technical landscape, reducing costs related to it, and providing stable operations for the business. The role of the enterprise architect in this phase is focused primarily on maintaining technical proficiency and good IT operations management, and the enterprise architect is tasked with the technical deployment and maintenance of IT infrastructure.
- **Business-IT alignment: Harmonizing goals**
  The alignment of business strategies with IT capabilities represents a mature phase in the evolution of enterprise architecture. Here, the architecture doesn't just support business operations; it actively enables and shapes them. This stage involves a strategic partnership between business leaders and IT executives to ensure that the technological infrastructure directly contributes to achieving business objectives. This also impacts the role of the enterprise architect, who becomes a strategic partner involved in shaping business strategies by working closely with business leaders to ensure that IT initiatives support and drive business objectives.
- **Digital transformation: Toward a future-ready enterprise**
  The next stage in the evolution is digital transformation. This phase extends beyond mere alignment, envisioning a fundamental transformation of business models,

cultures, and customer experiences through digital technologies. Digital transformation involves a comprehensive rethinking of how an organization uses technology in creating value for its stakeholders. In this stage, the enterprise architect evolves to become a visionary leader, guiding the organization through digital innovation and transformation by focusing on leveraging technology to reinvent business processes and create new business models.

- **AI revolution: The next frontier**
  Building on the momentum of digital transformation, the AI revolution represents the next frontier in enterprise architecture. This phase is characterized by the integration of AI and machine learning technologies into all facets of business operations. The objective is to harness the power of AI to enhance decision-making, automate processes, and deliver personalized customer experiences at scale. During this phase, the enterprise architect's role expands to include the orchestration of AI-driven initiatives, ensuring ethical AI deployment, data governance, and continuous innovation. AI-based tools will revolutionize the discipline of enterprise architecture by enabling more dynamic and predictive modeling, real-time analytics, and automated compliance and risk management.

The evolution of enterprise architecture represents a shift from a reactive, IT-centric approach to a proactive, business-driven strategy that fully leverages technological innovations. In Chapter 17, we'll explore emerging trends in enterprise architecture that are likely to influence the role of the enterprise architect. However, it is clear from our discussion that the enterprise architect will assume an increasingly critical role in the fast-evolving business landscape of today.

SAP's Definition of Architecture Roles and the Enterprise Architect

In the context of SAP, an enterprise architect's role is the following: an enterprise architect orchestrates the development of a single holistic enterprise architecture model for the entire organization from a business, applications, data, and technology perspective that is closely aligned with the customer's business strategy and operating model. Besides the enterprise architect, who works across domains, SAP defines roles for the business architect, application architect, data architect, and technology architect.

Therefore, the enterprise architect is responsible for orchestrating all architecture domains, ensuring strategic alignment and seamless interfacing between them. This role includes leading the overall design of the architecture and managing its complexity from design to value realization. Enterprise architects also guide and influence customer relationships and architecture teams toward specific outcomes and goals. Additionally, they oversee the setup and management of architecture practices. Key competencies for this role include leadership, effective communication, orchestration, change management, and a deep understanding of the customer's business environment. Enterprise architects must also be adept in value engineering and have a sales-oriented mindset.

1.4 Challenges and Obstacles When Implementing Enterprise Architecture

Let's consider a hypothetical example. AlphaTech—a well-established market leader in industrial automation solutions—ventured boldly into a major business transformation, aiming to overhaul its enterprise architecture to better support rapid scaling and integration of emerging technologies. Fueled by ambition, it launched an aggressive timeline to adopt cloud-based infrastructure and advanced data analytics. However, the transformation took a disastrous turn. Among its enterprise architecture team, a lack of a detailed understanding of its existing as-is architecture, insufficient experience with planning complex architectures, and limited conceptual knowledge of architecture methods and tools led to frequent mistakes. Critical systems failed to communicate, data breaches occurred, and project milestones were missed. As costs skyrocketed and the initial excitement turned to frustration, AlphaTech's transformation came to a halt.

While the example of AlphaTech is fictional, it underscores a genuine and critical reality: implementing enterprise architecture involves navigating a complex landscape of challenges and obstacles. Each stage of enterprise architecture implementation, from initial assessment to full deployment, requires a careful orchestration of skills, knowledge, and strategic alignment. This narrative highlights the fact that, in practice, enterprises must prepare to address countless hurdles that can significantly impact the success of their enterprise architecture initiatives.

Figure 1.4 highlights several key challenges and obstacles our customers encounter when implementing an enterprise architecture practice.

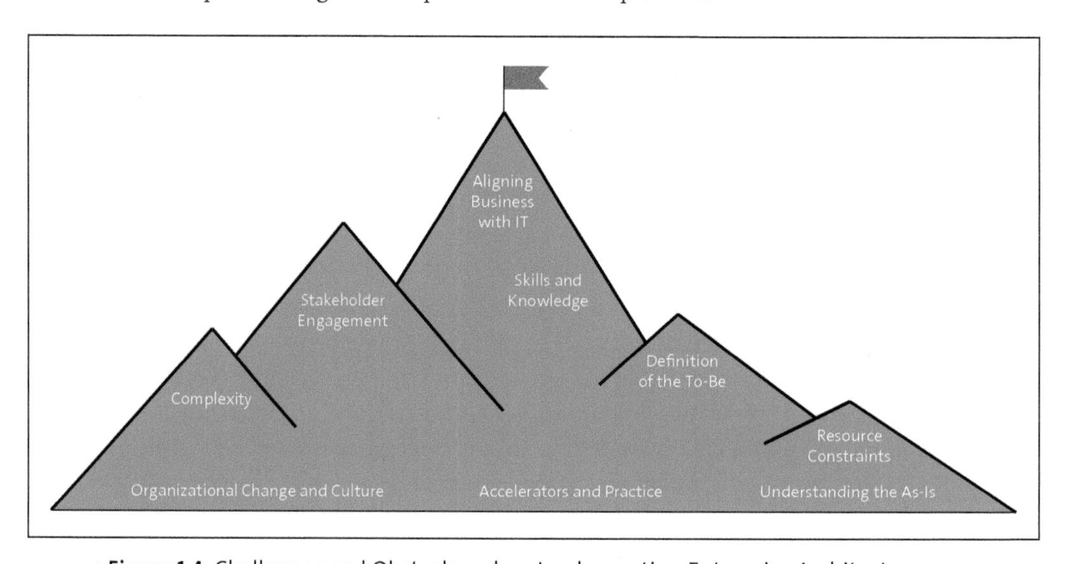

Figure 1.4 Challenges and Obstacles when Implementing Enterprise Architecture

Let's briefly describe each of these challenges:

- **Skills and knowledge**
  The foundation of any successful enterprise architecture implementation lies in the skills and knowledge possessed by the team. Enterprise architects must not only be adept in technical domains but also possess a solid understanding of business architecture and other enterprise architecture domains. Any gap in necessary skills—ranging from technical expertise to strategic thinking—can significantly hinder the adoption of enterprise architecture practices.
- **Accelerators and practice**
  Accelerators—such as predefined frameworks, reference content, and tools—can streamline the enterprise architecture implementation process. However, effectively integrating these accelerators into daily practices poses its own set of challenges, particularly when it comes to customization and scalability to meet specific organizational needs.
- **Understanding the as-is**
  A thorough understanding of the *as-is* (the current state of IT and business processes) is crucial. This involves detailed documentation and analysis, which can be resource intensive and difficult, especially in organizations where information is siloed or outdated.
- **Definition of the to-be**
  Defining the *to-be* state involves envisioning the future state of the enterprise architecture that aligns with business goals. This future state must be realistic and attainable yet ambitious enough to drive significant improvement. Balancing these aspects can be a major challenge, especially when dealing with competing visions and expectations from different business units.
- **Aligning business with IT**
  One of the recurrent challenges in enterprise architecture implementation is aligning business objectives with IT capabilities. Misalignment can lead to projects that are technically sound but don't deliver business value, resulting in wasted resources and missed opportunities.
- **Stakeholder engagement**
  Stakeholder engagement is critical for success. Gaining buy-in from all levels of the organization, from C-level executives to IT staff, can be challenging due to differing priorities and perspectives. Using effective communication and negotiation skills is essential to navigating these waters.
- **Resource constraints**
  Resource constraints often pose significant barriers, whether they be financial, human, or time related. Balancing limited resources while aiming to achieve optimal architectural changes requires meticulous planning and prioritization. This

applies not only to constraints on architecture resources (enterprise architects, solution architects, data architects, and so on) but frequently also to key stakeholders (e.g., business process experts) who at the same time have a multitude of projects and tasks to handle.

- **Complexity**
  The inherent complexity of changing existing systems and processes can be intimidating. Organizations often face technical debt and legacy systems that are deeply embedded in their operations, making transformations risky and complex. Changing a well-running and fine-tuned architecture that has supported the business over many years requires thorough planning and careful adoption of changes.
- **Organizational change management and culture**
  Organizational change management (OCM) is a critical aspect of enterprise architecture implementation. An organization must manage change on a cultural level, ensuring its workforce is ready to adopt new processes and technologies—because as the saying goes, "culture eats strategy for breakfast." Cultural resistance to change is a common obstacle that can derail even the best-planned enterprise architecture initiatives but unfortunately, most organizations overlook the OCM aspect of enterprise architecture or narrow it down to training and enablement. Therefore, an organization must embed stakeholder analysis, communication strategies, change impact analysis, and other tools of OCM from the beginning in any transformational program.

Navigating the myriad challenges in implementing enterprise architecture requires a robust strategy backed by a deep understanding of the organization's current capabilities and future needs. Addressing these challenges head-on with a comprehensive plan that includes skilled resources, stakeholder buy-in, and effective change management can pave the way for a successful transformation that aligns IT infrastructure with business goals, driving significant long-term value.

In Part IV, we'll explore the practice of enterprise architecture and illustrate how to establish an effective enterprise architecture practice using SAP EA Framework.

1.5 Summary

This chapter delved into the realm of enterprise architecture, positing it as an indispensable cornerstone of planning modern SAP landscapes and integral to aligning business with IT. The narrative opens with a twist on the tale of the Leaning Tower of Pisa, asserting that just as the tower's fate was sealed by its shaky foundation, the success of any enterprise architecture similarly hinges on the robustness of its underlying base. This illustration drives home the message: enterprise architecture must be built on a solid foundation to avoid the risks of structural weaknesses that can compromise the entire architecture.

The discussion progressed to outline enterprise architecture's evolution from a purely technical framework to a strategic imperative crucial for navigating today's rapidly changing business and technological environments. It emphasizes the importance of closely aligning enterprise architecture with business strategies, ensuring it can swiftly adapt to ongoing technological shifts and evolving business demands. This approach to enterprise architecture is comparable to the dynamic and responsive planning seen in urban development, but it requires even greater flexibility.

This chapter explored key challenges to implementing enterprise architecture, stressing the critical need for a thorough understanding of the baseline (the as-is) and a clearly envisioned target state (the to-be). These steps are vital to ensuring that the enterprise architecture not only aligns with but actively supports business objectives, thus enabling smooth and transformative changes within organizations.

This chapter highlights the role of the enterprise architect as increasingly vital within companies because it blends deep technical expertise with strategic and business acumen while steering through the complexities of modern IT and business landscapes. This evolving role mirrors the shifting priorities of organizations as they respond to new market pressures and technological opportunities.

Additionally, this chapter tackled prevalent misconceptions about enterprise architecture, such as its supposed ivory tower approaches and theoretical nature, countering these views by comparing enterprise architecture to city planning. This analogy helps demystify the discipline, making it more relatable and understandable, though it also acknowledges the limitations of this comparison in capturing the full breadth of enterprise architecture's adaptability and scope.

In conclusion, this chapter established a foundational understanding of the critical role of enterprise architecture in contemporary business, especially in SAP-centered landscapes. It sets the stage for further discussions of enterprise architecture while clarifying the challenges and vital nature of this discipline in fostering business agility and growth in an ever-evolving technological environment.

We'll continue with an in-depth look at SAP EA Framework in the next chapter.

Chapter 2
The SAP Enterprise Architecture Framework

With its own enterprise architecture framework, SAP empowers businesses to gain comprehensive control over their IT landscape, facilitates agile decision-making, and accelerates digital transformation journeys. Delve deeply into this chapter, where we explore SAP EA Framework and its five primary components: methodology, reference content, tools, practice, and services.

SAP offers its own enterprise architecture framework, SAP EA Framework, which we introduce in this chapter. With many frameworks already available in the market, one might wonder why SAP introduced yet another one. SAP EA Framework stands out as a comprehensive framework encompassing methodology (with a practitioner focus), reference content, tooling, and practice. In this chapter, we'll explore all the components of this framework.

In general, we define an *enterprise architecture framework* as a well-structured, holistic approach that provides a comprehensive view of the key components of an organization, their interrelationships, and the principles guiding their design and evolution. It's essentially a blueprint that defines the structure and operation of an organization, integrating strategic, business, and technology planning. An enterprise architecture framework provides a standardized methodology for organizing the complex systems and processes of an enterprise into a coherent and understandable structure, enabling effective decision-making and strategic alignment.

The need for an enterprise architecture framework arises from the complexities inherent in managing large-scale organizations. With the rapid pace of technological advancements and growing business demands, organizations often grapple with the challenge of integrating and aligning their IT systems with business goals. An enterprise architecture framework helps mitigate these challenges by providing a clear view of the organization's structure and operations. It helps identify redundancies, inconsistencies, and gaps in the organizational structure and processes, facilitating their timely resolution.

Looking back, the importance of enterprise architecture frameworks has grown with the increasing complexity of business and IT landscapes. In the past, many organizations suffered from poor decision-making and inefficient business processes due to their lack of a clear understanding of their organizational structures and operations. Without an enterprise architecture framework, it was difficult for organizations to align their IT systems and processes with their business goals, leading to wastage of resources and missed opportunities. Moreover, the absence of a standardized methodology for organizing and managing the organizations' structure and operations often resulted in chaos and confusion.

Moving forward, the requirements for an enterprise architecture framework are set to increase even further. With the advent of digital transformation, cloud computing, AI, and other technological advancements, managing an organization is expected to get dramatically more complicated. Having an enterprise architecture framework will be crucial in helping organizations navigate this complexity by providing a clear, comprehensive, and standardized view of their organizational structures and operations. It will play a key role in aligning IT systems and processes with business goals, facilitating strategic planning and decision-making, and optimizing resource utilization. By doing so, enterprise architecture frameworks will enable organizations to stay agile, competitive, and relevant in the fast-paced business environment of the future.

2.1 Overview of Enterprise Architecture Frameworks

There are several main enterprise architecture frameworks available on the market, and each has its own strengths and areas of focus (see Figure 2.1).

	TOGAF 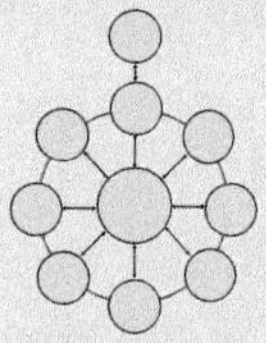	Zachman 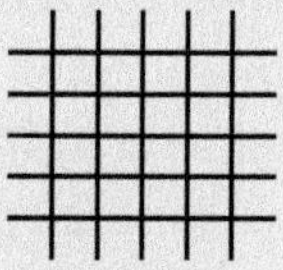	Gartner 
Scope and Coverage	●●●●○	●●●○○	●●●○○
Flexibility and Adaptability	●●●●○	●●○○○	●●●●○
Maturity and Adoption	●●●●○	●●○○○	●●●○○
Reference Models and Content	●●●○○	●○○○○	●○○○○
Ease of Use and Practicality	●●●○○	●○○○○	●●●●○

Figure 2.1 Comparison of Important Enterprise Architecture Frameworks

The Zachman Framework is one of the earliest and most influential. It provides a structured way of viewing and defining an enterprise, focusing on six different perspectives (planner, owner, designer, builder, subcontractor, and functioning enterprise) across six different categories (data, function, network, people, time, and motivation).

The Open Group Architecture Framework (TOGAF), on the other hand, provides a more detailed approach, offering a methodology for designing, planning, implementing, and governing an enterprise's IT architecture. Other notable enterprise architecture frameworks include the Federal Enterprise Architecture (FEA), which is used by the US federal government, and the Gartner Enterprise Architecture Framework, which emphasizes the strategic role of enterprise architecture frameworks in achieving business outcomes. Each of these frameworks offers its own unique approach to managing enterprise architecture, and the choice among them often depends on the specific needs and context of the organization.

Note that the choice of a particular enterprise architecture framework often depends on the specific requirements of the organization, the nature of its IT infrastructure, and the expertise of its staff. It is notable that most of SAP's customers who seek an enterprise architecture framework have embraced TOGAF.

However, SAP has introduced its own framework: SAP EA Framework. This framework doesn't aim to reinvent the wheel or propose a completely different approach to enterprise architecture, compared with established frameworks such as TOGAF. On the contrary, SAP EA Framework aligns closely with the TOGAF framework, incorporating many of its terminologies and concepts. Rather than focusing heavily on theoretical concepts, SAP EA Framework emphasizes practicality. It's specifically designed for practitioners who wish to conduct enterprise architecture planning activities within their SAP-focused architecture.

One of the significant advantages of using SAP EA Framework is its tight integration of SAP reference content and SAP enterprise architecture tools within the enterprise architecture framework. This is a distinct advantage over frameworks like TOGAF and others that introduce the concept of reference content and tooling to be used in the development process of enterprise architecture artifacts. Unlike these other frameworks, the SAP EA Framework provides out-of-the-box content that can be used immediately, offering practical utility to its users.

Figure 2.2 illustrates the five foundational elements of SAP EA Framework: methodology, reference content, tools, practice, and services.

The *methodology* encompasses the core principles of SAP EA Framework, such as the architecture domains, elements of the meta model, terminologies, and artifacts. The *reference content* primarily concentrates on strategy, business architecture, and solution architecture. With SAP's recent acquisition of LeanIX, the *tooling* component has become even more integral to SAP EA Framework, with SAP Signavio and SAP Cloud ALM emerging as key resources for the enterprise architect. Enterprise architecture *practice* is organizational implementation by adopting the methodology and setting up appropriate governance and change management processes. Finally, the *services* outline how the methodology, reference content, and tools can be utilized in real-world enterprise architecture planning activities and can be sourced internally or externally.

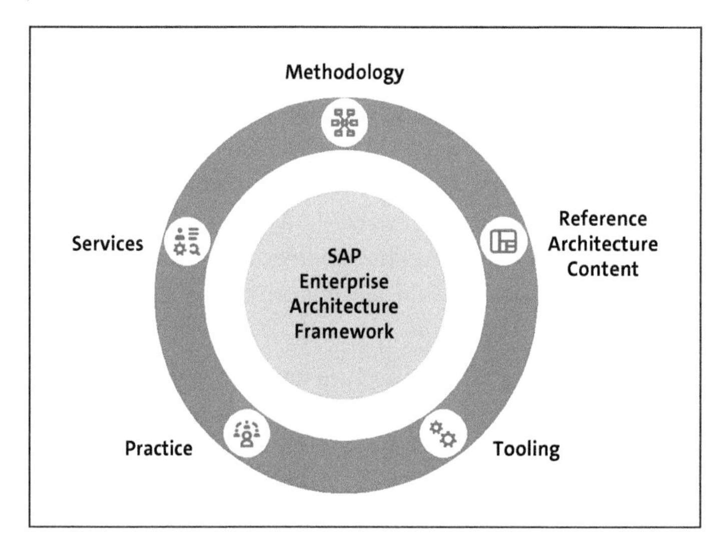

Figure 2.2 SAP EA Framework and Its Components

Continuously growing the amount of reference architecture content and its availability in the end-to-end transformation toolchain will make the adoption of SAP EA Framework more attractive, as reference content will be able to blend with customer-specific content.

The following sections will delve deeper into each of these five foundational elements of SAP EA Framework.

2.2 Enterprise Architecture Methodology

The first pillar of SAP EA Framework is the methodology for developing enterprise architecture artifacts. Figure 2.3 highlights this overall approach and the relationships among the different domains, including some high-level meta model connections between selected enterprise architecture artifacts like the business strategy model and the business capability model.

Let's take a closer look at each section of the methodology:

1. **Architecture vision**
   The *architecture vision* domain outlines the expected outcomes of the architecture work. It includes identifying stakeholders, gathering business and IT contexts, and defining the scope and requirements of the next project phases.
2. **Strategy and motivation**
   The *strategy and motivation* domain derives objectives from company priorities and long-term goals. It aligns enterprise architecture with business strategy and clearly communicates the benefits and rationales behind architectural decisions to motivate stakeholders.

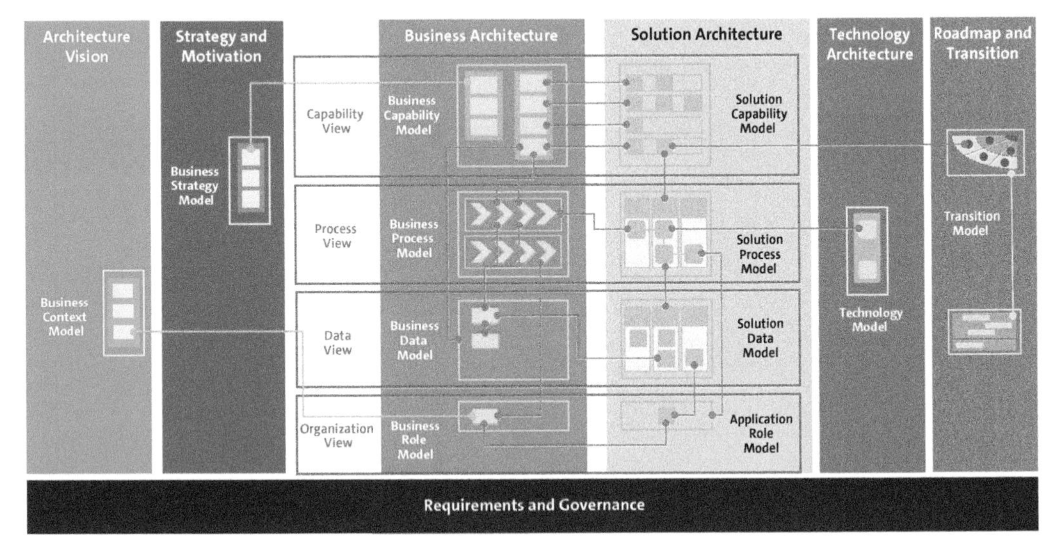

Figure 2.3 SAP EA Framework Methodology Overview

3. **Business architecture**
   The *business architecture* domain includes all business aspects, including capabilities, processes, data, and organizational structure. It facilitates stakeholder discussions and decisions based on agreed-upon business terms.
4. **Solution architecture**
   The *solution architecture* domain focuses on aligning technology solutions with business goals and requirements. It serves as a bridge between the business and technical sides of an organization to ensure effective technology implementations.
5. **Technology architecture**
   The *technology architecture* domain documents the delivery of target solution architecture through technology components like operating systems, hardware, and networks. It details the deployment of IT systems in specific data center locations.
6. **Roadmap and transition**
   The *roadmap and transition* domain uncovers project growth opportunities and manages system or process transitions. It involves planning migrations and executing changes to achieve project objectives efficiently.
7. **Requirements and governance**
   The *requirements and governance* domain encompasses governance artifacts like risk catalogs and architecture principles. It captures important architecture decisions and manages requirements across all architecture domains.

Within each distinct enterprise architecture domain, SAP EA Framework outlines the development of specific artifacts. These artifacts are similar in function to some of the artifacts in TOGAF, but SAP EA Framework consistently emphasizes practicality in the context of SAP during artifact development. SAP EA Framework also concentrates solely on the critical artifacts necessary for architectural planning. Figure 2.4 provides a detailed overview of the artifacts as per the enterprise architecture domain.

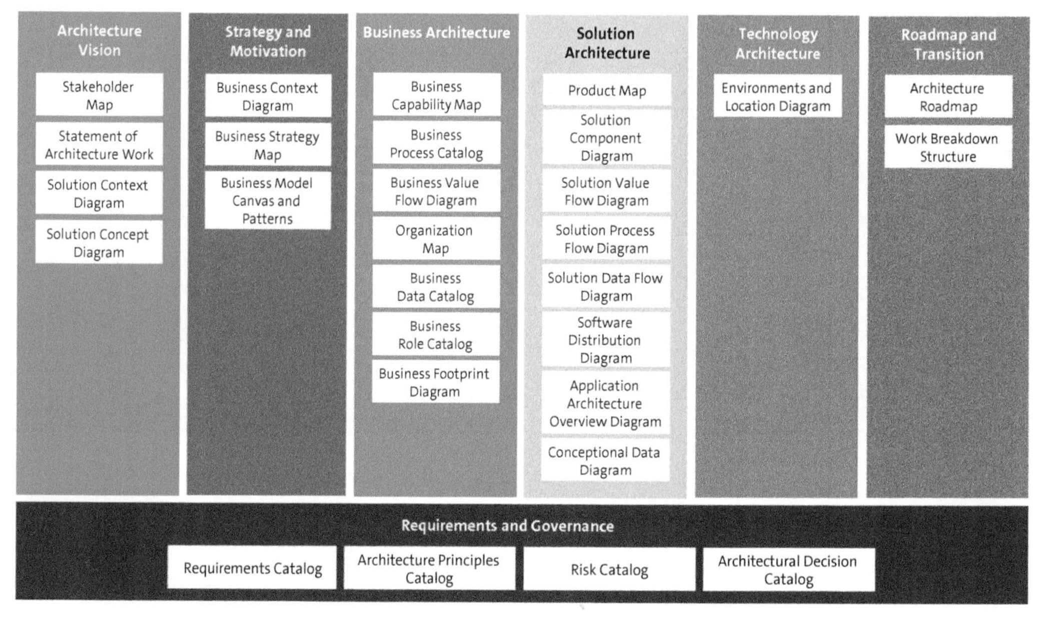

Figure 2.4 SAP EA Framework Artifacts Overview

We'll explain each of the artifacts in the following sections. In addition, we'll demonstrate in this book how the artifacts are developed as part of use cases. For that, refer to Chapter 5 through Chapter 10.

Meta Model of SAP Enterprise Architecture Framework

SAP EA Framework includes an internal meta model that defines all the main entities used in its methodology and illustrates their interconnections. However, the complexity of this meta model can be overwhelming, meaning it can provide little clarity for the reader trying to comprehend the subsequent content. While the meta model ensures consistency in navigating between artifacts and aids in maintaining coherence in reference content and tool development, it's not essential for practitioners to grasp every detail of the metal model. The primary focus for users should be on leveraging the framework's practical applications rather than delving into the intricacies of the meta model.

Chapter 11 offers an overview of the SAP LeanIX meta model, which is leaner and easier to comprehend but still aligns with the overall SAP EA Framework meta model. Those who are interested in understanding the concept of a framework meta model should refer to Chapter 11.

2.2.1 Architecture Vision

The architecture vision domain provides an initial sketch of the desired outcomes of the architecture work. It focuses on identifying relevant stakeholders and gathering essential context from both business and IT perspectives. This domain includes already formulated business and IT strategies, which will be refined for the selected scope. Ultimately, it produces the statement of architecture work, which defines the scope and requirements for the subsequent phases of the project.

We'll walk through the key components of the architecture vision in the following sections.

Stakeholder Map

Performing a stakeholder analysis is a crucial preliminary step in enterprise architecture planning because it can help an organization identify and prioritize the needs, expectations, and concerns of various stakeholders within the organization. By comprehensively understanding the stakeholders involved—including executives, employees, customers, and partners—an organization can align its architectural efforts with strategic objectives and ensure that the resulting architecture meets the diverse requirements of all involved parties. This analysis enables architects to anticipate potential challenges, mitigate risks, and foster stakeholder buy-in throughout the architectural planning and implementation processes.

Creating a *stakeholder map* is a key part of stakeholder analysis. As illustrated in Figure 2.5, a stakeholder map is typically divided into four distinct quadrants that categorize stakeholders as promoters, enthusiasts, resisters, and opponents.

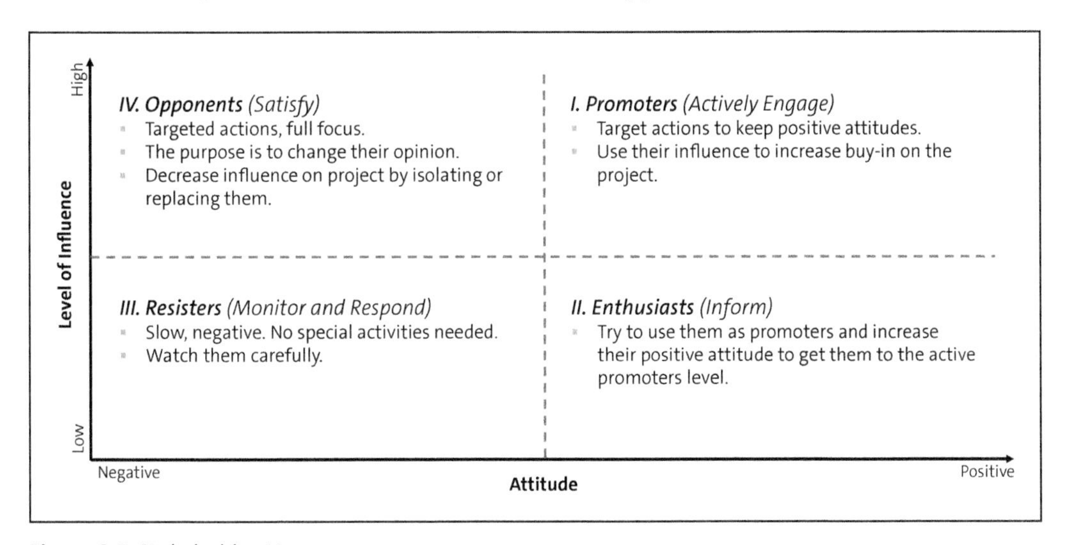

Figure 2.5 Stakeholder Map

The initial stage of creating a stakeholder map involves consulting the organizational chart to pinpoint relevant stakeholders. These individuals often include those most impacted by architectural changes, those executing significant influence over the architecture, and those with a vested interest in the success of the architectural development process. Subsequently, stakeholders are analyzed in accordance with the defined quadrants of the stakeholder map. This analysis informs the formulation of tailored action plans to effectively manage each stakeholder group. Stakeholder management then emerges as an essential, ongoing endeavor throughout the entire architectural development cycle, demanding nuanced strategies tailored to the unique dynamics of each stakeholder cohort.

Statement of Architecture Work

The *statement of architecture work* (SoAW) plays an important role in the enterprise architecture planning process by providing a clear and comprehensive outline of the objectives, scope, and constraints of the architectural initiative. By defining the purpose and goals of the enterprise architecture effort up front, the SoAW serves as a guiding document that aligns architectural endeavors with the overarching strategic vision of the organization. The SoAW provides clarity that helps stakeholders understand the intended outcomes of the architecture planning and implementation processes and ensures that efforts remain focused and coherent throughout.

Following the TOGAF framework, a SoAW typically contains the following elements:

- Title
- Architecture project request and background
- Architecture project description and scope
- Overview of architecture vision
- Specific change-of-scope procedures
- Roles, responsibilities, and deliverables
- Acceptance criteria and procedures
- Architecture project plan and schedule
- Approvals

The SoAW facilitates effective communication and collaboration among stakeholders by establishing a common understanding of the architectural initiative. By clearly articulating the scope, deliverables, timelines, and resource requirements, the SoAW enables stakeholders to align their expectations and commitments accordingly. This alignment fosters a sense of ownership and accountability among stakeholders, promoting engagement and buy-in throughout the enterprise architecture planning process. Additionally, the SoAW serves as a baseline for measuring progress and assessing the success of the architecture, providing a framework for iterative improvements and adjustments as the architectural initiative unfolds.

Solution Context Diagram

The *solution context diagram* serves as a comprehensive visualization of the intended solution within enterprise architecture endeavors. Its primary objective is to offer a clear and accessible depiction of the aspired-to solution, ensuring that it's easily understandable by both technical and nontechnical stakeholders. By encapsulating the required business capabilities to be fulfilled by the architecture, the diagram provides a strategic roadmap for aligning the solution with organizational objectives. Key prerequisites for developing the solution context diagram include the creation of a stakeholder map and an SoAW, which help contextualize the solution within the broader enterprise landscape. Particularly for stakeholders from the business domain, the solution context diagram serves as a vital tool for conveying how the envisioned solution interfaces with various organizational units, roles, and business functions. Its importance lies in guiding architectural decision-making, fostering consensus among stakeholders, and promoting alignment between the solution and business objectives throughout the architecture planning and implementation phases.

To create the solution context diagram, follow these steps and use the template from Figure 2.6:

1. **Translate learning into business capabilities**
   Begin by translating your understanding of the aspired-to solution into business capabilities. These capabilities describe what the solution can do in terms of main functions or features, expressed in business terminology. Aim for a list of five to ten main capabilities that capture the essence of the solution's intended functionality.

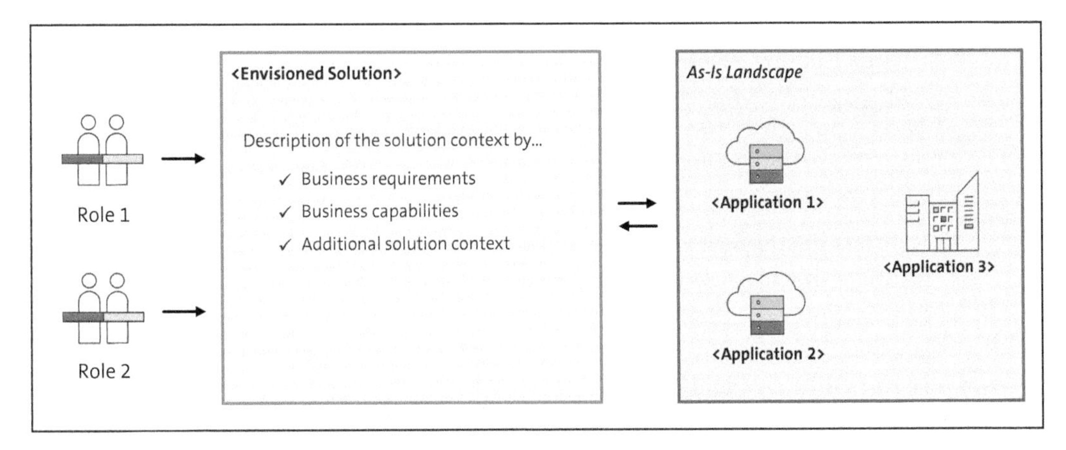

Figure 2.6 Solution Context Diagram

2. **Identify architectural components and organizational elements**
   After defining the business capabilities, identify the architectural components needed to support them. Consider related organizational units, business roles, and existing applications. Use the stakeholder map to reference users and roles. This ensures that the solution context diagram reflects both technical and organizational

aspects, providing a comprehensive view of the solution's ecosystem and alignment with stakeholder needs.

3. **Iterate and refine**
   Iterate on the solution context diagram based on feedback from stakeholders and validation against business requirements. Refine the diagram to ensure clarity, completeness, and accuracy in representing the intended solution and its alignment with organizational goals.

Solution Concept Diagram

The *solution concept diagram* serves as an important tool for crafting a high-level portrayal of the proposed solution. It offers a concise yet comprehensive overview of the solution, capturing its key components, functionalities, and interrelationships. This diagram is instrumental in aligning the solution with the strategic objectives and requirements of the architecture engagement. By distilling complex solution elements into a visual representation, it helps stakeholders gain a clear understanding of the proposed approach and its intended outcomes. Furthermore, the solution concept diagram serves as a communication tool, facilitating discussions among stakeholders and fostering consensus on the proposed solution's viability and alignment with business goals. Overall, this diagram plays a crucial role in guiding decision-making processes and ensuring that the proposed solution effectively addresses the needs and objectives of the enterprise architecture initiative.

To create a solution concept diagram like the one illustrated in the template in Figure 2.7, you must follow several key steps:

1. **Identify high-level architectural building blocks**
   Begin by identifying the high-level architectural building blocks required to fulfill the key objectives and business capabilities outlined in the solution context diagram. These building blocks should represent the fundamental components necessary to realize the desired solution.
2. **Translate business-oriented capabilities into architecture building blocks**
   Translate the business-oriented capabilities described in the solution context into corresponding architecture building blocks. This involves mapping each business capability to the architectural elements necessary to support it, ensuring alignment between business requirements and architectural components.
3. **Outline relationships among building blocks**
   Establish the relationships among the identified building blocks within the solution concept diagram. These relationships may encompass various types, such as request-response interactions or information flows, and should reflect the dependencies and interactions among architectural components.
4. **Incorporate users, roles, and organizational units**
   Integrate users, roles, and organizational units into the solution concept diagram, illustrating their relationships with the architecture building blocks. This step

ensures that the diagram reflects the organizational context and depicts how different stakeholders interact with and utilize the solution's components.

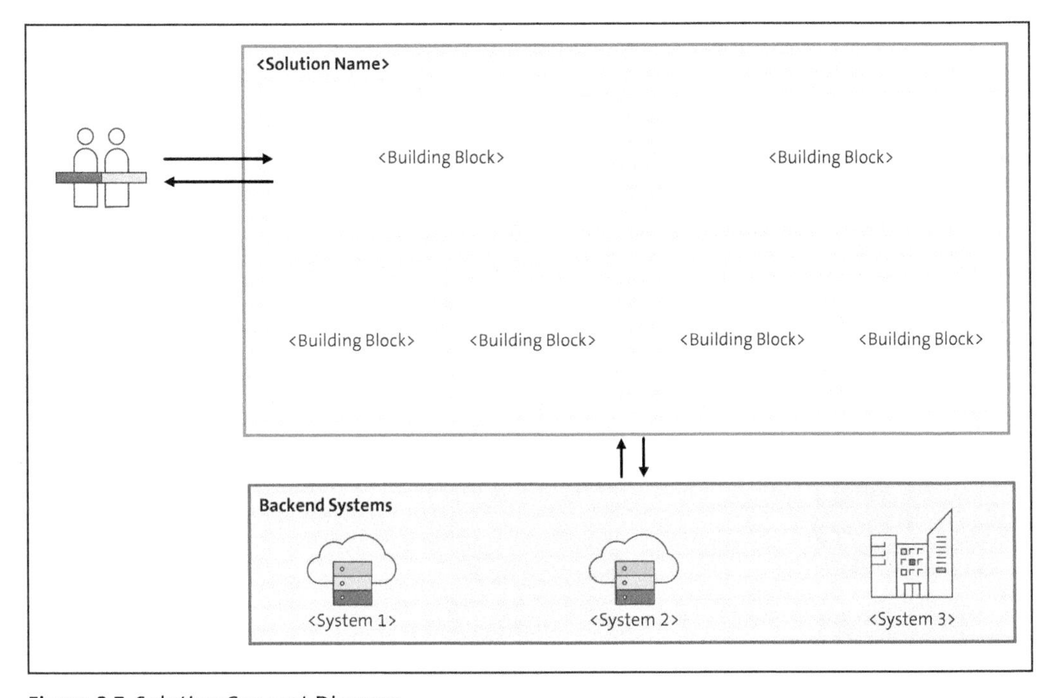

Figure 2.7 Solution Concept Diagram

2.2.2 Strategy and Motivation

This enterprise architecture domain is crucial for deriving objectives based on the company's strategic priorities and long-term goals, and it forms the first part of the business architecture. It ensures that the enterprise architecture aligns with the overall business strategy. Additionally, it plays a vital role in motivating stakeholders by clearly communicating the benefits and rationale behind architectural decisions and transformations, thereby establishing a compelling vision for the company's future.

Let's dive into the key components of the strategy and motivation domain next.

Business Context Diagram

A *business context diagram* serves as a crucial artifact in enterprise architecture planning due to its ability to offer a comprehensive and accessible depiction of how a business interfaces with its external environment. By visually illustrating the various entities that interact with the business—including customers, suppliers, partners, and regulatory bodies—the diagram provides a holistic view of the organization's ecosystem. Additionally, it describes the flows of information, resources, and materials between the business and these external entities, shedding light on the complex networks and dependencies that underpin business operations.

The initial step of constructing the business context diagram entails defining the organizational entities—such as customers, suppliers, competitors, and financial institutions—as named business actors. Subsequently, high-level business flows among these actors must be outlined. These include product flow, payment flow, and information flow, among others. This process provides a structured framework for capturing the interactions and exchanges that occur within the business environment. A standard template for the business context diagram, exemplifying these defined entities and flows, is depicted in Figure 2.8. This template serves as a visual reference point for organizing and representing the complex web of relationships and activities essential to the business's functioning.

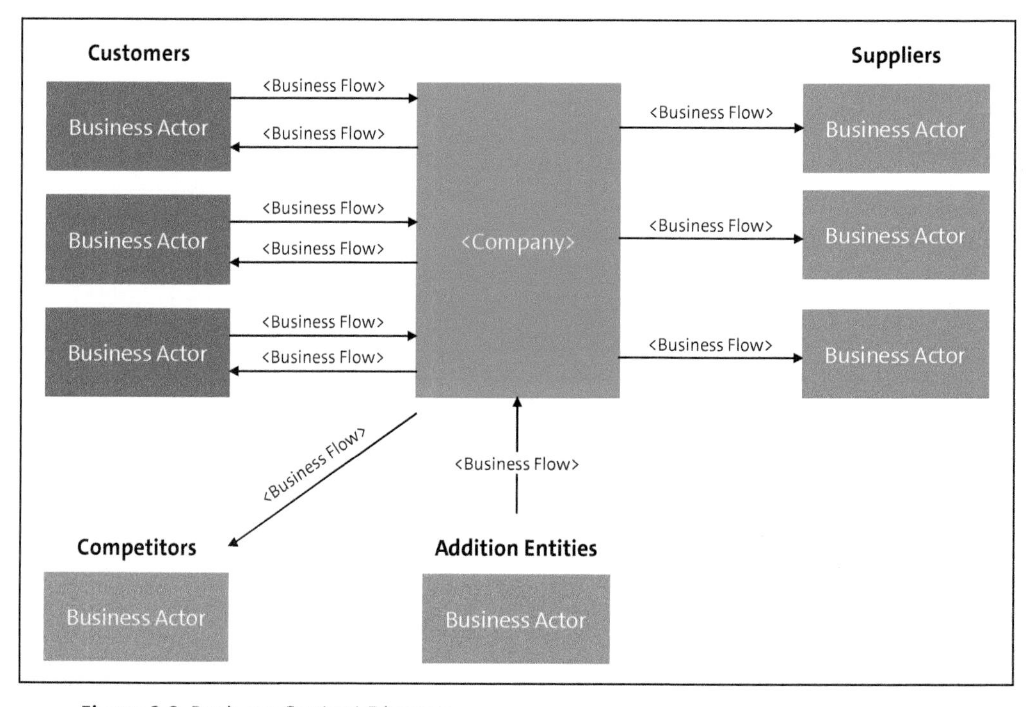

Figure 2.8 Business Context Diagram

A business context diagram offers several benefits, including providing insights into the complex networks of relationships that generate value through dynamic exchanges among individuals, groups, and organizations. By visually representing these relationships and flows of information and resources, and by highlighting the most critical business activities for value creation, the diagram promotes strategic clarity. This clarity enables stakeholders to prioritize efforts and resources in their efforts to optimize key processes and relationships, which helps them ultimately drive efficiency and effectiveness in operations. Additionally, the diagram serves as a catalyst for innovation by identifying intersections where information and material flows converge. This in turn presents the organization with opportunities for the development of new products, services, and processes that enhance its competitive advantage.

Strategy Map

In enterprise architecture planning, a *strategy map* is a crucial tool for effectively translating business strategy into actionable business architecture. It aids in establishing a clear connection between the strategy map and business capabilities. Moreover, it provides a comprehensive view of the organization's readiness to adopt new or improved business changes. Significantly, it helps align business and IT priorities, ensuring that business-critical priorities are adequately addressed in the IT architecture and roadmap design. Thus, a strategy map serves as a pivotal component in achieving strategic alignment and facilitating business transformation.

The main components of a strategy map and what they do are as follows:

- *Strategic priorities* are the catalysts that drive change within an organization. They serve as the guiding principles that shape the organization's direction and decision-making processes.
- *Business goals*, on the other hand, represent the end state that stakeholders aim to achieve. They provide a clear vision of what success looks like for the organization.
- *Objectives*, also known as *value drivers*, are measurable short-term or midterm targets that have direct relevance to the business. They come with defined target values and schedules, acting as stepping stones on the way to reaching the business goals.
- Lastly, *business capabilities* reflect the organization's potential to carry out business activities successfully, attain its objectives, and deliver value to its customers. They underline the competencies and resources of the organization in executing its strategic priorities and achieving its goals.

The principal procedures for constructing a strategy map like the one in Figure 2.9 include the following:

1. **Understand the strategy**
   Familiarize yourself with the strategic priorities of your company or business unit. These are often outlined in strategic documents specific to the corporation or business division, which may be accessible internally or publicly. Additionally, conduct interviews with key stakeholders who play a role in shaping the strategy of the targeted organizational unit. This can provide valuable insights and information. To enhance your understanding, compare this information with external analysis from public sources, such as industry trends.
2. **Clarify the objectives**
   From the strategic priorities, extrapolate the specific business objectives of your organization. These objectives provide more detail on the strategic priorities. They may be numerous, and sometimes, you may need to arrange them in a hierarchical structure. To maintain a manageable scope, we suggest aligning no more than three to five objectives with each strategic priority.

3. **Identify value drivers**
   Link the appropriate value drivers to the established objectives. Value drivers, or key performance indicators, make the business objectives quantifiable and are typically associated with a timeframe. If you can't determine value drivers from the strategy documents available, or if there's a need to expedite the strategy mapping process, you can bypass this step and directly associate business capabilities with business objectives.

4. **Map business capabilities**
   Identify the enabling and core business capabilities that relate to value drivers. In this phase, you bridge the gap between strategy and operations by aligning business capabilities with value drivers (or with business objectives, if you omitted the previous step). Concentrate on the most crucial and distinctive business capabilities, rather than attempting to map all necessary capabilities. Consequently, you'll have an initial business capability map that supports the organization's strategy, which can serve as a foundation for further development of the business capability map within the business domain.

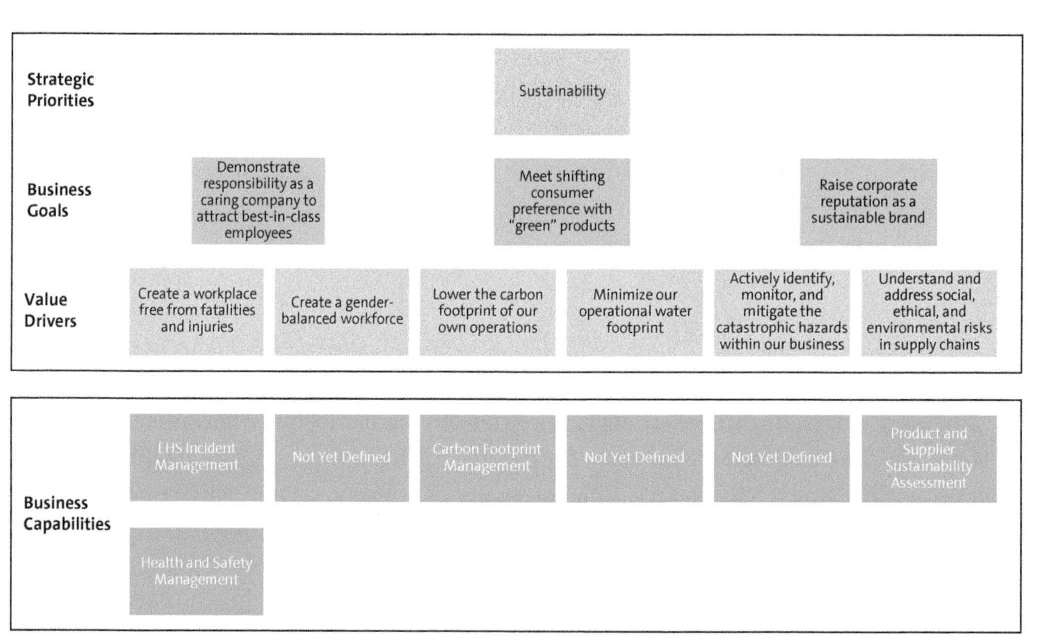

Figure 2.9 Strategy Map

Business Model Canvas

The *business model canvas* (BMC) is a strategic management tool that provides a structured framework for visualizing and analyzing the key components of a business model. It consists of nine building blocks, as shown in Figure 2.10: key partnerships, key activities, key resources, value propositions, customer relationships, channels, customer segments, cost structure, and revenue streams. Each building block represents a fundamental aspect of the business, and using these building blocks to create the BMC allows stakeholders to understand how different elements interact and contribute to the overall business model. The BMC enables organizations to articulate and refine their business strategies, identify areas for innovation and improvement, and align various business functions with common goals. It promotes a holistic view of the business model, fostering collaboration and informed decision-making among different departments and stakeholders.

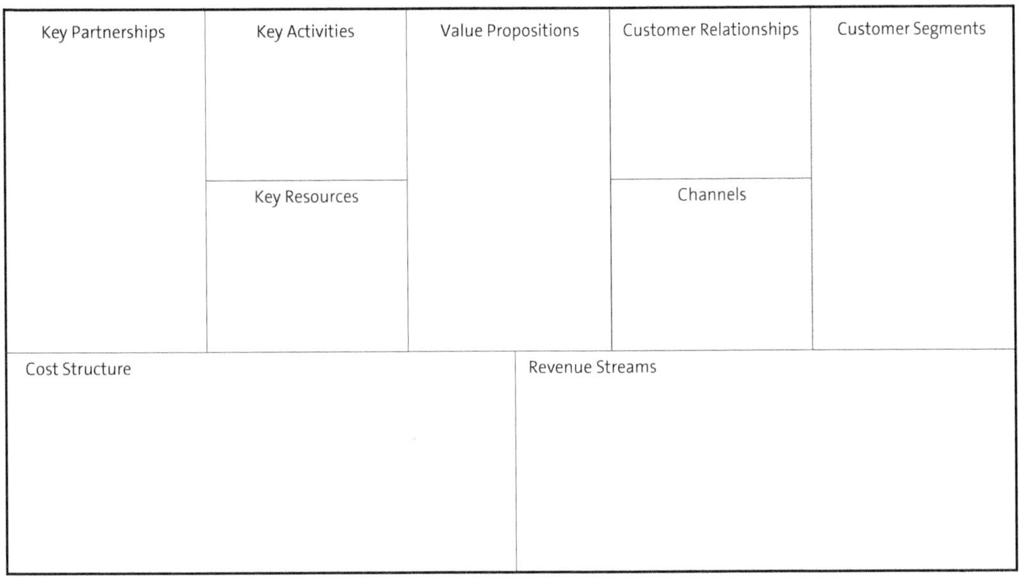

Key Partnerships
Key Activities
Key Resources
Value Propositions
Customer Relationships
Channels
Customer Segments
Cost Structure
Revenue Streams

Figure 2.10 Business Model Canvas

In the context of enterprise architecture, the BMC serves as a crucial artifact. Firstly, it provides a structured framework for enterprise architecture practitioners to analyze and understand the business architecture of an organization. By mapping out the various components of the business model, enterprise architecture teams can identify dependencies, synergies, and potential gaps between different business functions and processes. This understanding enables them to develop coherent and aligned architecture blueprints that support the organization's strategic objectives. Secondly, the BMC helps enterprise architecture practitioners to communicate complex business concepts and strategies in a visual and easily understandable format. This facilitates stakeholder engagement and buy-in, as it allows for more meaningful discussions about the organization's business model and how it intersects with technology and IT infrastructure. Ultimately, the BMC serves as a bridge between business strategy and enterprise architecture, helping organizations to translate their strategic priorities into actionable architectural decisions.

2.2.3 Business Architecture

The business architecture domain focuses on creating a comprehensive description of the organization, covering all business aspects, including capabilities, processes, data, and organizational structure. It facilitates business-led discussions with stakeholders and aids in making decisions based on agreed-upon business terms. This domain enables clear communication of business values and the impact of architecture work, ensuring that all stakeholders are on the same page.

Let's walk through the key components of the business architecture domain.

Business Capability Map

An organization's *business capabilities* constitute its ability to effectively execute its business activities, accomplish its business aims, and provide value to its stakeholders. A capability represents a distinct skill or capacity that a business might possess or leverage to attain a particular objective or result. Business capabilities lie at the heart of business architecture as they articulate *what* the business accomplishes and not *how* it's accomplished. These capabilities can encompass existing functions or those needed to support a novel direction or strategy. The idea of business capability has evolved into a key element in crafting business architecture. With numerous organizations grappling with the complexity of their operations, they seek to address queries about their advancement toward achieving strategic objectives. Business capabilities offer a simplified representation of the business landscape, facilitating discussions among stakeholders and furnishing a practical framework for both business and IT endeavors.

Business capabilities are best represented in a *business capability map* like in Figure 2.11.

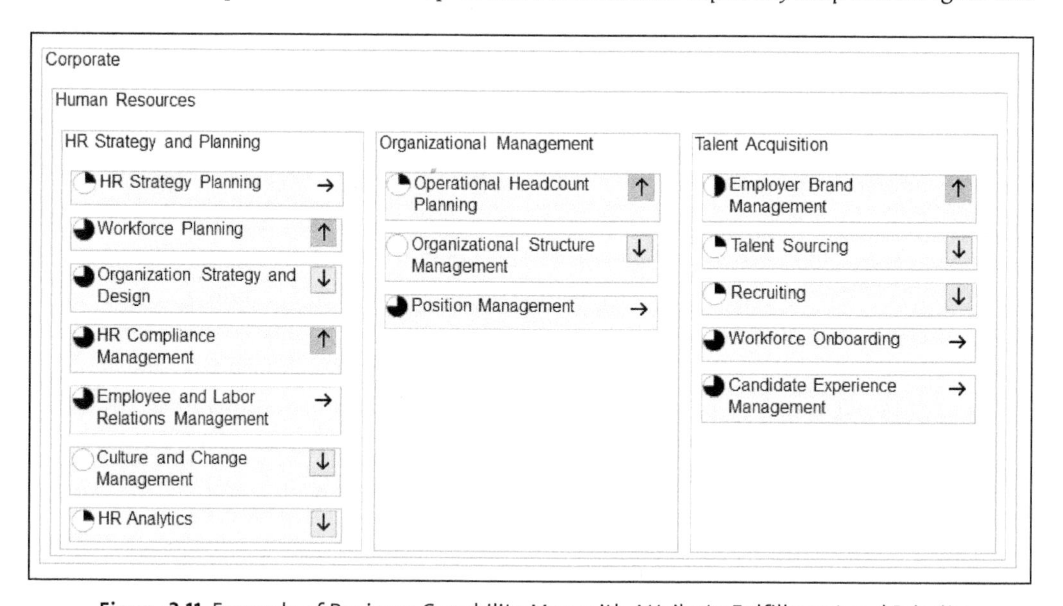

Figure 2.11 Example of Business Capability Map with Attribute Fulfillment and Priority

The main steps involved in creating a business capability map are as follows:

1. **Identify business functions**
   Begin by identifying the primary functions or activities within your organization. These could include sales, marketing, customer service, product development, etc.
2. **Break down functions into capabilities**
   Break down each business function into discrete capabilities. These are specific abilities or capacities that the business must possess to perform the function effectively. For example, under the "sales" function, capabilities might include lead generation, prospecting, negotiation, etc.
3. **Maintain attributes and assess capabilities**
   Adding attributes to business capabilities can provide additional context and information, enhancing the understanding of each capability. Evaluate each business capability to assess its current state and performance. Potential attributes could be description, strategic importance, maturity level, requirements, stakeholder, organizational or business unit, etc.
4. **Iterate and maintain**
   Business capabilities evolve over time, so it's important to regularly review and update the diagram to reflect changes in the organization's strategy, structure, and operations. Continuously iterate on the diagram to keep it relevant and useful for decision-making.

Utilizing reference content is crucial when developing a business capability diagram as it provides valuable insights and guidance based on established industry standards and best practices. Drawing from reference content allows organizations to leverage the collective knowledge and expertise of their industry peers, ensuring that their business capability diagram reflects relevant and proven approaches to structuring and organizing business functions. SAP offers a wealth of reference business capability models tailored to specific industries. These models serve as invaluable resources, offering predefined frameworks and templates that align with industry-specific requirements and nuances. Refer to Section 2.3 to learn more about SAP's reference content.

Business Process Catalog

A *business process catalog* is a comprehensive repository that documents all business processes within an organization. It serves as an organized inventory of the various processes that are critical to the functioning of the enterprise. This catalog includes detailed descriptions, objectives, inputs, outputs, key stakeholders, and any interdependencies between processes. Essentially, it provides a high-level view of how different business activities are structured and interconnected, offering a clear understanding of the operational landscape of the organization.

The business process catalog is crucial to enterprise architecture because it provides or aids in the following:

- **Holistic view**
  It provides a complete overview of the organization's business processes, facilitating a better understanding of how different functions and activities are interrelated.
- **Alignment**
  It ensures that business processes are aligned with the organization's strategic goals and objectives. This alignment is essential for achieving organizational coherence and effectiveness.
- **Decision making**
  It aids in informed decision-making by providing a clear picture of existing processes, which helps in identifying areas for improvement, optimization, and innovation.
- **Integration**
  It supports integration efforts by highlighting interdependencies and interfaces between processes, ensuring smooth collaboration among various departments.

The business process catalog typically takes the form of a structured document. It may include the elements shown in Table 2.1.

Business Catalog Element	Business Catalog Description
Process ID	A unique identifier for each process
Process name	The name of the process
Description	A brief overview of the process
Objectives	The goals and intended outcomes of the process
Inputs	The resources, data, and prerequisites required to execute the process
Outputs	The deliverables or results produced by the process
Organizational unit	Organizational or business units
Process owner	Key business processes where the owner is responsible
Stakeholders	Key individuals or groups involved in or affected by the process
Interdependencies	Relationships and dependencies with other processes
Performance metrics	Key process performance indicators (PPIs) and metrics used to measure the effectiveness and efficiency of the process

Table 2.1 Example of Business Process Catalog

Level of Business Processes

Business processes are often organized into five levels of depth to provide a clear and structured approach:

1. **Business area**
   Represents the broadest categories, aligning with the main functions of the organization
2. **Process group**
   Groups related processes within a business area
3. **Business process**
   Defines specific processes that achieve certain business objectives
4. **Process variants**
   Defines variations on the same process, based on different conditions or data inputs
5. **Process steps**
   Provides the most detail, outlining individual tasks or activities within a process

Developing a business process catalog involves several key steps:

1. **Identify processes**
   Conduct workshops and interviews with stakeholders to identify all significant business processes within the organization on level 1 to level 3.
2. **Document processes**
   Capture key information about each process, including descriptions, objectives, inputs, outputs, stakeholders, and interdependencies.
3. **Standardize information**
   Ensure that the documentation follows a consistent format and standard nomenclature for easy understanding and comparison.
4. **Validate information**
   Engage with process owners and stakeholders to review and validate the documented processes for accuracy and completeness.
5. **Categorize and organize**
   Group processes into logical categories and hierarchies to create a structured and easily navigable catalog.
6. **Implement tool**
   Choose an appropriate tool or platform to store and manage the catalog, ensuring it's accessible and user-friendly.

Be aware that in enterprise architecture, the business process catalog provides a higher-level view of processes compared to solution blueprints. The catalog offers an overview of all business processes, concentrating on levels 1 to 3, rather than detailed process descriptions. This higher-level focus, developed during the planning and analysis phases, aids strategic decision-making and ensures alignment with organizational goals. In contrast, solution blueprints provide detailed technical designs during the implementation phase, concentrating on how specific processes will be executed within a particular system. Thus, while the catalog serves as a foundational tool for understanding and optimizing business operations, the detailed process descriptions are part of the actual solutioning phase, when solutions are built and implemented.

Business Value Flow Diagram

A *business value flow diagram* is integral to SAP Enterprise Architecture Methodology (SAP EA Methodology). It depicts the integration of value-adding business activities aimed at achieving specific goals and delivering valuable outcomes. These flows are pivotal for enterprise architecture planning, offering insights into how value is generated and delivered across the organization. By mapping these activities, stakeholders identify opportunities for improvement, innovation, and strategic alignment. The diagram provides visibility into the current and desired future state of the organization, ensuring alignment between business and IT priorities. Ultimately, understanding business value flows empowers organizations to enhance their efficiency, agility, and competitiveness.

Business processes and business value flows and serve distinct yet interconnected purposes within an organization. *Business processes*, represented in Business Process Model and Notation (BPMN), detail the sequential steps and activities required to accomplish a specific task or achieve a particular outcome. They provide a granular view of the operational workflow, highlighting the sequence, interactions, and decision points involved in executing a process. On the other hand, *business value flows* focus on the flow of value-adding activities aimed at achieving strategic business objectives and delivering value to stakeholders. While business processes focus on operational efficiency and effectiveness, business value flows provide a higher-level perspective, emphasizing the alignment of activities with overarching business goals and priorities. In essence, business processes offer a detailed blueprint of operational activities, while business value flows offer a strategic roadmap for value creation and delivery within the organization.

Business Process Model and Notation

BPMN is a standardized graphical representation of specific business processes. It provides a set of symbols and notations that are used to create detailed process maps. Key elements of BPMN include the following:

- **Flow objects**
  These are the core elements—such as events (circles), activities (rectangles), and gateways (diamonds)—that define the actions and decisions in the process.

- **Connecting objects**
  These are arrows that connect flow objects, showing the sequence and direction of the process flow.
- **Swimlanes**
  These divide the diagram into lanes or pools to represent different participants or departments involved in the process.
- **Artifacts**
  These constitute additional information, such as data objects or annotations, that provide more context to the process.

To illustrate a business value flow, we can examine Figure 2.12, which presents a business value flow extracted from SAP's reference content (see Section 2.3 for more details).

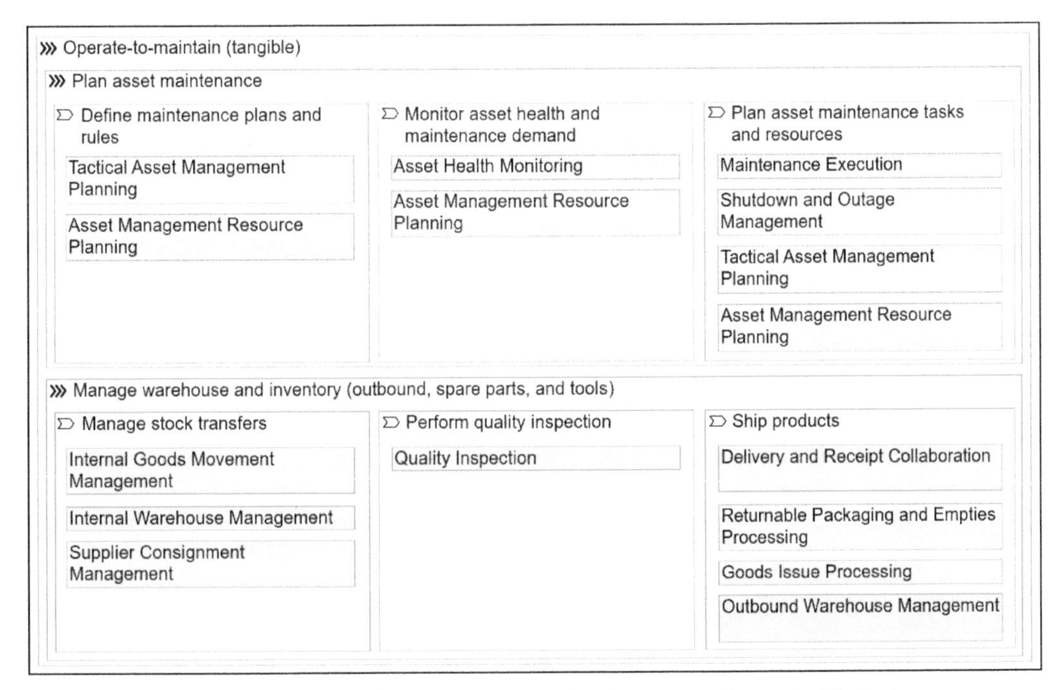

Figure 2.12 Example of Value Flow for Operate-to-Maintain Process Mapped with Business Capabilities

Usually, you can create the business value flow diagram by performing the following steps:

1. **Identify end-to-end business value flows**
   Begin by identifying the high-level, end-to-end business value flows that encompass the organization's primary objectives or goals. These level 1 business value flows provide a top-down perspective on how value is created and delivered within the organization.

2. **Identify industry-specific variants**
   Determine whether there are any industry-specific variations or nuances that need to be considered for each business value flow. This step involves identifying how industry-specific factors may impact the execution of business activities and the overall flow of value within the organization.
3. **Explore further levels of business value flows**
   Delve deeper into each level 1 business value flow to identify and define more detailed levels of value flow, down to level 4. These lower levels break down the level 1 flows into specific business activities, providing a more granular understanding of the value creation process.
4. **Map business capabilities to business activities**
   Identify the business capabilities that are associated with each business activity. This helps you to understand which capabilities are necessary for the organization to execute each activity and how they contribute to value creation within the organization.
5. **Review and iterate**
   Once you have created the initial diagram, review it with stakeholders to ensure its accuracy and completeness. Make adjustments as necessary based on feedback and further analysis.

Organization Map

An *organization map* is a comprehensive representation of the various entities that make up an organization. It details the hierarchical arrangement of and relationships among different components such as groups, business units, regions, countries, sales organizations, and production sites. This map provides a clear understanding of how the organization is structured, showing the distribution and coordination of roles, responsibilities, and operations among different levels and areas (see Figure 2.13).

<table>
<tr><td colspan="10">Group</td></tr>
<tr><td colspan="7">Business Unit 1</td><td>Business Unit 2</td><td>Business Unit 3</td><td>Business Unit 4</td></tr>
<tr><td colspan="3">EMEA</td><td colspan="2">APJ</td><td colspan="2">LA</td><td>EMEA</td><td>EMEA</td><td>NA</td></tr>
<tr><td>Germany</td><td>South Africa</td><td>England</td><td>China</td><td>India</td><td>Brazil</td><td>Mexico</td><td>Italy</td><td>France</td><td>USA</td></tr>
</table>

Figure 2.13 Organization Map

The organization map is a crucial artifact in enterprise architecture because it provides an overview of the organizational structure, serving as a foundational framework for structuring business processes and capabilities. It ensures alignment between the enterprise architecture and the organizational hierarchy, supporting strategic goals and operational needs. Additionally, the map aids in understanding business needs by clearly defining roles and responsibilities among various entities, thus facilitating effective planning and resource management. In summary, the organization map helps you navigate through the enterprise architecture of the organization.

Developing an organization map involves several key steps:

1. **Identify key entities**
   Determine the main groups, business units, regions, countries, sales organizations, and production sites within the organization.
2. **Document relationships**
   Map out the relationships and reporting lines between different entities.
3. **Create the visual representation**
   Use diagramming tools to create a visual representation of the organizational structure.
4. **Validate with stakeholders**
   Review the draft map with stakeholders to ensure its accuracy and completeness.

Business Data Catalog

A *business data catalog* serves as a comprehensive repository of metadata, encompassing various dimensions of an enterprise's information assets. It houses definitions of data objects and fields, alongside crucial business system attributes, such as business rules, process relevance, data activities, and access controls. This catalog offers a clear and up-to-date depiction of *data reality* from both business and IT perspectives, fostering collaboration and facilitating impact analysis for new data requirements. Crucially, it aids in understanding the creation, storage, transmission, and reporting of enterprise data entities, thus contributing to enterprise architecture planning. By providing a holistic view of the entire organizational landscape, the catalog transcends specific business functions, enabling its utilization in diverse business operations.

The business data catalog can encompass diverse attributes such as the following:

- Data object and field definitions
- Business rules and process relevance
- Data activities such as data quality management
- People and organizations involved in data management
- Locations of data, including master and secondary systems
- Access controls and security limitations

- Compliance requirements and regulations
- Timing and events related to data processing
- Components of the data lifecycle (creation, maintenance, and deletion)

For instance, we can consider Table 2.2, which offers just a glimpse of the entire business data catalog.

Data Object	Data Type	Business Definition	Process Relevance	Data Owner
Sales order	Transactional data	An order placed by a customer for a product	Order-to-cash	Sales manager
Material	Master data	A unique identifier for a product in inventory	Plan-to-produce	Inventory manager
Customer	Master data	The name of the customer placing an order	Lead-to-order	Customer service manager
Order quantity	Transactional data	The quantity of a product ordered by a customer	Lead-to-order	Production manager
Delivery date	Transactional data	The scheduled date for delivering the order	Lead-to-order	Logistics manager
Payment status	Transactional data	An indication of whether payment for the order is received	Order-to-cash	Finance manager

Table 2.2 Example of a Business Data Catalog

The key steps for creating a business data catalog involve two main phases—catalog definition and catalog delivery:

1. **Catalog definition**
   - Engage in discussions with data architecture stakeholders to gather requirements.
   - Agree on the structure and level of detail for the initial version.
   - Define usage scenarios for the metadata catalog.
   - Establish a roadmap for content delivery and evolution.
2. **Catalog delivery**
   - Name priority-1 data objects along with their business and IT attributes.

- Begin populating the catalog with content based on the identified priorities.
- Expand the baseline version to encompass all data objects.
- Include data objects with specifically identified priority-1 business and IT attributes.
- Incorporate nonpriority objects to complete the overall version.
- Ensure all data objects and attributes are captured in the metadata catalog.

Business Role Diagram

A *business role diagram* is a detailed representation of the various roles within an organization and their relationships to business activities and processes. This artifact outlines the responsibilities, duties, and authorities of different roles, providing a clear understanding of who is involved in what activities within the enterprise. It typically includes roles such as executives, managers, specialists, and other key positions, and it maps these roles to specific business functions and activities.

The business role diagram typically includes the following elements:

- **Roles**
  Definitions of various roles such as executives, managers, specialists, administrators, and clerks
- **Responsibilities and duties**
  A description of the responsibilities and duties associated with each role
- **Relationships to business activities**
  Mapping of roles to specific business activities and processes
- **Attributes**
  Seniority level, required skills, and other role-specific characteristics

Business roles are intricately related to other entities within the organizational and process frameworks. They are directly connected to business activities, indicating who is responsible or accountable for each activity, as defined in the extended RACI model (RACI meaning responsible, accountable, consulted, and informed). Additionally, business roles have relationships with other roles, often forming hierarchical structures where roles like learning manager report to roles such as chief learning officer. Business roles also interact with application roles, requiring specific software access to perform their duties. They are linked to business capabilities, which define the skills and resources needed to carry out business activities. Moreover, business roles can be associated with job roles and personas, providing a detailed breakdown of responsibilities within the organization and aligning roles with specific personal profiles or job positions. These relationships ensure that business roles serve as a bridge between organizational structure, business processes, and individual responsibilities, creating a cohesive and comprehensive framework for enterprise architecture.

Business Footprint Diagram

A *business footprint diagram* is a graphical representation that illustrates the various components, relationships, and interactions within an organization's business environment. It typically includes elements such as business goals, organizational units, business functions, applications, and technical components. This diagram serves as a comprehensive map, providing clarity on the structure and functioning of the enterprise. Understanding the business footprint is crucial for effective enterprise architecture planning as it enables stakeholders to visualize the current state of the organization, identify areas for improvement, and devise strategies for alignment with business goals. By mapping out the business footprint as in Figure 2.14, organizations can make informed decisions regarding resource allocation, technology investments, and process optimization, ultimately enhancing their operational efficiency and driving business success.

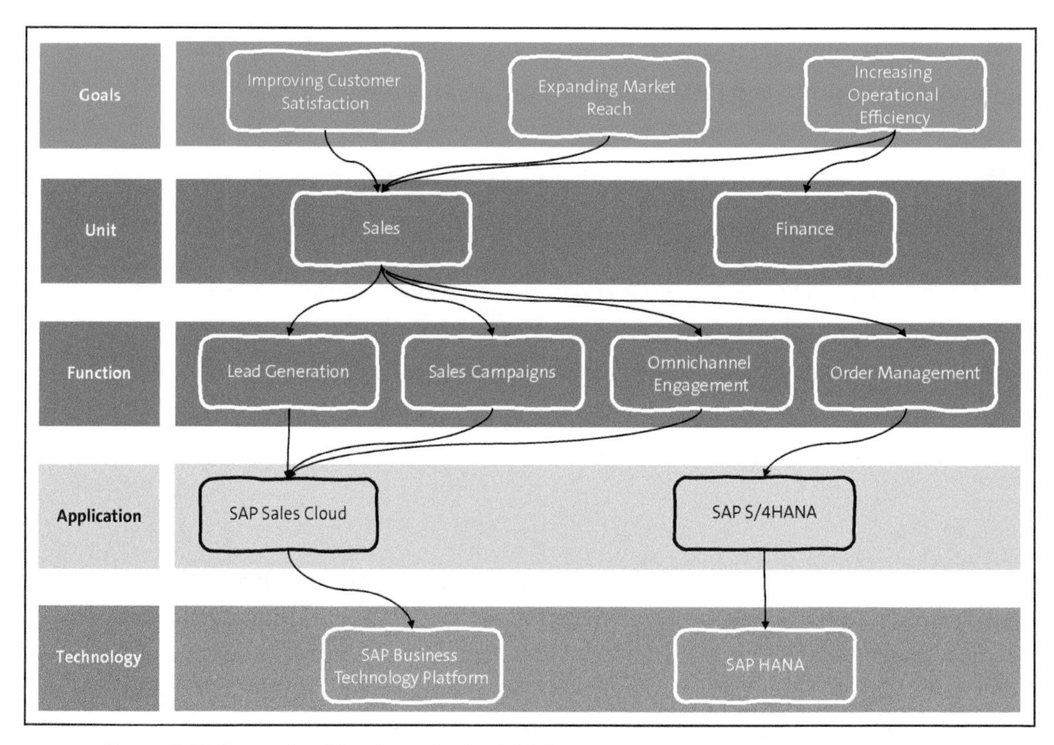

Figure 2.14 Example of Business Footprint Diagram

You can create a business footprint diagram by performing the following steps:

1. **Gather requirements**
   Understand the business goals, organizational structure, and key functions of the company. Identify stakeholders and gather input on their needs and requirements.

2. **Define components**
   List the organizational units, business functions, services, and technical components relevant to the business. This may involve conducting interviews and workshops or reviewing existing documentation.
3. **Establish relationships**
   Determine the links between business goals, organizational units, business functions, and services. Map out how these functions rely on each other to achieve the company's objectives.
4. **Map technical components**
   Identify the technical components (e.g., software applications, hardware infrastructure) that support the business functions. Connect these components to their corresponding services and functions.
5. **Create visualization**
   Use a diagramming tool or software to visually represent the relationships and connections you've identified in the previous steps. Arrange the components in a logical layout that is easy to understand and annotate the diagram as needed to provide clarity.

2.2.4 Solution Architecture

The solution architecture domain covers reference, base, target, and transition architectures, focusing on capabilities, processes, data, and organizational structure. Its primary value lies in aligning technology solutions with business goals and requirements. Acting as a bridge between the business and technical sides of an organization, it ensures that technology solutions are designed and implemented in a way that meets business needs, providing a cohesive approach to integrating technology with business operations.

We'll review the key components of solution architecture in the following sections.

Product Map

A *product map* describes the functional ability of a single or multiple solution components to implement and support a specific business capability. The product map captures solution capabilities structured by business domains, business areas, and possibly business capabilities or solution components for reference purposes. Figure 2.15 shows an example of a product map for asset management. Benefits of a product map include the following:

- It lists relevant solution capabilities structured by business needs.
- It helps to identify relevant solution components and the role they can play in fulfilling the business scope.
- It supports the comparison and value proposition of solution components.

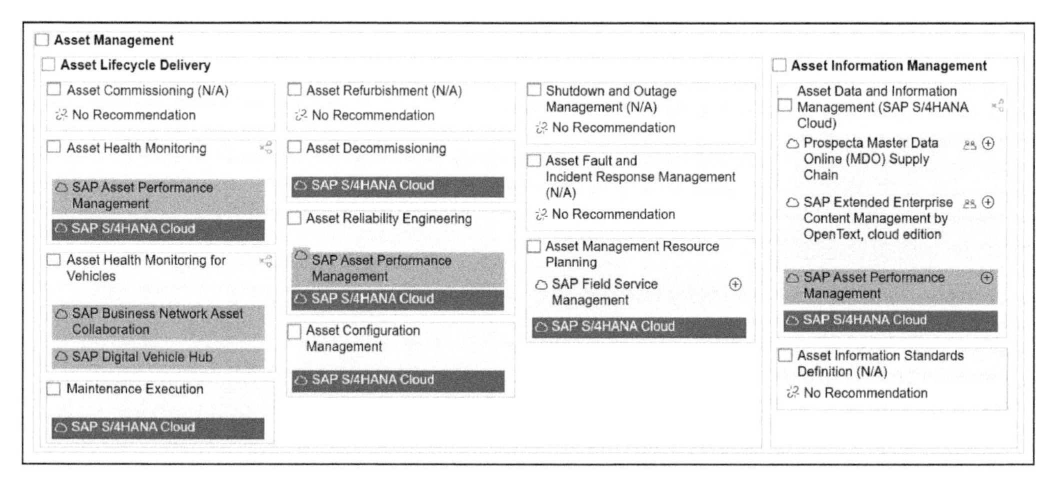

Figure 2.15 Example of Product Map for Asset Management

Creating a product map involves a structured approach to aligning business needs with technical solutions. Initially, you identify the business scope by structuring the product map into relevant business domains and business areas, possibly further refined by relevant business capabilities at a granular level 3. Once you identify the business scope, you assign supporting solution capabilities to address the identified business areas or business capabilities, thus potentially offering multiple alternative solutions for flexibility. Subsequently, for each solution capability, you should list the required and optional solution components, clearly describing the components that are necessary to fulfill the capability and those that offer additional features or enhancements. Using this structured methodology ensures that the product map effectively captures the alignment between business requirements and technical solutions and thus facilitates informed decision-making and solution design processes.

You can depict product maps in different layouts and different heat mapping, depending on the context in which the product map will be used. For example, in the context of application rationalization, you can establish a product map on the tolerate, invest, migrate, or eliminate (TIME) methodology or the rehost, replatform, repurchase, refactor, retire, and retain (6R) methodology. When using SAP LeanIX, this comes out of the box, and you can immediately apply it to visualize the product map. (For additional details on SAP LeanIX, see Part III.)

Solution Component Diagram

A *solution component diagram* serves as a graphical representation of the essential solution components mandated for a specific solution variant. It accurately outlines the deployment possibilities of these components and explains the complex integration points among them, referencing corresponding implementation artifacts like application programming interfaces (APIs) and integration flows. Its significance lies in its capacity to clarify the collection of solution components for addressing a given business scenario, thus providing stakeholders with a comprehensive understanding of the solution's architecture. Additionally, the diagram outlines the necessary integration channels among these solution components, offering insights into their interdependencies and interactions that are critical for seamless operation. Moreover, it aids in the identification of crucial implementation artifacts such as APIs, communication scenarios, and recommended middleware content, thus empowering organizations to streamline their implementation processes and ensure optimal performance.

Solution components are modular, software-based units that are essential for constructing an entire solution and are tailored to fulfill specific business needs. These components, which can manifest as configured applications or services, are designed to perform distinct functions within the solution architecture. They can be further categorized into more granular components or grouped into coarser-grained units to describe the overall structure of the solution. For instance, in Figure 2.16, there are eight top-level solution components (such as SAP SuccessFactors HXM Suite and SAP Master Data Integration), and each represents a distinct functional aspect of the solution. Moreover, these solution components may be hierarchically organized, with higher-level components (like SAP SuccessFactors HXM Suite) encapsulating finer-grained components like SAP SuccessFactors Employee Central Payroll. Additionally, deployment units are demarcated by a cloud icon, distinguishing them from other solution components within the architecture. Overall, solution components serve as the building blocks of a solution, delineating its structure and functionality in a modular and scalable manner.

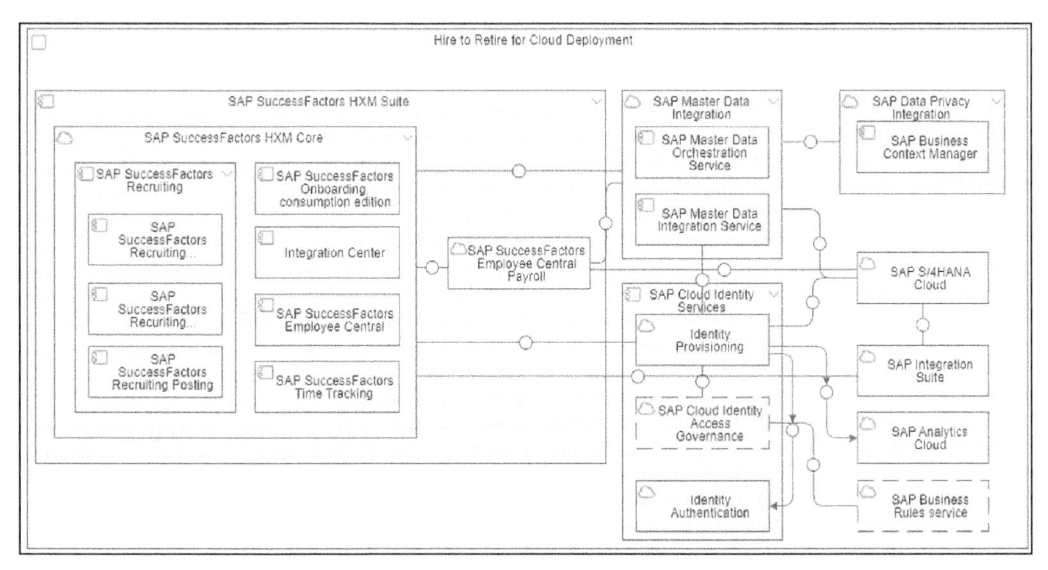

Figure 2.16 Example of Solution Component Diagram for Solution Process Hire-to-Retire (Refer to https://api.sap.com/)

Creating a solution component diagram involves a methodical process comprising several key steps:

1. **Outline the solution variant**
   Get clear about the business scenario (e.g., the business process variant). The solution will address and determine the name of the solution variant.
2. **Identify and structure solution components**
   Identify the solution components required to address the scope of the solution variant as well as relevant external components and understand the role they play within the solution variant. Then, group or detail the solution components to transfer required details in a clearly arranged manner.
3. **Introduce communication channels**
   Introduce schematic communication flows between the solution components.
   Note that the communication channels between solution components must be integrated with each other.
4. **Consider details**
   Take the details (e.g., associated communication scenarios, APIs, message flows) into account as required.

Solution Value Flow Diagram

A *value flow* is a sequence of activities and interactions that deliver value to stakeholders. It differs from a business process, which is typically depicted in a BPMN diagram that details specific steps and tasks. Value flows can be nested in several hierarchies, but on the lowest level, the main entity is the business activity that is the main step of the value flow. In addition, each business activity can be mapped to corresponding business capabilities. Refer to Section 2.3.1 for additional details.

The artifact known as a *solution value flow* encapsulates an abstract representation of a solution process, showcasing a collection of business activities aimed at generating additional value for stakeholders. These addressed business activities are linked with relevant solution components and solution capabilities that are responsible for their implementation. Typically depicted through solution value flow diagrams, these visualizations offer insights into how value is created and propagated throughout the solution architecture. Additionally, these diagrams may incorporate the display of associated solution components and solution capabilities, providing a comprehensive view of the underlying mechanisms driving value creation within the enterprise architecture. Overall, solution value flows serve as invaluable tools for understanding and optimizing the flow of value across business processes and technical solutions.

Figure 2.17 illustrates an example of a solution value flow diagram for the **Market products and services (generic)** business segment. Within this diagram, specific business activities such as **Execute promotional activities** or **Perform customer profiling** are correlated with corresponding solution capabilities like **Digital Asset Management** or

Customer Profile Management, and they are also associated with solution components such as **SAP Customer Data Cloud**.

»» Market products and services (generic)	Execute promotional activities	Perform customer profiling	Analyze and respond to customer insight
SAP Customer Data Cloud		Customer Profile Management (CustData Platform, CustData CLD)	
SAP Customer Data Platform		Customer Profile Management (CustData Platform, CustData CLD)	Customer Data Analytics (CustData Platform)
SAP Digital Asset Management Cloud by OpenText	Digital Asset Management (OpenText CLD)		
SAP Emarsys Customer Engagement	Marketing Campaign Management (Emarsys) Social Media Management (Emarsys)		Social Media Management (Emarsys) Marketing Analytics (Emarsys) Recommendation Management (Emarsys)
SAP Omnichannel Promotion Pricing	Promotion Execution (OPP CLD)		
No Recommendation	Marketing Collaboration (N/A)		

Figure 2.17 Example of Solution Value Flow Diagram

Creating a solution value flow artifact follows a structured process to elucidate value-adding processes within the solution architecture:

1. **Specify business activities**
   Depict all addressed value-adding business activities in a logical order that is conducive to easy explanation of the solution process. Note that the actual execution sequence may differ from the depicted order.
2. **Identify solution capabilities and solution components**
   Assign business activities to relevant solution capabilities and solution components that are responsible for realizing them.

Solution Process Flow Diagram

The *solution process flow diagram* provides a comprehensive depiction of the solution process, illustrating its sequential execution and interactions within an ecosystem. Functioning as a behavioral diagram, it articulates the solution process through a tangible flow. Leveraging the framework of BPMN 2.0 collaboration or process diagrams, it incorporates SAP EA Methodology extensions to delineate integration architecture nuances. Within this framework, the rectangular areas (which we call *pools*) symbolize deployment units and encapsulate solution components, while message flows elucidate the collaboration and integration dynamics among these units. Lanes, on the other hand, serve to delineate subsolution components or application roles, offering a granular view of the process landscape. Through this structured representation, the solution process flow diagram serves as a vital tool for comprehending, analyzing, and optimizing complex solution architectures. Figure 2.18 depicts an example of a solution process flow diagram for performing quality inspections.

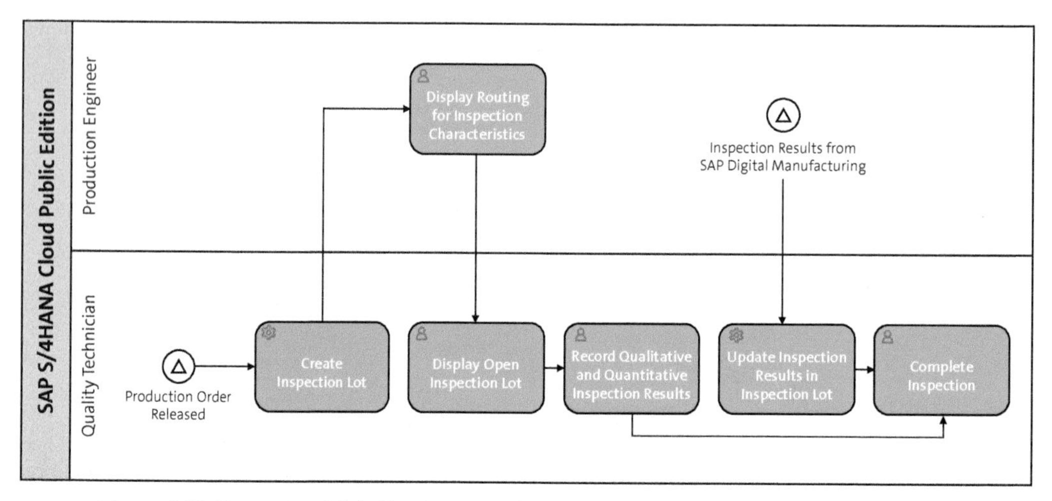

Figure 2.18 Example of Solution Process Flow Diagram for Performing Quality Inspection Process

Creating a solution process flow diagram involves several key activities aimed at capturing the essence of the solution process and its components. Here are the main activities:

1. **Specify flow of solution activities**
   Depict all relevant solution activities, each of which represents one or a part of one or multiple business activities and the ordering constraints between them. Define the flow of solution activities, including parallel and optional process flow paths.
2. **Identify solution components**
   Assign solution activities to solution components, realizing them in the context of the given solution process flow. You can assign a given solution activity to exactly one solution component, which is represented as a pool in the diagram.
3. **Specify integration**
   Define how integration of relevant solution components is realized in the context of the given solution process flow. Use message flows to depict different integration points between any two solution components.

Solution Data Flow Diagram

The *solution data flow diagram* serves as a visual representation of the data flow within a particular solution, revealing the exchange of information among solution components. At its core, the diagram illustrates the interconnectedness of solution components and their associated solution data objects, showcasing how data travels through the system. Through the depiction of data flows, it highlights the journey of data originating from the system of records and its propagation among various solution components. By providing a structured view of data movement, the solution data flow diagram aids in your understanding of the data exchange dynamics within the solution architecture, thus facilitating effective analysis and optimization of data flow processes. Figure 2.19 illustrates a sample solution data flow diagram for the hire-to-retire process within a cloud deployment scenario.

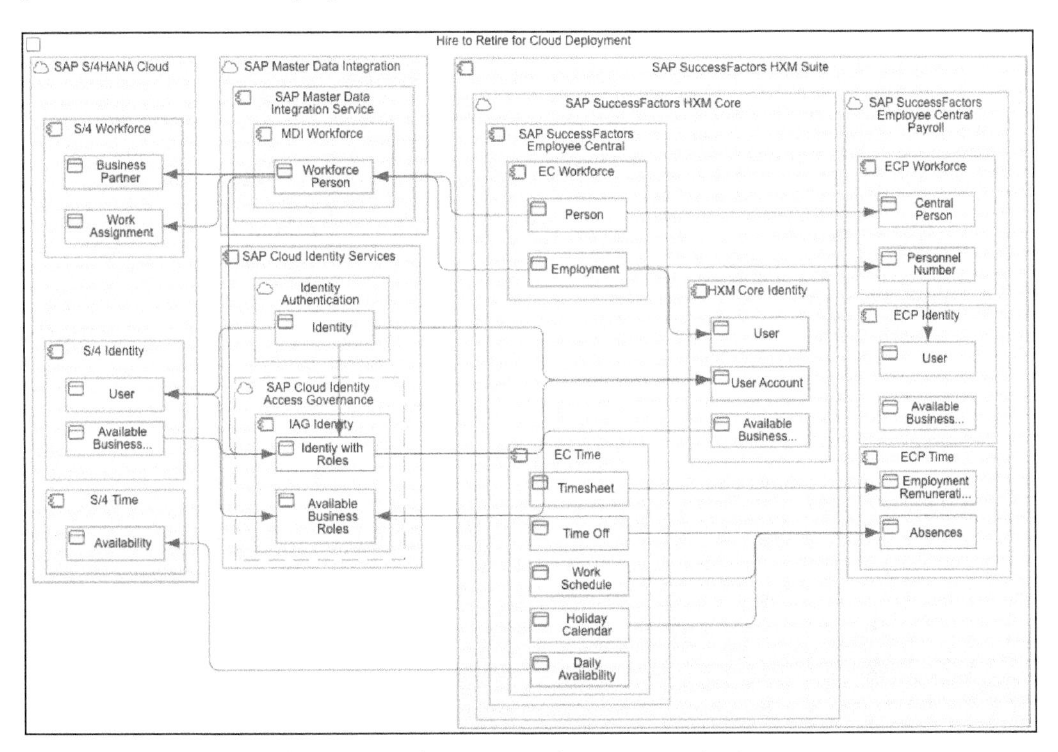

Figure 2.19 Example of Solution Data Flow Diagram for Hire-to-Retire Process (Refer to https://api.sap.com/)

Creating a solution data flow diagram involves a series of systematic steps aimed at capturing the intricacies of data flow within the solution architecture. Here are the main steps:

1. **Identify solution data objects**
   Specify all solution data objects within the relevant solution components, emphasizing their relevance to integration processes. These objects serve as specific realizations of business objects, representing parts or combinations of business objects tailored to the solution components.

2. **Specify data structure**
   Define the structure of the identified solution data objects, particularly focusing on

their attributes and relationships. In the context of SAP, the structure is typically outlined by SAP One Domain Model, which provides a standardized framework for data representation.

3. **Specify data flows**
   Establish dependencies between identified solution data objects by defining the flow of data among relevant solution components. Data flows are the pathways by which data moves among solution data objects situated in different solution components. Each endpoint of a data flow signifies a solution data object, indicating the direction of data transmission from the sending object to the receiving object.

SAP One Domain Model

SAP One Domain Model is a standardized data model that ensures consistent data integration across various SAP applications. It aims to eliminate data silos by providing a unified view of business data, facilitating seamless integration and interoperability. SAP One Domain Model serves as a common language that enables different SAP solutions to communicate effectively, ensuring that data remains consistent, accurate, and accessible across the enterprise. This model is crucial for businesses looking to harmonize their data landscape and drive digital transformation.

Software Distribution Diagram

The *software distribution diagram* provides a high-level overview of how individual solution components or building blocks are structured and distributed across the IT infrastructure and underlying technology foundation. Its level of detail is contingent upon the granularity of the solution components under consideration. Typically, the diagram categorizes environments into on-premise data centers, private clouds (designated private areas within an IT partner's data center), and public clouds (services hosted in public hyperscaler data centers). This visualization aids in understanding the potential impact, risks, and connectivity challenges associated with the selected solution components during the application and data architecture phase. While the diagram primarily focuses on these aspects, additional details pertaining to the physical locations of data centers, their connectivity (such as virtual private or interconnections), and specifics regarding the operational software layer (e.g., virtualization, container technology) are addressed within the technology architecture domain (Section 2.2.5).

Creating a software distribution diagram like the one in Figure 2.20 involves several systematic steps aimed at illustrating the distribution of solution components among various deployment models and environments.

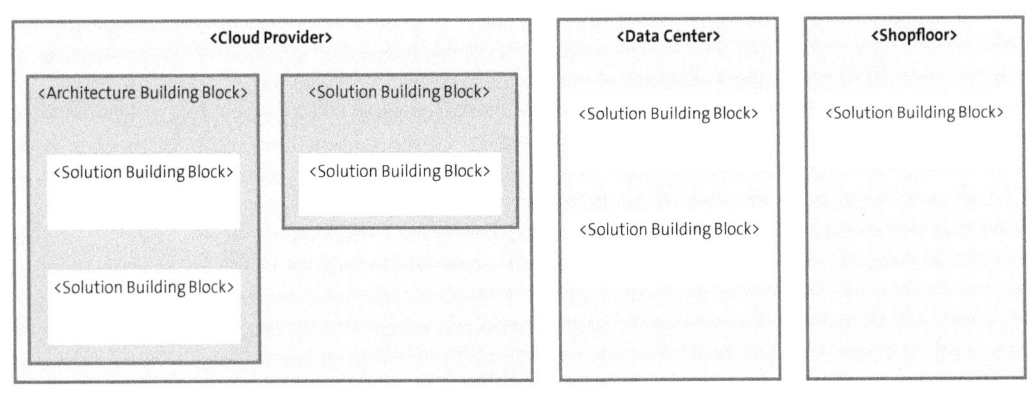

Figure 2.20 Template for Software Distribution Diagram

Here are the typical steps:

1. **Identify solution components and deployment models**
   Begin by identifying the key building blocks of the solution variant. This entails understanding the specific deployment types or deployment models involved. Utilize content from the related solution component diagram, if available, to inform this process.
2. **Depict relationships among the identified solution components**
   Visualize the communication relationships among the solution components within the diagram. This step is crucial for understanding potential latency considerations or network requirements associated with the solution architecture.
3. **Identify hosting and deployment environments as well as infrastructure providers**
   Map the identified solution components to their respective hosting and deployment environments, taking into account the infrastructure providers involved. This step helps to establish a clear understanding of where each component resides within the overall IT infrastructure landscape.

Application Architecture Overview Diagram

The *application architecture overview diagram* serves as a comprehensive visualization of the solution components and their interconnections within an enterprise architecture. It encapsulates both SAP and third-party solution components, offering a holistic perspective on the architecture landscape. Unlike the formalized structure of solution component diagrams, this overview diagram presents a more illustrative and accessible representation, making it easier to grasp the architecture's essence at a glance. While it potentially encompasses a broader scope than the solution component diagram suggested by SAP Reference Solution Architecture, it must align with and complement the architecture description outlined in the corresponding solution component diagram. The application architecture overview diagram thus plays a crucial role in providing stakeholders with a clear, cohesive understanding of the enterprise architecture's composition and relationships.

Figure 2.21 is an application architecture overview diagram example showing different solutions like SAP Ariba, SAP Business Technology Platform (SAP BTP), and SAP S/4HANA, including their components like product compliance (in SAP S/4HANA) and SAP Ariba Sourcing (as part of SAP Ariba).

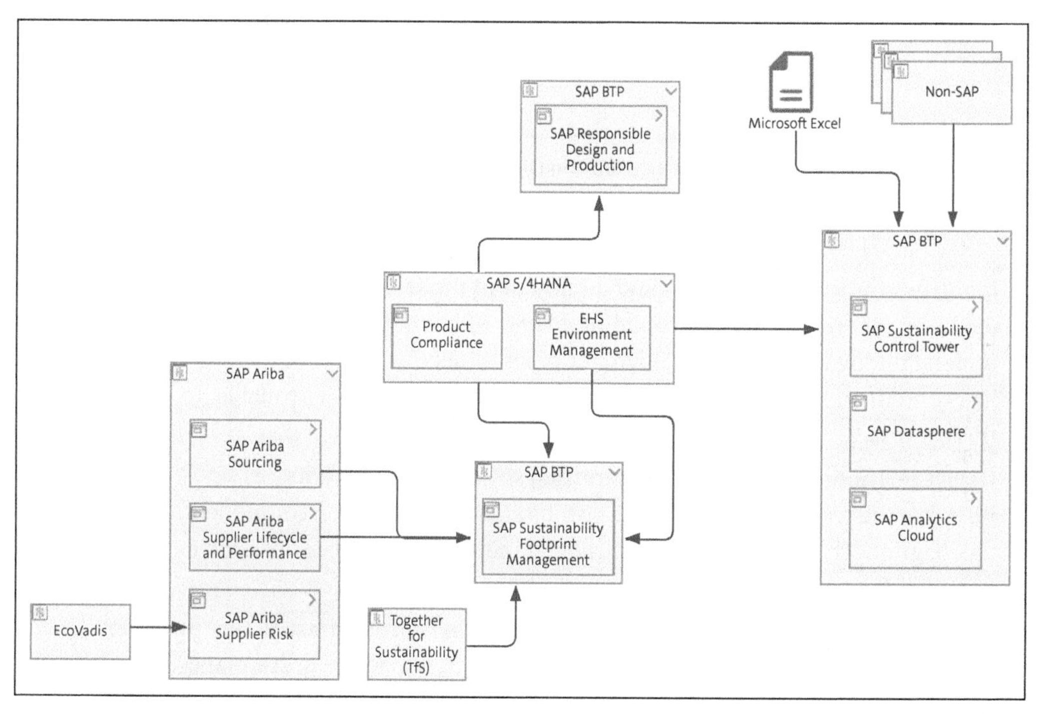

Figure 2.21 Example of Application Architecture Overview Diagram with Focus on Sustainability

Creating an architecture overview diagram involves a structured approach that begins with the identification and organization of solution building blocks. This process may utilize existing artifacts, such as a solution concept diagram if available, to define the key components of the solution variant. Additionally, you can highlight areas lacking SAP solution components (which we call *white spots*) to indicate potential gaps in the architecture. To enhance clarity, you can employ icons and color-coding to distinguish between SAP and non-SAP components. It is therefore essential that you understand the dependencies between these components. Leveraging SAP reference architecture facilitates this task by mapping out the required solution components. By comprehending these dependencies, such as application interfaces, the diagram effectively captures the intricate relationships and dependencies within the architecture, providing stakeholders with invaluable insights into the system's structure and functionality.

Conceptional Data Diagram

A *conceptual data diagram* is a visual representation that outlines the data entities involved in a solution building block and the relationships among them. This artifact is fundamental for understanding the data architecture of an enterprise because it illustrates how different data entities are structured and interconnected. It provides a high-level view of the key data elements and their associations, serving as a blueprint for data management and integration within the organization.

Developing a conceptual data diagram like the one in Figure 2.22 involves the following key steps:

1. **Identify relevant data entities**
   Explore the business process and activities to identify the primary data entities involved.
2. **Define attributes**
   For each data entity, identify and name key attributes with appropriate data types.
3. **Establish relationships**
   Identify the relationships between data entities, such as one-to-one, one-to-many, or many-to-many, and represent them graphically.
4. **Create the visual representation**
   Use diagramming tools to create a visual representation of the entities and their relationships.

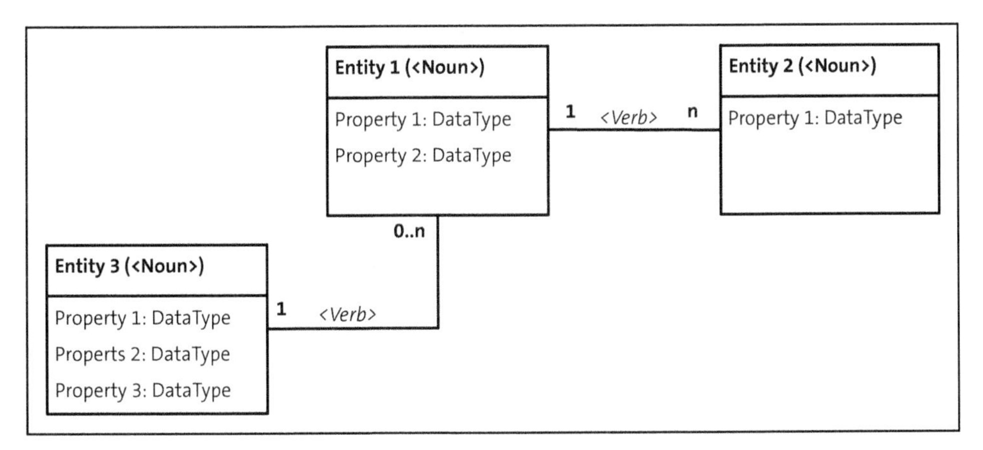

Figure 2.22 Conceptual Data Diagram

This artifact is also related to several other key artifacts. It supports the business value flow by detailing the necessary data for process execution and aligning data dependencies with value streams. It complements the organization map by defining role-based data access and stewardship, ensuring data security and governance. Additionally, it connects with other enterprise architecture artifacts like the business process catalog, business role diagram, solution value and process flows, and business capability models, thus ensuring that data architecture aligns with business architecture and organizational structure.

2.2.5 Technology Architecture

This domain is concerned with documenting how the target solution architecture building blocks are delivered through various technology components, such as operating systems, virtualized environments, hardware, and networks. It provides a detailed depiction of the deployment of the organization's IT systems in specific data center locations, ensuring that the technological infrastructure supports the overall architecture effectively.

There's one key diagram available for technology architecture, and we'll discuss it in this section.

Environment and Location Diagram

The *environment and location diagram* serves as a comprehensive visual representation of the deployment and operation of solution building blocks in various data center locations. It defines the required physical connections between these blocks, explaining the infrastructure supporting them. Moreover, it factors in the geographic distribution of users and external systems that are essential for interacting with the solution components. This integrated perspective extends to delineating data exchange pathways over public lines and identifying areas necessitating VPN security enhancements. By encapsulating these elements, the diagram provides a holistic view that is vital for orchestrating a robust and secure architectural framework. Figure 2.23 shows a template that you can use in the SAP context for an environment and location diagram.

Creating the environment and location diagram involves the following steps:

1. Begin with the previously created software distribution diagram and advance it into the environment and location diagram.
2. Identify the deployment environments where your building blocks are operational, then enhance the landing zones of the software distribution diagram by explicitly specifying data center providers, data center locations, and infrastructure details for each landing zone.
3. Align the solution building blocks, as outlined in the solution realization diagram, with the corresponding deployment environments.
4. Visualize the relationships, such as request-response or information flow, among the building blocks. This visualization aids in your understanding of network requirements.

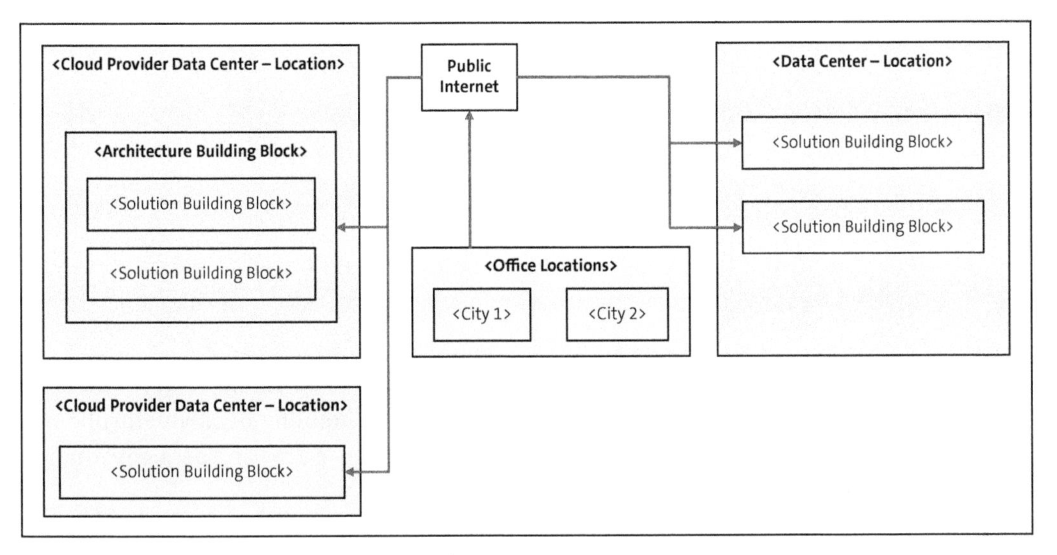

Figure 2.23 Template for Environment and Location Diagram

2.2.6 Roadmap and Transition

The roadmap and transition domain focuses on identifying and planning required initiatives and managing the transition of systems or processes. It involves pinpointing potential advantages, planning migrations, and executing changes to achieve the project's objectives efficiently. This domain ensures that transitions are smooth and that the organization can adapt to new systems or processes without significant disruptions.

We'll walk through the key components of the roadmap and transition domain in the following sections.

Architecture Roadmap

The *architecture roadmap* in enterprise architecture encompasses two primary artifacts that serve distinct yet interrelated purposes:

- The *business architecture roadmap* offers a visual depiction of business objectives over varying timeframes and functional groupings. It provides stakeholders with a clear understanding of the anticipated business outcomes and how they align with strategic goals.
- The *application architecture roadmap* delineates the deployment sequence of individual solution building blocks relative to business lines or value drivers on a specified timeline. It guides the implementation journey by outlining the order of steps needed to transition from a baseline architecture to a desired target architecture.

These roadmaps work in tandem, offering complementary perspectives and facilitating a cohesive approach to architectural planning and execution within the enterprise.

Creating a comprehensive architecture roadmap involves considering several influencing factors spanning various dimensions of the organization. These factors include the following:

- Delivery scope, which encompasses the extent and boundaries of the project's delivery, as well as the specific business segments, geographical entities, and lines of business (LoBs) affected
- Business architecture, which entails mapping out the business operating model, strategy, pain points, and prioritization of capabilities based on value and change drivers
- Application architecture considerations, which involve detailing the as-is and to-be architecture, aligning with SAP product strategies, defining deployment options, sequencing implementations, ensuring user experience (UX) and data governance, and integrating intelligent capabilities
- Technical architecture considerations, which span platform strategies, sizing, intelligent technologies, and hyperscaler strategies
- Transformation programs, all within the context of boundary conditions, ongoing initiatives, and timelines

The first step in crafting an architecture roadmap typically involves assembling an *initiatives catalog* like the one in Figure 2.24. This entails creating a list of initiatives derived from the target architecture. Moreover, where feasible, it's advantageous to connect these initiatives with customer goals to establish an *initiative/goals map*. This map provides a clear picture of how each initiative relates to specific customer objectives, thereby enhancing the strategic planning process. Each initiative can be assessed using a *value-feasibility matrix*, which takes into account the initiative's ease of implementation and the value it generates. This process aids in prioritizing initiatives that offer maximum value while considering the implementation effort required. The emphasis should be on initiatives with high value and low implementation effort as quick wins, with others of high value targeted for later stages. Additionally, it's advisable to map dependencies among initiatives within the same catalog of initiatives.

No.	Initiative	Description	Goal Contribution	Dependencies on Other Initiatives	Value Assessment	Ease of Implementation
A1	Conversion Assessment	Conduct a conversion assessment study for a brownfield conversion of the SAP ERP system PRD to SAP S/4HANA.	Goal 2	Initiative B1 and D4	High	Medium
A2	Data Cleansing Concept	Develop a detailed concept on data cleansing requirements as a preproject of the planned SAP S/4HANA conversion project.	Goal 2	Initiative A1 and D3	High	Low
A3	...	...	...		...	..

Figure 2.24 Initiatives Catalog

Utilizing the initiatives catalog as the cornerstone, you can formulate the roadmap. Initially, you can group the primary initiatives based on business or IT initiatives, LoB-specific endeavors (such as finance or procurement), end-to-end business processes, SAP products, or a combination of diverse criteria. Further structural elements may include phases of transformation programs (e.g., foundation, quick wins, innovation) or alignment with the organization's key strategic objectives (e.g., supporting a new business model). During the creation of these clusters, it's imperative to consider dependencies and analyze and determine sequences. Ultimately, you can illustrate the roadmap in various formats, one example of which is provided in Figure 2.25.

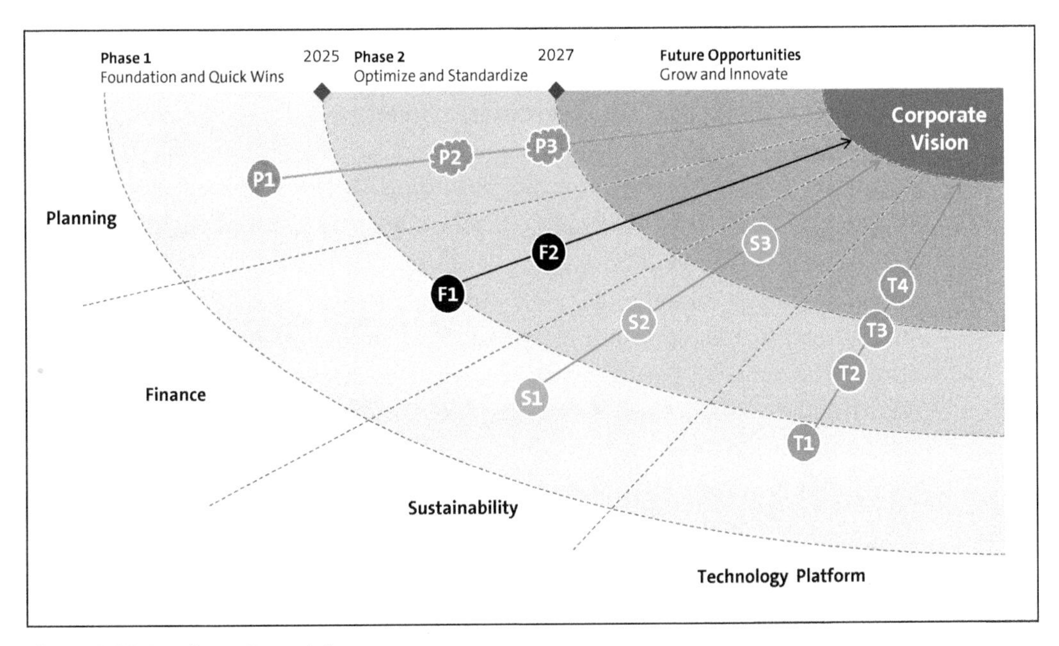

Figure 2.25 Roadmap Template

The architecture roadmap, alongside the application architecture overview diagram, typically stands out as one of the most crucial artifacts produced during planning activities because it synchronizes all elements from various artifacts on a timeline. However, it's important to note that you should not treat the roadmap as a detailed project plan at this stage, as doing so would entail including additional levels of detail that are not intended during the architecture planning phase. Nonetheless, when coupled with the work breakdown structure discussed in the next section, the architecture roadmap serves as the primary foundation for all subsequent program or project-related planning endeavors.

Work Breakdown Structure

The *work breakdown structure* (WBS) is a crucial artifact in project management because it functions as a hierarchical framework that dissects a project into smaller, more manageable components. Its primary purpose is to provide a clear definition, organization, and control mechanism for the entirety of a project's lifetime. By structuring initiatives into discrete sections and deliverables, the WBS enables teams to better allocate resources and track progress effectively. Moreover, it serves as a foundational tool for estimating both the effort and the cost associated with each initiative within the project. Typically, collaboration between enterprise architects and the implementation team lead facilitates the development of the WBS, effort, and cost estimations, marking a pivotal transition from the architecture planning phase to the implementation phase of the project.

You can create the work breakdown structure by following these steps:

1. Define major deliverables and milestones and break down the initiative in scope into deliverables, alongside project phases, milestones, and activities. First, identify the major outcomes or deliverables that need to be achieved throughout the project. Then, break down the entire project scope into smaller, more manageable deliverables, aligning them with project phases, milestones, and specific activities required to accomplish them.
2. Estimate resource requirements and estimate relevant resources per line item (internal, external, license, third-party costs, etc.). Once the deliverables are defined, estimate the resources needed to complete each line item effectively. Consider both internal and external resources, licensing needs, and any third-party costs that may arise during the project.
3. Align with key stakeholders and review and agree with them on the breakdown structure, effort, and cost. Engage key stakeholders throughout the WBS creation process to ensure alignment with project goals and expectations. Present to stakeholders the breakdown structure, along with effort and cost estimations, for their review and approval. Collaboratively refine as needed to gain consensus and ensure everyone's understanding and agreement.

2.2.7 Requirements and Governance

The requirements and governance domain is not linked to a specific phase of an architecture engagement, but it includes relevant governance artifacts such as risk catalogs and architecture principles. This domain captures important architectural decisions as *architecture decision records* (ADRs) and manages requirements that can emerge in any architecture domain. It ensures that governance is maintained throughout the project, supporting consistency and compliance with architectural principles and policies.

Let's explore the key components of requirements and governance in the next sections.

Requirements Catalog

The *requirements catalog* is a structured document that captures and organizes both the functional and the nonfunctional requirements for a specific project, system, or product. It plays a crucial role in identifying and collecting high-level business needs, which helps to clearly define the objectives of the business, leading to a more focused and aligned approach to project development. Through a refinement process, the initial high-level requirements undergo thorough analysis, resulting in clear and actionable requirements that can be effectively implemented. By classifying requirements, the catalog enables easier review and tracking, ensuring that each requirement is considered by the appropriate owner and well documented for subsequent analysis. This process ultimately enhances overall project management and communication.

To create a requirements catalog, like the one in Figure 2.26, you can perform the following steps:

1. Document all identified high-level business needs and requirements.
2. Classify requirements as either functional or nonfunctional.
3. Further classify nonfunctional requirements using classes of *nonfunctional requirements*.
4. Categorize requirements based on business capabilities, distinguishing between strategic business requirements and operational business requirements.
5. Maintain additional attributes for business requirements such as status, priority, and owner.

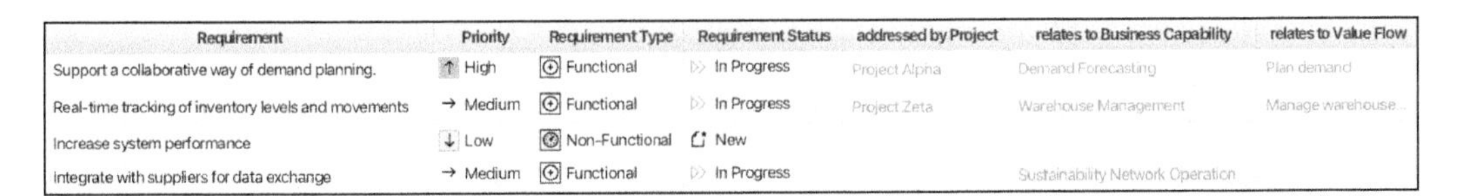

Requirement	Priority	Requirement Type	Requirement Status	addressed by Project	relates to Business Capability	relates to Value Flow
Support a collaborative way of demand planning.	High	Functional	In Progress	Project Alpha	Demand Forecasting	Plan demand
Real-time tracking of inventory levels and movements	Medium	Functional	In Progress	Project Zeta	Warehouse Management	Manage warehouse...
Increase system performance	Low	Non-Functional	New			
integrate with suppliers for data exchange	Medium	Functional	In Progress		Sustainability Network Operation	

Figure 2.26 Example of Requirements Catalog

Architecture Principles Catalog

Principles serve as foundational guidelines in enterprise architecture planning, expressing overarching, long-term objectives that drive organizational transformation. They are high-level and enduring, rooted in strategic business and IT goals rather than reactive to day-to-day demands. An example principle, "Subscribe before buy before build," highlights this strategic focus by emphasizing the prioritization of subscription-based solutions over immediate purchase or in-house development.

In contrast, *policies* are actionable rules that guide the application of enterprise architecture. While they may be influenced by current trends, policies refrain from specifying particular products or protocols. For instance, a policy dictating the adoption order of cloud solutions—such as software as a service (SaaS), platform as a service (PaaS), and infrastructure as a service (IaaS)—aligns with strategic objectives without naming specific vendors or platforms.

Finally, *standards* precisely define the products, best practices, and protocols that support policies. While they aim for longevity, standards may change more frequently than policies, ensuring alignment with evolving technology landscapes. An example standard, such as Amazon Web Services (AWS) as the designated cloud IaaS provider, illustrates the specificity of standards compared with the broader scope of policies and principles.

Table 2.3 gives an example of the definition of a principle.

Principle	Simplify and standardize processes.
Statement	Streamline and standardize business processes across the organization to optimize efficiency and agility.
Rationale	Simplifying and standardizing processes in SAP S/4HANA facilitates seamless integration, reduces complexity, and enhances scalability. This principle promotes operational consistency, accelerates deployment timelines, and mitigates risks associated with fragmented or redundant processes.
Implications	Adhering to this principle entails conducting thorough process assessments, identifying opportunities for consolidation and optimization, and defining standardized processes aligned with industry best practices. It also requires change management efforts to ensure organizational buy-in and adoption of standardized processes.

Table 2.3 Example of Architecture Principal

Architecture principles are typically formulated by enterprise architects in collaboration with key stakeholders and subsequently endorsed by the architecture board. They draw upon existing principles at the enterprise level, if applicable, ensuring coherence and alignment with broader organizational goals. These principles are meticulously crafted to provide clear guidance for decision-making processes and are tailored to steer the architecture and implementation of the target architecture in harmony with business strategies and visions. The development of architecture principles is chiefly influenced by various factors, including the enterprise's mission, strategic plans, and organizational structure, as well as ongoing strategic initiatives and external constraints such as market dynamics and regulatory requirements. Additionally, consideration is given to the current technology landscape within the enterprise, encompassing existing systems, infrastructure, and emerging industry trends that may impact the enterprise environment. Through a comprehensive assessment of these factors, architecture principles are crafted to form a robust framework for driving architectural decisions and fostering strategic alignment across the organization.

Risk Catalog

The *risk catalog* serves as a comprehensive repository for systematically capturing, categorizing, and managing risks throughout the enterprise architecture process. This artifact encompasses various types of risks, including technical, operational, strategic, and regulatory, among others. By documenting risks alongside their associated impact, likelihood, and mitigation strategies, the risk catalog enables stakeholders to proactively identify and address potential threats to the successful execution of architectural initiatives. Additionally, it facilitates informed decision-making by providing a centralized reference point for assessing risk exposure and prioritizing risk mitigation efforts. Through regular updates and reviews, the risk catalog evolves dynamically, reflecting changes in the organizational landscape and ensuring that risk management remains an integral aspect of the architectural governance framework.

Creating a risk catalog for architecture planning like the one in Figure 2.27 involves the following steps:

1. **Identification**
   Collaborate with stakeholders to identify potential risks relevant to the architectural initiatives, considering factors like technological dependencies, stakeholder requirements, and regulatory compliance.
2. **Documentation**
   Document each identified risk along with its perceived frequency of occurrence and potential effect on architecture objectives.
3. **Assessment**
   Track the assessment status of each risk to monitor its evolution over time, facilitating proactive risk management.
4. **Ownership**
   Assign a risk owner to each identified risk to ensure accountability and effective implementation of mitigation strategies.
5. **Scope alignment**
   Align identified risks with the scope of the architecture program to contextualize their relevance and prioritize mitigation efforts accordingly.

Risk	Frequency	Risk Effect	Risk Assesment Status	Risk Owner	in scope of Program
Data migration failure	Likely	Critical	assessed	Data Architect	Demo Program
Integration Challenges with legacy systems	Occasionally	Critical	in evaluation	Integration Lead	
Resource constraints	Seldom	Marginal	new	Project Manager	
Security vulnerabilities	Unlikely	Catastr...	in evaluation	Security Lead	
Budget overruns	Seldom	Marginal	assessed	Finance Manager	

Figure 2.27 Example of Risk Catalog

Architectural Decisions Catalog

The *architectural decisions catalog* supports architecture planning as a comprehensive record of the key decisions made throughout the architectural process. This artifact encapsulates the rationale, considerations, and outcomes of significant architectural choices, providing valuable insights into the evolution of the architectural framework. Typically, the architecture board is responsible for making these decisions—but by documenting architecture decisions, stakeholders gain clarity on the reasoning behind specific design choices, helping them to gain better understanding of and alignment with overarching architectural goals and objectives. Moreover, this artifact serves as a reference point for future decision-making, facilitating consistency, transparency, and traceability within the architectural process.

To create an architectural decisions catalog like the one outlined in Figure 2.28, follow these steps:

1. **Identify architectural decisions to be taken**
   Begin by listing all the significant architectural decisions that need to be addressed.
2. **Assign priority**
   Prioritize each architectural decision based on factors such as impact, urgency, and complexity. This helps you focus on the most critical decisions first.
3. **Describe the decision taken**
   For each decision, provide a description. This should include what the decision is, why it was made, and the implications it has.
4. **Document the criteria that led to the decision**
   Outline the criteria and rationale that guided the decision-making process. This could include business requirements, technical constraints, performance considerations, and any other relevant factors.
5. **Optional: Cross-reference decision accelerators**
   If any decision accelerators (e.g., reference content from SAP) were used in making a decision, document them as well. Provide cross-references to the relevant resources for future reference.

Ref	Decision	Priority	Decision Description	Criteria	Reference to Decision Accelerator
1	SAP EWM	High	SAP EWM will be deployed embedded in SAP S/4HANA	Business processes integration, maintenance, latency, risk, effort	Deployment options for SAP EWM
2					
3					
4					
...	...	...	...	...	...

Figure 2.28 Example of an Architectural Decisions Catalog

2.3 SAP's Reference Content

Throughout the diverse activities involved in enterprise architecture planning, *reference content* plays a crucial role, providing invaluable guidance and insights. Whether enterprise architects are analyzing current systems, shaping future architectures, or aligning technology with business objectives, reference content serves them as a reliable resource. It offers a structured foundation for decision-making, ensuring architects can leverage established best practices, standards, and organizational knowledge effectively. Reference content simplifies the daily tasks of an enterprise architect by providing a ready repository of knowledge and insights. With access to established best practices, architects can quickly find solutions to common challenges and industry standards. Also, by leveraging reference materials, architects can reduce the time they spend on research, ensure consistency across projects, and mitigate risks effectively. Ultimately, reference content empowers architects to focus their energy on innovation and strategic planning, rather than constantly reinventing the wheel. It thus enhances productivity and drives the success of architectural initiatives.

The reference content within SAP EA Framework provides enterprise architects with distinct advantages over other frameworks like TOGAF and Zachman. It provides enterprise architects with a wealth of resources, including best practices, methodologies, and industry-specific insights tailored to SAP environments. With access to such comprehensive reference materials, architects can expedite their decision-making processes, enhance the quality of their designs, and ensure alignment with SAP-specific standards and principles. Furthermore, the richness of reference content within the SAP framework enables architects to navigate complexities more effectively, addressing challenges with greater confidence and precision. Ultimately, this advantage translates into streamlined architectural endeavors, improved implementation outcomes, and enhanced value realization for organizations leveraging SAP solutions.

Figure 2.29 shows the current publicly available or planned reference content that SAP provides as part of SAP EA Framework.

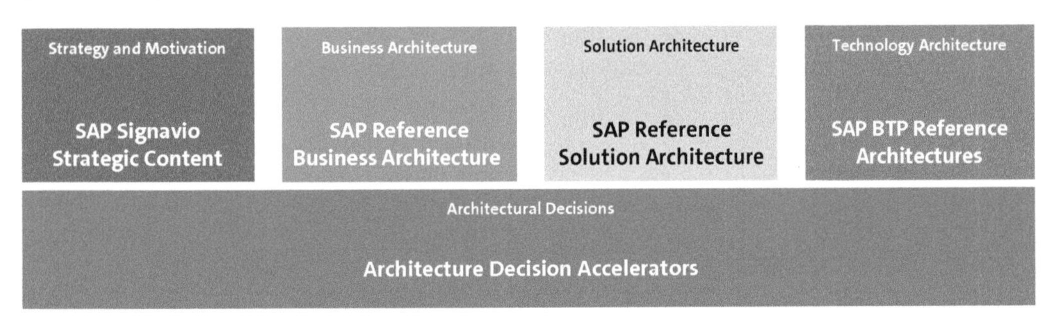

Figure 2.29 Overview of Available and Planned Reference Content for SAP EA Framework

SAP's reference architecture content primarily emphasizes business architecture and solution architecture, exemplified by SAP Reference Business Architecture and SAP Reference Solution Architecture. Moreover, SAP extends its offerings through SAP Signavio, providing strategic content for various industries from an outside-in perspective. Within the domain of technology architecture, SAP furnishes best-practice content tailored to diverse SAP BTP architectural use cases. Additionally, at the solution implementation level, SAP offers best-practice business process content in BPMN format, accompanied by configuration guides and test scripts. However, since this content pertains to solution implementation rather than enterprise architecture, it's not considered part of the enterprise architecture domain.

SAP reference architecture content offers a comprehensive approach to exploring and understanding SAP's solution offerings, guided by the structure and format of SAP EA Methodology and aligned with industry standards like TOGAF and the American Productivity & Quality Center (APQC). The APQC Cross-Industry Process Classification Framework (APQC Cross-Industry PCF) is a comprehensive model used to standardize business processes across different industries. The APQC Cross-Industry PCF includes a wide range of business processes, categorized into functional areas such as finance, human resources (HR), supply chain, and customer service, among others.

Central to the SAP EA Framework reference content are SAP Reference Business Architecture and SAP Reference Solution Architecture. The former provides standardized content for business architecture, and the latter maps recommended solutions to SAP's portfolio. They ensure consistency and clarity in communicating the architectural landscape of enterprises, thus facilitating informed decision-making and alignment with organizational goals.

In subsequent sections, we'll delve deeper into the key repositories of reference content provided by SAP for use with SAP EA Framework.

2.3.1 SAP Reference Business Architecture

SAP Reference Business Architecture encompasses diverse collections of architectural reference content focused on the business domain. Both SAP Reference Business Architecture and other reference sources adhere to a unified meta model that is akin to the artifacts within SAP EA Framework mentioned earlier. This alignment streamlines artifact development, ensuring coherence through a shared language. The primary components of SAP Reference Business Architecture include the following:

- Business capability model
- Business process model
- Business data model (planned)
- Business organizational model (planned)

As depicted in Figure 2.30, SAP Reference Business Architecture is organized in alignment with distinct enterprise domains, offering a range of end-to-end business process models and business capability models tailored to each domain. Moreover, SAP Reference Business Architecture incorporates industry-specific business capability and process models to cater to diverse sectors. Consequently, choosing an end-to-end business process model such as idea to market, with associated business capabilities in areas like research and development (R&D), engineering, and product management, can yield varied outcomes depending on the chosen industry. Additionally, alongside industry-specific iterations of the content, a generic perspective remains accessible in instances where a specific industry is not yet covered by SAP Reference Business Architecture.

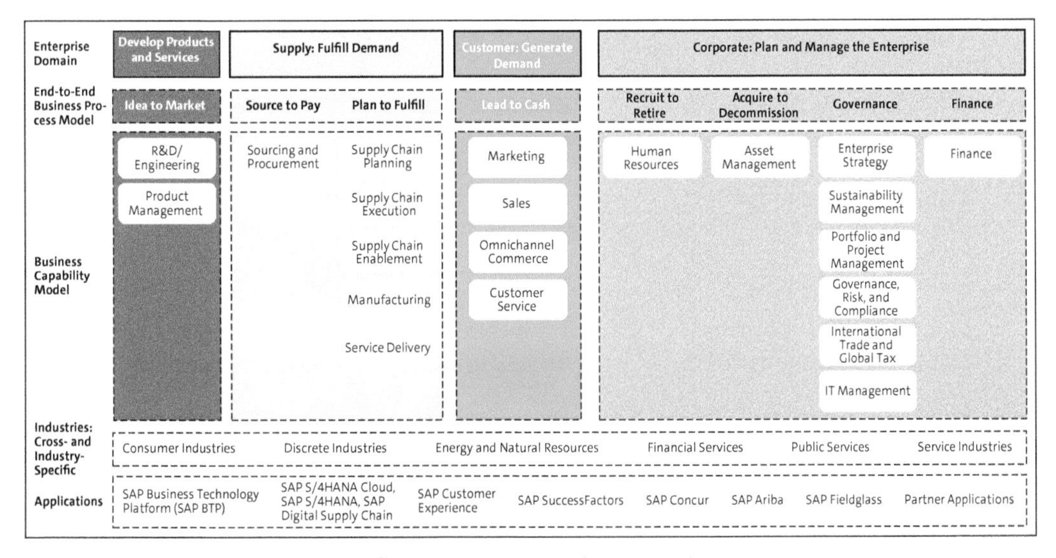

Figure 2.30 Introduction to SAP Reference Business Architecture

Figure 2.31 lists the core and supporting processes of SAP Reference Business Architecture. Each of the listed end-to-end business processes (e.g., source to pay) is further detailed into modular business processes (e.g., source to contract, procure to receipt).

Further exploration into detailed business activities beneath those modular business processes is consistently feasible, as illustrated in Figure 2.32. Under the modular business process **Source-to-contract**, designated business segments such as **Source products and services** are established as elements, providing structure to underlying business activities. For each business activity, such as **Prepare RFX** (request for proposal/information/quotation) or **Process RFX**, corresponding business capabilities are also mapped, including **Supplier RFX Preparation**, **Supplier RFX Execution**, and **Sourcing Collaboration**.

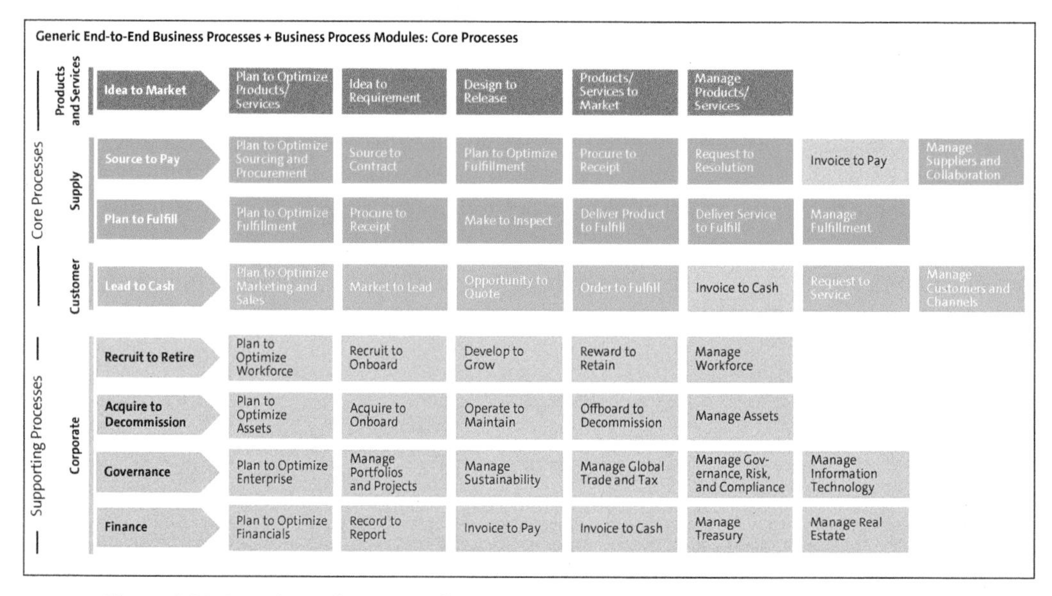

Figure 2.31 Overview of Supported SAP Reference Business Architecture Processes

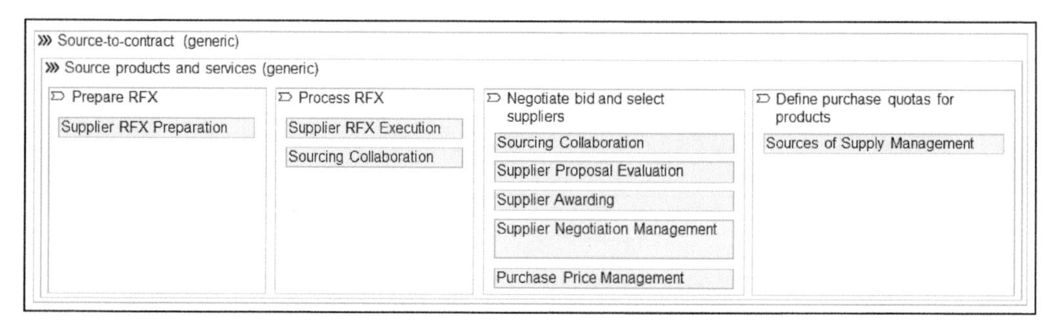

Figure 2.32 Source to Contract Drilldown in SAP Reference Business Architecture

SAP Reference Business Architecture not only supports general business processes but also accommodates industry-specific variations known as *business process variants*. For instance, within the source to pay domain, there exist specialized variants such as source to pay for direct physical products, source to pay for subcontracting, and source to pay for services, among others.

Furthermore, SAP Reference Business Architecture enables detailed scoping and exploration of business activities and capabilities, offering a comprehensive model encompassing both generic and industry-specific business capabilities. This model spans business domains (level 1), business areas (level 2), and detailed business capabilities (level 3), as illustrated on the business domain level in Figure 2.33 and the complete drill down from the corporate business domain to business area asset management and its related business capabilities in Figure 2.34.

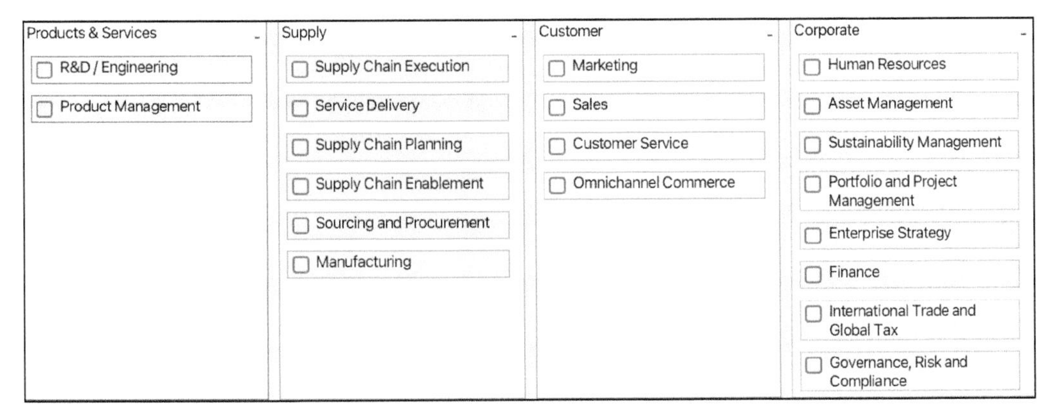

Figure 2.33 Business Domains from SAP Reference Business Architecture

Asset Management

Asset Management Strategy and Planning
Asset Policy Management
Asset Objectives Management
Asset Demand Management
Asset Strategy Management
Strategic Asset Management Planning
Operations and Maintenance Expenditure Planning
Tactical Asset Management Planning

Asset Lifecycle Delivery
Maintenance Execution
Asset Refurbishment
Asset Commissioning
Asset Decommissioning
Asset Reliability Engineering
Asset Configuration Management
Shutdown and Outage Management
Asset Fault and Incident Response Management
Asset Management Resource Planning
Asset Health Monitoring

Asset Review and Risk Management
Asset Review and Audit
Asset Costing and Valuation
Asset Performance Analysis
Asset Risk Assessment
Asset Contingency Planning and Resilience Analysis
Asset Stakeholder Management

Digital Asset Operations
Digital Asset Usage Analysis
Digital Asset Operations Controlling
Issue Resolution
Unplanned Downtime Prevention
Digital Asset Management Resource Planning
Operations Expenditure Planning
Digital Asset SLA Monitoring
Digital Asset Performance Analysis

Figure 2.34 Drilldown from Business Domain (Level 1) to Business Area (Level 2) to Business Capabilities (Level 3)

The primary benefit of SAP Reference Business Architecture lies in its provision of both generic and industry-specific business process and capability models, with a strong alignment between the two. This facilitates seamless navigation between business processes and capabilities in both directions. Additionally, SAP Reference Business Architecture is agnostic to solutions and applications, allowing it to map the business architecture domain without being tied to any particular software provider.

2.3.2 SAP Reference Solution Architecture

SAP Reference Solution Architecture is the counterpart to SAP Reference Business Architecture, and it outlines the application, data, and technology aspects of SAP and partner solutions in alignment with business architecture assets. It provides recommended implementation guidance utilizing SAP software components, including endorsed partner components. Through SAP Reference Solution Architecture, stakeholders gain access to an implementable SAP reference architecture that illustrates how the components interact to support business processes. Furthermore, organizations can leverage this content to define their own target architecture based on their specific needs and objectives. The primary components of SAP Reference Solution Architecture include the following:

1. Solution capability model
2. Solution process model
3. Solution data model (planned)
4. Solution organizational model (planned)

SAP Reference Solution Architecture can be seen as the specialized counterpart of SAP Reference Business Architecture. It facilitates the transition from a solution-agnostic perspective in the business architecture to an SAP-specific solution architecture, revealing the necessary solution components to uphold the chosen business process or capability.

Figure 2.35 illustrates the correlation between SAP Reference Business Architecture and SAP Reference Solution Architecture. The transition from the business domain to the solution domain is facilitated through the alignment of business capabilities with solution capabilities.

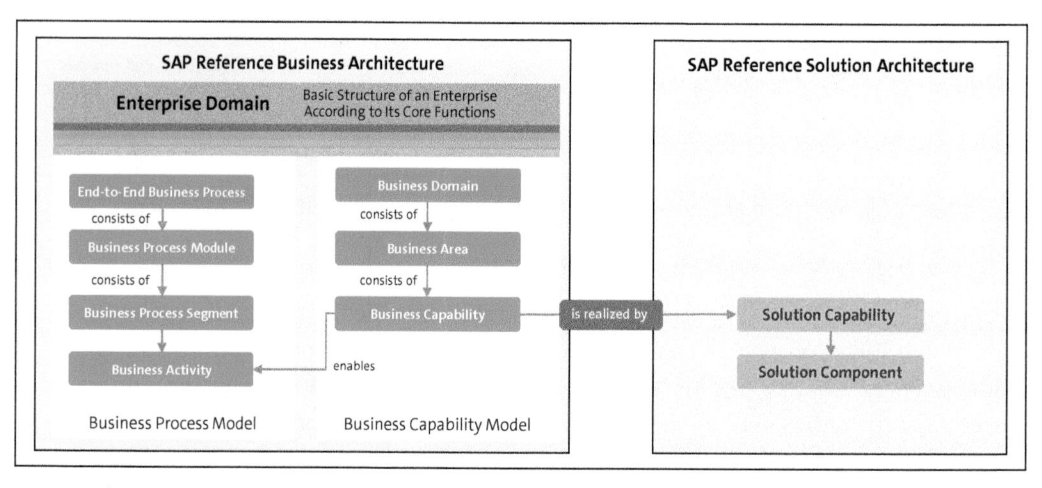

Figure 2.35 Connection of SAP Reference Business Architecture to SAP Reference Solution Architecture

In the SAP reference model, the linkage between business processes at the business activity level and business capabilities, as well as between business capabilities and solution capabilities, enables the mapping of required solution components from either the business process level or the business capability level.

This relationship is displayed in Figure 2.36. We revisit the **Source-to-contract** example depicted in Figure 2.32, this time examining it from a solution component perspective, wherein the necessary solution components are aligned with business activities through solution capabilities.

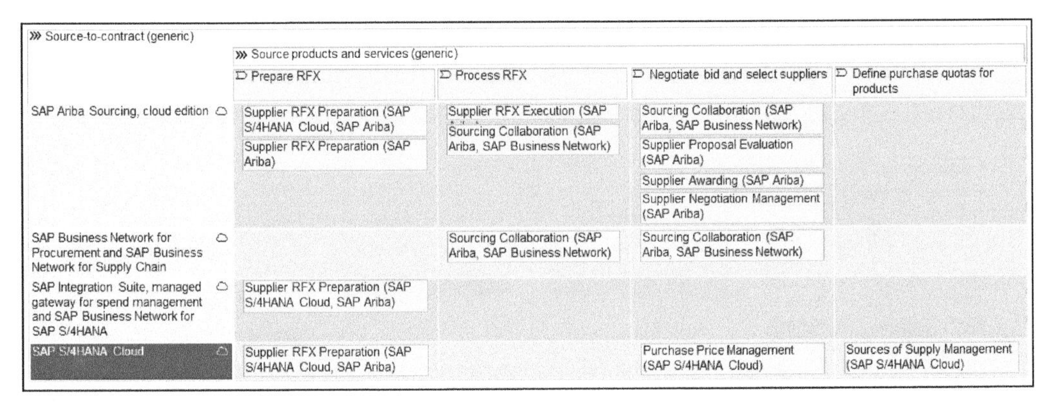

Figure 2.36 SAP Reference Solution Architecture for Business Process Source to Contract (Selection)

Within the business process segment focusing on sourcing products and services, and within its associated activities such as **Prepare RFX**, **Process RFX**, **Negotiate bid and select suppliers**, and **Define purchase quotas for products** (which are drawn from SAP Reference Business Architecture), SAP Reference Solution Architecture automatically establishes connections between solution capabilities and business activities. For instance, the business activity **Prepare RFX** corresponds to the solution capability **Supplier RFX Preparation (SAP S/4HANA Cloud, SAP Ariba)**. Through this alignment, the reference content facilitates an automated mapping from solution capabilities to the required or optional solution components. For example, **Supplier RFX Preparation** is linked to SAP S/4HANA Cloud and SAP Ariba Sourcing. Furthermore, the content encompasses mappings for SAP partner solutions and solutions supporting specific use cases.

If the business capability is intended as the starting point for defining the business architecture, SAP Reference Solution Architecture facilitates not only the mapping from the business process level to solution components but also mapping directly from business capabilities to solution components. An example illustrating this is provided in Figure 2.37.

In the business domain of **Asset Management** and specifically within the business area of **Asset Lifecycle Delivery**, the underlying business capabilities are linked to solution components. For instance, the business capability **Asset Health Monitoring** is supported by solution components such as **SAP S/4HANA Cloud** and **SAP Asset Performance Management**. This method allows us to first identify the necessary business capabilities and then delve into the required solution components when constructing the solution target architecture based on the business architecture. With the reference content already providing mappings for SAP solutions, this approach significantly expedites the process within any SAP-centric environment. However, since most landscapes comprise not only SAP solutions but also third-party alternatives, we have the capability to map business capabilities to any third-party non-SAP solution as well. This mapping can be easily executed using the tools detailed later in this chapter and is an integral part of every service engagement for architecture planning.

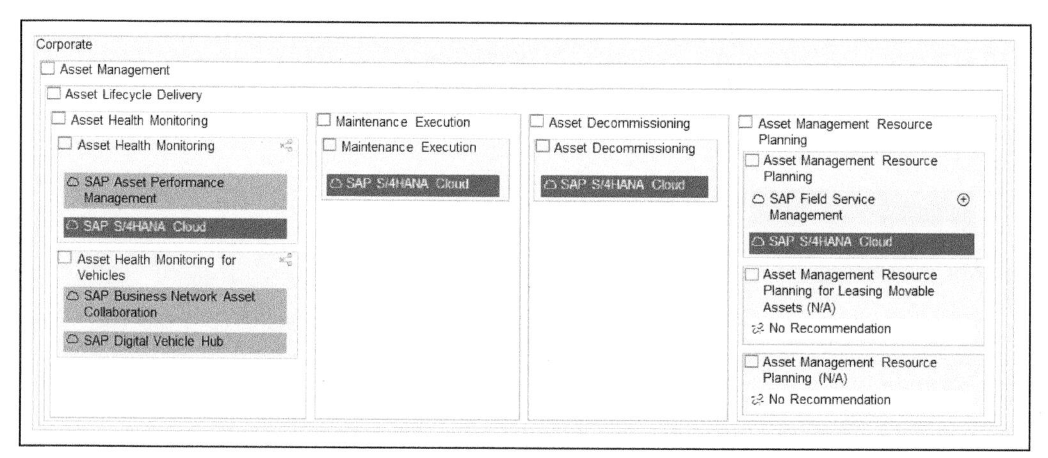

Figure 2.37 SAP Reference Solution Architecture: Mapping Business Capabilities to Solution Components

2.3.3 Additional Reference Content

In addition to SAP Reference Business Architecture and SAP Reference Solution Architecture, SAP offers supplementary reference architecture sources tailored to various use cases. However, some of these additional sources are highly specific to particular use cases and lack the broad scope of SAP Reference Business Architecture and SAP Reference Solution Architecture. Other sources, such as SAP S/4HANA scope items at the BPMN level, extend beyond the domain of enterprise architecture and are more suited to solution implementation levels. Hence, they are not included in this chapter. Additionally, we refrain from discussing reference content exclusive to SAP and inaccessible to external enterprise architects.

SAP Signavio Process Explorer and SAP Signavio Process Navigator

SAP Signavio, as a suite of tools, can't be classified solely as a source of reference content. SAP Signavio Process Explorer and SAP Signavio Process Navigator do offer SAP

Reference Business Architecture and SAP Reference Solution Architecture as part of their content sources (along with other implementation-focused reference materials such as best practice scope items at the BPMN level). However, it is not sufficient to consider SAP Signavio solely from a tool perspective where reference content can be accessed.

SAP Signavio also provides outside-in content for the strategic domain of SAP EA Framework. This content includes strategic objectives and business goals that can be aligned with required business processes from SAP Reference Business Architecture. In Chapter 6, we'll demonstrate a use case illustrating how you can connect the strategy domain to the business domain by leveraging strategic content.

You can access SAP Signavio Process Navigator at *https://me.sap.com/processnavigator*.

SAP Business Technology Platform

SAP BTP Guidance Framework is a comprehensive resource designed to help organizations effectively navigate and leverage the capabilities of SAP BTP. As shown in Figure 2.38, it provides structured guidance, best practices, and tailored recommendations to ensure successful adoption and implementation of the platform's various services and tools.

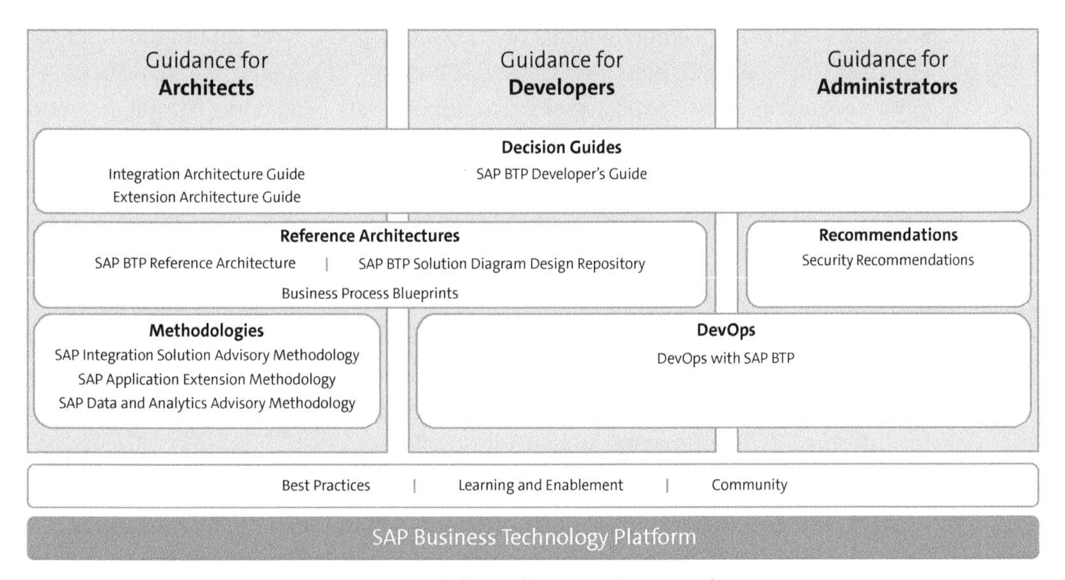

Figure 2.38 SAP BTP Guidance Framework (http://s-prs.co/v586300)

The key elements of SAP BTP Guidance Framework as depicted in Figure 2.38 are decision guides, reference architectures, methodologies, recommendations on implementation options, and DevOps principles. We describe these elements as follows:

- **Decision guides**
  The *Integration Architecture Guide* is designed to help enterprise and integration architects modernize their existing integration architectures using SAP BTP. It provides insights into SAP's current technology offerings and highlights the integration services available to meet the diverse needs of heterogeneous IT landscapes. The *Extension Architecture Guide* provides guidance on extending SAP applications on SAP BTP using modular, flexible approaches. The *Developer's Guide* is a comprehensive resource for developers building, deploying, and managing applications on SAP BTP. It offers tools, services, and best practices to maximize development efficiency and effectiveness.
- **Reference architectures**
  The *SAP BTP reference architectures* provide a comprehensive catalog of best-practice architectural patterns to guide organizations in designing and implementing solutions on SAP BTP. SAP BTP's *solution diagram design repository* is a collaborative space that offers prebuilt solution diagrams, enabling architects and developers to visualize and design their SAP BTP implementations effectively. Meanwhile, the *business process blueprints* offer detailed process descriptions for key business processes, ensuring alignment with industry standards and best practices.
- **Recommendations**
  SAP BTP's *security recommendations* provide a comprehensive set of guidelines for ensuring the security of your SAP BTP environment. These recommendations cover key areas such as identity and access management, data protection, compliance, and secure application development.
- **Methodologies**
  SAP Integration Solution Advisory Methodology provides a structured approach for guiding organizations in designing and implementing effective integration strategies using SAP solutions. *SAP Application Extension Methodology* focuses on extending SAP applications on SAP BTP, offering best practices for scalable and maintainable extensions. Meanwhile, *SAP Data and Analytics Advisory Methodology* offers a framework for maximizing the value of data and analytics within SAP ecosystems, ensuring informed decision-making and business insights.
- **DevOps**
  DevOps with SAP BTP provides a comprehensive guide for implementing DevOps practices on SAP BTP. It covers key principles like continuous integration, continuous delivery, monitoring, and automation, tailored specifically to SAP environments. The guide helps organizations streamline their DevOps processes, ensuring faster delivery, improved quality, and more efficient collaboration between development and operations teams. It also offers tools and best practices to support DevOps initiatives within SAP BTP.

SAP Business Accelerator Hub

SAP Business Accelerator Hub, formerly known as SAP API Business Hub, is an online platform provided by SAP that offers access to a wide variety of reference and best-practice content. You can access SAP Business Accelerator Hub at *https://api.sap.com/*.

For example, the following types of reference content can be found there:

- Solution value flow diagrams
- Solution component diagrams
- Solution data flow diagrams
- Solution process flows
- Application programming interfaces (APIs)
- Events
- Core data services (CDS) views

Architecture Decisions Accelerators

In SAP-centric enterprise architecture landscapes, you must make specific decisions pertaining to enterprise architecture when designing the new to-be architecture. Drawing from our experience, we know that numerous decisions can be standardized using predefined criteria that facilitate decision-making. As a result, the SAP Transformation Hub team has introduced *architecture decision accelerators* as reference content. These accelerators offer predefined architectural patterns, best practices, and implementation guidelines, enabling architects to make informed decisions efficiently and accelerate the design and implementation of robust architectures. By leveraging architecture decision accelerators, enterprise architects can reduce the time and effort required to address recurring architectural issues, enhance the quality and consistency of their solutions, and ultimately drive innovation within their organizations.

As examples, there are architecture decision accelerators for the following decisions:

- Financial analytics deciding between SAP S/4HANA, SAP Analytics Cloud, and SAP Disclosure Management
- SAP Extended Warehouse Management (SAP EWM) deployment options to help make a decision between embedded and de-centralized warehouse management
- Solution options for manufacturing operations management
- SAP S/4HANA event management standalone or add-on deployment options

These decision accelerators include decision trees that simplify the process of choosing the right architecture. An example of a decision tree for manufacturing performance management is illustrated in Figure 2.39.

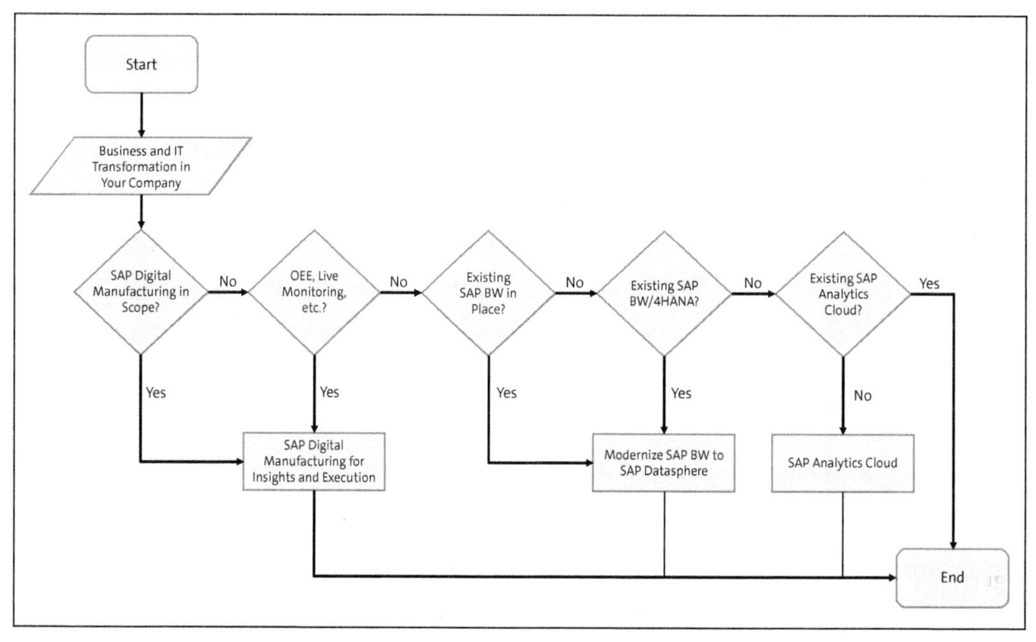

Figure 2.39 Example Decision Tree for Manufacturing Performance Management

These decision accelerators can be applied within enterprise architecture work and documented within the architecture decision catalog artifact, serving as references to justify why a particular decision was made.

You can find the decision accelerators at *http://s-prs.co/v586301*.

2.4 Preview of Tools and Practice

Enterprise architecture tools are crucial for maintaining consistency and managing the complexity of business and IT objects because you can't efficiently achieve these goals with basic tools like PowerPoint and Visio. In this section, we aim to provide a high-level overview of the intended integrated toolchain within SAP EA Framework. For more information on enterprise architecture tools in general, refer to Chapter 4. For detailed information on SAP LeanIX, refer to Part III of this book.

Figure 2.40 illustrates the vision of integration of various tools within SAP EA Framework to streamline enterprise architecture and process management.

At the top left, SAP Signavio Process Explorer and SAP Signavio Process Manager are highlighted. *SAP Signavio Process Explorer* allows organizations to visualize and explore their business processes, facilitating better understanding and optimization. *SAP Signavio Process Manager* complements this by enabling detailed process modeling and management, ensuring that processes are documented and can be continuously improved.

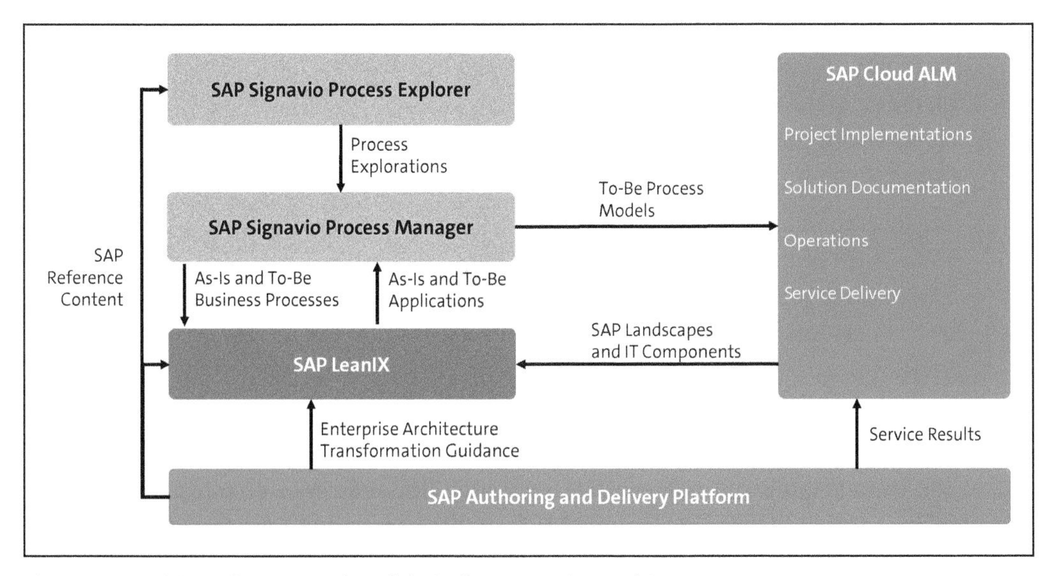

Figure 2.40 Vision of Integrated Toolchain for Enterprise Architecture

The middle section of the diagram features *SAP LeanIX* enterprise architecture management. This tool is crucial for managing the enterprise architecture by providing insights into the IT landscape and helping organizations align their IT strategy with business goals. It supports the documentation, planning, and analysis of the enterprise architecture, fostering better decision-making and strategic alignment.

The section on the right shows the integration with *SAP Cloud ALM*, which supports various aspects of project implementations, solution documentation, operations, and service delivery.

Together, these tools are supported by a cohesive SAP authoring and delivery platform that ensures consistency in reference content usage and service results documentation. This integrated toolchain enables organizations to manage their enterprise architecture and processes comprehensively, promoting alignment, transparency, and continuous improvement.

Chapter 15 provides a comprehensive guide to setting up an enterprise architecture practice, emphasizing the organizational implementation of enterprise architecture. Establishing an enterprise architecture practice involves adopting methodologies and governance processes, including change management. The success of an enterprise architecture practice depends on the organizational context, management support, and desired outcomes. The practice can be embedded within an organization in various ways, such as in a dedicated organization, in federated groups, or as part of initiatives, and it's also crucial to define the expected outcomes and deliverables.

We introduce a maturity model to assess and plan the progression of the enterprise architecture practice, suggesting that the highest level of maturity is not always necessary. We highlight governance as a critical element, requiring a balance between open collaboration and defined rules and standards. Key aspects include a stakeholder map, capabilities, processes, data management, applications, technologies, and transition planning. Small and midsized companies are advised to document their implicit principles and architecture approaches to maintain a shared understanding among key stakeholders.

Organizational dimensions such as geography, size of the business, lines of business, and business functions significantly impact the organization of an enterprise architecture practice. We recommend that enterprise architecture practices should not be split by business function but should be aware of specific functional needs, like local regulations in finance and HR. We discourage organizing enterprise architecture practices based on application platforms, to avoid competitive conflicts. Instead, a common approach to all functions ensures better integration and synergies.

You can explore various organizational setups for enterprise architecture practices, such as linking them to business and IT, making them part of the project organization, or making them separate from the project management office (PMO). Different setups like permanent teams versus rotating assignments and organizational units versus virtual teams have their pros and cons. You can examine the maturity of an enterprise architecture practice through a five-level maturity model that emphasizes reaching a defined level where the practice has a structured operating model, defined roles, processes, and tools. Aligning enterprise architecture practices with strategic transformations and involving business support and collaboration from the outset are essential for success and acceptance.

SAP Trainings and Certifications

SAP offers various internal and external training courses and certifications for SAP EA Framework. Instead of detailing all the offerings, especially internal ones, we focus on an external option: SAP IEA10 (SAP Intelligent Enterprise – Architecture Foundation). This course provides foundational knowledge and certification in SAP enterprise architecture, helping participants understand and apply the SAP EA Framework effectively. It covers creating a company-specific roadmap to an intelligent enterprise, taking into account business strategies and priorities. The course includes modules on architecture vision, business architecture, technology architecture, and creating an architecture roadmap. It's designed for enterprise architects, solution architects, and business architects, and it requires a foundational knowledge of TOGAF. For more details, you can visit the SAP Training page at *http://s-prs.co/v586302*.

2.5 SAP's Enterprise Architecture Services

SAP provides a range of enterprise architecture services to its clients, with variations in service scope and commercial terms depending on the specific organizational unit within SAP. The main teams responsible for delivering these services include SAP Customer Advisor (formerly SAP Presales), Business Transformation Services, and SAP Transformation Hub.

Despite operating within different organizational structures and contractual frameworks, all of these teams adhere to SAP EA Framework as outlined in this chapter. This serves as a common framework and foundation for their activities. Employing SAP EA Framework offers these teams numerous benefits, such as consolidating efforts to develop methodologies, reference materials, and tools. Additionally, it facilitates seamless transitions in architecture deliverables through standardized artifacts and a shared language of enterprise architecture.

In this section, we provide a brief overview of the service approach of the SAP Transformation Hub team, focusing on the North Star service the team offers to SAP premium engagements customers, who represent some of the largest and most complex clients within SAP.

2.5.1 SAP's Enterprise Architecture Service Approach

SAP Transformation Hub offers a range of enterprise architecture services, categorized into two main types: holistic scope and dedicated scope.

The North Star service is tailored to provide a comprehensive approach to enterprise architecture planning, covering all domains outlined in the SAP EA Framework. Conversely, SAP Transformation Hub offers architecture point of view (APoV) services, which focus on specific architecture issues without considering the broader landscape. For instance, an APoV service might address the transition from warehouse management in SAP ERP to SAP EWM without delving into other architectural aspects such as financial transformation or reporting capabilities.

In addition to APoV and the North Star service, the SAP Transformation Hub provides other services like the target application assessment, which supports the review of already developed to-be architectures.

Overall, the service approach of the SAP Transformation Hub can be viewed as a general strategy for enterprise architecture planning: starting with a smaller scope and addressing specific architecture questions (the APoV approach), or beginning with a broader scope covering multiple enterprise architecture domains, or reviewing existing architectural artifacts. Regardless of the chosen option, all services adhere to the same methodology, ensuring seamless handover points.

2.5.2 Patterns and Modularization

As discussed in the preceding section, the North Star service adopts a comprehensive approach to architecture planning, encompassing various enterprise architecture domains. Within each domain, multiple modules are available, and each contains one or more artifacts from SAP EA Framework, along with detailed descriptions of the artifacts' development. Figure 2.41 illustrates the sets of modules that are associated with the various enterprise architecture domains.

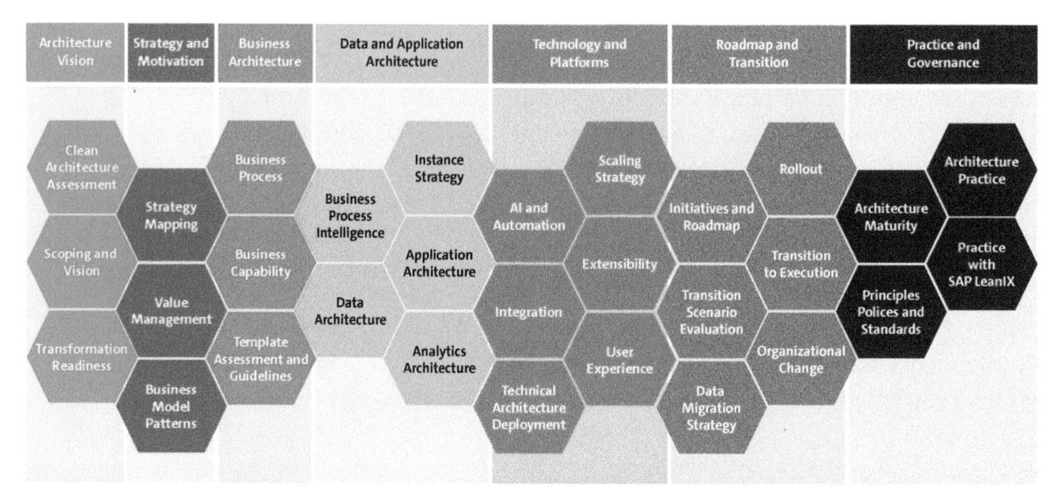

Figure 2.41 Overview of Modules in Relation to Enterprise Architecture Domains

The primary advantage of organizing artifacts from the SAP EA Framework into modules lies in their ability to outline a logical progression for artifact creation, complete with detailed scoping and delivery instructions. Each module establishes necessary prerequisites from other modules and identifies potential successor modules. This modularization approach furnishes a comprehensive toolbox for addressing all topics related to enterprise architecture.

However, while the modularization approach aims for a holistic view of the landscape, the intention is not to deliver all modules simultaneously. Instead, the goal is to define meaningful iterations of module flows. Additionally, we adhere to the TOGAF nomenclature regarding levels of depth, providing definitions of informal, light, and core levels of delivery depth for most modules.

To further enhance this approach, we introduce the concept of *patterns*, which are predefined flows of modules at predetermined levels of depth. Patterns are designed to address specific architecture problem statements, and example of patterns are as follows:

- Target architecture and roadmap (including assessment)
- Landscape consolidation
- Business transformation
- Rollout strategy assessment
- Mergers, acquisitions, and divestitures
- Two-tier target architecture

By incorporating patterns, we integrate all previously mentioned concepts, including modules, iterations, and levels of depth. Figure 2.42 offers an overview of this concept.

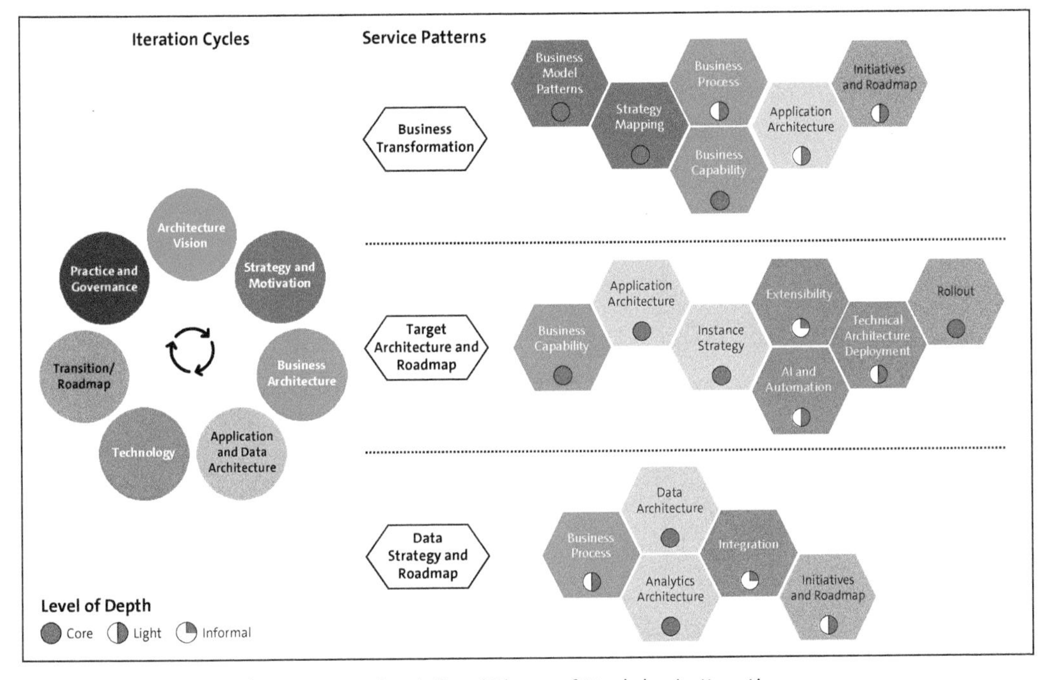

Figure 2.42 Concept of Patterns as Predefined Flows of Modules in Iterations

In this example, the initial iteration would concentrate on the pattern of business transformation. Within this pattern, specific modules take priority. These modules include business model patterns, strategy mapping, business capabilities and processes, target application architecture, and initiatives and roadmaps. However, we need other modules to devise a comprehensive enterprise architecture. To do this, we employ iterations, initially focusing on strategic and business-related artifacts.

During a second iteration, the target application architecture, already developed at a light level, can undergo further refinement. We can also introduce additional modules, such as instance strategy, extensibility, and AI and automation. In a potential third iteration, we can shift our attention to developing the required data strategy. This iterative approach accelerates architecture planning activities by initially prioritizing the most crucial patterns and progressively refining the overall target state with additional modules or deeper levels of detail for already delivered modules.

Iteration Types

Iterations can happen in different types. One way, as explained in Figure 2.42, is to run iterations over different patterns (i.e., in each iteration, you focus on a particular pattern). However, sometimes a pattern as such is already too complex for a single iteration. In those cases, you can plan iterations in the following way for an individual pattern:

- **Option 1**
  Define the iterations over a subset of modules inside the pattern. For example, you can focus iteration 1 on strategy mapping and business capabilities, iteration 2 on application architecture, and iteration 3 on initiatives and roadmap.
- **Option 2**
  Define the iterations on the level of depth of the selected modules inside the pattern. For example, you can run all modules first in an informal first iteration and then add the following iterations at additional levels of depth.
- **Option 3**
  Define the iterations on functional domain focus. For example, you can start the first iteration with central functions, continue the second iteration with regional functions, and so on.

There are also other options that you can apply to structure the iterations (e.g., by business areas, by geography, etc.).

2.5.3 Applying the Service Methodology

As demonstrated in the preceding section, modularization and patterns are robust tools for defining the scope of an architecture engagement. The pattern-based scoping approach enables a streamlined selection of modules, drawing from the collective experience of previous engagements addressing similar issues. However, if a particular pattern is not applicable, you can still select modules through an assemble-to-order approach, ensuring that necessary input and output relationships among modules are satisfied.

In essence, the following approach is employed to define and deliver the North Star service, leveraging patterns and modularization:

1. Gain a comprehensive understanding of the architectural scope by reviewing existing documents such as current architecture, planned or ongoing initiatives, and business operating models.
2. Determine whether a predefined pattern can be applied or if modules need to be selected using an assemble-to-order approach, considering the required depth of analysis. This results in a list of modules for service delivery.
3. Define meaningful iterations of the selected modules and ascertain if the necessary requirements for each module can be fulfilled.
4. Based on the delivery instructions for each module, establish the team composition (e.g., additional solution or data architects required), estimate the effort in person-days, and outline an initial agenda for delivery.

5. Execute the delivery according to the module instructions and generate a comprehensive report detailing the developed artifacts and recommendations as the outcome of the architectural work. A management summary serves as an effective way to communicate a coherent and persuasive narrative at the executive level.

2.6 Summary

SAP EA Framework helps businesses gain comprehensive control over their IT landscape. It aids in making agile decisions and accelerates digital transformation journeys. The framework includes four primary components: methodology, reference content, tools, and best practices.

The methodology provides a holistic view of the key components of an organization, their interrelationships, and the principles guiding their design and evolution. The reference content offers best practices, methodologies, and industry-specific insights tailored to SAP environments. The tools component has become even more integral to the enterprise architecture framework, with SAP Signavio and SAP Cloud ALM emerging as key resources for the enterprise architect. Lastly, best practices are meticulously crafted to provide clear guidance for decision-making processes.

SAP EA Framework aligns closely with TOGAF, incorporating many of the latter's terminologies and concepts. One of the significant advantages of using SAP EA Framework is its tight integration of SAP reference content and SAP enterprise architecture tools within the enterprise architecture framework, which provides out-of-the-box content that can be used immediately.

SAP Reference Business Architecture and SAP Reference Solution Architecture are key repositories of reference content provided by SAP for use with SAP EA Framework. SAP Reference Business Architecture offers a range of end-to-end business process models and business capability models tailored to each domain, while SAP Reference Solution Architecture outlines the application, data, and technology aspects of SAP and partner solutions in alignment with business architecture assets.

SAP also provides a range of enterprise architecture services to its clients, including the SAP North Star service it offers to SAP premium engagements customers. These services adhere to SAP EA Framework, consolidating efforts to develop methodologies, reference materials, and tools.

In conclusion, SAP EA Framework is a comprehensive tool set that offers a clear, comprehensive, and standardized view of an organization's structure and operations. This enables effective strategic planning and decision-making, optimizes resource utilization, and accelerates digital transformation.

In the next chapter, we'll discuss how enterprise architecture impacts and integrates with other business areas.

Chapter 3
Functional Integration of Enterprise Architecture

In this chapter, we'll explore the relationship of enterprise architecture to "neighboring" functions. Some enterprise architects might think that their function is self-contained and the most strategic one for their company—and such a view could be perceived as ivory tower thinking. This is why we promote integration with other functions.

Even though enterprise architecture is recognized as a dedicated function and profession, mostly due to its strategic relevance, it's important to understand related and partially overlapping functions. Typically, these related functions are managed by other groups and require close alignments. If these alignments are not implemented, it can be difficult to harvest the benefits of good enterprise architecture work. At worst, it can lead to conflicting results and different groups "fighting" for their perspectives.

This chapter will go into multiple other functions, assess perspectives from those functions, and also assess perspectives from an enterprise architecture viewpoint. The question is whether "the other" function is included in, essential to, supportive of, or mostly not relevant to enterprise architecture.

Figure 3.1 shows that there are three large groups of functions. What they are, what they do, and the differences among them are as follows:

1. The architecture functions highlighted on the left belong to the realm of either "enterprise" or "system" architecture. There's no clear-cut difference between the two. We'll elaborate on this aspect in Section 3.4.
2. Horizontally, we distinguish among process, data, application, and technology management functions, which largely map to enterprise architecture domains. The richness of the subpractices in the blue circles on the right explains why it doesn't make sense to fully integrate these functions with enterprise architecture. In enterprise architecture, we work with the broader business domains, which include organization, function, and processes. The horizontal blue arrows within each of the domains indicate that domain-specific architectures are seen as integral parts of enterprise architecture.
3. The lower green bar of the diagram lists additional functions—including requirements management, risk management, innovation management, and portfolio

management, which we'll investigate in Section 3.5. These functions have touch points with both the domain-specific and the cross-domain functions.

SAP-oriented practices might differ because they address everything from adopting standard software to departments where entire systems are developed from scratch.

We see a pattern in some companies where the enterprise architecture is managed for IT's non-SAP departments as opposed to the SAP organization, which manages its own sphere—often without establishing an enterprise architecture practice. From that angle, SAP-oriented architecture teams should collaborate with functional practices other than non-SAP global architecture teams.

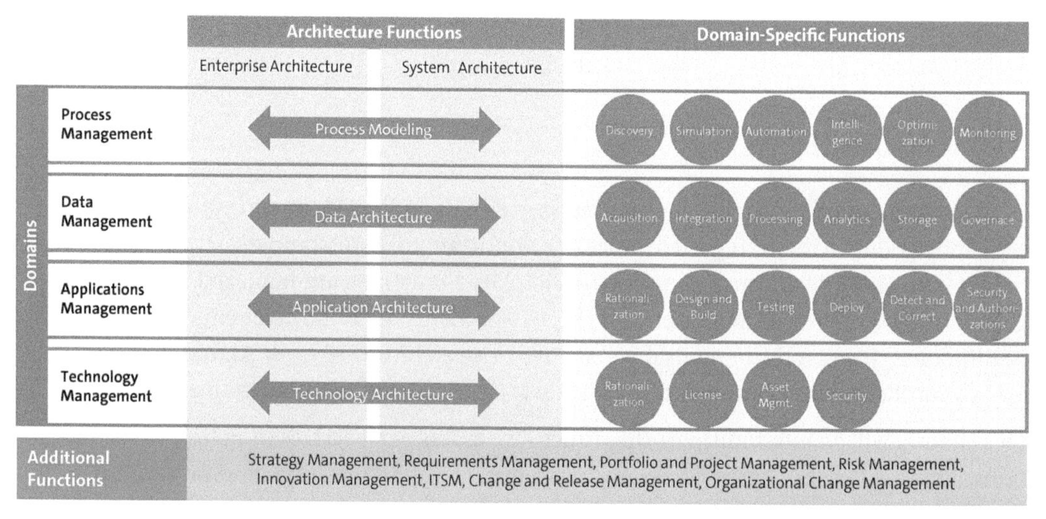

Figure 3.1 Interrelated Functions and Practices

In Chapter 4, we'll discuss the need for tooling—not only for enterprise architecture but also for the functions described in this chapter and what integrations between them are needed.

Not all companies will need these functions or have them established explicitly. Their need to do so will depend on the size and maturity of their business. Organizational considerations, such as whether functions are to be performed by an explicit organizational unit or integrated with project organizations, will be discussed from an enterprise architecture perspective in Chapter 15.

3.1 Business Process Management

Business processes are essential parts of enterprise architecture as described in Chapter 2. In this framework, we link specific process elements of different levels to strategy, business capabilities, and business roles within the business architecture. In the solution architecture, we link activities to solution components and process flows to interfaces. Enterprise architecture business process modeling can be seen as a subdiscipline of enterprise architecture, and most enterprise architecture tools include process-modeling features that range from depictions of simple value flows to multilevel nested Business Process Model and Notation (BPMN) diagrams.

In our view, enterprise architectures that lack any process focus are incomplete. To help you not get lost in deep process model levels, it can make sense to narrow down the process perspective to value flows, which are higher-level process representations that don't include message flows and branching as in BPMN.

This often implies that enterprise architecture uses business processes primarily to link other enterprise architecture domain objects such as goals, capabilities, and applications with a business context. Besides providing a catalog of business processes and value flows, business process modeling often is still understood as a part of enterprise architecture.

Figure 3.2 shows the different process-related disciplines and how they relate to each other.

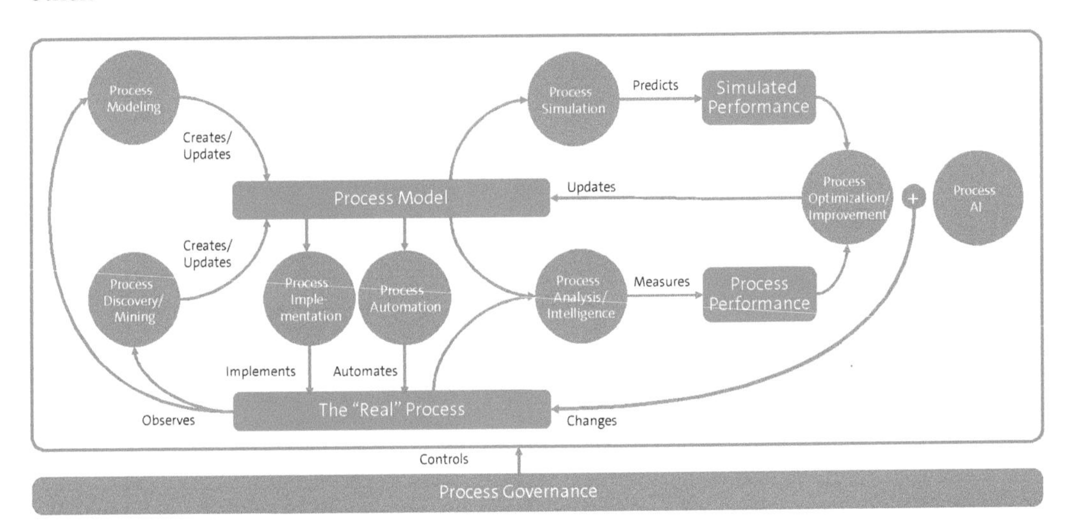

Figure 3.2 Business Process Practices

The term *business process management*, which we use as umbrella term, includes discovery, implementation, simulation, monitoring, modeling, and process governance. We define these terms as follows:

- **Process governance**
  Process governance is the control layer. In it, each enterprise defines its methodology and standards, level of details, process ownership, and tools. Without clear governance, process management activities will be one-off, meaning they can't be sustained over time. Typical elements of process governance are as follows:

  - Process framework
  - Modeling standards
  - Process ownership
  - Approval of model changes
  - Conformance check for process models
  - Process classification frameworks for different governance levels (e.g., corporate/global, divisional/regional, local)
  - Process owner network
  - Aligned change notifications
  - Alert thresholds for process monitoring
  - Process documentation updates
  - Business process resilience

- **Process modeling**

  Process modeling de-facto standards are BPMN and value stream modeling. Note that the terms *value stream* and *value flow* are often used interchangeably—for a description of them, see Chapter 2. Two additional techniques that we mention here for the sake of completeness are simple flow charts and petri nets, but explaining these practices in detail is beyond the scope of this book.

  Figure 3.3 shows a sample value flow from SAP Reference Business Architecture content that depicts the generic end-to-end lead-to-cash business process. The subprocesses in this value flow are color coded according to their primary business domain. Hence, the invoice-to-cash subprocess is coded in yellow, which signals that finance is part of the corporate domain.

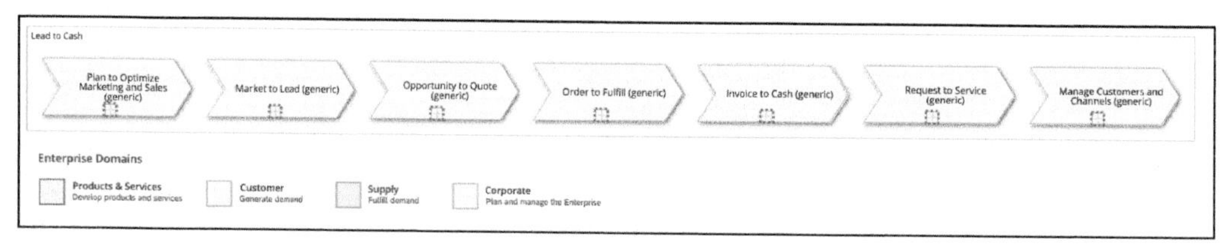

Figure 3.3 Lead-to-Cash Value Flow

- **Process implementation**

  Process implementation is an aggregate of all other activities, meaning that it's equal to running an entire development and implementation project. It's important that initial models are adjusted and refined during the implementation phase if need be. If not, process models may not depict reality on the go-live date.

- **Process automation**

  Process automation is a specialized area of process implementation with a focus on eliminating steps that require human intervention. If process automation is achieved on the basis of the as-is process model, updating the model is an important subtask of any automation activity. Process execution times (cycle times) are key performance indicators, and these times are typically higher if processes include manual tasks. Needless to say, manual tasks engender high failure rates. In this context, we often use the term *robotic process automation* (RPA), which adds the perspective of intelligent automation steps. It's critical that the process be understood so well that the automation result is not only fast but at least as good as or better than the former manual process.

- **Process discovery and process mining**
  The quality of process models that are based solely on human modeling activities will depend on the observation skills of the person doing the modeling and the effort they put into it. A superior approach is one where a process model is fully or partially generated. This is typically achievable in environments where process steps leave trace points in the system. An example is a change record that traces when and why documents change their state. By aggregating changes over time, a good process discovery engine won't only depict different process variants but will also visualize how often these are run, what their cycle times are, and other process indicators.

  Figure 3.4 shows a process mining example from SAP Signavio Process Intelligence. In the depicted procurement process, the initial steps haven't been executed according to the model (the dashed line is bypassing these steps). The magenta bars above the executed steps indicate the corresponding cycle times.

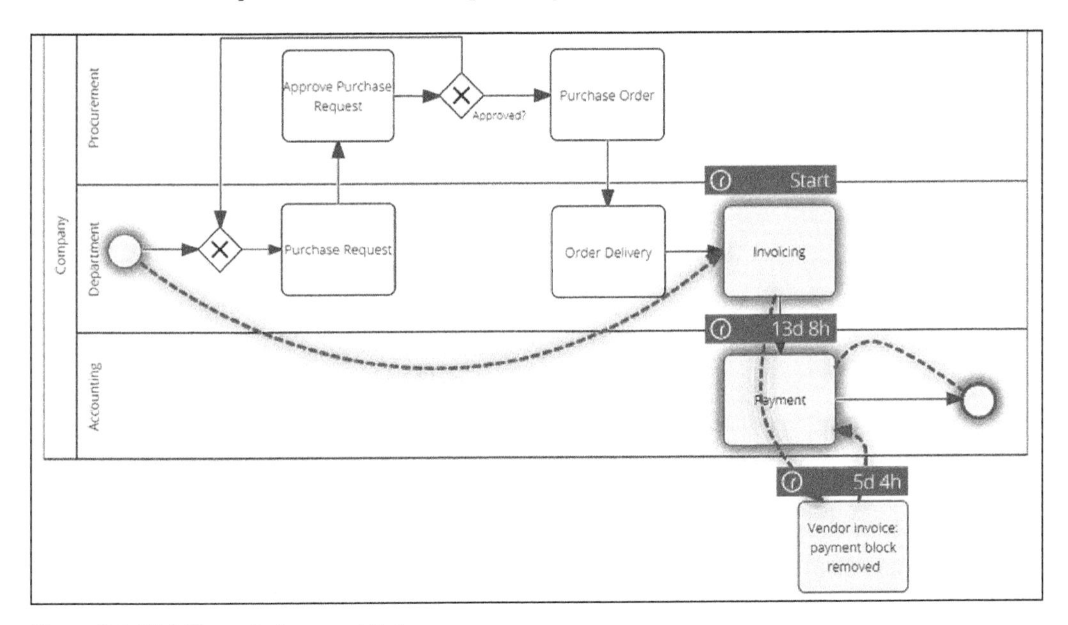

Figure 3.4 SAP Signavio Process Mining

- **Business process simulation**
  Simulating a business process enables an impact analysis of process changes. The

prerequisite is that the process is modeled in a computer system and matches the real process as closely as possible. Multiple parameters will be changed over many simulation runs, and the results of the simulations are sets of key performance indicators (KPIs) indicating judgments on the different simulated variants. Selecting the best-performing variant can then trigger a process adjustment or automation. Obviously, models are no more than approximations of real processes, so results need to be validated by process experts.

- **Process monitoring and alerting**
  A simulation will measure the simulated performance, whereas *process monitoring and alerting* measure the real process. Out-of-the-box solutions like SAP S/4HANA often include alerts regarding business KPIs, meaning that they only need to be supplemented in case of additional needs.
- **Process analysis/process intelligence**
  If process mining, process monitoring, or process simulation shows that there's room for improvement in terms of process performance (in cost, time, etc.), then the processes need to be improved. You can improve them by changing the system configuration or, in some cases, by simply retraining the individuals running the processes. Process changes are more invasive, especially if they require automation of manual steps (see process automation) or—in the extreme—a new IT system. In the latter case, this will result in a partial or full process reimplementation. (See our earlier coverage of process implementation.)
- **Process AI**
  The use of a specialized large language model (LLM) or large process model (LPM) that is fed by best practices and real process data can open up a multitude of new capabilities. These may range from detecting matching KPIs to giving recommendations to adjusting the BPMN process model. We'll discuss this in more detail and in the context of other trends in Chapter 17.

Modeling company-specific business processes conflicts with the adoption of predefined best-in-class processes, so you'll need to prioritize those process areas for which a deep modeling of the baseline and the target process is both wanted and needed.

We see the relationship to enterprise architecture in providing context to other enterprise architecture building blocks like applications and capabilities. The more this context is exposed, the more effort you'll need to spend keeping detailed enterprise architecture and business process management content aligned. Of the business process management subfunctions we listed previously, business process modeling has the strongest linkage to enterprise architecture. Here, we see different process variants that can often require significantly different business capabilities and can therefore end up in different architectures.

Possible reasons for an exceptionally strong process focus are as follows:

- Industries requiring a properly detailed process description
- Harmonization needs after a merger or acquisition (see Chapter 8)
- Obviously poor process performance in some areas

It's obvious that not all subpractices can and should be subsumed under the enterprise architecture umbrella. This is all the more true when it comes to simulation, monitoring, and alerting, in which the required toolsets are quite different and often bundled into business process management tool suites. SAP Signavio is such a process management suite and covers most of the functions mentioned previously.

When establishing an in-depth process management practice, integrating it with enterprise architecture artifacts such as data flow diagrams is critical. Figure 3.5 shows such an integration scenario. It is also essential to replicate nonprocess objects like applications and business capabilities toward a process management suite and to reverse-replicate value streams, value flows, and processes from a process management suite into an enterprise architecture repository.

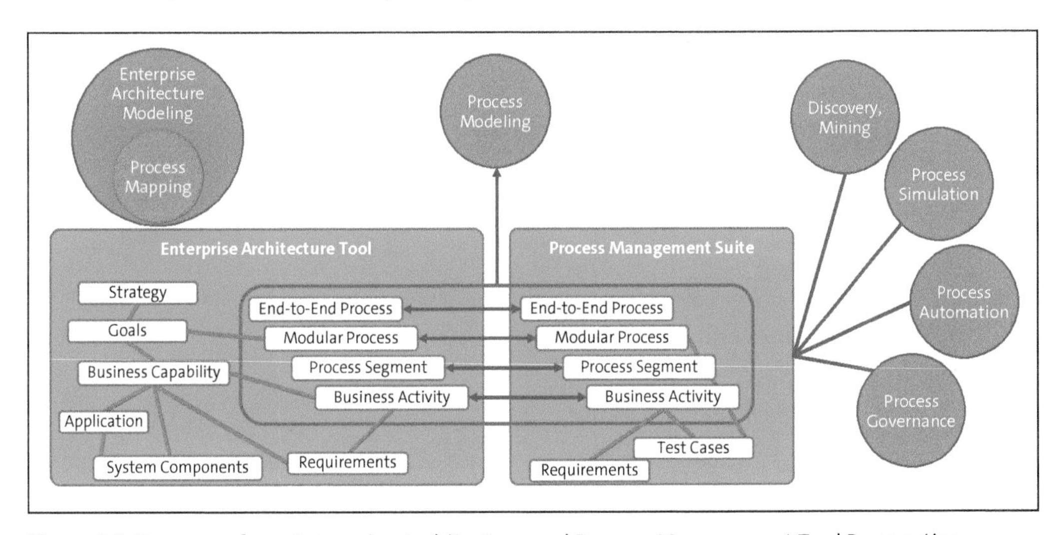

Figure 3.5 Processes from Enterprise Architecture and Process Management Tool Perspective

3.2 Data Management

For years, many companies have been making strong moves toward becoming *data-driven enterprises*. They seek to break down their data siloes and harmonize their data structures, and this requires significant changes from an organizational as well as a responsibility perspective.

Data warehouses and analytical solutions are long-established parts of IT landscapes. Due to their importance, accountability for them is bundled in central teams. Managing the data lifecycle—from creating, receiving, and harmonizing data right down to archiving and deleting data in operational and nonoperational systems—is also a well-established practice that surpasses the typical scope of enterprise architecture.

Within SAP, the SAP Data and Analytics Advisory Methodology depicted in Figure 3.6 is part of SAP BTP Guidance Framework, which we introduced in Chapter 2. It puts *data products* in the center, and it includes reference architectures for both business and technical use cases. For more information, see *https://community.sap.com/t5/technology-blogs-by-sap/new-release-of-sap-data-and-analytics-advisory-methodology/ba-p/13725453.*

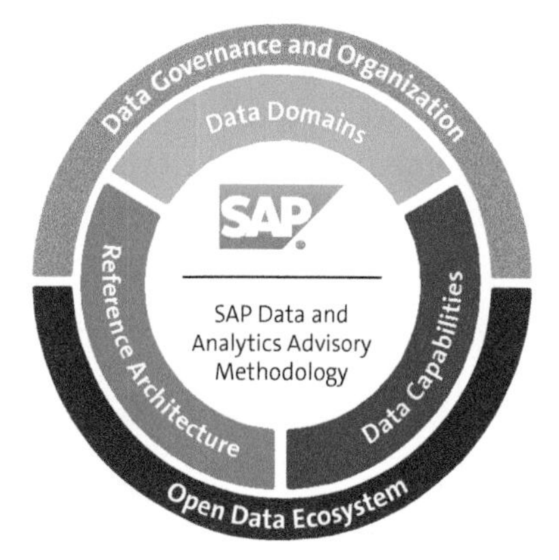

Figure 3.6 SAP Data and Analytics Advisory Methodology

The elements of this methodology are as follows:

- **Data governance and organization**
  - Consider the organizational impact to acquire or develop new skills or competencies required by new solutions.
  - Apply best-practice data governance processes to ensure consistent data management.
- **Data domain reference model**
  - Establish a common understanding of SAP data across the enterprise.
  - Reuse the domain structure of SAP Reference Business Architecture to organize SAP data.
  - This is partially covered by SAP One Domain Model.
- **Data and analytics capability model**
  - Create harmonized and solution-agnostic data and analytics capability terms and definitions.
  - Simplify definition of solution architecture building blocks.

- **Reference architecture**
  - Implement faster design of target solution architecture by identifying reference architectures through business and technical use case patterns.
  - SAP BTP reference architectures can be rapidly adopted, tested, and validated.
- **Open data ecosystem**
  - This is open for third-party technologies and solutions.
  - Partner solutions are considered in SAP BTP reference architectures.

Figure 3.7 divides data management into different categories and functions. This is a kind of capability model for data, and SAP is refining its categorization approach so that it will be able to elevate such content as reference content in the future.

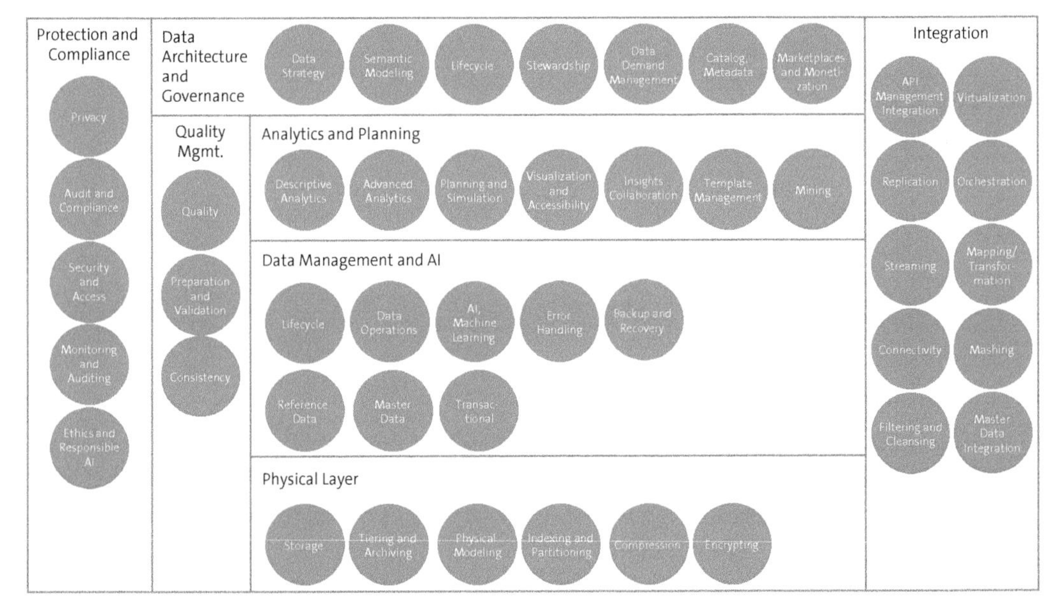

Figure 3.7 Data Management Functions

Some companies establish separate data organizations under the lead of a chief data officer (CDO) who is not a part of the regular IT organization. Such a CDO will be accountable for identifying the value of the data assets but could also be in charge of defining data standards and mastering and deploying data management practices. See *https://www.gartner.com/smarterwithgartner/understanding-the-chief-data-officer-role* for more information.

As data is the core of all processes and applications, a CDO's organization will never be able to work in isolation, so we recommend defining an overarching data strategy like the one in Figure 3.8. This will help ensure that to start with, you *collect* requirements and then *organize* and prioritize them by domains and use cases. This will clarify the input you need to *structure* the required technical capabilities, which will in turn guide you to the final step, in which you *define* a target architecture matching your strategic goals.

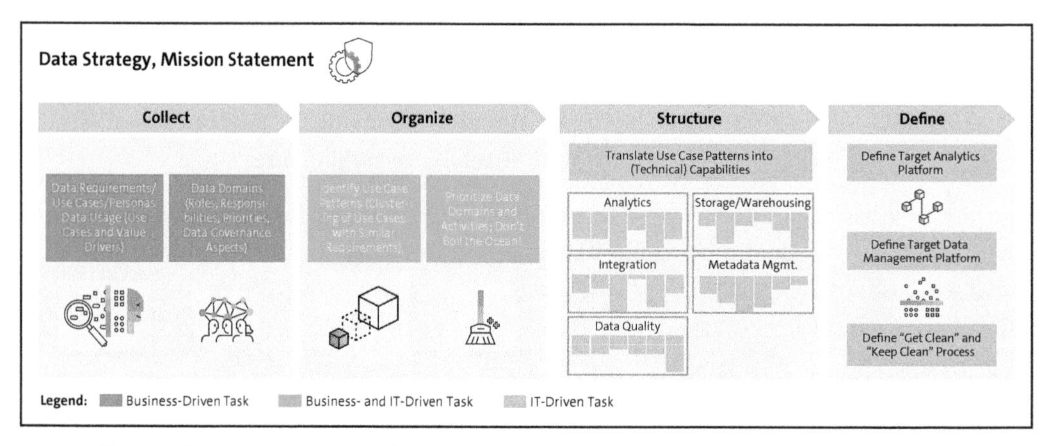

Figure 3.8 Data Strategy: Architecture Approach and Methodology

This data strategy can initially be driven from within the enterprise architecture function, but a dedicated team will become necessary as maturity in this field increases and more roles arise around data ownership, data quality, data retention, data distribution, and data security. This can but doesn't necessarily have to go hand in hand with the work of longer-established groups that are accountable for analytics and data warehousing.

In Figure 3.9, we refine the business input needed to drive a holistic data strategy. This includes the data needed for end-to-end process as well as a well-structured data domain model, which differentiates reference and configuration data from master data and transactional data domains.

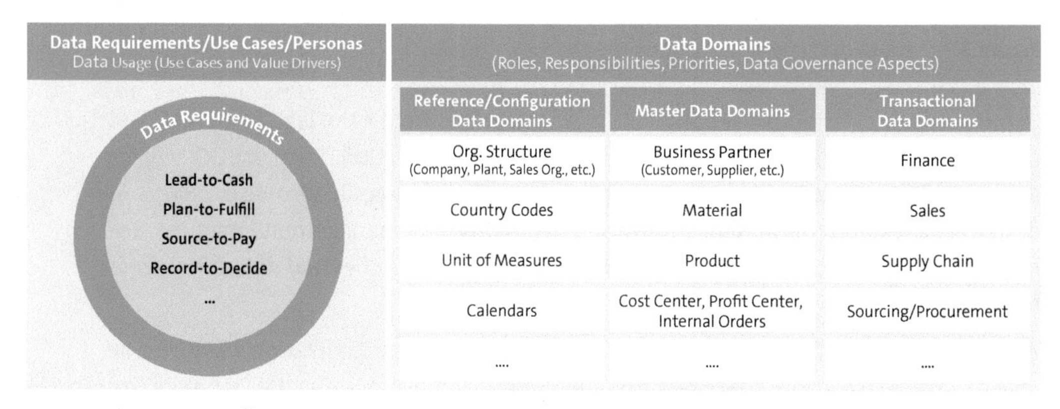

Figure 3.9 Collecting Business Input for Overall Data Strategy

3.3 Application Management

Application management focuses on the entire lifecycle of an application, meaning it takes a bird's eye view of an application's lifecycle. The term *application* itself can be ambiguous in this context. From a reference architecture perspective, SAP S/4HANA and SAP Integration Business Planning for Supply Chain (SAP IBP) are applications. In real landscapes, there will often be more than just one SAP S/4HANA application, so the common approach is to call the specific application instance an *application* and the generic application a *solution component*.

The related term *application portfolio management* is also used frequently, so the approaches described in Section 3.5.1 equally apply.

Application lifecycle management and *application management* are often used interchangeably too. They cluster all activities from portfolio, requirements, design, implementation, development, testing, and change management to operations. *Application rationalization*, which is the discipline of *not* risking the operation of an uncontrolled set of partially outdated and often overlapping applications, can be seen as part of application portfolio management. This leads to the following three functional clusters:

- **Application rationalization**
  Managing the lifecycle of applications and, ideally, reducing their number is clearly a central objective of any good enterprise architecture practice. It's critical when mergers and acquisitions lead to overlapping and/or redundant applications. Enterprise architecture approaches such as capability mapping are useful for assessing the functional fit. In Chapter 12, we'll see the tool support provided by this product.
- **Design/system architecture**
  The SAP-centric view tends to center on the adoption of standard solutions, where only the extensions (and not entire systems) are designed. Predesigned extensibility options often do away with specific system design.

  From a broader perspective, new solutions can be architected and designed entirely from scratch. This brings many more considerations into the equation, two of them being which platform to use and which programming language to use. Companies will typically not develop their own enterprise resource planning (ERP) from the ground up, but in the case of a highly specific business need, they may find that a custom-developed solution is better than an ill-fitting standard solution.
- **Integration management**
  Dedicated integration organizations are often formed to both realize an integration and operate it. This is especially desirable for complex, time-critical integrations involving multiple business partners.

SAP Solution Manager and SAP Cloud ALM are two solutions that SAP provides for managing the entire SAP product range. Many functions also support non-SAP solutions.

While in the past, SAP Solution Manager was the dominant IT for IT offering from SAP, SAP Signavio and SAP LeanIX spread process management and enterprise architecture management functions across an end-to-end transformation suite, which we will look at in depth in Chapter 11.

Let's compare SAP Cloud ALM to SAP Solution Manager:

- **SAP Cloud ALM**
  SAP Cloud ALM started by supporting customers with a public cloud-only footprint, and it's not yet as comprehensive as SAP Solution Manager. It's the strategic platform for SAP, so most companies use SAP Cloud ALM and SAP Solution Manager in parallel and will migrate by function once they become available in SAP Cloud ALM.
- **SAP Solution Manager**
  SAP Solution Manager 7.2 is widely adopted in the SAP install base because customers get a custom-fit ALM suite with it as part of the SAP Enterprise Support contract. SAP Solution Manager is also integrated with best-in-class solutions for specific functions (e.g., Tricentis for test automation, Redwood Workload Automation for job scheduling).

3.4 System versus Enterprise Architecture

Differentiating between enterprise architecture and system architecture is a challenge because there aren't any clear boundaries. Often, people make an architecture request without specifying whether they want it to address "enterprise" or "system" concerns. We derive policies and standards from enterprise architecture principles that will then be used in system architecture while defining a component architecture, for example.

The *Zachman Framework* depicted in Table 3.1 differentiates three layers—business model, system model, and technology model—and represents them in a matrix. Even though the Zachman Framework as a whole is seen as an enterprise architecture framework, the business model represents the enterprise architecture perspective while the system and technology models are closer to system architecture.

	Data (What)	Function (How)	Network (Where)	People (Who)	Time (When)	Motivation (Why)
Business Model (Conceptual)	Conceptual data model	Business process model	Business logistics	Workflow model	Master schedule	Business plan

Table 3.1 Zachman Framework

	Data (What)	Function (How)	Network (Where)	People (Who)	Time (When)	Motivation (Why)
System Model (Logical)	Logical data model	Application architecture	Distribution of system architecture	Human interface architecture	Processing structure	Business rule model
Technology Model (Physical)	Physical data model	System design	Technology architecture	Presentation architecture	Control structure	Rule design

Table 3.1 Zachman Framework (Cont.)

To illustrate the differences in more detail, we'll reuse the structure introduced in Figure 3.1. Figure 3.10 shows multiple domain topics and their implications for enterprise architecture and/or the system architecture level. Without proper governance, architecture work will be conducted from a system architecture perspective and possibly only much later from enterprise architecture perspective. Corrective actions need to be triggered to reduce the architecture landscape's overall complexity and/or to comply with architecture principles and standards.

	Enterprise Architecture	System Architecture
	• Broad Cross-Application Concerns • Frameworks: TOGAF	• Specific System Design • Frameworks: UML, Model-Driven Architecture
Domains: **Process Management**	• Value Streams • Business Roles	• Multilevel BPMN • Workflow Design
Domains: **Data Management**	• Conceptual Data Domain Models • Interapplication Data Flow Diagrams • Integration Pattern and Standards	• Technical Data Models, including Partitioning • Data Retention and Archiving • Analytics Dashboard Design
Domains: **Application Management**	• Application Components • UI Technologies, Principles, and Standards • User Roles	• Application Components and Detailed APIs • Detailed WRICEFs • User Administration
Domains: **Technology Management**	• Technology Standards • Overall Cloud Architecture	• Configuration Management Database (CMDB) • Hardware Configuration
Security	• Policies • Overall Cloud Architecture	• Identity Management • Technical Security
Change and Release	• Release and Rollout Planning	• Daily Change Control
Risk Management	• Enterprise Risks • Cross-Concerns	• System-Specific Risks

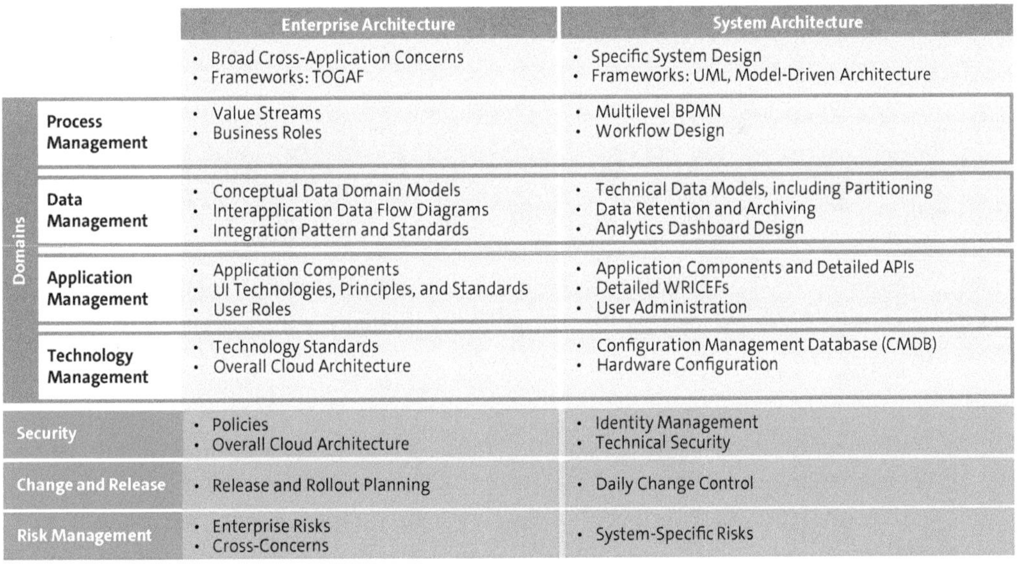

Figure 3.10 Differentiation between Enterprise and System Architecture

Within the application management layer of system architecture, traditional workflows, reports, interfaces, conversion, enhancements, and forms (WRICEF) development tasks should be assessed with a cloud-friendly extensibility strategy. The SAP BTP Guidance Framework introduced in Chapter 2 gives guidance for this in the following ways:

- Via SAP Application Extension Methodology
- On integration via SAP Integration Solution Advisory Methodology
- On overall developer guidance via the SAP BTP Developer's Guide

3.5 Additional Functions and Practices

So far, we've reflected on several enterprise architecture-connected functions. The more loosely related functions that we'll look at in the following sections don't cover the entire spectrum and merely reflect what we see as most important. At the end of the chapter, we'll look further with a short assessment of the relevance of these functions to enterprise architecture.

3.5.1 Portfolio Management

Portfolio management can be seen as a generic function depending on what types of portfolios are to be managed. The term encompasses the full set of activities that are necessary for driving the lifecycle of all elements in the portfolio. Portfolio entries are constantly assessed across their entire lifespan, from ideation through close. The type of portfolio will typically determine the assessment criteria. A roadmap describes the planned portfolio changes over time, and the granularity of roadmaps can vary in the different domains.

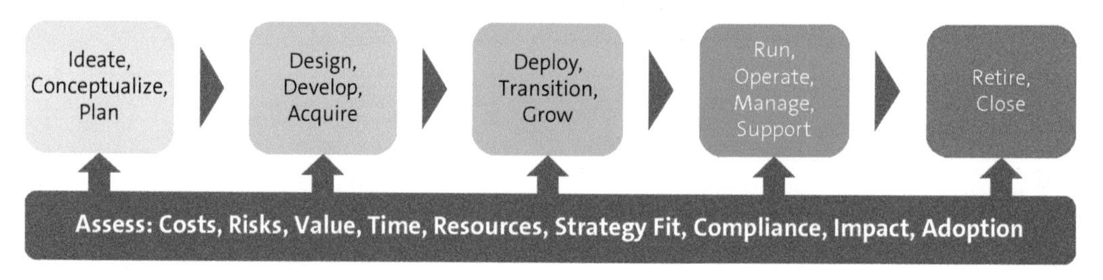

Figure 3.11 Generic Portfolio Lifecycle and Assessment Criteria

You can define almost any collection of physical or conceptual objects as a portfolio. For example, you can think of the shirts in your wardrobe as your shirt portfolio. In our enterprise architecture context, however, we focus on the following portfolios:

- Tangible product portfolios (sellable products)
- Financial portfolios (stocks, currencies, bonds, etc.)
- Real estate portfolios
- Service portfolios (ITSM/ITIL)
- Project portfolios

- Application portfolios
- IT assets (servers, laptops, etc.)

Project and program management will always look at the important dimension of time, while project scheduling can be simplified by defining successor relationships.

Companies shouldn't do portfolio- and time-related assessment solely for projects and programs. They should also include many other subjects of enterprise architecture, the most important of which are applications, solution components, application releases, business capabilities, and solution capabilities. Adding successor relationships for not only projects but also applications, releases, and solution capabilities can help them visualize different progressions of an enterprise architecture, as follows:

- **Applications**
  When managing application portfolios (Section 3.3), corporations can plan when solutions need to switch from one life cycle stage to the next. Each stage has a validity period and can therefore be depicted on the roadmap. Good portfolio management will also show the financial perspective, as shown in Figure 3.12.

Application Portfolio Management

Live Applications by Business Capability (30 days)		Application run cost broken down by Business Capability (30 days)	
Total live Applications	85	Application Run Cost	$11M
Customer Relationship	11	Inventory	$2.4M
HR	10	Information Technology	$1.7M
Information Technology	10	HR	$1.4M
Corporate Services	9	Customer Relationship	$1.3M
Marketing	9	Finance	$1.1M
Inventory	8	Marketing	$1M

Figure 3.12 Application Portfolio Management in SAP LeanIX

- **Application releases**
  During its lifetime, an application typically runs through several release cycles. For the different deployment models (on-premise, private cloud, and public cloud), it's important to understand the various ways in which releases can be applied to custom and standard solutions:
  - *Custom solutions*: These are self-developed solutions that can follow a fully customer-driven release strategy. If a custom solution is developed for deployment as an extension to a standard software package, release planning makes a lot of sense. This should be aligned with the solution provided by the software vendor.
  - *Standard software packages (on-premise and private cloud)*: Releases have their own lifecycles that determine the time windows in which the releases can be

used. Using a release longer than its officially designated lifetime can pose a critical risk, as the user will typically no longer be able to rely on the vendor to provide essential patches and legal adjustments. There are, however, cases in which software providers extend the maintenance liability for an extra service charge (i.e., provide extended support).

  - *Public cloud solutions like SaaS*: Customers have no influence over the timing of a software release switch. This enables an automatic, constant flow of innovations. For critical SaaS solutions, some testing is still recommended when larger SaaS versions are upgraded. It can make sense to document the official SaaS release cycles as well, especially if critical solution capabilities needed to support parts of the business are added during implementation and rollout.

  The Product Availability Matrix publishes maintenance windows for SAP software components (see *http://support.sap.com/pam*). SAP LeanIX enables access to the lifecycle catalog as documented at *http://s-prs.co/v586303*.

- **Initiatives and projects**

  The dimensions that should typically be assessed are value and ease of implementation. Strategic investment planning will benefit from a close look at how different investment decisions will impact the various portfolios. It's critical to take into account that the option of "doing nothing" (i.e., not investing in a transformation initiative) can have an impact on the running of critical applications (e.g., if the support of underlying software components is discontinued).

  Managing projects in a traditional and agile way leads from portfolio management to project management. In other words, you'll not get from an idea to a running solution without a project.

We'll return to application portfolio management in Chapter 14, where we'll take a closer look at SAP LeanIX, for which application portfolio management is a core use case.

3.5.2 Requirements Management

Requirements management aims to capture needs of all kinds. Needs may arise not only from IT but also from the development of a tangible product or even the acquisition of a new building.

Within the TOGAF Architecture Development Method (ADM), requirements are depicted at center stage and linked to all phases of the architecture development process. This implies that requirements are not only broken down from corporate strategies.

Managing requirements can be equated with managing a requirements portfolio (Section 3.5.1).

Requirements management can be handled in large software suites like Jira/Confluence or in smaller, light web solutions. These solutions typically combine several features to structure and assess the complexities of requirements (e.g., story points in scrum to weigh user stories). A more traditional approach is *function point assessment*, which is a standardized way (developed by the International Organization for Standardization [ISO]) to measure the amount of business functionality provided to a user. It is rooted in work at IBM in 1979.

Different types of requirements can come up throughout all phases of a transformation initiative. They can differ in the following ways:

- Abstract strategic requirements will typically be classified as goals and will belong to the strategy management function (see the next section), rather than to requirement management.
- Functional requirements will typically link to business capabilities, while nonfunctional requirements will rather link to already implemented applications.

Instead of collecting all requirements from the bottom up by starting with a blank sheet of paper, it's best practice to assume an SAP standard/reference configuration as the baseline and only document gaps it doesn't address as requirements.

3.5.3 Strategy Management

Each company has an explicit (or implicit) strategy. *Strategy management* therefore includes all efforts to create such strategies. Typically, we see the entire company strategy as the business strategy and the IT strategy as focused on the IT function—but ideally, the IT strategy should be in support of the overall business strategy.

We look at strategy management from the following three perspectives:

- **Enterprise architecture perspective**
  Strategy maps are artifacts within the business architecture. They link strategic priorities to goals and value drivers. In a typical architecture development cycle, the real business strategy input is most likely given by third parties and not developed from scratch. An important next step is to link goals to the enabling capabilities and processes so that they can be mapped to IT components in turn.
- **Strategy management perspective**
  IT strategies are clearly owned by the CIO, who often will leverage enterprise architecture as a supportive practice. In the overarching company business strategy, IT typically is only seen as an enabler, and therefore, enterprise architecture methodology won't often be used. Only specific notations like the business model canvas, which is seen as a business architecture artifact by the enterprise architecture function, might be used.

- **Integrative perspective**
  We recommend that the enterprise architecture function shouldn't lead during a strategy development cycle. Instead, it should consult to assess the strategy's impact. At a minimum, it should communicate strategic decisions to the enterprise architecture function in a timely manner so that the strategy can be broken down into manageable goals, capability gaps, and transformation programs.

3.5.4 Innovation Management

Innovations enable change, and the idea of *innovation management* is to accelerate this process in a controlled way. The motto "fail early, fail often" encourages tryouts of new approaches, technologies, and (in simple terms) "change" without the promise of success. Managing the innovation process implies that failure is an inherent part of innovation, that it's free of negative connotations, and that it's ideally detected early on.

To enable the necessary innovation speed, not many restrictions or fixed boundaries are set for the innovation process. This means that early-stage innovations can be prototyped without you having to work within narrow boundaries.

Innovative ideas that look successful may still be problematic when it comes to scalability. Promising early prototypes, for example, can inflate expectations and then fail when integration, scalability, and other functional and nonfunctional requirements come into play.

Success criteria for any innovation should be defined up front. This ensures that the right follow-up steps will be taken to scale the innovation and harvest its full potential once the idea's positive outcome becomes apparent. This can be done through subsequent additional funding to build a business case for scaling. In the last phase of the process, the scaling, the innovation should be looked at through the enterprise architecture lens.

Possible approaches to the innovation process are as follows:

- **Contract innovation from outside**
  Large companies can acquire small startups to boost their innovation pipeline. On a smaller scale, companies can collaborate with universities and use student master theses or employ working students to develop new innovations.
- **Innovation groups and departments**
  Companies can set up and fund dedicated business and IT innovation groups. Such groups should be directly linked to corporate strategy and may collaborate with

other parts of the company in some respects, as it won't be possible to have all skills needed on the team. The benefit of this approach is that the dedicated innovation group can refine the assessment criteria and invest in the innovation processes. Such a team will also be well positioned to make use of the "contract innovation from outside" approach on top of their internal projects.

- **Central funding**
  Central innovation funding is funding for which multiple groups can compete. If such a program is set up, the guardrails should be clear: How many innovations will be funded? Will all get the same amount of money, or will it vary? Clearly communicating all dates and steps for the process is essential if all submitters are to have a fair chance.

  Central funding can be established without any organizational changes. A committee of people from different parts of the organization can decide if funding is to be granted, and the size and distribution of the committee will depend on the amount of funding available.
- **Grassroots and uncontrolled innovation**
  Decentralized innovation will be organically adopted if it brings any benefits. It will occur naturally in every organization, but it has two downsides. First, any innovation's full potential might not be harvested, and second, multiple similar and/or overlapping innovations might be invented in different parts of an organization.

The need for innovation is obvious, not only for the business as a whole but also for IT. To safeguard an innovation process' momentum, it's important for enterprise architecture teams to not act as gatekeepers who seek to stop any change and maintain the status quo. Rather, enterprise architecture teams should act as enablers who provide guidance on how to scale innovations and properly integrate them with the rest of the IT landscape.

The best organizational innovation setup for an enterprise architecture team might be a dedicated innovation department, as scalability and enterprise architecture alignment can be planned jointly once a given innovation has been deemed promising. Figure 3.13 illustrates a possible collaboration model between enterprise architecture and innovation: while the innovation function focuses on managing the innovation funnel, enterprise architecture helps with qualifying and scaling innovations to ease broad adoption of the innovation.

When following totally decentralized innovation approaches, innovation can push through many different channels and may often be overlooked in its early stages. In such cases, the risk that these innovations will fail to scale is significantly higher.

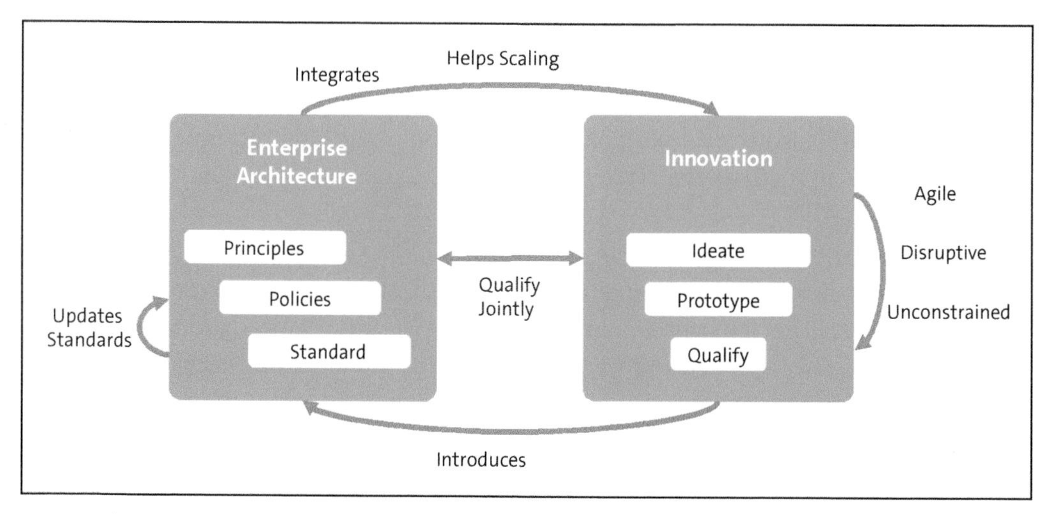

Figure 3.13 Interaction between Innovation Team and Enterprise Architecture

3.5.5 Risk Management

As geopolitical turbulence increases, there is more awareness of the need to manage risk in both business and IT. The task of *risk management* is to establish a practice that identifies risks before they materialize, so that risk mitigation activities can be planned and put into place in time. Risk management follows five general steps: risk identification, likelihood and impact analysis, mitigation planning, risk monitoring, and acting on the mitigation plan. Table 3.2 shows a risk register containing the risk areas that are relevant to enterprise architecture.

Technology Risks	Organizational Risks	Operational Risks	Geopolitical and Regulatory Risks
▪ Technology obsolescence (nonsupported components) ▪ Complexity ▪ Vendor lock-in ▪ Lack of agility ▪ Technical depth/ legacy management ▪ Technology dependencies	▪ Available skill set and talent retention ▪ Complex governance structures ▪ Misalignment of divisions ▪ Misalignment between IT and business ▪ Resistance to change	▪ Security vulnerability ▪ Insufficient data governance ▪ Availability ▪ Business continuity and disaster recovery ▪ Cost overruns and total cost ownership (TCO) ▪ Licensing issues	▪ Sanctions ▪ War ▪ Climate impact ▪ Compliance (data, legal, etc.)

Table 3.2 Enterprise Architecture-Related Risk Area Register

Risks are like requirements; they can occur at any stage of the development lifecycle and in any architecture layer. The risk management process within enterprise architecture is not different from the processes in any other business functions where risk management is established.

As enterprise architecture deals with many concepts, it's best practice to link identified risks to the related elements (applications, processes, and organizations) in the risk register. When using tools, these links must be bidirectional so that users can immediately navigate from the architecture elements to the risks in the register and vice versa.

3.5.6 Value Management

Building a strong business case for a business transformation isn't an architecture-only job. Impacted business units and finance departments need to be involved.

The task of tracking the promised value of such complex business transformations is not simple either, and it can only be accomplished if different business functions are involved and commit to identified savings. Some projects like release changes recur again and again, and they should ideally be planned after the solution is implemented. It can be difficult to find business cases for each upgrade project, especially when solutions reach the end of their lifespan and/or are modified. In this case, upgrades need to be accompanied by comprehensive retesting if not reengineering.

Two major value management approaches are as follows:

- Top-line growth (increasing revenue)
- Cost reduction via automation, which ideally contributes to reducing the workforce and leads to business savings

There are strong arguments in favor of both of these approaches, but an organization will typically only commit to them if the transformation has a large organizational impact (e.g., when introducing centralized shared services or within the framework of a company acquisition). Two further value management areas that organizations can look into are increasing productivity (e.g., in the area of usability) and targeting better data quality to reduce repetition of work.

3.5.7 Organizational Change Management

For large transformation initiatives, both enterprise architecture and organizational change management (OCM) are largely accepted as musts. While exploring the changes for both business and IT from an enterprise architecture perspective is a challenge in itself, managing the changes from an organizational or people aspect can be an even bigger task.

Let's take a look at an example. A company wishes to extract procurement and HR from country-specific organizations and move them to a global shared service center. Enterprise architecture will document the processes that will change as well as the additional capabilities the company needs, and it will go on to select the necessary IT solutions. Enterprise architecture focuses documenting the changes to the business architecture of the organization, working in the knowledge that enabling and adapting the organization to make these changes often is a hard task. For example, positions may move to other countries, which requires recruiting new talent and possibly letting go of existing talent.

Other transformations might "just" affect processes and IT systems, but it's essential, even in these situations, to engage impacted employees early on if the transformation program is to succeed. They should have the opportunity to get acquainted with new technologies and processes as soon as possible, and their needs should be listened to up front. This will increase acceptance and smooth the transition phase.

3.6 Isolated versus Integrated Functions

So far, this chapter has dealt with a large range of functions. Now, we'll give you our opinionated perspective on how a traditional enterprise architecture function might relate to other functions and vice versa. Figure 3.14 shows how these functions view themselves and each other. When the functions (depicted as circles) don't intersect, it means that they assume that they are completely independent of each other. If another function is fully incorporated into the enterprise architecture function, it's often seen as a subordinate, and if another function's circle is smaller than the enterprise architecture function, it's considered to be less relevant.

The matrix in Figure 3.14 shows three perspectives:

- In the first column, the IT-centric enterprise architecture perspective often assumes the inclusion of the neighbor functions within enterprise architecture. The enterprise architecture function therefore steers the "included" functions with an IT centric mindset.
- In the second column, the other function's perspective assumes that the enterprise architecture function is not needed or at most should be run in isolation. If the enterprise architecture function is not shown at all, it means that the respective other function sees enterprise architecture as irrelevant.
- In the third column, the hybrid perspective, which is the one we endorse, sees enterprise architecture as a function shared by both business and IT. Most of the other related functions in this scenario are run by business and IT too, in what we call a hybrid mode.

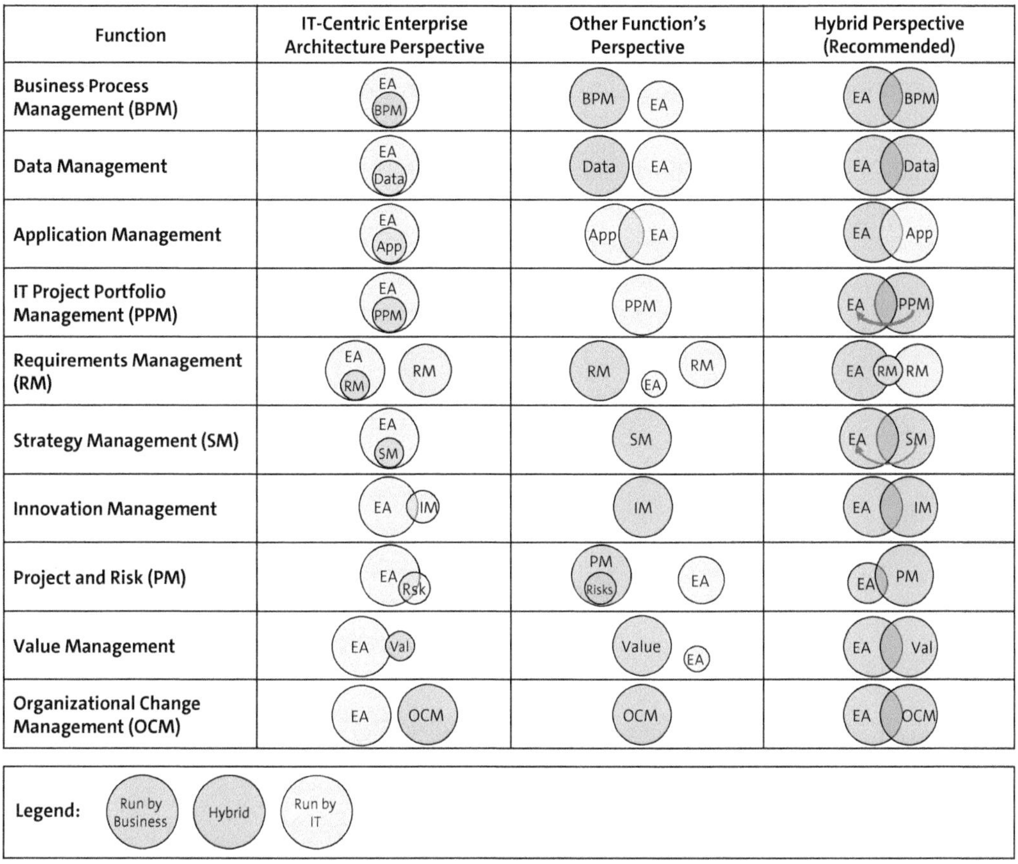

Function	IT-Centric Enterprise Architecture Perspective	Other Function's Perspective	Hybrid Perspective (Recommended)
Business Process Management (BPM)	EA BPM	BPM EA	EA BPM
Data Management	EA Data	Data EA	EA Data
Application Management	EA App	App EA	EA App
IT Project Portfolio Management (PPM)	EA PPM	PPM	EA PPM
Requirements Management (RM)	EA RM RM	RM EA RM	EA RM RM
Strategy Management (SM)	EA SM	SM	EA SM
Innovation Management	EA IM	IM	EA IM
Project and Risk (PM)	EA Rsk	PM Risks EA	EA PM
Value Management	EA Val	Value EA	EA Val
Organizational Change Management (OCM)	EA OCM	OCM	EA OCM

Legend: Run by Business | Hybrid | Run by IT

Figure 3.14 Perspectives of Function Dependencies

3.7 Summary

In this chapter, we've gone through a large variety of functions that can all play significant roles in the enterprise architecture context. However, enterprise architecture shouldn't be seen as the strategic function that orchestrates all others. Most enterprise architecture teams won't have the capacity to cover all the concerns mentioned in this chapter, so a prioritization and adoption plan should be established. It's beyond the scope of this book to cover all relevant functions, but we've listed some supporting functions in Figure 3.1.

In Chapter 15, we'll cover important aspects for setting up or strengthening an enterprise architecture practice. In addition to the core capabilities, we'll revisit some of the functions discussed in this chapter and discuss their importance when setting up a good practice.

Depending on the maturity of the enterprise architecture function or of the partner function the enterprise architecture function needs to collaborate with, a stepwise adoption plan should be developed. It would be too difficult to implement or improve all in one step.

In the next chapter, we'll turn our attention to what makes a good enterprise architecture tool.

Chapter 4
Tool Capabilities for Enterprise Architecture

While Chapter 2 introduced tooling in the context of the SAP EA Framework, this chapter will discuss the capabilities tools must have to support the SAP EA Framework and the adoption of reference content. The overview provided in this chapter is complemented by Part III, which is a deep dive into SAP LeanIX.

Can you design a good enterprise architecture by using only PowerPoint and/or Visio? Yes, possibly, but with increasing complexity, you'll need to depict different perspectives on the same business and IT objects. Executing simple operations like deleting or renaming an object in a consistent way won't be supported when relying solely on office tools.

Therefore, the need to ensure consistency among a large amount of data and visualizations is the primary need driving the use of an enterprise architecture tool. A pure content management system like a Microsoft SharePoint site can be established to hold enterprise artifacts, but from our perspective, it would never qualify as enterprise architecture tool.

A model-based approach and an aligned visualization approach form the key way to achieve data consistency. In this process, multiple graphical representations (reports and diagrams) of the same object are linked to one single model entry, thus automatically propagating renaming, deletion, etc.

Another important characteristic is that semantics in diagrams and reports should always originate in the model and not have their own implicit semantics. Ignoring this design principle would downgrade the tool to a pure drawing program (like PowerPoint), thus losing consistency support.

In this chapter, we'll dive deeper into the value of tools for enterprise architecture in an effort to demonstrate their benefits over nonspecialized, distributed solutions. We'll see how enterprise architecture tools are structured, we'll help you understand a meta model approach, and we'll walk through key artifacts and features.

4.1 Generic Reference Architecture

Before we dive into specific tool aspects, we'll introduce a generic reference architecture in Figure 4.1. This can be seen as an overarching enterprise architecture solution concept. We'll discuss each of the areas in subsequent chapters, but for now, here's a preview of each:

- **Architecture model**
  We've already laid out why taking the model-based approach and operating on the basis of tools makes more sense than working with single isolated "drawings." In Section 4.2, we'll discuss the relevance of a meta model-driven approach.
- **Architecture artifacts**
  Artifacts are visual representations of the architecture model, and they allow exploration that helps you gain insights and typically expose interactive user interface elements to help you modify the model state.
- **Integrations**
  In Chapter 3, we spoke about a number of enterprise architecture-related functions. As many of them will come with their own specialized tool support, there's a strong need for the enterprise architecture tools to not lie in closed repositories but rather embrace bidirectional integrations. In the SAP context, we'll also highlight the importance of landscape autodiscovery as well as the use of SAP reference architectures as described in SAP EA Framework.
- **Features and capabilities**
  Comparing features and functions of various enterprise architecture tools can be a daunting exercise. Therefore, our focus will be on extracting and evaluating a core set of concepts that we see as critical.

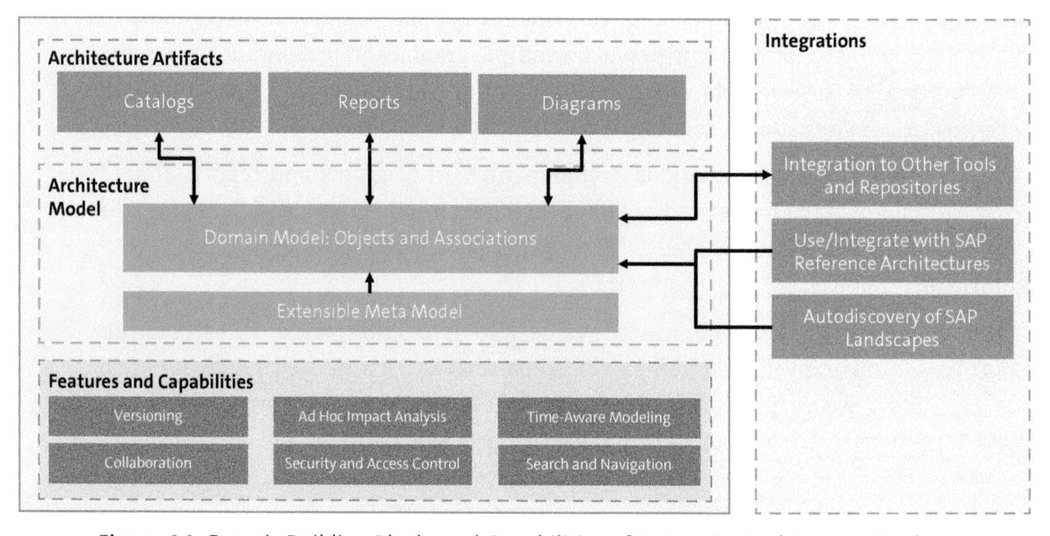

Figure 4.1 Generic Building Blocks and Capabilities of Enterprise Architecture Tools

Exploring and Extracting SAP's Reference Architecture Content

The availability of reference architecture content on multiple architecture layers can help accelerate enterprise architecture progress. Reference architectures can have multiple sources.

As described in Chapter 2, the use of SAP reference content is a key accelerator for SAP transformations, and we recommend that you actively use it in daily enterprise architecture work.

Chapter 13 shows current SAP reference content use cases for SAP LeanIX. This topic will gain an even higher priority in the future.

4.2 Architecture Model

The difference between a model-driven approach and free charting in Visio or PowerPoint is that the meta model predefines the types and attributes as well as the relationships different concepts can have with one another. If you depict, for example, an application in PowerPoint as a text box on multiple slides, there's no way to manually ensure consistent naming and coloring on multiple slides and large numbers of office documents.

When you're selecting a tool, the fit between the meta model and the purpose should be an important consideration. More concepts are not necessarily better:

- The Open Group Architecture Framework (TOGAF) 9.2 Core Content Model includes 21 concepts, and with the content extensions, the number increases to 35.
- ArchiMate is even more finely granular and holds up to 60 concepts.
- The SAP LeanIX meta model, in contrast, consists of only 12 base concepts. When counting the 21 specific subtypes (instead of the generic main types), we reach 28 concepts (12 – 5 + 21 = 28).

A simpler meta model generally gains a higher degree of acceptance from the user base. Also, adding more concepts can exponentially increase the complexity and therefore the cost of maintenance because each added concept will typically connect to multiple other concepts. Additional catalogs of objects will, in other words, require the maintenance of a growing network of interconnected concepts.

We will introduce SAP LeanIX in detail in Part III of this book.

4.3 Architecture Artifacts

Many enterprise architecture tools offer the possibility of adding free graphical objects (rectangles, circles, etc.), but it's a best practice to always link them to real model objects. For example, if an arrow between two objects represents a relationship in the model and the relationship is deleted at a later point in time, the arrow should either be eliminated from the visual diagram automatically at that time, or there should at least be an alert that the model-linkage has been "lost." Likewise, if a new capability is added to a business capability model or if the sequence of capabilities is changed, then all graphical representations of that part of the model should reflect the change.

Tools (such as SAP LeanIX) often distinguish among the following three representation types:

- **Data-driven visualizations (reports)**
  These can be hierarchies, matrix reports, circular interface reports, or just simple lists or tables. We define a *report* as a fully data-driven output describing a given model state. A report can be interactive in the sense that it allows interactive exploration, and it can also be a user interface for changing the underlying architecture model. One example would be a report in which dragging one object into another would trigger the establishment of a relationship between the two objects. Other operations could be adding and deleting objects, changing object attributes, etc.
- **Manually designed visualizations (diagrams)**
  Often, these are flow diagrams for components. Even though good graphical layout algorithms are widely available, good design skills are sometimes required to represent the integration flows of a complex landscape (often with the goal of having no or just a few crossing integration lines while keeping semantically related objects close to each other).
- **Object lists (catalogs)**
  Object lists offer the simplest way to expose architecture content. Typically, they can be filtered by type and by any other attribute or relationship. Catalogs are often used for data entry or for branching to visualizations of these objects.

In the generic depiction of an enterprise architecture in Figure 4.1, we used the concepts of a report and a diagram, where the report has the connotation of a generated and then a static output while the diagram is something like an architecture design that can be manipulated by the user. In our opinion, the distinction between reports and diagrams is somewhat fuzzy, as both follow conventions, and the visualization will depend on the parameters set by the user. This is why we prefer to use the term *view*. A view can be specified by relations and attributes, as we'll see in the following sections.

4.3.1 Representing Relationships in Artifacts

Relationships between concepts can be depicted by one of the following options described in Figure 4.2:

- **Association via lines and arrows**
  Users often connect objects with lines when analyzing large association networks.

ArchiMate heavily uses shapes connected with arrows to depict the relationships within a model, and when visualizing organizational structures in diagrams, lines are often used to denote hierarchies as well. The downside of using lines for depicting many different types of associations is that the readability of such diagrams usually decreases. A manual line layout can reach its limit when you want to expand nodes interactively during dynamic exploration sessions.

- **Association via inclusion**
  You can easily depict hierarchical relationships by nesting rectangular shapes. Typically, you can use mouse operations like drag and drop to change the associations, add new objects, or change their natural order.
- **Association via a matrix or grid**
  In simple matrix diagrams, a filled cell indicates a relationship between the object in the row header and the one in the column header. Such cells can also be used to show attributes, which are defined by the relationship between row and column objects. In a more complex matrix diagram, the cells can contain one or multiple architecture objects that have an association with the respective row and column headers. The example in Figure 4.2 is a matrix in which the rows show applications, the cells show provided business capabilities implemented by the application, and the columns show the parent business capabilities.

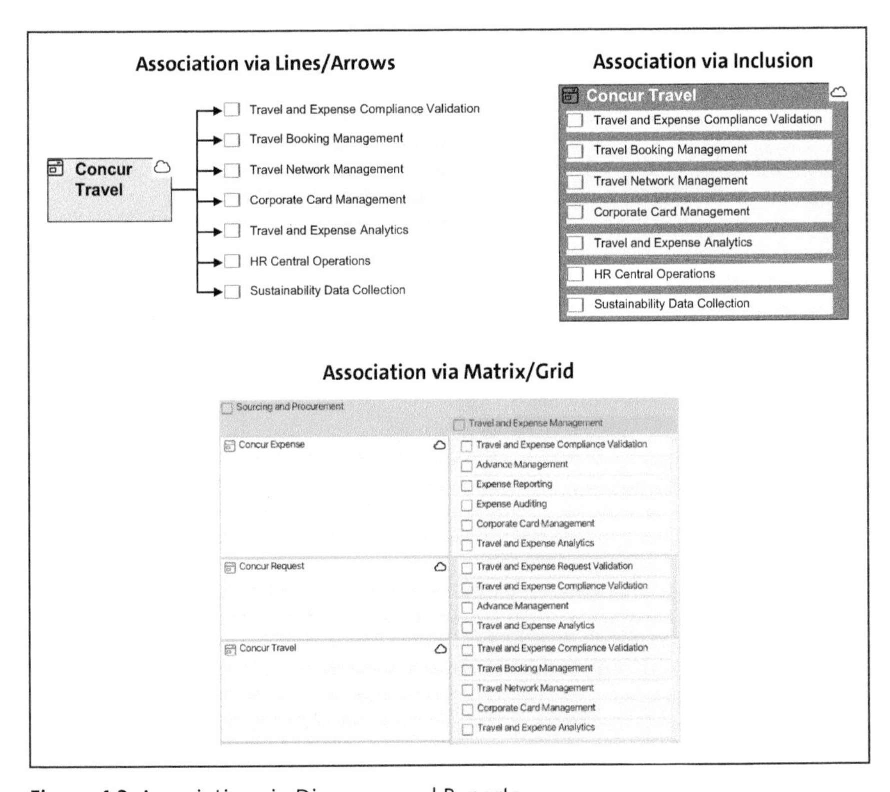

Figure 4.2 Associations in Diagrams and Reports

4.3.2 Representing Attribute Values in Artifacts

The base entities in an enterprise architecture tool form different catalogs that are interrelated via relationships. Objects of different types have different attributes. Capabilities might have the name, description, criticality, current maturity, and future maturity attributes, to name some examples. Applications may have attributes such as name, description, number of users, deployment model, and lifecycle.

Tools will typically offer specific editing user interfaces to allow changing the attributes' values via typical user interface controls like text boxes, dropdown lists, radio buttons, sliders, etc.

There are multiple ways to show attribute values in reports and diagrams and to use them to drive a report layout. A good visualization depends on the number of attributes (dimensions) that need to be discernible at the same time. Showing multiple attributes at the same time enables insights into multidimensional assessments.

Figure 4.3 shows five simple ways of depicting a single value: heatmapping, icons, cluster/box diagrams, tagging, and text appends.

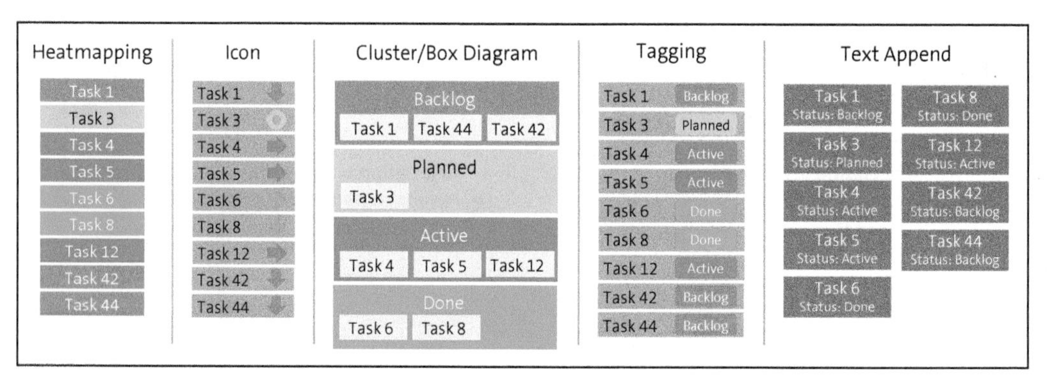

Figure 4.3 Five Ways of Depicting an Attribute Value

If you wish to depict multiple attributes, you can mix and match these ways or use a more complex report design like the following:

- Two attributes: Matrix or mixed use of cluster + heatmapping or text appends
- Three attributes: Matrix + heatmap or radar + heatmap
- Four or more attributes: Any combination of the above

Note that a radar report can structure objects in segmented concentric rings.

Comparison to The Open Group Architecture Framework

TOGAF is the concept of an architecture repository, and it includes a meta model, artifacts, and linked or included repositories or libraries for references, standards, governance, and more. Artifacts in TOGAF are defined as follows:

- **Catalog**
  A *catalog* can be understood as a list of objects with a set of predefined attributes that specify them.
- **Matrix**
  A *matrix* expresses relationships between objects. We see the relationships part of the architecture model layer (the green model layer in Figure 4.1), and we would see a specific matrix as an editable report. However, relationships can be exposed in other reports, diagrams, and the catalogs described previously.
- **Diagram**
  A *diagram* is the visual representation of an element of the architecture model (called a *building block* in TOGAF).

In this high-level comparison, the TOGAF way of thinking is very close to the concepts introduced earlier.

4.4 Features and Capabilities

In this section, we'll explore the features and capabilities introduced in Figure 4.1 that play an important role in supporting the enterprise architecture practice. Let's walk through each of them:

- **Versioning**
  Building up knowledge over time often is not a linear approach. A tool should be capable of reverting to older states and be able to do so for defined subsets, as in when you need to use enterprise architecture reports to present an architecture decision proposal to the enterprise architecture board. Decisions are taken based on insights and facts known at a specific point in time, so if you like to keep track of enterprise architecture decisions, it's crucial for you to have at hand the "original" artifact or deliverable that the enterprise architecture board has used for decision making. If more insights are gained over the course of time, the "live" reports will change, and the board needs to be informed of those changes because they may use them to arrive at a revised decision. When you're modeling target or transition states, the tool should also enable you to design alternative transition or target states.
- **Ad hoc exploration**
  Static depiction won't be sufficient if the amount of data is building up. The tool will need to allow visual ad-hoc explorations, which could be done via network traversal or zooming through different levels.
- **Time-aware modeling**
  Enterprise architecture tools should not only be snapshots of a single point in time but should also be able to look back to former states and depict or compare multiple

future aspects. One way to enable this is by ensuring that each object and each relationship has a validity period (a start-to-end date) defined.

When you're investigating a tool, we strongly recommend, with regard to the time dimension, that you check whether it's implemented in an easy-to-understand way.

- **Collaboration**
  Sharing, chatting, assigning tasks, and directing collaborative working sessions are modern ways of working. We don't postulate that enterprise architecture tools should replicate the full functionality of a specialized solution like Microsoft Teams, but having a relevant set of day-to-day features such as the following will increase productivity:
  - Commenting on objects, diagrams, and reports
  - Assigning tasks and following up via status
  - Synchronous modeling in diagrams (like in Mural or PowerPoint)
- **Search and navigation**
  Assuming that an enterprise architecture management tool will grow exponentially, at least in the early phase, getting on target will require powerful search capabilities and a well-structured approach to aggregating, categorizing, and interlinking artifacts. When in the tool selection phase, we recommend that you focus on ease of use while still enabling a rich set of features.
- **Security and access control**
  We believe that enterprise architecture should be done in a very interactive, collaborative manner, whereas defining a finely granular security concept (which limits access and editing rights) can decrease productivity. A tool that internally keeps track of who changed what and when, and that also allows you to undo specific edits, provides a solid basis for arguing in favor of not being too rigid in the security model. On the other hand, if a tool allows hundreds of users full editing or even administration rights, the risks of unintentional edits or deletions will increase.

4.5 Integrations

Chapter 3 discussed the need for collaboration with related functions and supporting integrated processes. As many of these functions will be implemented in dedicated environments, we draw a hypothetical abstract integration architecture in Figure 4.4.

Especially in cases where related information can take different channels to flow toward the same targets, a carefully designed integration concept needs to be in place.

You, as architect, may well ask, for instance, if you need dedicated tools for all nine topics listed in Figure 4.4. There are two questions that you need to consider here:

- **Can other functions be included in the enterprise architecture tool?**
  The answer to this will depend on the maturity or existence of those other functions. As an example, SAP LeanIX is well suited to planning project portfolios and depicting them in Gantt charts. However, complex project management planning tasks like resource assignment, baselining, and different optimization techniques need to be managed with external dedicated tools like Jira that can integrate with SAP LeanIX.
- **Can enterprise architecture work also be conducted in the configuration management database?**
  There are well-established configuration management database (CMDB) tools that claim to cover enterprise architecture work, but we tend to be very careful around such claims. From a configuration management perspective, a CMDB may be well equipped to maintain comprehensive catalogs of business capabilities or other enterprise architecture concepts effectively, but other enterprise architecture-specific artifacts may not be as easy to find. These include the following:
  - Simulating different versions
  - Fully flexible visual reporting
  - Representation of time

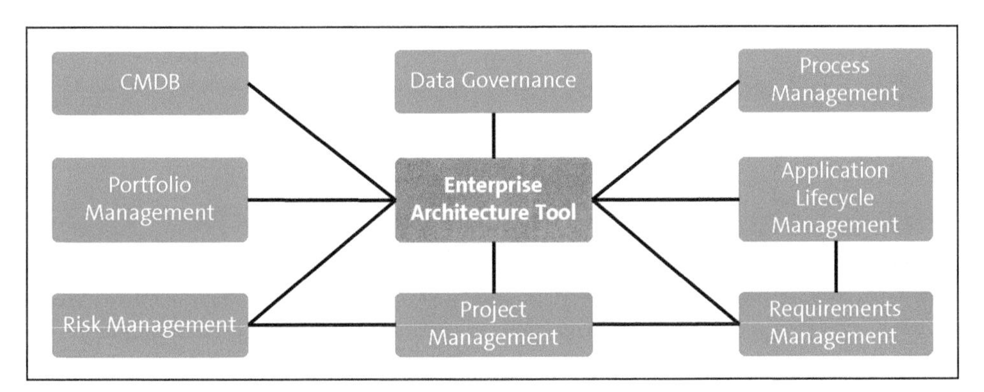

Figure 4.4 Generic Integration Architecture for Enterprise Architecture Management

4.6 Comparing Different Types of Enterprise Architecture Tools

Enterprise architecture tools can be generic or follow a very targeted set of rigid approaches. When implementing an enterprise architecture tool, you should check whether the tool will support your needs. Table 4.1 introduces three classes of tools: rigid, open, and guided. These are by no means standardized, official categories, and there may be tools that fall into none of them. Our intent in introducing them here is to reflect on their key strengths and weaknesses.

	Definition	Pros	Cons
Rigid Tools	A rigid tool comes with opinionated, preconfigured use cases. Not only the domain, but also the way of working and (with that) the methodology are embedded in the tool. Visualization can be advanced but is often specific.	The benefit of an opinionated rigid tool is that it guides you through a specific methodology and notation. Less conceptual work is needed when implementing the tool.	As the focus of enterprise architecture can range from application management via " classical" enterprise architecture to system architecture, you'll need to carefully assess whether your current and possible future needs will be covered. The extensibility options of rigid tools are limited.
Open Tools	This category is very flexible and highly customizable. The complexity of customization might be high and, if so, will require intensive care. The flexibility of this tool category is intended to help you configure additional domains within the tool, rather than integrate it with other tools.	They have a high degree of flexibility and extensibility.	Be aware of whether key aspects like versioning and time-dependent modeling are supported. The costs of customization and operation can be high, too.
Guiding Tools	These tools are more general purpose, with a preconfigured core to support all domains, but they're not as sophisticated as the tools in the rigid and the open categories. The meta model is built around few concepts, and the category doesn't aim to cover "neighboring domains."	Core use cases can be supported without extra effort. Hurdles to adoption are low because these tools are easy to learn.	For advanced use cases, integration with other specialized repositories will be needed. Implementing a comprehensive methodology will be difficult.

Table 4.1 Classification of Enterprise Architecture Tools

Let's look at a common dilemma that is not specific to enterprise architecture tools: feature-rich versus easy-to-use products. Figure 4.5 gives a framework for comparing ease of use with feature completeness. This matrix resembles the ease-of- implementation versus value assessment for initiatives.

One option for preventing occasional users from being overwhelmed by the tool's advanced functionality is to provide different expertise levels that expose the respective user groups to just those features and functions that they need to perform their specific jobs.

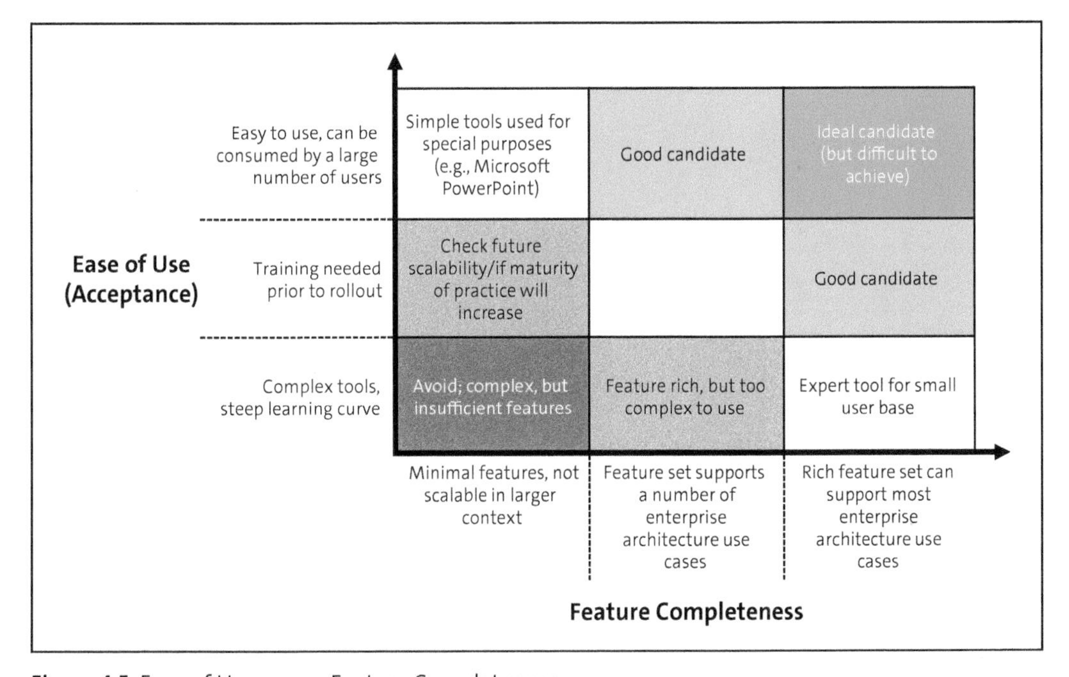

Figure 4.5 Ease of Use versus Feature Completeness

After you've read Part III or when you're reflecting on your own experience, you may wonder where SAP LeanIX fits in here. We place SAP LeanIX in the class of guiding tools, and we see them as easy to use and covering many to most enterprise architecture use cases. That is to say that in the matrix, SAP LeanIX falls into the upper-right, bright green box: ideal candidate (but difficult to achieve).

Some might see this as an opinionated SAP "internal" classification, so we would like to invite you to draw your own conclusions and share them.

4.7 Summary

Even the best tool doesn't guarantee good enterprise architecture. Rather, this will require a good methodology, good content governance, and, primarily, a mature enterprise practice. We'll discuss the latter in Chapter 15.

When selecting a tool, make sure it supports your current needs but also think ahead: once your maturity improves, you might want to explore how to represent more complex concepts like representation of time or the ability to model and judge alternative architectures. At the other end of the spectrum, an easy-to-understand, less powerful tool that will be adopted more easily by a larger user base may have an overall higher impact than a niche comprehensive tool.

Let's continue on to Part II and walk through some use cases that demonstrate enterprise architecture in real-world scenarios. We'll start with cloud transformation in the next chapter.

PART II

SAP Enterprise Architecture Use Cases and Patterns

Implementing SAP software, like SAP S/4HANA, may allow a business to get by with an organizational structure that maps to the standard software package. However, this strategy typically only works in the short term. Eventually, new business challenges arise and force change, requiring a broader focus on enterprise architecture transformation with SAP.

This part discusses a variety of enterprise use cases and the architecture patterns that can be used to solve common design problems that may arise in each scenario. The patterns used in these scenarios can often be applied to additional use cases. By working through these use cases, you'll learn about the ongoing activities involved in enterprise architecture transformation.

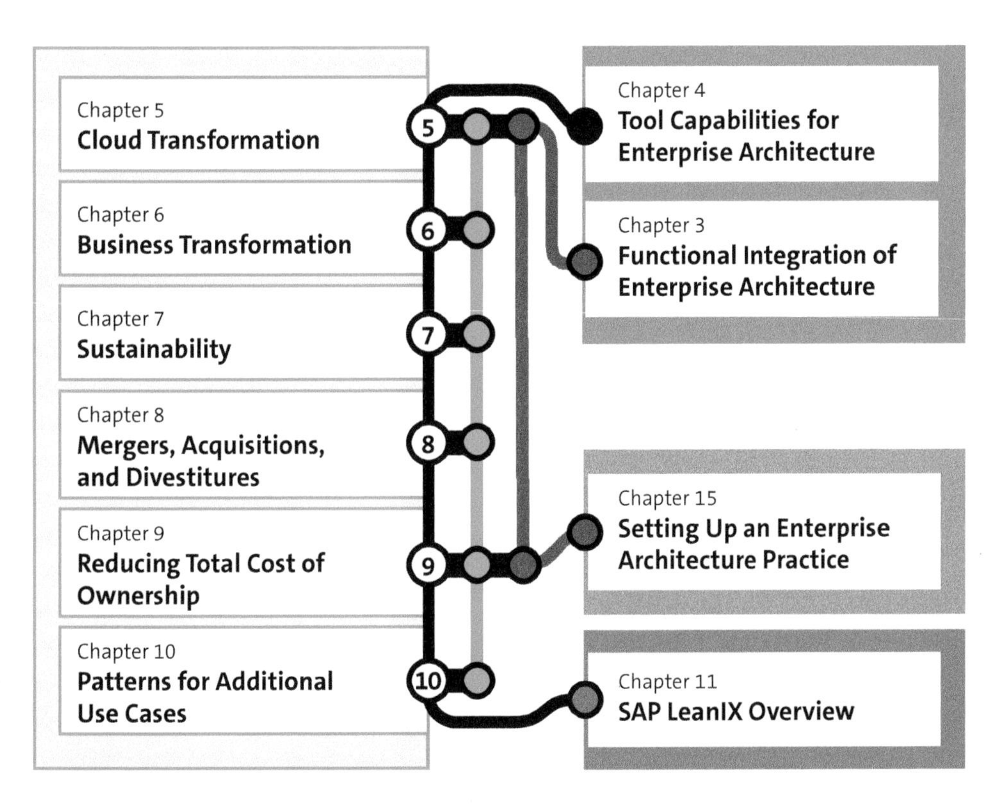

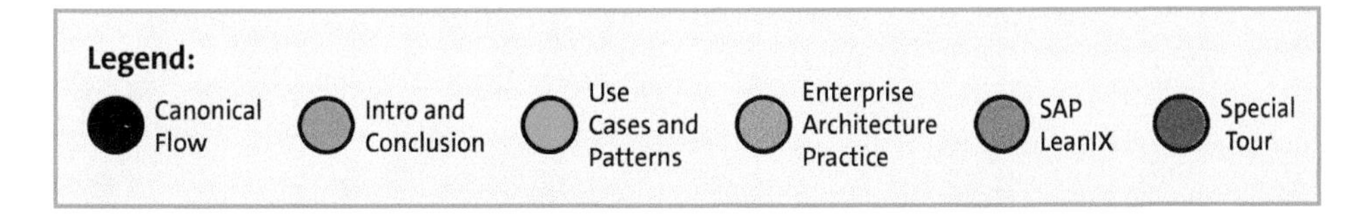

Chapter 5
Cloud Transformation

In this chapter, we'll introduce you to the practical application of SAP cloud transformation, employing SAP EA Framework as a guiding tool. Through a detailed use case, we navigate the structured approach to architecture transformation in the cloud, considering essential factors such as clean core, two-tier architecture, and organizational readiness.

In the ever-evolving landscape of business technology, the shift toward cloud computing has emerged as a defining transformation. This transition is not merely a change in infrastructure but a fundamental reimagining of how businesses operate, innovate, and compete. SAP systems lie at the heart of this transformation for many enterprises; the systems form the critical backbone supporting numerous business processes spanning several SAP and non-SAP applications. The transformation of SAP-centric landscapes into cloud architecture represents an essential opportunity to achieve agility, scalability, and digital innovation. Additionally, the concept of *shadow IT*—in which business departments independently adopt cloud solutions outside the traditional IT infrastructure—can promote a culture of self-empowerment.

For decades, SAP systems have been the cornerstone of enterprise resource planning (ERP), managing everything from supply chain logistics to customer relationships. Historically rooted in on-premise infrastructure, these systems were synonymous with significant capital investment, complexity, and rigid operational silos. However, as the digital economy accelerates, the limitations of traditional SAP landscapes have become increasingly apparent. Businesses face the imperative to adapt swiftly to market changes, scale on demand, and harness the power of data analytics for strategic decision-making. Herein lies the transformative potential of cloud architecture.

This chapter explores the intersection of cloud technology and SAP landscapes, demonstrating the path to successful cloud migration and adoption. It aims to provide a comprehensive guide for enterprise architects, IT professionals, and business leaders navigating this transition. This chapter illustrates the cloud journey through a fictional use case inspired by real-life transformation examples of SAP customers. This use case gives you an example of how to assess the readiness of your organization and current SAP environment for the cloud, build the required business and IT architecture, and ensure a smooth transition path to the cloud.

Before we dive into the details of the use case, we'll need to frame the broader business and IT context of cloud transformation. We'll discover how shifting to the cloud aligns with broader business objectives such as digital transformation, operational flexibility, and market responsiveness. In addition, we'll explore the reasons why the clean core paradigm is a transformative approach to modernizing SAP landscapes in the cloud: it reduces complexity, lowers total cost of ownership, and accelerates the journey toward becoming a truly intelligent enterprise. This foundation will set the stage for understanding why cloud adoption is not just an IT initiative but a strategic business move.

A critical aspect of successful cloud transformation involves strategic planning and structured execution. To this end, we'll delve into how to apply SAP EA Framework to facilitate this process. This discussion will include a detailed examination of customer use cases, showcasing the main artifacts created during the transformation journey, such as target architecture, a transformation roadmap, and governance structures. By illustrating how these elements come together within the framework, we'll help you gain insights into how to craft a coherent and effective strategy for migrating SAP landscapes to the cloud.

Finally, the move to the cloud is not just a technical upgrade; it's a strategic transformation that can unlock new opportunities for growth and innovation. By leveraging cloud capabilities, SAP customers can enhance operational efficiency, foster innovation, and deliver exceptional experiences to their customers and employees. However, this journey requires careful planning, execution, and ongoing management to realize its full potential. As you navigate through the chapter, we aim to equip you with the knowledge, strategies, and insights needed to successfully transform your SAP landscape through cloud architecture.

5.1 Business and IT Context

Cloud transformation represents a fundamental shift in how businesses approach their operational, infrastructural, and technological strategies. From a business perspective, cloud transformation promises to help organizations achieve enhanced scalability, agility, and cost efficiency. It allows businesses to rapidly adapt to market demands and customer needs without the heavy up-front investment traditionally associated with IT infrastructure expansion. Moreover, it offers a pathway to innovation by leveraging the latest in artificial intelligence (AI), data analytics, and other technologies, facilitating new product development and improved service delivery.

From an IT perspective, cloud transformation is pivotal in simplifying infrastructure management, enhancing security through centralized controls, and providing the flexibility to use a wide range of services and tools that improve efficiency and productivity.

Taking this dual view (from a business and an IT perspective) will help you ensure that cloud transformation is not just a technological upgrade but a strategic enabler that aligns IT capabilities directly with your business's objectives, fostering a more resilient and competitive organization in the digital age.

However, cloud transformation also comes with its own set of risks. From a business perspective, there's the challenge of working within standardized cloud environments that often offer limited flexibility to adjust the software to specific business needs. This can lead to constraints in customization and may require businesses to adapt their processes to fit the cloud solutions rather than the other way around. From an IT perspective, migrating to the cloud can expose organizations to new security vulnerabilities and compliance challenges, as sensitive data may be stored off the premises and be subject to different regulatory environments. Furthermore, the complexity of managing a hybrid cloud environment can strain IT resources and expertise, potentially leading to misconfigurations or service outages.

We'll explore these business and IT perspectives in depth next.

5.1.1 Business Context

In the following sections, we'll examine the common business drivers of cloud transformation. We'll also explore the organizational impacts of cloud transformation and delve into the concept of a bimodal cloud transformation.

Business Drivers

In the digital age, businesses are rapidly transforming their operations by leveraging the cloud to stay competitive and meet evolving market demands. Cloud transformation is driven by multiple business imperatives, each of which contributes to the organization's overall growth, efficiency, and innovation. Let's explore the primary business drivers of cloud transformation:

- **Agility and flexibility**
  At the heart of cloud transformation is the pursuit of agility and flexibility. In a market where customer preferences and industry trends can pivot overnight, the ability to adapt swiftly is invaluable. Cloud computing empowers businesses with this agility. For example, in the cloud, a retail company can instantly access additional computing power during the holiday season to manage increased online shopping traffic and can then just as quickly scale down once the rush subsides. This level of operational agility was exemplified during the global shift to remote work, when companies on the cloud could seamlessly transition their workforce to out-of-office employment.
- **Cost optimization**
  The financial implications of cloud transformation are profound. Traditional IT

infrastructure requires significant up-front capital investment and ongoing maintenance costs. By migrating to the cloud, businesses transition from a capital expenditure (CapEx) model to an operating expenditure (OpEx) model, which is more predictable and scalable. A startup, for example, can avoid the hefty initial investment in servers by utilizing cloud services, thereby freeing up capital for other areas like research and development (R&D) and marketing.

- **Scalability and elasticity**
  Scalability and elasticity are perhaps the most tangible benefits of the cloud. In traditional settings, with highly modified and large ERP implementations, a sudden surge in demand could overwhelm IT infrastructure, leading to service outages and the loss of customers. The cloud addresses this by allowing businesses to scale resources in real time. A media company streaming a popular event can elastically add more bandwidth to ensure a seamless viewer experience, demonstrating how cloud services provide the necessary scalability to manage unpredictable demands.
- **Innovation and speed to market**
  Innovation is the lifeblood of growth, and the cloud is the catalyst for rapid innovation. By reducing the time it takes to procure, install, and configure IT resources from weeks to mere minutes, the cloud enables businesses to experiment and iterate faster. A fintech firm can roll out new features on its app much quicker than competitors stuck with lengthy deployment cycles, highlighting how cloud services can shorten the innovation loop.
- **Collaboration and communication**
  The modern workplace is no longer confined to a physical office. Cloud services offer a suite of collaboration tools that enable real-time communication and project management, regardless of geographical barriers. Consider a multinational team working on a product launch; cloud platforms allow them to coordinate and work together as if they were in the same room, fostering a collaborative culture.
- **Insights to action**
  In the data-rich environment of today's digital economy, the ability to translate information into insights is crucial. The cloud provides advanced analytics tools that can process large volumes of data to uncover patterns and predictions. A marketing team can use these tools to gain insights into customer behavior and subsequently tailor campaigns for better engagement.
- **Sustainability**
  As environmental concerns become increasingly central to corporate responsibility, the cloud offers a greener alternative to private data centers. Cloud providers typically utilize the latest, most energy-efficient technology at scale, which individual companies may find cost prohibitive. By leveraging the cloud's shared resources, a company not only reduces its own carbon footprint but also aligns with broader sustainability goals—a move that should be well received by ecoconscious consumers.

- **Artificial intelligence**
  Finally, AI in the cloud is a game-changer for businesses seeking to innovate. Cloud-based AI services allow companies to implement smart technologies without the need for specialized hardware. For instance, a health care provider can use AI for predictive analytics in patient care, enabling early intervention and personalized treatment plans.

Organizational Impact

The organizational impact of cloud transformation extends far beyond the IT department, impacting almost every facet of a business. This profound change redefines the operational, cultural, and strategic aspects of an organization. As businesses embrace the cloud, they witness a transformation that unlocks new potential while challenging traditional norms.

Let's explore the five main aspects of a business that cloud transformation impacts:

- **Organizational structure**
  Cloud transformation often leads to a restructuring of teams and roles within an organization. The IT department evolves from providing a support function to filling a central, strategic role, driving business innovation and enabling other departments to leverage technology more effectively. Traditional roles may shift, with new positions such as cloud architects and DevOps engineers emerging to manage and orchestrate cloud resources.
- **Organizational culture**
  Organizational culture plays a significant role in the success of cloud adoption. Fostering a culture of innovation, decentralized decision-making, collaboration, and continuous learning encourages employees to embrace change, experiment with new technologies, and drive digital transformation initiatives.
- **Leadership and governance**
  Clear leadership and governance structures are essential for guiding the cloud adoption process. Establishing executive sponsorship, defining roles and responsibilities, and implementing governance frameworks ensure alignment with organizational goals, policies, and compliance requirements.
- **Change management**
  Effective change management practices are critical for managing resistance to change and facilitating smooth transitions during cloud migration. Providing training, communication, and support to employees helps alleviate concerns, build confidence, and promote adoption of cloud technologies.
- **Skill development**
  Cloud adoption requires the development of new skill sets and capabilities within the organization. Investing in employee training and development programs to

build expertise in cloud technologies, DevOps practices, and agile methodologies ensures that teams are equipped to leverage the full potential of the cloud.

- **Cross-functional collaboration**
  Collaboration among different departments and teams is essential for successful cloud adoption. Breaking down silos, fostering cross-functional collaboration, and promoting knowledge sharing enable organizations to align cloud initiatives with business objectives and address diverse stakeholder needs.
- **Vendor management and partnerships**
  Establishing strategic partnerships with cloud service providers and technology vendors is key to leveraging their expertise, resources, and support throughout the cloud adoption journey. Effective vendor management ensures alignment of services, contracts, and service-level agreements (SLAs) with organizational requirements.

Bimodal Cloud Transformation

In the evolving narrative of business modernization, the concept of bimodal cloud transformation has emerged as an essential concept. It focuses on how companies can maintain the delicate balance between operational stability and the imperative to innovate. This approach is not simply a technological shift but a strategic reorientation that acknowledges the distinct rhythms at which different parts of a business must operate to harmonize sustained efficiency and disruptive innovation. Bimodal cloud transformation leverages the stability and reliability of mode 1 for core operations while embracing the agility and innovation of mode 2 to drive differentiation and rapid adaptation in competitive markets.

At its core, bimodal cloud transformation recognizes that not all business areas can—or should—move at the same speed. The bimodal strategy is about enabling businesses to be stable where necessary and agile where possible. It's a tale of two tempos, which we refer to as *modes*:

- **Mode 1: Stability**
  In mode 1, the business focuses on predictability and reliability. Here, cloud transformation is methodical and risk averse, and it supports essential business functions that form the backbone of the enterprise. This mode aligns with core systems that require stringent controls and gradual enhancements—such as financial systems, supply chain logistics, and core database applications. The emphasis is on robust cloud services that offer high availability and compliance with regulatory standards. Businesses can't afford missteps in these areas, as they represent the integrity of the enterprise and are key to maintaining the trust of customers.
- **Mode 2: Agility**
  Mode 2 is the business's experimental heart, where cloud transformation allows for rapid growth and innovation. In this mode, cloud services are leveraged for their

scalability, elasticity, and capacity to support iterative development. This is where businesses experiment with new customer engagement models, digital product enhancements, and market explorations. Mode 2 thrives on the cloud's ability to pilot new initiatives with minimal up-front investment and to scale successful experiments into broader business offerings.

The bimodal approach demands a thoughtful orchestration of resources. Budget allocation, for instance, becomes a delicate art—ensuring that steady-state operations are well funded while also investing in the agile initiatives that fuel growth and transformation. HR must also be managed bimodally, with some teams dedicated to the continuous improvement of reliable systems and others empowered to innovate and disrupt. Governance in a bimodal world takes on a nuanced dimension. Traditional risk management must coexist with a tolerance for the calculated risks inherent in rapid innovation. The two modes require different success metrics, leadership styles, and cultural attitudes.

For a successful bimodal transformation, businesses must ensure that mode 1 and mode 2 are not isolated from each other. Instead, they should be interlinked so that insights and innovations can flow back into core operations and stability can underpin exploratory ventures. It's a symbiotic relationship; the agility of mode 2 often sparks enhancements to mode 1, and the robustness of mode 1 provides a stable foundation for the ventures of mode 2.

Examples of a Bimodal Approach

We can illustrate a bimodal approach through several well-known examples:

- **Amazon**
  Beyond its core e-commerce platform (mode 1), Amazon constantly innovates with new business models and services, such as Amazon Web Services (AWS), Amazon Prime, and its foray into artificial intelligence with Alexa (mode 2).
- **Nike**
  Alongside its established business of designing and selling athletic footwear and apparel (mode 1), Nike integrates digital innovation through its Nike+ ecosystem, custom online services, and direct-to-consumer sales strategies, enhancing customer engagement and brand loyalty (mode 2).
- **Disney**
  With its traditional media and theme parks as mode 1, Disney has embraced mode 2 with the launch of its streaming service, Disney+, which directly competes with other streaming platforms. This move represents Disney's strategy to innovate while maintaining its core business models.

The journey of bimodal cloud transformation is ongoing and iterative. It requires business leaders to maintain a dual focus: ensuring that today's business thrives while simultaneously planting the seeds for tomorrow's success. By embracing the bimodal approach, businesses can commit to a transformation that is both grounded and visionary, leading to a future where the organization can swiftly adapt to ever-changing market landscapes and customer expectations.

While the bimodal cloud transformation model offers a structured pathway for balancing stability and innovation, it can also present significant drawbacks. One of the primary criticisms is that it may unconsciously hamper a quick and comprehensive move to the cloud by further investing in a modified and non-cloud-ready system of records. By maintaining mode 1's focus on stability and incremental improvements to legacy systems, organizations might find themselves entrenching outdated infrastructure and processes rather than fully embracing the transformative potential of cloud-native solutions. This dual investment in both legacy and innovative systems can lead to fragmented IT strategies, increased complexity, and higher costs. Moreover, the bifurcation of efforts can create silos within the organization, where mode 1 and mode 2 teams operate with divergent goals and methodologies, potentially leading to misalignment and reduced overall efficiency. Ultimately, this approach can slow down the holistic adoption of cloud technologies, limiting the organization's ability to fully capitalize on the agility, scalability, and cost efficiency that cloud transformation promises.

5.1.2 IT Context

This section will concentrate on the IT perspective on cloud transformation. We'll begin by discussing typical IT drivers and their organizational impacts, and we'll then introduce the clean core concept, which is a crucial paradigm in the context of SAP cloud transformations.

IT Drivers

As organizations face pressures to innovate, differentiate themselves from the competition, and achieve operational efficiency, IT departments increasingly look to the cloud as a catalyst for change. The move toward cloud infrastructure is driven by a host of compelling IT drivers, each of which addresses critical aspects of technology management—from modernizing legacy systems to optimizing costs, enhancing scalability, ensuring business continuity, and embracing a DevOps culture. The cloud also offers a path to leaner operations, access to advanced technologies, and enhanced security measures. Collectively, the following drivers form the cornerstone of an IT department's cloud transformation strategy and facilitate a more resilient, agile, and forward-thinking organization:

- **Modernization**
  Modernization is the process of updating IT systems and applications to conform to newer, more current standards. In the context of cloud transformation, this means

moving away from legacy systems that are often inflexible, expensive to maintain, and unable to meet the dynamic needs of today's business environment. By modernizing their IT landscape through the cloud, organizations can benefit from improved performance, reduced operational risks, and greater innovation potential.

- **Cost optimization**
  Cost optimization is about achieving the right balance between spending and performance efficiency. Cloud services help in this regard by offering a pay-as-you-use model, eliminating the need for large up-front investments in infrastructure. IT departments can monitor and adjust their usage based on current needs, and this allows for more precise budgeting and spending and leads to a reduction in overall operational costs.
- **Scalability and elasticity**
  Scalability is the degree to which a system can handle growing amounts of work by having more resources added to it. *Elasticity* is the degree to which a system can adapt to workload changes by provisioning and deprovisioning resources automatically, so that at each point in time, the available resources match the current demand as closely as possible. Cloud environments are highly scalable and elastic, allowing IT departments to effortlessly scale their infrastructure up or down based on the business demands without incurring downtime or expensive changes to their existing setup.
- **Business continuity**
  Business continuity involves planning and preparation to ensure that an organization can continue to operate in case of serious incidents or disasters and can recover to an operational state within a reasonably short period of time. Cloud services provide reliable data backup, disaster recovery, and redundant systems that can be activated when needed to ensure minimal disruption to operations.
- **Transition to DevOps**
  The *transition to DevOps* is a cultural shift that merges development with operations, aiming to shorten the development lifecycle and provide continuous delivery with high software quality. Cloud transformation is inherently suited to DevOps due to its emphasis on automation and monitoring throughout the software development lifecycle. This synergy improves collaboration and enhances the rate at which products can be brought to market.
- **Lean operations**
  Lean operations focus on streamlining processes, eliminating waste, and ensuring efficiency. In IT, this translates to more automated, simplified, and efficient processes, with less redundancy and more focus on value-adding activities. Cloud transformation facilitates lean operations by offering tools and services that automate routine tasks, reduce manual intervention, and help IT departments focus on strategic activities that add business value.

- **Advanced technologies**
  The availability and accessibility of *advanced technologies* like AI, the Internet of Things (IoT), machine learning, and big data analytics are made more feasible through cloud platforms. IT departments can leverage these technologies to drive innovation, make data-driven decisions, enhance user experiences, and stay ahead in their market. The cloud acts as a gateway to these emerging technologies without the need for heavy investment in specialized infrastructure.
- **Enhanced security**
  Enhanced security in the cloud includes data encryption, identity and access management, physical security, and cybersecurity measures. While security is a concern for all IT departments, cloud providers are uniquely positioned to offer robust security measures that may be out of reach for individual organizations due to cost or complexity. Cloud providers invest heavily in security, benefiting from economies of scale that individual businesses can utilize.

Organizational Impact

As organizations embrace cloud transformation, the impacts on the IT department are multifaceted. This transition redefines not only the technological framework but also the operational, cultural, and strategic aspects of the IT landscape. Let's explore the five main impacts of cloud transformation on the IT department:

- **Redefinition of IT roles and skills**
  Cloud transformation necessitates a shift in the IT skill set from traditional infrastructure management to cloud-based competencies. There's an increased need for skills in cloud architecture, security, and services management, and roles become more focused on integration, orchestration, and digital innovation. IT professionals are required to upskill and embrace continuous learning to stay abreast of rapidly evolving cloud technologies. This shift often leads to a more strategic role for IT, in which it contributes directly to business value through technological innovation.
- **Enhanced collaboration among departments**
  The centralized nature of cloud services fosters greater collaboration among different departments. IT becomes a pivotal player in cross-functional teams, working closely with business units to understand their needs and deliver solutions swiftly. The barriers between IT and other departments fall, enabling a more seamless flow of information and a unified approach to achieving business goals.
- **Shift toward agile and DevOps practices**
  Cloud transformation accelerates the adoption of agile and DevOps methodologies within the IT department, focusing on rapid deployment, continuous improvement, and high operational efficiency. This shift demands changes in workflows, processes, and mindsets. The IT department transitions from being a service provider to being a partner in development, enabling faster innovation and responsiveness to market demands.

- **Realignment to operational expenses**
  The adoption of cloud services transforms the CapEx model into an OpEx model, and this realignment impacts budgeting and financial planning within the IT department. It introduces a pay-as-you-go structure, which allows for more predictable and flexible allocation of resources based on actual consumption. This in turn enables IT to optimize costs effectively and allocate resources to more strategic initiatives.
- **Redefinition of security and compliance**
  In the cloud, security becomes a shared responsibility. The IT department must adapt to new security frameworks that cover data protection, privacy laws, and regulatory compliance. This shift involves implementing and managing identity and access management, encryption, and threat detection measures that are inherent to cloud services. The IT department's role expands to continuous monitoring and adjustment of security postures in alignment with the organization's risk management strategies and compliance requirements.

Cloud Transformation Is Not Hosting

Cloud transformation represents a strategic overhaul that goes far beyond the concept of traditional hosting. While *hosting* often involves the simple relocation of infrastructure, applications, or services to be managed off-premise, *cloud transformation* is a comprehensive shift to a new way of leveraging technology. It embraces the dynamic allocation of resources; delivers services through advanced cloud computing models like SaaS, PaaS, and IaaS; and incorporates modern approaches to development, operations, and delivery. This transformation enables agility, scalability, and innovation at a speed and scale not possible with traditional hosting, which is generally more static and limited in scope. Cloud transformation is not just a change of location for IT assets but a change in how an organization thinks about and uses technology to drive business outcomes.

Clean Core

The concept of a *clean core* has attained paramount importance for SAP and its customers, particularly those transitioning from complex and heavily customized landscapes. As businesses expand and evolve, they often find themselves entangled in a web of modifications and enhancements that, while initially beneficial, become burdensome over time, hampering agility and innovation. A clean core approach is essential in addressing these challenges, as it enables organizations to return to simplified, standardized, and streamlined ERP operations. This shift is crucial for maintaining competitiveness in a fast-paced digital market, where the ability to quickly adapt to new technologies, regulatory demands, and business processes is essential. By fostering a clean core environment, SAP and its customers can minimize disruptions during upgrades, reduce maintenance overhead, and unlock the flexibility needed to drive forward-looking business strategies with a clear focus on future growth and efficiency.

The clean core concept embodies a strategic philosophy and operational framework that is underpinned by robust governance and clear guidelines and that is essential for crafting an agile and future-proof ERP system. It advocates for a robust method of enhancing functionality in a manner that is stable, safe from obsolescence during upgrades, and fully transparent, while simultaneously providing a platform for innovation and differentiation. Embracing a clean core facilitates more rapid software deployments and simplifies the integration of SAP's latest innovations as well as compliance with regulatory updates. It introduces novel solutions for meeting business requirements without accruing unnecessary technical debt, thereby enabling organizations to capture the full strategic value of their transformation efforts and keep the associated costs under control.

We can examine the concept of a clean core through multiple dimensions that we list as follows and illustrate in Figure 5.1:

❶ **Processes**
You should implement strong process governance and process management practices and document processes according to defined principles (e.g., focus on differentiating processes). Leverage SAP standard solutions, implement SAP Best Practices where applicable, and measure as-is process execution and performance in comparison to documented processes and defined key process performance indicators.

❷ **Extensibility**
You should minimize the use of extensions whenever feasible, but if necessary, you should create cloud-compliant extensions that adhere to a strong governance framework. It's important to develop custom extensions that are compatible with upgrades and leverage released application programming interfaces (APIs). This allows you to create separate extensions that can fully utilize the extensibility capabilities of both on-stack resources (e.g., inside the SAP S/4HANA platform) and SAP Business Technology Platform (SAP BTP), while also making informed decisions to avoid accruing technical debts.

❸ **Data**
You should define and govern data requirements including accuracy, completeness, consistency, timeliness, validity, and uniqueness, while also emphasizing the need to control data volume for optimal memory and disk usage. Additionally, it's important to ensure that data doesn't include outdated, unused, or redundant information, and personal master data should only be stored for justifiable purposes.

❹ **Integration**
You should use standard APIs whenever possible and leverage the capabilities of SAP Integration Suite for API integration. For loosely coupled integrations, we

recommend an event-driven design based on standard events, and you should avoid traditional APIs like remote function calls (RFCs) and IDocs along with their classical extension options. To ensure proper monitoring and error resolution capabilities, you can utilize SAP Application Interface Framework.

5 Operations
You should integrate clean core in the end-to-end operations concept, with release management serving as the basis. We recommend that you follow SAP Best Practices for housekeeping activities and adhere to the agreed distribution of roles and responsibilities.

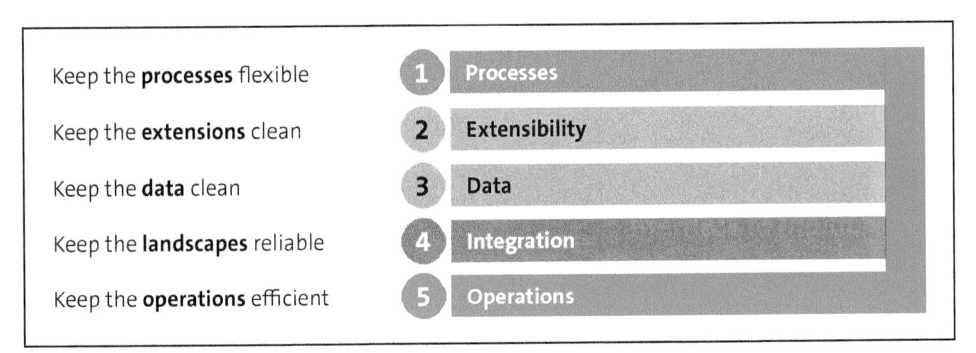

Figure 5.1 Clean Core Concept

SAP's Five Golden Rules

SAP's five golden rules provide a strategic framework that helps ensure successful cloud transformations by emphasizing the importance of standardization, agility, and adopting best practices. The following list summarizes these golden rules:

1. Foster a cloud mindset by adhering to fit-to-standard and agile deployment detailed in SAP Activate, based on SAP Best Practices, preconfigured processes, and the SAP Fiori user experience (UX).
2. Build your solution on data quality (configuration, master, and transactional) to leverage innovation by ensuring data accuracy, completeness, privacy, and consistency while managing data volumes and resource utilization.
3. Use modern integration technologies by following SAP Integration Solution Advisory Methodology and building integration scenarios on SAP BTP leveraging published APIs.
4. Use modern extension technologies by following SAP Application Extension Methodology and building extensions on-stack or side-by-side with SAP BTP while retiring unused custom code.
5. Ensure transparency on deviations from these golden rules by establishing a solution standardization board and leveraging SAP Cloud ALM.

For more information, refer to *http://s-prs.co/v586304*.

SAP customers embarking on the journey toward a clean core are often confronted with numerous challenges. This transition can be intimidating, as it requires loosening complex, deeply ingrained customizations and adopting new, standardized processes. Organizations may face resistance to change from within, as well as technical difficulties in aligning their existing systems with the streamlined architecture of a clean core. Establishing data cleanliness and integration reliability while maintaining operational effectiveness and adapting to new models are just some of the hurdles that can seem insurmountable. Moreover, the need to reskill employees to help them thrive in a clean core environment adds another layer of complexity. Despite these challenges, for organizations using SAP, steering toward a clean core is an indispensable strategy. The imperative to stay agile, competitive, and ready for the future in a digital-first landscape leaves no other viable option but to embrace the discipline and clarity that a clean core offers.

5.2 Problem Statement

This section delves into the cloud transformation journey of the fictitious company PumpTech. PumpTech's story is based on real-world experiences and challenges that various SAP customers have faced as they've navigated the process of cloud transformations. We'll begin by presenting an overview of the company and outlining its strategic objectives for the cloud transformation initiative. Then, we'll explore PumpTech's existing IT infrastructure, along with the current and anticipated challenges it faces. Subsequently, we'll discover the organizational elements that are relevant to PumpTech's cloud transformation vision.

5.2.1 Company Profile

In our hypothetical scenario, PumpTech Innovations has been a leader in the manufacturing of high-quality pumps for over three decades. With a strong focus on R&D, PumpTech has pioneered many advancements in pump technology, catering to a wide range of industries including water treatment, chemicals, pharmaceuticals, and oil and gas. The company prides itself on its engineering excellence, sustainable practices, and exceptional customer service.

Traditionally, PumpTech's business model has been centered on the sale of physical pump units, coupled with maintenance services. This model has served the company well, allowing it to establish a prominent market position and a loyal customer base. However, the evolving industrial landscape and the rise of digital technologies have prompted PumpTech to reassess its strategy.

PumpTech envisions a future where its pumps are not just products but integral components of its customers' success stories. The company plans to leverage cloud technology to transition into a subscription-based model, offering "pumping as a service."

This innovative approach will allow customers to pay based on usage, ensuring they have access to the latest pump technology without the need for up-front investment. The cloud infrastructure will also enable PumpTech to offer real-time monitoring, predictive maintenance, and analytics services, enhancing its value proposition to its customers.

The following list outlines the key strategic priorities and objectives of PumpTech's cloud transformation:

1. **Develop a scalable cloud platform**
   PumpTech aims to develop a robust, secure, and scalable cloud platform that will support the pumping-as-a-service model. This platform will facilitate equipment monitoring, data analytics, and customer interaction seamlessly.
2. **Implement IoT solutions**
   By integrating IoT devices into their pumps, PumpTech plans to collect operational data in real time, enabling predictive maintenance, performance optimization, and a personalized customer experience.
3. **Ensure seamless integration without disruption**
   Critical to the transformation is the goal to integrate the subscription-based model in a manner that doesn't hamper the traditional product-based business model. This new model should coexist with the existing one, providing an additional revenue stream while the company still expects the traditional business model to drive significant revenue. The company will carefully balance resources, marketing, and customer support to ensure that both models thrive without cannibalizing each other's success for the time being. Over the long term, the new business model is supposed to replace the traditional one.
4. **Implement strategic resource allocation**
   PumpTech will allocate resources in a way that supports growth and innovation in the subscription model while continuing to invest in the product-based model. This dual investment approach will ensure that advancements in the product-based model drive further innovation in the subscription model, creating a synergistic effect that will enhance overall business performance and customer satisfaction.

5.2.2 IT Landscape

PumpTech operates in an environment shaped by emerging players who, unburdened by legacy systems, are quick to adopt cloud technologies and agile methodologies, thus challenging established companies to accelerate their own transformation efforts. For traditional manufacturing companies like PumpTech Innovations, this represents both a challenge and an opportunity to redefine their value proposition by leveraging cloud transformation to offer more adaptable and customer-centric solutions in the market.

PumpTech Innovations operates within an ecosystem of SAP and non-SAP systems (depicted in Figure 5.2) to support its extensive manufacturing operations and business processes.

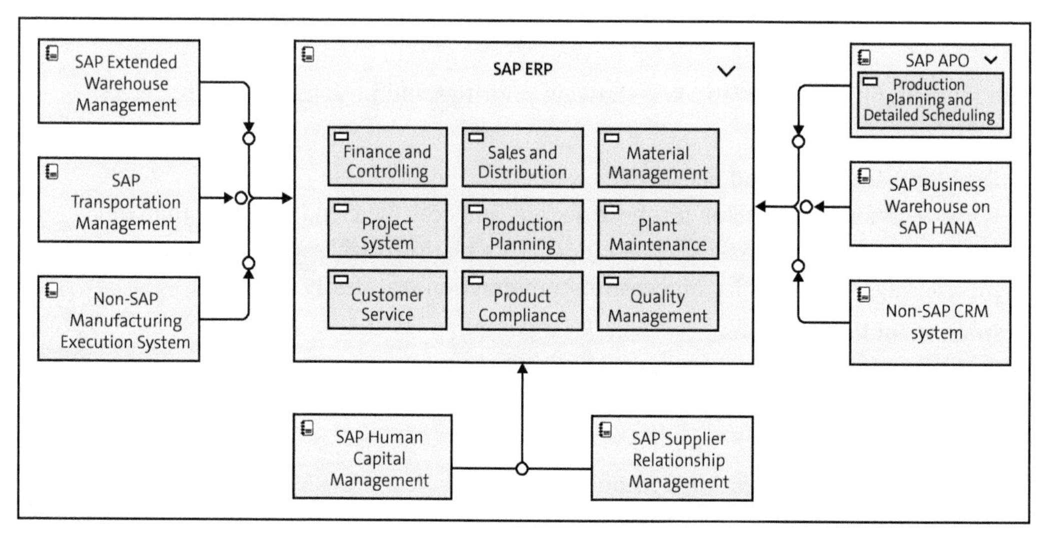

Figure 5.2 As-Is Landscape of PumpTech Innovations

At the core of this ecosystem is a single global instance of SAP ERP, which is the backbone ERP system that integrates the following key modules:

- Finance (FI) and Controlling (CO) manage the financials, providing real-time tracking of cost and revenue flows.
- Sales and Distribution (SD) streamlines order processing, billing, and customer interaction.
- Quality Management (QM) ensures that products meet stringent quality standards.
- Product Compliance oversees adherence to regulatory requirements and industry standards.
- Project System facilitates project management, from planning through to execution.
- Materials Management (MM) optimizes procurement and inventory management.
- Customer Service enhances after-sales support and service management.
- Plant Maintenance (PM) ensures that production facilities are maintained and operate efficiently.
- Production Planning (PP) is pivotal for manufacturing processes and scheduling.

To complement SAP ERP, PumpTech utilizes SAP Extended Warehouse Management (SAP EWM) to manage complex warehousing operations and SAP Transportation Management (SAP TM) to optimize logistics and distribution. The SAP Human Capital

Management (SAP HCM) suite manages employee lifecycle processes, while SAP Supplier Relationship Management (SAP SRM) streamlines procurement and collaboration with suppliers.

Advanced planning capabilities are furnished by SAP Advanced Planning and Optimization, with production planning and detailed scheduling (PP/DS) ensuring that production schedules are efficient and resources are utilized optimally.

SAP Business Warehouse (SAP BW) on SAP HANA empowers PumpTech with real-time analytics and the ability to handle large volumes of data, offering insights that drive strategic decision-making and operational performance.

However, not all of PumpTech's systems are SAP based. A non-SAP customer relationship management (CRM) system is in place, and it was likely chosen for its specific features that cater to PumpTech's unique customer engagement requirements. Additionally, a non-SAP manufacturing execution system (MES) interfaces with the production hardware and machinery on the shop floor, managing and monitoring the production process in real time.

PumpTech is currently confronting the following challenges with its legacy systems:

1. **Roadblocks to upgrades**
   PumpTech struggles with a substantial number of custom modifications within its SAP ERP environment, which historically has led to challenging and often postponed upgrades.
2. **Innovation blockers**
   The burden of supporting an outdated and heavily tailored SAP ERP environment impedes PumpTech's ability to pursue innovations it needs to support new business models.
3. **Integration constraints**
   The current landscape is impacted by integration limitations, impeding seamless communication between SAP ERP and other critical business systems.
4. **Costly customization maintenance**
   Maintaining the unique customizations in SAP ERP not only is cost prohibitive but also diverts vital technical resources away from innovation.
5. **Data accessibility issues**
   Legacy system structures create data silos, complicating the extraction and consolidation of data for analysis and decision-making.
6. **Limited scalability**
   The existing SAP ERP system shows restricted scalability, hindering PumpTech's ability to expand and adapt to new business opportunities and market demands.
7. **Compliance and security risks**
   As regulatory requirements evolve, the legacy SAP ERP system may fall short in meeting modern compliance standards and safeguarding against current cybersecurity threats.

8. **Operational inefficiencies**
   Fragmented processes across the legacy landscape result in inefficiencies, redundancy, and increased potential for human error.
9. **Downtime and reliability concerns**
   The age and customization of the current systems pose risks of increased downtime and reduced reliability, impacting business continuity.

IT as an Enabler of Innovation

At PumpTech Innovations, the current IT landscape, once the foundation of the company's operations, has become a stark example of how technology can transition from an asset to an impediment in the face of cloud innovation. With its heavily customized legacy systems, IT at PumpTech has inadvertently erected barriers that slow adaptation to agile, cloud-based solutions, which are essential for maintaining competitive advantage in today's fast-paced market. In the future, PumpTech IT needs to evolve as part of the planned cloud transformation to become a catalyst for innovation.

PumpTech stands as an example of the challenges faced by many SAP customers, whose deeply entrenched legacy systems and extensive customizations present significant hurdles to embracing the agility and innovation offered by modern cloud solutions. In the journey of cloud transformation, it's crucial to realign IT as a catalyst for innovation, ensuring that technology once again becomes a driving force behind business growth and adaptation in the digital era.

5.2.3 Organizational Aspects

PumpTech's current organizational aspects, as inferred from the provided context, likely include the following key structural and cultural elements that have both supported its traditional business operations and may now influence its transformation journey:

- **Hierarchical structure**
  The company may have a traditional hierarchical organizational structure, with clear reporting lines and established protocols that facilitate order but may slow down innovation.
- **Departmental silos**
  As in many established manufacturing companies, departmental silos may be prevalent at PumpTech, creating isolated working environments that can hinder cross-departmental collaboration and communication.
- **Risk-averse culture**
  Given its success with the existing business model, PumpTech may possess a risk-averse culture that favors proven methods over novel, untested approaches. This can be a barrier to adopting cloud technology.

- **Centralized decision-making**
  Decision-making processes are likely to be centralized, with strategic decisions being made at the upper levels of the company, potentially delaying response times to technological advancements.
- **Specialized workforce**
  The workforce is probably highly specialized in their respective fields, with deep expertise in pump technology and manufacturing but possibly less familiarity with cloud technologies and modern digital tools.
- **Customer-focused approach**
  A customer-focused approach is ingrained in the company. It drives the company's commitment to quality and service but may also necessitate careful consideration of how cloud transformation will affect customer relationships.
- **Resource allocation**
  Resources have been traditionally allocated to areas directly tied to product development and manufacturing. This may mean that investment in IT and digital transformation requires a shift in budgeting and resource prioritization.
- **Performance metrics**
  Current performance metrics likely focus on production efficiency, product quality, and customer service. This may need to be expanded to include metrics related to digital initiatives and cloud integration.

Understanding these organizational aspects is critical because they'll influence how PumpTech approaches its cloud transformation. These elements will define both the challenges and the opportunities for change within the company, shaping the strategies for managing the organizational impact of the company's new cloud-based business model.

Cloud Transformation Doesn't Only Affect IT

A cloud transformation is fundamentally an enterprise-wide endeavor that impacts every facet of a business, not just its IT framework. It requires a holistic approach that includes process reengineering, cultural change, and business model redefinition. If driven solely by the IT department, a cloud transformation is at risk of falling short because it may not align with the broader business objectives or secure buy-in from all stakeholders. Success in cloud transformation demands collaborative efforts among departments—marketing must understand and convey the benefits of the cloud, sales must adapt to the new capabilities it provides, and executive leadership must champion the transition to ensure strategic alignment and resource allocation. Without this cross-functional engagement and a unified vision, a cloud transformation initiative is likely to fade and be unable to deliver its full potential and drive the organization forward.

5.3 Applying the Enterprise Architecture Transformation Approach

The enterprise architect has been assigned the task of developing an initial architectural plan that aligns with PumpTech's overarching goals as outlined in Section 5.2: to develop a scalable cloud platform, ensure seamless integration without causing disruptions, implement IoT solutions, and allocate resources strategically. After thoroughly analyzing PumpTech's detailed strategy and realigning with various business and IT stakeholders, the next step for PumpTech's enterprise architect is to select the relevant modules that will be included in the architectural planning activities.

As presented in Chapter 2, we advocate for a modular and iterative strategy when practically applying SAP EA Framework. Reflecting PumpTech's specific goals and needs, we've chosen the modules shown in Figure 5.3 for the initial phase. This selection doesn't imply that other modules lack importance; rather, for the initial phase of architectural planning, we suggest concentrating on those modules that are most critical and significant. This use case will address solely the modules selected for the first iteration. For information on additional modules not discussed in this chapter, refer to the chapters detailing other use cases.

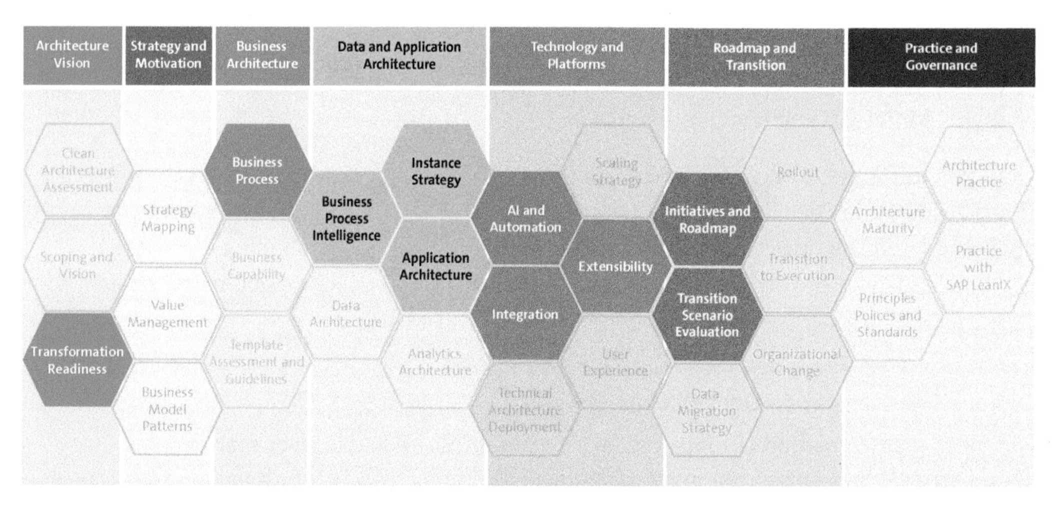

Figure 5.3 Modules in Scope for PumpTech

Let's apply these modules to our PumpTech use case:

- **Transformation readiness**
  This module assesses PumpTech's readiness for cloud transformation, considering the organization's cultural, operational, and technological preparedness. It evaluates the existing infrastructure and applications to identify areas that would benefit from cloud capabilities, focusing on scalability, security, and compliance needs. It also highlights the skills and mindset shift needed among teams to embrace cloud technologies.

- **Business process**
  This module focuses on reevaluating and optimizing PumpTech's business processes for the cloud environment. It involves identifying processes that can be streamlined or automated using cloud-based solutions, leading to improved efficiency and reduced costs. It emphasizes aligning business processes with cloud capabilities to leverage scalability and innovation opportunities.

- **Business process intelligence**
  This module uses cloud-based analytics and intelligence tools to gain insights into PumpTech's business processes. It identifies bottlenecks, inefficiencies, and opportunities for optimization. The cloud enables real-time data analysis and predictive insights, facilitating smarter business decisions and enhancing process outcomes.
- **Application architecture**
  This module defines the future-state application architecture in the cloud, considering PumpTech's new business model and the need to support legacy systems. It involves selecting cloud-native services and solutions that align with business objectives, ensuring interoperability, scalability, and resilience. It also addresses the migration paths for existing applications to the cloud environment.
- **Instance strategy**
  This module develops a strategy for cloud instances, focusing on how PumpTech will manage its cloud environment. It includes choosing among single, multi-instance, and hybrid cloud approaches, based on the company's operational requirements and business goals. This strategy ensures that PumpTech optimizes its cloud resource usage and costs while meeting performance and regulatory requirements.
- **AI and automation**
  This module explores how PumpTech can leverage AI and automation within the cloud to enhance business operations. It focuses on identifying tasks and processes that can be automated, using AI to drive efficiency, innovation, and new capabilities. It includes implementing AI-driven analytics, process automation, and intelligent applications to transform customer experiences and operational efficiency.
- **Integration**
  This module addresses the seamless integration of cloud services and solutions with PumpTech's existing systems and data sources. It ensures that data flows smoothly across cloud and on-premise environments, supporting a unified operational view. It also focuses on leveraging cloud integration tools and platforms to enable interoperability and real-time data exchange.
- **Extensibility**
  This module focuses on how PumpTech can extend its core applications and systems in the cloud to meet specific business needs. It involves using cloud platforms to build, deploy, and manage custom applications and extensions that enhance functionality and support new business models. It also ensures that the cloud environment supports rapid innovation and flexibility.

- **Initiatives and roadmap**
  This module creates a strategic roadmap for PumpTech's cloud transformation journey, outlining key initiatives, milestones, and timelines. It aligns cloud transformation efforts with business objectives, ensuring a phased and manageable approach to migrating to the cloud. It also helps prioritize actions and resources, ensuring alignment with the overall business strategy.
- **Transition scenario evaluation**
  This module evaluates different scenarios for transitioning to the cloud, considering the impact on PumpTech's operations, costs, and risk profile. This module involves scenario planning to identify the most effective path for cloud migration, including assessing the legacy landscape's transition alongside deploying new cloud-based solutions.

Role of the Enterprise Architect in Cloud Transformation

An enterprise architect plays a crucial role in cloud transformation, acting as a strategic planner and a bridge between the organization's business vision and its technology capabilities. They are responsible for designing an overall architecture that aligns with business goals, assessing the impact of cloud technology on business processes, and ensuring that the transition to cloud services is smooth, secure, and cost-effective. They navigate complex technical landscapes to recommend the right mix of cloud solutions and legacy integrations, paving the way for a future-proof, scalable, and agile enterprise environment.

The following list highlights the three essential skills an enterprise architect must possess to implement successful cloud transformations:

- **Strategic thinking**
  Ability to envision the big picture and strategically plan the integration of cloud solutions into the business model, considering current and future industry trends
- **Technical proficiency**
  Deep understanding of cloud computing technologies, architectures, and services, alongside expertise in enterprise IT infrastructure, to guide sound technical decisions
- **Change management**
  Strong skills in leading change, managing stakeholder expectations, and facilitating the organization's transition to new processes and technologies

The next actions for the enterprise architect would be to verify the fulfillment of all requirements for each module and to ascertain the availability of the necessary documentation. Following this, the enterprise architect should gather and scrutinize the relevant documents. Key documents that would be crucial for PumpTech's planning cycle are as follows:

- A comprehensive inventory of existing IT infrastructure, applications, and services
- Documents providing insight into organizational culture and readiness for change
- Documentation of compliance and security standards currently in place
- Detailed mapping of current business processes
- A list of business stakeholders and their objectives
- Business continuity and disaster recovery requirements
- Performance requirements and SLAs
- A current-state analysis of IT systems and data flows
- Requirements for custom applications and system extensions
- Policies for custom development in a cloud environment
- A list of strategic business objectives to be achieved through cloud adoption
- Prioritized initiatives and corresponding outcomes
- A timeline of milestones and deliverables
- Criteria for evaluating the impact of transition scenarios on operations and finances
- Baseline maturity assessment of current architectural practices
- A documented set of enterprise architecture principles that align with the business vision and goals
- Standards for architecture practices, including design, documentation, and review processes

As a further step, the enterprise architect should conduct an initial stakeholder analysis to identify main parties with a vested interest in PumpTech's cloud transformation initiative. This analysis should map out the various stakeholders based on their influence on and interest in the transformation program. The enterprise architect should evaluate each stakeholder's current position, potential impact on the transformation, and strategy for engagement. An example of a stakeholder map might categorize stakeholders into four quadrants as depicted in Figure 5.4.

The chief information officer (CIO) of PumpTech has a long-standing record of dedicated service to the company. Recently, however, the company's chief executive officer (CEO) established a new position on the board—the chief digital officer. To fill this role, the CEO selected an individual who has a background in startups, with the specific intent that the chief digital officer will copilot the cloud transformation initiative alongside the CIO. The enterprise architect who is reporting to the CIO must therefore be aware of potential conflicts as the CIO is rather skeptical about the planned cloud transformation and more interested in the maintenance and incremental improvement of the technologies at PumpTech. On the other hand, the chief digital officer is pushing for rapid innovation, disruptive technologies, and the prioritization of agility and speed-to-market over the more methodical processes favored by the CIO.

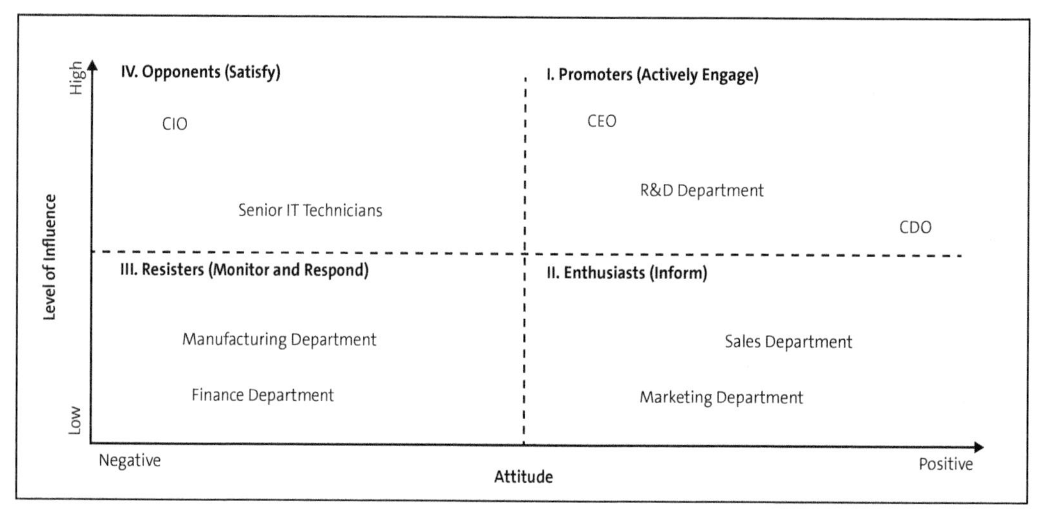

Figure 5.4 Stakeholder Map at PumpTech

The enterprise architect must determine the prevailing attitude of each of the following stakeholders toward the cloud transformation and must also consider strategies for engagement with the stakeholder throughout the architectural work:

- **Chief digital officer**
  As a driving force behind the digital shift, the chief digital officer will likely champion cloud adoption. The enterprise architect should collaborate closely with the chief digital officer to ensure the architecture aligns with digital goals, facilitating a smooth transformation.
- **Chief executive officer**
  The CEO's vision for the company's future includes this digital transformation. The enterprise architect should ensure that the architecture supports the CEO's strategic objectives and communicate how the cloud services will achieve long-term business goals.
- **Research & development (R&D)**
  The R&D department is likely excited about how cloud computing can further innovation. The enterprise architect should involve R&D in designing the architecture so that it enables new product developments and accelerates the research processes.
- **Sales department**
  Sales will be eager to sell the new service model. The enterprise architect should inform sales staff about the benefits and capabilities of the cloud services, helping them to understand and thus sell the concept more effectively.
- **Marketing department**
  Marketing will play a critical role in communicating the change to customers. The enterprise architect should provide marketing staff with clear information on the technical advantages and customer benefits of the cloud platform.

- **Chief information officer**
  Given the CIO's skepticism, the enterprise architect should address their concerns directly, demonstrating how the cloud can enhance, not hinder, existing operations, and offering assurances about security, data integrity, and transition planning.
- **Senior IT staff**
  Senior IT staff may view cloud transformation as a threat to their current way of working. The enterprise architect should engage them by highlighting how their roles can evolve and grow with the cloud transformation, ensuring that they feel included and vital to the process.
- **Manufacturing department**
  Manufacturing staff will likely be worried about disruptions, so they'll need reassurance. The enterprise architect should explain how cloud technology can streamline operations and address specific concerns about process continuity.
- **Finance department**
  Financial stakeholders may be resistant due to the cost implications. The enterprise architect should present a clear cost-benefit analysis, emphasizing the financial efficiency and scalability of cloud solutions.

Following the comprehensive scoping efforts, the enterprise architect is now able to craft a detailed strategic plan for the forthcoming weeks and months. This plan will articulate the architecture project's trajectory, defining a clear project timeline and listing the necessary inputs to and outputs from each module within the scope. The plan will outline the schedule for workshops, specifying the essential participants to involve and providing a precise inventory of the deliverables expected from these sessions. Additionally, the plan will include periodic feedback loops with relevant stakeholders, ensuring the architectural work aligns with PumpTech's broader strategic objectives. This iterative consultation process is designed to promptly identify and tackle any impediments, ensuring they are resolved at the appropriate organizational level.

5.4 Key Deliverables and Outcomes

Building on the groundwork laid in the initial scoping stage, we'll now explore in greater depth the key outcomes associated with each chosen module. As we previously touched upon, one could make a case for the relevance of modules such as strategy mapping and business model patterns to PumpTech's shift to cloud services, given that this shift represents a change in the business model itself. Nonetheless, we've dedicated Chapter 6 to business model transformation to underscore the business-centric elements of such a change. In this particular scenario, our emphasis will be on examining the business processes and their corresponding architectural components. The business model perspective won't be our focus here. For insights into that aspect, we encourage you to review the case study provided in Chapter 6.

5.4.1 Transformation Readiness

The *transformation readiness assessment* module is a preemptive analytical tool that merges customer experience data with operational data to furnish a holistic view of a company's present condition. This rigorous examination is akin to taking an X-ray of the company, and it covers various transformation drivers in both the business and the IT realms. The main deliverable is a detailed report containing findings, recommendations, and proposed actions that lay the groundwork for a successful transformation. This report is instrumental for organizations before starting a cloud transformation because it identifies *white spots*—areas that haven't been fully considered or are underprepared for change. The module is data driven, combining experience data (X-data) and operational data (O-data) to shape a tailored report that provides clear, actionable items. It ensures value retention in business transformation by preempting potential roadblocks.

O-data includes metrics and measurements related to business performance that is extracted from operational systems. Examples of O-data would include the following:

- **SAP Readiness Check data**
  Results from automated checks that assess the readiness of an SAP ERP system for conversion to SAP S/4HANA
- **SAP EarlyWatch Alert reports**
  System health checks and performance analysis reports on the SAP systems in place
- **ABAP Test Cockpit (ATC) results**
  Analysis of custom code quality and adherence to SAP Best Practices
- **SAP Signavio Process Insights reports**
  Data on business process performance, efficiency, and compliance with standard practices

X-data captures the human factors of business operations—the beliefs, emotions, and sentiments of stakeholders involved in or affected by the business processes. This type of data is typically qualitative and is collected through methods that seek to understand people's experiences, opinions, and feedback.

In the transformation readiness module, X-data is primarily collected through preconfigured surveys that are distributed to a wide range of stakeholders, including both IT and business personnel within the organization. The goal of these surveys is to gather subjective insights into the experiences, expectations, and sentiments of the individuals who will be affected by the transformation to a cloud-based infrastructure. The questions asked in these surveys would typically be designed to shed light on important cloud transformation drivers, such as the following:

- **Cloud and digitalization experience**
  This might include gathering feedback on users' experiences with current cloud and digital tools, their comfort level with digital changes, and their attitudes towards potential shifts to cloud-based systems.
- **Strategy and challenges**
  Here, X-data could be collected to help leadership understand stakeholders' perceptions of the company's strategic direction, their understanding of the challenges ahead, and their alignment with the proposed transformation strategy.
- **Digital readiness**
  This could involve assessing the workforce's preparedness for working in a more digitized environment, their digital competencies, and any gaps that may exist in digital skills.
- **Architecture, application, and IT security**
  Gathering experience data would be relevant to understanding user experiences with current IT architecture, their trust in application performance and usability, and their confidence in the organization's IT security posture.
- **Process data and analytics**
  X-data collection might focus on how users interact with and perceive current business processes, their experiences with data analytics, and their feedback on process efficiency and effectiveness.
- **Project method experience**
  This would be about collecting insights into past project management experiences, preferences for certain methodologies (e.g., agile versus waterfall), and perceptions of past project successes or failures.

Figure 5.5 presents part of the conducted transformation readiness assessment for PumpTech.

The transformation readiness assessment for PumpTech reveals critical insights into the company's innovation and digital transformation journey. Cloud technology adoption, which is pivotal for achieving PumpTech's aspiration to become a leader in digital innovation, is lagging in key business areas due to misalignment of strategic goals with actual technology usage. Employee feedback highlights frequent disruptions and a perception that an inflexible, heavily customized ERP system is impeding innovation. Additionally, the assessment uncovers a disconnect in the integration of cloud services with existing on-premise ERP systems, leading to inefficiencies and the formation of data silos. The company's management also recognizes the need for better cloud service integration to enhance processes. Recommendations include clarifying the link between strategic objectives and cloud initiatives, establishing a clean core to mitigate customization excess, and focusing on areas that will drive quick wins to build momentum for the cloud integration process. This holistic view, combining O-data and

X-data, emphasizes the need for alignment of strategy, process improvement, and technological modernization to support PumpTech's transformation objectives.

Findings: O-Data Finding	Findings: X-Data	Key Insights	Recommendation
Cloud technology adoption in key business areas is not aligned with PumpTech's strategic goal of becoming a market leader in digital innovation.	Surveys indicate mixed understanding among staff regarding the role of cloud technology in achieving strategic business objectives.	• O-data shows a misalignment between technology adoption rates and strategic objectives, with slow progress in areas critical to innovation leadership. • X-data suggests that the workforce is not fully aware of or aligned with the strategic vision, hindering effective execution of cloud initiatives.	Clarify the connection between cloud initiatives and strategic goals through targeted internal marketing campaigns. Facilitate workshops to align departmental objectives with the overall cloud strategy and create visible leadership sponsorship for cloud projects.
Analysis of PumpTech's systems reveals extensive customizations with a high number of modifications deviating from standard ERP configurations.	Employee feedback indicates frequent issues with process interruptions and a sentiment that system rigidity hinders innovation.	• O-data points to a "dirty core;" an ERP system bogged down by excessive customizations, which complicates updates and maintenance. • X-data suggests that employees are frustrated by the broken processes resulting from these customizations, feeling that this environment stifles innovation and agility.	Initiate a clean core initiative aimed at reducing system customizations, bringing the ERP environment closer to standard configurations. This includes establishing a governance process for any future customizations to ensure they are truly necessary.
The integration of cloud services with existing on premise ERP systems is incomplete.	There is a significant desire among middle management for better integration of cloud services.	• O-data demonstrates that the current hybrid cloud and on premise architecture is creating data silos and inefficiencies. • X-data suggests that management sees the potential of cloud integration to streamline processes.	Prioritize the completion of cloud services integration, focusing on areas that will provide quick wins to build momentum and management buy in.

Figure 5.5 Transformation Readiness Results for PumpTech

5.4.2 Business Process

In the analysis of PumpTech's business processes as part of the enterprise architecture transformation, we must take a dual-focused approach. On the one hand, we must conduct a thorough analysis of the existing business processes that underpin the traditional business model—processes that form the foundation of PumpTech's operations and are deeply embedded within the current as-is landscape. This will involve a review of their current effectiveness, pain points, and closeness to SAP standards and best practices, as well as the potential impact of transitioning to new models.

On the other hand, we must give equal attention to the emerging business process requirements that originate from the innovative pumping-as-a-service business model. These new processes are essential for future-proofing the company and will require a redesign of the IT landscape. The challenge lies in achieving a balance between these two paradigms, ensuring a seamless transition that sustains operational continuity while propelling the company toward its envisioned future state.

Business Processes for the Traditional Business Model

We suggest the following approach to examine the as-is business processes at PumpTech:

1. **Catalog existing processes**
   The first step is to catalog all existing business processes. For PumpTech, this would likely involve compiling a list of processes from the as-is landscape, such as production order management, quality inspection, and inventory management, to name a few.
2. **Determine relevance and impact**
   The next step is to assess each process for its relevance to the cloud transformation initiative and its potential impact on the business. PumpTech could do this by looking at factors such as the volume of transactions the process handles, its criticality to ongoing operations, and the degree to which it's expected to change in the new business model.
3. **Select processes for deep dive**
   Then, PumpTech would select key processes from the catalog for a deep dive. Candidates for selection could include processes that have known pain points, are less aligned with best practices, and are expected to change significantly under the new model.

In this section, while it's not feasible for us to perform a complete analysis of all business processes within PumpTech's extensive enterprise architecture, we've selected the **Operate manufacturing (discrete)** business process segment of the **make-to-inspect** end-to-end process for an in-depth review, as shown in Figure 5.6. We've singled out this process due to its centrality to PumpTech's core operations, its pivotal role in the company's cloud transition, and its substantial potential for innovation and improvement. It's heavily integrated with various systems and directly influences product quality, compliance, and customer satisfaction.

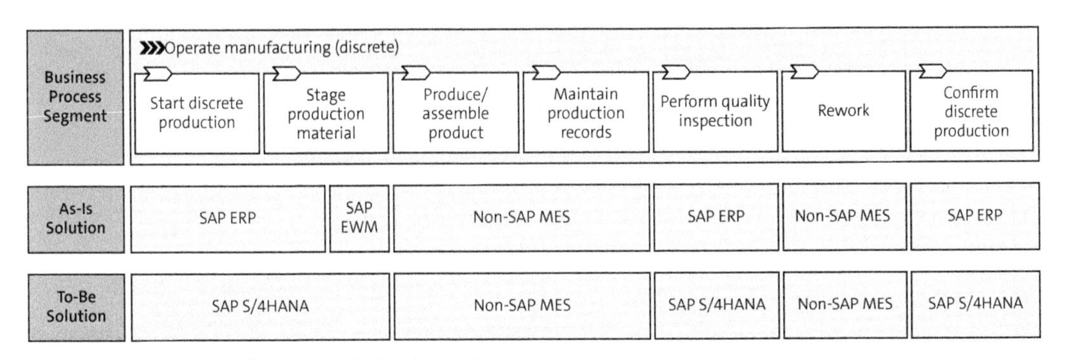

Figure 5.6 Mapping of Systems to Business Process

For the processes we've selected for the deep dive, we carry out the following activities:

1. **Document current state and target state**
   Once we've defined the scope, we document each selected business process in its current state. This includes mapping out the process flow and detailing the roles and systems involved (as-is and to-be).

As an illustration, Figure 5.6 presents the mapping of the business process segment to the current as-is solution, encompassing the activities involved and the prospective to-be solutions targeted for implementation.

2. **Measure effectiveness**
   We measure the effectiveness of each process against key performance indicators. For example, we might measure production order management by its accuracy, timeliness, and flexibility.
3. **Identify pain points**
   Through a combination of stakeholder interviews, workshops, and process performance data, we identify pain points within each process. These could be inefficiencies, sources of errors, or areas where the process doesn't meet business needs.
4. **Assess conformance to best practices**
   We assess each process against SAP standards and industry best practices. This may involve benchmarking against best-in-class organizations or leveraging the enterprise management layer for SAP S/4HANA for best practices.
5. **Analyze improvement opportunities**
   We identify improvement opportunities for processes that are not effective or are misaligned with best practices. This may involve process redesign, system enhancement, or changes in policy or procedure.

Figure 5.7 depicts the analysis that focuses on pain points and adherence to SAP standards within the **Operate manufacturing (discrete)** business process segment. The primary areas of concern identified are quality inspection and rework.

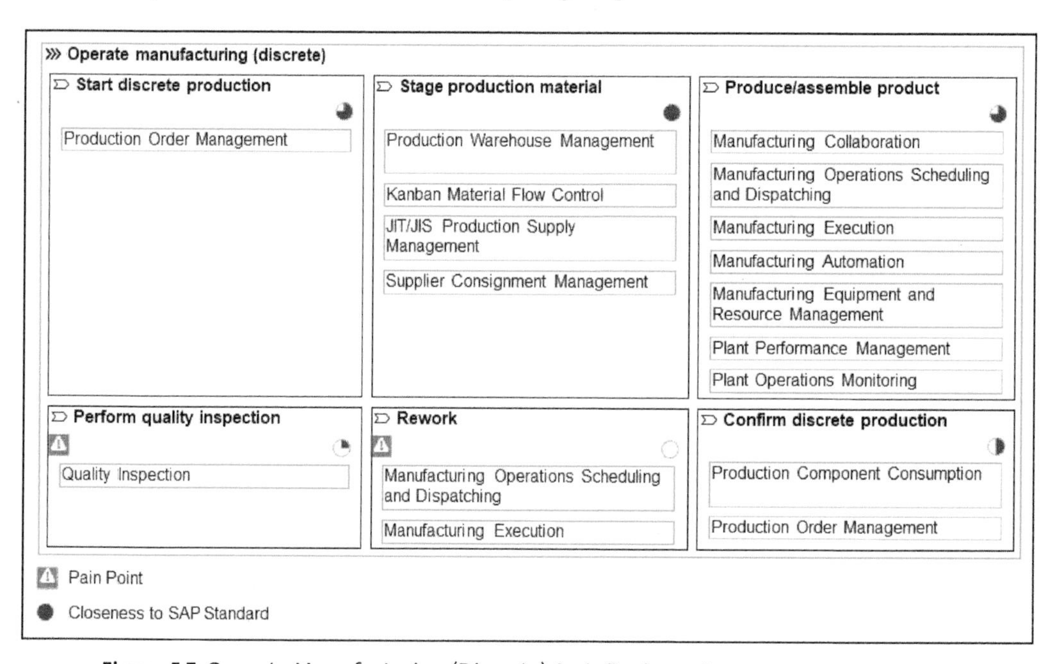

Figure 5.7 Operate Manufacturing (Discrete) As-Is Business Process

Currently, quality inspection is conducted through the SAP ERP system, which requires complex interfaces with a non-SAP MES to relay critical data during the **Produce/assemble product** activity. Additionally, the quality inspection process has been significantly customized with additional business logic to meet PumpTech's high-quality standards, reflecting enhancements that were not part of the original SAP system implementation. Rework activities, also managed by the non-SAP MES, require complex interfacing with SAP ERP. Due to the complexity of these integrations, rework activities are sometimes inaccurately reflected in SAP ERP, leading to the need for manual intervention to update production orders accordingly. This represents a significant challenge, as it undermines the efficiency and data integrity of the manufacturing operations.

When moving to SAP S/4HANA, PumpTech, facing pain points in quality inspection and rework processes, should consider the following:

- **Process simplification and standardization**
  PumpTech should evaluate its current customizations and interfaces for quality inspection and rework processes. SAP S/4HANA offers the opportunity to standardize and simplify processes, minimizing custom code and complex interfaces.
- **Integration capabilities**
  SAP BTP has enhanced integration capabilities that could potentially reduce the complexity of interfacing with non-SAP systems. PumpTech should consider how to leverage these capabilities to streamline data exchange between SAP S/4HANA and MES systems.
- **Enhanced functionality**
  SAP S/4HANA provides improved functionality that could negate the need for some customizations. PumpTech should explore these native features to address business needs that were previously met through enhancements.
- **Quality management**
  SAP S/4HANA's advanced quality management features could offer a more robust solution for PumpTech's quality inspection processes. The company should look at how the new system's features can support its stringent quality requirements.
- **Innovation opportunities**
  Look for opportunities to innovate, using SAP S/4HANA as a platform for digital transformation. These can include IoT, advanced analytics, and machine learning to further enhance manufacturing operations.

Business Processes for the New Business Model

In this section, we focus on PumpTech's strategic shift toward its innovative pumping-as-a-service model, underscored by the implementation of SAP Reference Business Architecture. We pinpoint five key business processes that are integral to the success of this model, selecting each for its vital role in operationalizing and enhancing the new business model offering:

- **Subscription management and billing**
  This process is essential for managing recurring revenue streams and providing flexible billing options that are foundational to the pumping-as-a-service model.
- **Service configuration**
  This process allows for the customization of pumping services to meet the diverse needs of PumpTech's clientele, ensuring that customer preferences drive the service delivery.
- **Asset management**
  This process is critical for tracking and maintaining the physical pumps in use, which is central to ensuring the reliability and efficiency of the pumping-as-a-service offering.
- **Predictive maintenance and service**
  By predicting when pumps are likely to require maintenance, this process minimizes downtime and enhances the overall service quality.
- **Real-time analytics and reporting**
  This process provides deep insights into operational efficiency and customer usage patterns, enabling data-driven decision-making to continually refine the pumping-as-a-service model.

In the story of this section, we'll zoom in on asset management and particularly on the **Plan asset maintenance** and **Manage asset data and collaboration** business process segments (see Figure 5.8). These are examples of how processes converge to provide a competitive edge and enable PumpTech to deliver continuous value to its customers.

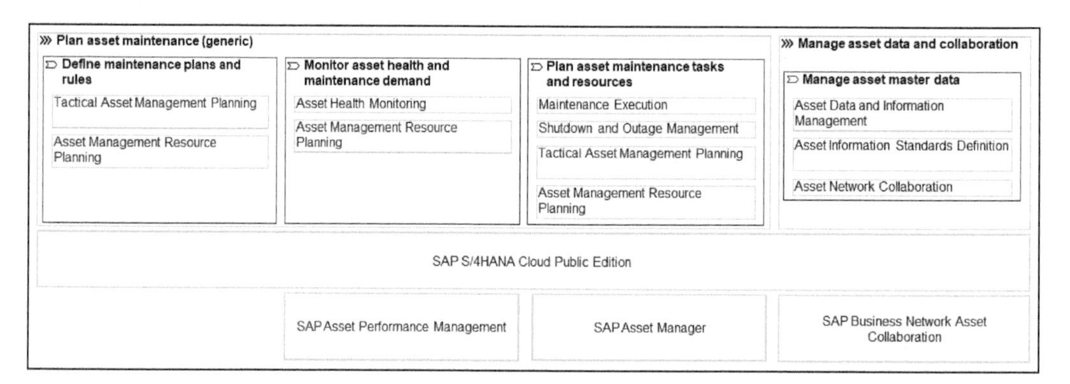

Figure 5.8 Example for Asset Management

Analyzing business processes while considering a new business model like PumpTech's transition to pumping as a service involves a structured approach. Here are the key steps for such an analysis:

1. **Align with strategy**
   Begin by aligning the business process analysis with the strategic priorities, goals, and objectives of the pumping-as-a-service model. This ensures that the processes support the overarching goals of the transformation.
2. **Map processes**
   Document and map these to-be processes in detail. Understand how they are executed within the SAP environment, who is responsible for each step, and what data flows are involved.
3. **Gather requirements**
   Collect new requirements driven by the pumping-as-a-service model. Engage with stakeholders in different functions to understand their needs and expectations from the new business model.
4. **Perform high-level fit-to-standard analysis**
   Identify discrepancies between the SAP standard and requirements. Develop a strategy for how to close potential gaps (see Section 5.4.8 on extensibility).

Figure 5.8 provides an illustrative example of how processes are aligned with SAP Reference Business Architecture, as well as their correlation with potential future solutions from SAP. As PumpTech is intent on developing its new business model within a cloud-based framework, the focus is placed squarely on public cloud solutions right from the outset, with a strategic commitment to harnessing the SAP standard to its full potential.

The use of SAP's cloud-based solutions such as SAP Asset Performance Management, SAP Asset Manager, and SAP Business Network Asset Collaboration is instrumental for the pumping-as-a-service model because it facilitates proactive asset health monitoring, mobile management of maintenance activities, and collaborative network operations. These integrated processes ensure the reliability and availability of pump services, enhance the efficiency of field service operations, and optimize maintenance execution—which are all critical for delivering uninterrupted, high-quality service to customers. The cloud infrastructure also provides the necessary scalability and real-time data analytics to support the dynamic needs of a service-centric business approach.

5.4.3 Business Process Intelligence

In the preceding section, we addressed the obstacles present in PumpTech's existing business processes. Our examination was informed by consultations with business stakeholders and the enterprise architect, who utilized value flow diagrams to document critical pain points and potential areas for improvement. However, we could also automate this analysis with the help of business process intelligence, focusing on process insights and process intelligence. These activities enable PumpTech to gain a clear, data-driven understanding of how their current processes are performing, which is essential for identifying inefficiencies, bottlenecks, and opportunities for streamlining operations.

Here are two factors that help PumpTech utilize business process transformation effectively:

- *Process insights* provide the company with the ability to observe the real-time execution of its processes, assess those processes' effectiveness, and pinpoint areas where it can make improvements.
- *Process intelligence* takes this further by leveraging AI to analyze vast amounts of process data, offering predictive insights and actionable recommendations for optimization.

This thorough understanding of existing processes ensures that PumpTech can make informed decisions about where to innovate, which processes to standardize or automate, and how best to align its process transformation with its strategic business objectives. Ultimately, these analyses are foundational for driving operational excellence and achieving a competitive edge in the market.

SAP supports business process transformation (see Figure 5.9) comprehensively with SAP Signavio Process Transformation Suite, offering a multifaceted platform for understanding, improving, and transforming all business processes efficiently and at scale. This suite provides AI-powered process analysis and mining, coupled with process modeling and collaboration tools that foster cross-departmental alignment and active collaboration. With the use of value accelerators, SAP Signavio facilitates rapid transformation by providing best practices, benchmarks, and prebuilt content that help the company avoid reworking and accelerate time to value. Additionally, it incorporates process governance and automated execution to ensure compliance and regular reviews, supporting a continuous improvement loop that enables agile and resilient businesses.

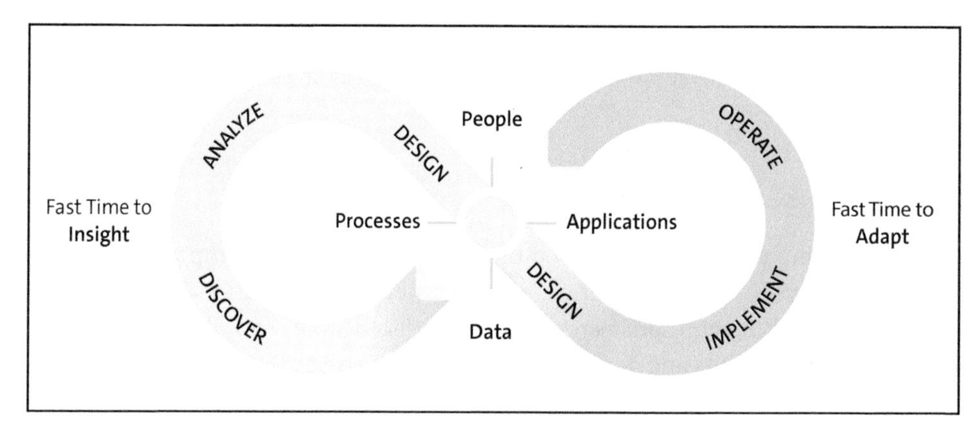

Figure 5.9 Business Process Transformation

PumpTech utilized SAP Signavio Process Insights to identify key areas within its source-to-pay process (see Figure 5.10) that required optimization during the transition to SAP S/4HANA.

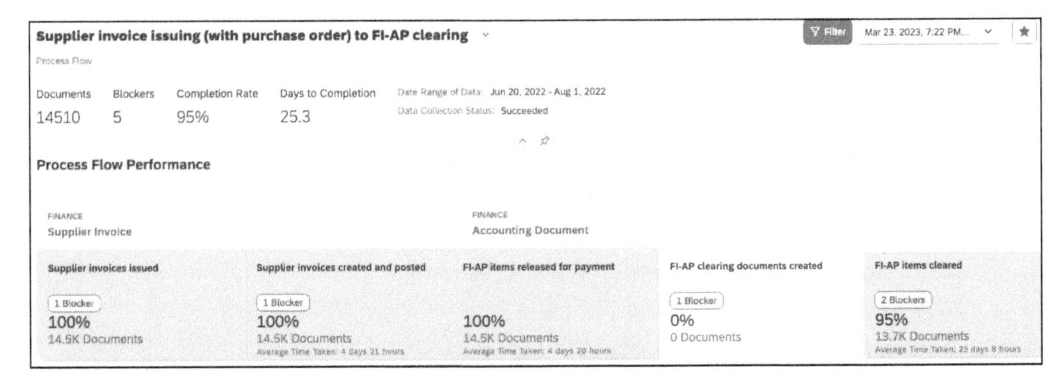

Figure 5.10 Example Analysis from SAP Signavio Process Insights for Source-to-Pay Process

A notable instance occurred during the company's supplier invoice issuing process. The insights provided by SAP Signavio Process Insights revealed several blockers in the process, leading to a long completion duration of over twenty-five days on average. The main blockers identified by SAP Signavio are the following:

- Supplier invoices created by a dialog user
- Supplier invoices created after the posting date
- Accounts payable items cleared after the net due date
- Open accounts payable items with a net due date in the past

For example, by analyzing the **Open FI-AP items with net due date in the past** blocker in more detail (see Figure 5.11), PumpTech can drill down further into the analysis of the blocker to either see what document types mostly contribute to the low performance or get a detailed list of the affected invoice documents.

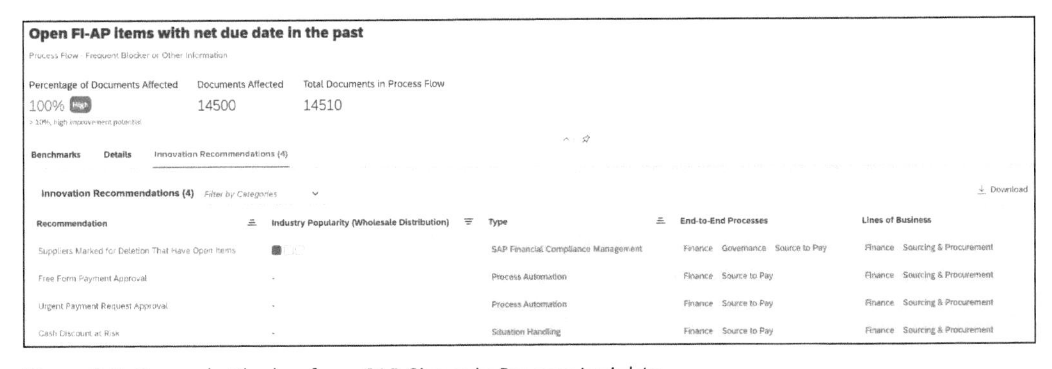

Figure 5.11 Example Blocker from SAP Signavio Process Insights

In addition, PumpTech can evaluate innovation recommendations given by SAP Signavio to improve process performance. In the given example, PumpTech can leverage process automation for a free-form payment approval or urgent payment request approval. The company could also use SAP Situation Handling and SAP Financial Compliance Management to innovate in that specific process area.

Armed with these insights, PumpTech was able to streamline its invoice and clearing processes and reduce manual interventions by implementing process automation. In addition, SAP Signavio pointed PumpTech to the corresponding SAP Best Practices processes that PumpTech then leveraged as a foundation for its back-to-standard endeavor. This preparatory step ensured a more seamless migration to SAP S/4HANA, with improved inventory accuracy and operational readiness for the new system's advanced capabilities.

5.4.4 Application Architecture

The target architecture that PumpTech Innovations plans to implement for its transition to a pumping-as-a-service model is an embodiment of a bimodal IT strategy. Centralized in the SAP S/4HANA platform in both private and public cloud variants, the architecture promises to support the existing business model of producing, selling, and servicing pumps (with SAP S/4HANA Cloud Private Edition) and also enable the new business model (with SAP S/4HANA Cloud Public Edition) without disruption. The integration of IoT technologies facilitates the collection of operational data for predictive maintenance in SAP Asset Performance Management—which is a cornerstone of the new service model. Because the service aspects play an even more important role in the new business model, the company implements solutions like SAP Field Service Management and SAP Asset Manager. SAP Business Network Asset Collaboration also enables data exchange with various parties in the supply chain.

SAP Analytics Cloud augments this by turning data into actionable insights, thereby enhancing customer value. Also, SAP Integration Suite stitches together various important systems to ensure a seamless user experience and continued operational excellence. Figure 5.12 depicts the final target architecture.

When developing a target architecture, it's essential to take into account various elements such as instance strategy (refer to Section 5.4.5), integration capabilities, data flows, security considerations, and user experience, among others. In the following discussion, we'll focus specifically on three key areas: deployment options, centralization of services, and innovation.

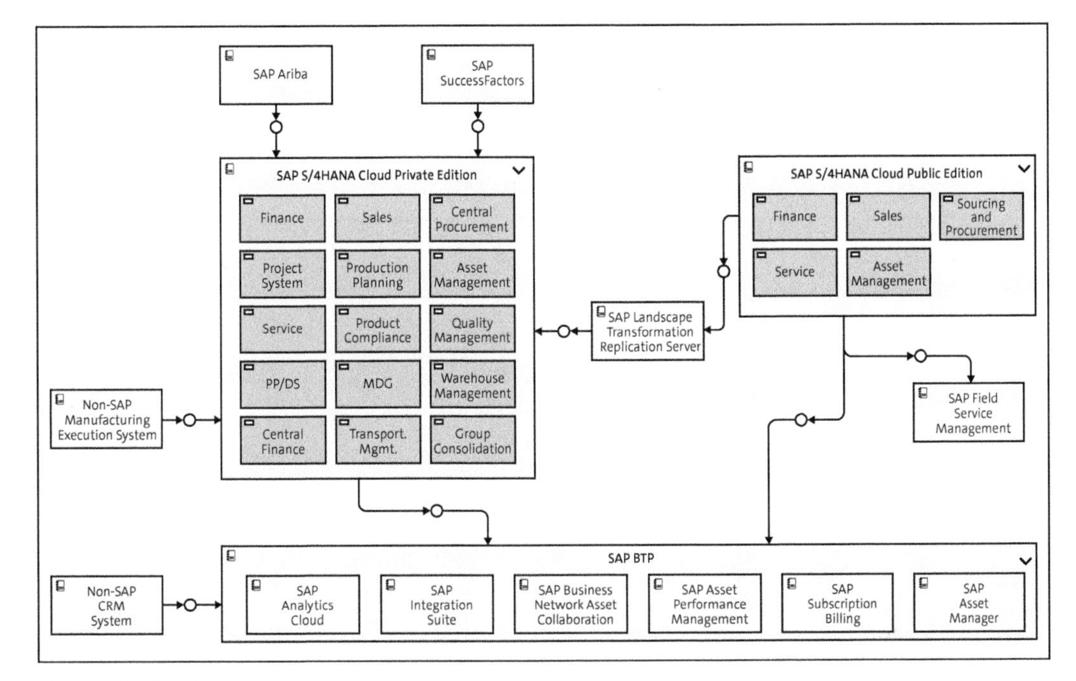

Figure 5.12 To-Be Landscape

Deployment Considerations

In the context of SAP, *deployment options* relate to the methods and platforms chosen for implementing SAP's software solutions. These range from traditional on-premise deployment, where the SAP systems are installed on the company's own servers, to cloud-based deployments on external servers managed by SAP or third-party providers and hybrid models that combine both. In addition, deployment can involve decisions to embed certain software components in the same installation of the software or operating them on a separate instance as a side-by-side solution. The deployment decision impacts various operational aspects like system performance, data accessibility, cost efficiency, maintenance responsibilities, and the ability to adapt to changing business needs.

In the next sections, we'll explain deployment considerations for the ERP system (SAP S/4HANA) and adjacent systems involved in our PumpTech use case.

SAP S/4HANA

SAP S/4HANA offers flexible deployment options to accommodate diverse business strategies and IT landscapes, as outlined in Table 5.1. The on-premise option allows businesses to host the software on their own servers, providing maximum control over the infrastructure, data, and security. This is ideal for organizations with stringent regulatory requirements. The private cloud option offers a dedicated cloud environment managed by SAP for technical operations, combining the control of the on-premise option with the scalability and managed services of cloud computing. Meanwhile, the public cloud option delivers SAP S/4HANA as a service that is fully managed by SAP, which facilitates rapid deployment, reduces operational costs, and offers a pay-as-you-go model. This makes the public cloud option suitable for businesses looking for agility and minimal IT overhead. Each of these options provides a different balance of control, cost, and management responsibility to meet specific business needs.

	SAP S/4HANA Cloud Public Edition	SAP S/4HANA Cloud Private Edition	On-Premise SAP S/4HANA
Implementation Types	New implementation.	System conversion, selective data transition, and new implementation.	
Innovations	The latest innovations are available (e.g., SAP Business AI, sustainability).		Not all innovations are available.
Extensibility	SAP S/4HANA extensibility framework; no classic extensibility is allowed.	SAP S/4HANA extensibility framework; classic extensibility is allowed but not recommended.	
Modifications	Not allowed.	Allowed but not recommended.	
Release Upgrades	Included and mandatory.	Customer owned, installation on request included.	Not included.
Upgrade Entitlement	Two times per year.	Every two years.	Every two years.
Third-Party Add-Ons	Certified public cloud add-ons are allowed.	A defined list of SAP S/4HANA add-ons is allowed.	Allowed.
License Model	Software subscription.		Perpetual software.
Technical Operations	SAP.	SAP.	Partner/customer/SAP.
Infrastructure	Hyperscaler and SAP.	Hyperscaler, SAP, and customer data center on request.	Customer data center, hyperscaler, SAP, premium supplier, and partner.

Table 5.1 Deployment Options for SAP S/4HANA

One crucial aspect to consider is the innovation opportunities available based on the deployment choice. Since SAP is pursuing a clear cloud strategy, not all innovations developed by SAP are or will be provided to SAP S/4HANA on-premise customers. This currently includes innovations in business AI, sustainability, and Central Finance, among others. Therefore, customers facing the deployment decision must carefully consider that a noncloud approach won't unlock the full potential of the ERP solution.

PumpTech has committed to a cloud-based approach, thus opting for a complete cloud deployment for its ERP needs. Its legacy business systems will shift to a private cloud, which will help the company capitalize on previous investments in ERP infrastructure while methodically progressing toward a more standardized framework and a streamlined core. This transformation will be gradual, so in parallel, the innovative business segment will be established in a public cloud setting. This strategy prioritizes adherence to SAP's standards and best practices right from the outset to ensure agility and a swifter realization of value.

SAP EWM and SAP TM

Organizations can choose between an embedded or side-by-side deployment of SAP TM and SAP EWM. The embedded option means that SAP TM or SAP EWM functionalities are integrated directly within the SAP S/4HANA landscape, offering a unified approach that simplifies the IT architecture and reduces integration complexity. On the other hand, the side-by-side deployment option allows SAP TM and SAP EWM to operate independently on a separate server, which can be advantageous for organizations with specialized or extensive warehouse and transportation operations that require dedicated systems.

PumpTech elected to implement embedded EWM and TM in its new SAP S/4HANA system to harness the benefits of streamlined processes and cohesive system architecture. By embedding these applications, the company aims to achieve tighter integration with core ERP functionalities, resulting in real-time data exchange and more efficient end-to-end processes. This approach not only simplifies the IT landscape but also reduces the complexity of maintaining separate systems, thereby lowering total cost of ownership. Furthermore, embedded TM and EWM facilitate quicker decision-making and improved visibility across logistics and inventory management, enabling PumpTech to optimize its supply chain operations with greater agility and precision.

Centralized Services

In the context of SAP S/4HANA, *centralized services* are business functions and processes that are consolidated from both an organizational and a systems perspective. By centralizing services, SAP S/4HANA acts as a central hub, offering streamlined access to real-time data, harmonized master data management, and integrated business processes across the organization. This enables the business to reduce redundancy, improve efficiency, and ensure a consistent information flow throughout the enterprise. Centralized services in SAP S/4HANA are particularly advantageous for large enterprises operating in multiple locations or regions, as they help to standardize operations, enforce compliance, and provide a single source of truth for decision-making and reporting.

PumpTech is transforming its organizational structure by centralizing the following critical services within its SAP S/4HANA environment, resulting in enhanced efficiency and coherence across the enterprise:

- **Central Finance**
  PumpTech is consolidating its financial data and centralizing its financial processes. This results in the creation of a unified financial platform that facilitates real-time analytics and a comprehensive view of the company's financial posture. This integration not only simplifies complex financial reporting and accelerates closing activities but also harmonizes financial operations across the entire organization.
- **Master data management**
  By centralizing master data management (MDM), PumpTech ensures uniformity and integrity of data across the business, enhancing the accuracy of business insights and the efficacy of operational processes.
- **Central procurement**
  Centralizing procurement brings together all procurement activities, capitalizing on bulk buying and standardizing procurement procedures to achieve cost savings and process uniformity.
- **Group reporting**
  The centralization of financial reporting across all divisions simplifies intercompany transactions and streamlines the group closing process, enabling faster and more transparent reporting at the group level.

Business Model Innovation

PumpTech's shift toward a subscription-based model leverages the following SAP solutions to facilitate the transition:

- **SAP Subscription Billing**
  Utilizing SAP Subscription Billing, PumpTech can efficiently manage recurring revenue streams, automate billing cycles, and provide flexible pricing models to meet the evolving demands of their customers.
- **SAP Business Network**
  By tapping into SAP Business Network, PumpTech can foster better collaboration with partners, streamline supply chain processes, and enhance customer engagement through integrated platforms.
- **SAP S/4HANA Asset Management**
  Implementing SAP S/4HANA Asset Management enables PumpTech to optimize the lifecycle of physical assets, ensuring maximum value extraction from their use in subscription services and enhancing overall service quality.

In conclusion, the target architecture we've outlined for PumpTech represents a robust and forward-thinking framework that underpins the company's strategic objectives. Anchored by the powerful SAP S/4HANA core, it's designed to seamlessly integrate a range of specialized modules and cloud solutions, establishing a flexible yet comprehensive system. This architecture not only promises to streamline current operations but also positions PumpTech to readily adapt to future demands and technological advancements. With its new, clear vision of a connected and scalable enterprise ecosystem, PumpTech is well on its way to achieving greater efficiency, agility, and continued innovation in its market space, ensuring that it remains competitive and responsive to the evolving business landscape.

5.4.5 Instance Strategy

The *instance strategy* module is designed to guide enterprises like PumpTech through the decision-making process for their SAP landscape architecture. It helps to determine the most suitable landscape option, considering the company's unique business strategy, operational needs, and technological infrastructure. The methodology is comprehensive, evaluating the pros and cons of various instance options both qualitatively and quantitatively, with a focus on SAP Best Practices criteria.

For PumpTech, this would involve a series of steps, starting with identifying the current instance strategy and any additional options the company is willing to consider. A set of tailored criteria would then be selected and weighted according to PumpTech's specific context. The next phase includes a collaborative assessment, engaging stakeholders in all business functions to evaluate each instance option against the selected criteria. The end goal is to propose a *best-fitting* instance strategy for PumpTech's SAP target landscape. This includes validating the technical and application feasibility of the options, assessing operational risks and costs, and ultimately minimizing future risks while reducing the costs of the target landscape.

For PumpTech, the instance strategy module explores the alignment of its SAP landscape with both current operations and strategic visions of the future. As depicted in Figure 5.13, three potential instance strategies have emerged as candidates after a thorough shortlisting process:

❶ The first option maintains the current single global instance, which supports both the traditional business model and the innovative pumping-as-a-service model. This offers a centralized approach but raises concerns regarding compliance with China's data security law.

❷ The second option proposes a separate instance specifically for China, creating a split system that ensures adherence to local regulations while allowing the rest of the global operations to remain on a single instance.

❸ The third option considers creating two separate instances based on business models: one for the traditional business model and another dedicated to the pumping-as-a-service model. This allows for tailored IT environments that can cater specifically to the distinct needs of each business line.

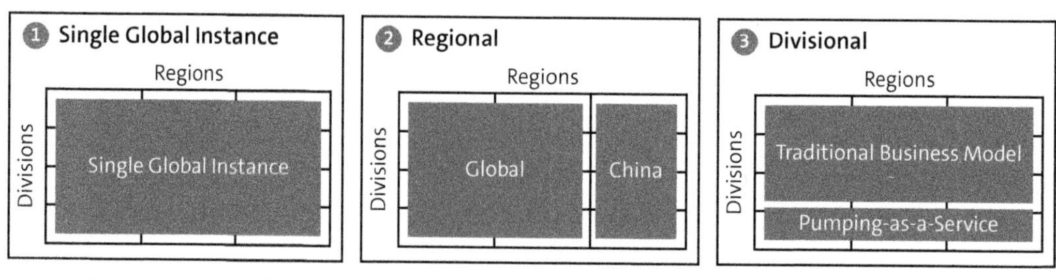

Figure 5.13 Shortlisted Options for PumpTech's Productive Instance Strategy

PumpTech assessed the instance strategy options using a variety of criteria, balancing business needs with IT capabilities. These criteria included the following:

- **Fit with corporate strategy**
  How each instance option supports business growth initiatives like direct-to-consumer channels, expansion into emerging markets, sustainability, resilience, speed to scale, continuous improvement, and financial targets
- **Business fit and flexibility**
  How well the instance options support a digital consumer experience, connected appliances, digital productivity tools, product modularization, digital manufacturing, digital supply chain, value-based sourcing, and template standardization
- **Risk**
  Adherence to data protection regulations, which is particularly relevant for operations in China, and how the options facilitate business continuity and minimize downtime impacts
- **Size scalability and performance**
  How the options would handle technical complexity, interfacing complexity, forecasted data growth, and system performance
- **Software change management**
  Facilitation of the corporate IT governance model, support for different availability needs in manufacturing and warehousing systems, simplification of the change management process, and the flexibility of these processes
- **Maintainability**
  The simplicity of IT operations and support for seamless integration of different IT applications, plus data consistency, security management, provision of real-time end-to-end views, and the impact on the operations team
- **Long-term IT cost**
  Infrastructure and maintenance costs, considering a cloud strategy and data lakes, replacement of end-of-lifecycle software, and adoption of common-use applications to keep things standardized
- **Cost and risk of the transition**
  The cost and duration of the transition, the complexity and risk associated with the transition, and how well the instance options fit into the existing rollout plan

Figure 5.14 displays the results of PumpTech's instance strategy assessment.

Criteria	Weight	Option 1	Option 2	Option 3
Fit-to-Corporate Strategy	3	9	6	15
Business Fit and Flexibility	2	6	4	10
Size, Scalability, and Performance	1	4	3	5
Software Change and Maintainability	1	4	3	3
Risk of Operation	2	4	10	4
Long-Term IT Costs	1	3	3	2
Cost and Risk of Transition	1	4	3	3
		34	32	42
Maximum Rating	55			

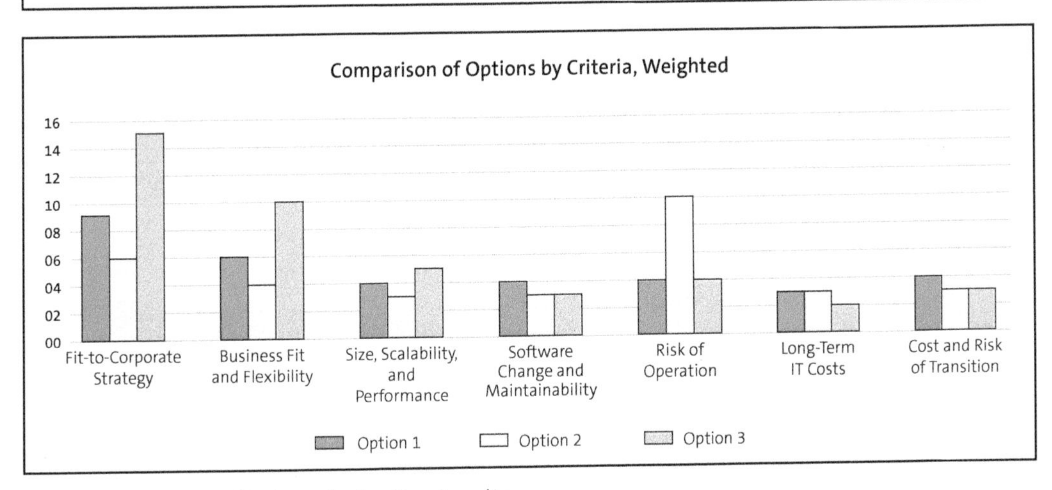

Figure 5.14 Instance Strategy Evaluation Results

The final evaluation of the instance strategy options for PumpTech, based on the weighted criteria, indicates that option 3 is the clear leader, with a total score of 42 out of a maximum of 55. It excels particularly in fit to corporate strategy and business fit and flexibility, reflecting its strong alignment with PumpTech's strategic objectives and adaptability to business needs. Option 1, while competitive, scores a total of 34, showing respectable adherence to corporate strategy but with room for improvement in scalability and maintainability. Option 2 is the least favored, with a total score of 32; despite its high rating in risk of operation and the fact that it addresses data security compliance in China, it falls short in crucial areas like corporate strategy fit and business flexibility.

After careful consideration and consultation with specialized law firms in China, PumpTech has determined that the risks associated with China's data security law are manageable. This assessment was crucial in guiding the company's decision-making process for the instance strategy. PumpTech places a high value on ensuring that its operational framework is in harmony with the corporate strategy and provides the best fit for its evolving business needs. The company has given these factors the greatest weight in its evaluation criteria, reflecting their central importance to PumpTech's long-term goals and operational efficiency. Consequently, despite the potential risks in China, PumpTech has prioritized strategic alignment and business adaptability as the driving forces in selecting its optimal instance strategy.

These results provide a clear direction for PumpTech, favoring a split instance strategy that differentiates between traditional operations and the pumping-as-a-service model. This ensures strategic alignment while managing operational risks and long-term costs.

5.4.6 Artificial Intelligence and Automation

Business AI, especially generative AI, is revolutionizing future business transformations by creating new capabilities that were previously unthinkable. Here, AI does more than just execute commands—it actively participates in problem-solving and innovation. Generative AI in particular stands out for its ability to generate novel content, summarize complex data, write code, and provide creative solutions, all while learning from business-specific data.

SAP Business AI, which is deeply integrated into the SAP product portfolio, offers a reliable, relevant, and responsible approach to embedding AI in business processes. It enables applications to learn, improve, and optimize outcomes, transforming core ERP areas, human capital management, spending management, and customer relationship management. With generative AI, SAP provides tools to help you modernize legacy code, accelerate cloud transformations, and drive business process efficiency. This integration not only makes processes more intelligent but also ensures that work is more efficient and user experiences are delightful, leading to significant savings in time and operational costs.

PumpTech approached the integration of AI and automation into its transformation by methodically assessing its business capabilities to uncover areas ripe for technological enhancement. The company started by assessing its automation potential by performing the following:

- **Business capability analysis**
  PumpTech evaluated each business capability to identify manual, repetitive, and time-intensive tasks.
- **Performance measurement**
  PumpTech analyzed key performance indicators (KPIs) and operational metrics to detect inefficiencies.
- **Process mapping**
  PumpTech documented existing workflows to spot stages that could benefit from automation.

- **Gathering employee input**
  PumpTech gathered insights from staff to determine where they face bottlenecks that AI could alleviate.

The company then did the following to match its assessment with SAP solutions that fit its use case:

- **Review SAP automation solutions**
  PumpTech looked at SAP's catalog of automation use cases to identify potential matches.
- **Capability-to-use case mapping**
  PumpTech aligned high-potential automation capabilities with corresponding SAP use cases.
- **Feasibility study**
  PumpTech conducted a thorough review to ensure that SAP solutions fit the identified capabilities.
- **Prioritization**
  PumpTech ranked the automation opportunities based on their potential impact and ease of implementation.

As an example, depicted in Figure 5.15, PumpTech has analyzed its financial business capabilities and mapped them to SAP's intelligent technologies to enhance efficiency and decision-making. For instance, business capabilities in the finance domain—such as budgetary accounting, predictive accounting, overhead cost accounting, and project accounting—have been mapped to intelligent technologies like Robotic Process Automation (RPA), machine learning capabilities, situation handling, and digital assistants.

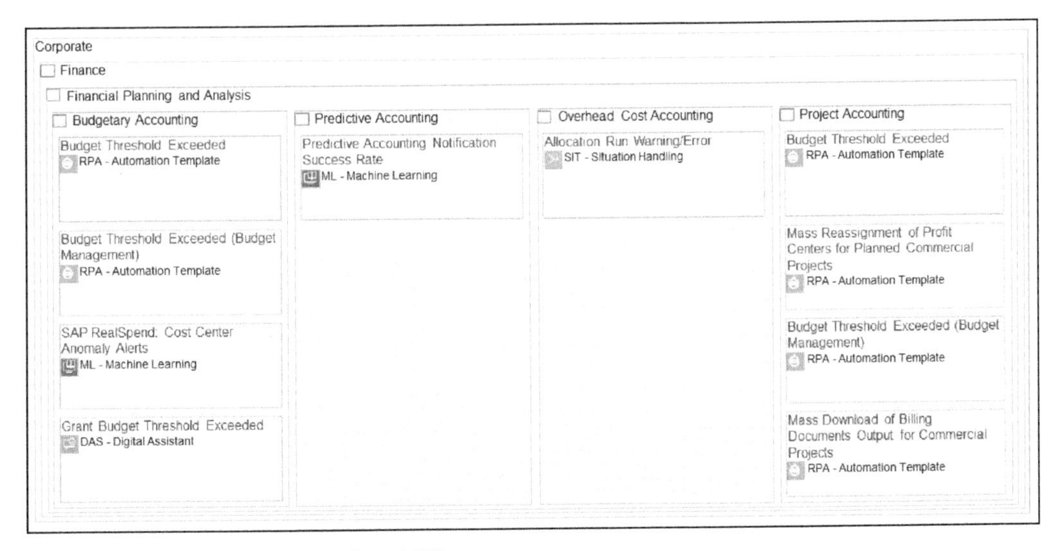

Figure 5.15 Intelligent Business Capabilities

Generative AI in Cloud Transformation: Future Impact Areas

The following list outlines key impact areas where generative AI will increasingly serve as a supportive technology for enterprise architects in planning cloud transformations:

1. **Data migration and transformation**
   Enterprise architects can use generative AI to automate the process of migrating and transforming data from legacy systems to SAP in the cloud. Generative AI can analyze data structures, map them to SAP-compatible formats, and generate scripts or code for seamless migration.
2. **Custom code management**
   AI-powered tools can analyze existing custom code repositories, identify redundant or outdated code segments, and generate recommendations for code refactoring or optimization. Additionally, generative AI models can automate code generation for custom extensions or enhancements, accelerating development timelines and improving code quality.
3. **Requirement analysis and prediction**
   Generative AI models can analyze requirements documentation, user feedback, and historical data to predict future needs and potential bottlenecks. This predictive analysis assists enterprise architects in designing a cloud architecture that can accommodate future growth and changes in demand.
4. **Architecture design and optimization**
   Generative AI tools can assist enterprise architects in designing and optimizing the architecture of cloud systems. By generating multiple design alternatives based on specified constraints and objectives, generative AI helps enterprise architects explore various architectural options and select the most suitable one for the cloud transformation project.
5. **Risk assessment and mitigation**
   Generative AI can help enterprise architects identify potential risks and vulnerabilities in the planned cloud architecture. By analyzing historical data and simulating different scenarios, generative AI models can predict potential failure points and suggest mitigation strategies to minimize risks during the transformation process.
6. **Cost estimation and optimization**
   Generative AI can assist enterprise architects in estimating the cost of cloud resources required for the transformation project. By analyzing factors such as workload patterns, resource utilization, and pricing models, generative AI models can generate accurate cost estimates and recommend cost optimization strategies to minimize expenses.

5.4.7 Integration

SAP Integration Solution Advisory Methodology serves as a guiding framework for enterprises looking to navigate the complexities of integration within hybrid IT landscapes. By offering a technology-agnostic approach, it aids in the meticulous planning, design, and implementation of integration strategies. Applying SAP Integration Solution Advisory Methodology can help PumpTech streamline its transition towards a cloud-centric model, ensuring seamless integration of legacy systems and new cloud-based architectures.

PumpTech faces several challenges in the realm of integration, stemming from a legacy of application team–centric integration efforts. This approach has led to a fragmented landscape of siloed data and processes with a lack of a coherent strategy for a unified system. The disparate methods, tools, and practices used across different teams haven't only created inefficiencies but also have hindered the company's ability to quickly adapt to new business models and scale its operations into the cloud. The absence of dedicated integration roles has further complicated the oversight and governance of integration tasks, impeding the transition to a more agile and interconnected IT infrastructure.

SAP Integration Solution Advisory Methodology offers a comprehensive framework to guide enterprise architects in developing and executing an integration strategy. It encompasses the following phases:

1. **Assessing integration strategy**
   Documenting and reviewing the current integration architecture to scope focus areas and adopt a use-case-driven approach
2. **Designing a hybrid integration platform**
   Mapping use case patterns to integration technologies and deriving organization-specific guidelines
3. **Defining integration best practices**
   Establishing architecture blueprints and decision criteria and emphasizing integration as a discipline
4. **Enabling a practice of empowerment**
   Democratizing integration delivery, defining self-service areas, and fostering a shared developer community

We'll walk through these phases in the following sections.

Assessing Integration Strategy

The first step, assessing integration strategy, is a phase within SAP Integration Solution Advisory Methodology in which an organization critically examines its current state of integration. For PumpTech, this would involve reviewing its legacy systems and integration methods, pinpointing inefficiencies, and identifying the extent to which its current approach aligns with the envisioned cloud-based model. This assessment paves the way for designing a robust integration architecture by highlighting areas for improvement, such as the need for new cloud-centric integration patterns or the elimination of redundant connections that may hinder the move to a more streamlined, service-oriented architecture.

In the assessment, PumpTech has examined the main integration domains as shown in Table 5.2.

	Relevance			Integration Services/ Technologies	
Integration Domain	Relevant	Not Relevant	Under Evaluation	As-Is	To-Be
Cloud to cloud	X			Not yet available	SAP Integration Suite
Cloud to on-premise	X			SAP Process Integration	SAP Integration Suite
User to on-premise			X	N/A	N/A
User to cloud			X	N/A	N/A
Thing to on-premise		X		Not yet available	Under investigation
Thing to cloud	X			Not yet available	Under investigation

Table 5.2 Integration Domains for PumpTech

PumpTech's assessment indicates that the primary integration domains for its forthcoming landscape are cloud-to-cloud and thing-to-cloud. While transitioning, cloud-to-on-premise remains essential until the complete cloud migration is achieved. Thing-to-on-premise is deemed nonessential, given that the IoT-driven business model is designed to be cloud native from inception. The entire user-centric integration landscape is currently under review at PumpTech.

Integration styles within SAP Integration Solution Advisory Methodology represent the archetypes of connectivity that organizations like PumpTech can leverage to align their various systems and services. These styles encapsulate the main categories of integration tasks, and each has unique requirements and tools. They provide a structured framework to identify the appropriate patterns and technologies. SAP Integration Solution Advisory Methodology identifies the following integration styles:

- **Process integration**
  Chaining business processes across applications
- **Data integration**
  Synchronizing data between applications

- **Analytics integration**
  Deriving and exposing analytical data for business insights
- **User integration**
  Integrating user-centric applications with business applications
- **Thing integration**
  Integrating real-world things (IoT) with business applications

PumpTech's primary focus on process integration is strategic; it aims to streamline and automate its business processes in various applications. This focus is critical as it enables seamless workflow transitions, enhances operational efficiency, and serves as the backbone of digital transformation. Relevant use case patterns for PumpTech would likely include application-to-application (A2A) integration, business-to-government (B2G) integration, master data integration, and business-to-business (B2B) integration. These patterns ensure that internal and external processes are efficiently connected, which is essential for supporting the pumping-as-a-service model.

Designing a Hybrid Integration Platform

In the second phase of SAP Integration Solution Advisory Methodology, the identified use case patterns are mapped to integration technologies and services. Following the integration style process integration, we can build a decision table as shown in Figure 5.16 to give guidance on integration technologies to be used.

Integration Guidelines Process Integration					
Integration	Recommendation Degree	Description/Use Case Patterns	Relevant Integration Domains		
			On-Premise to On-Premise	Cloud to On-Premise	Cloud to Cloud
SAP Integration Suite	General recommendation	Use for all new integration use cases and gradually for migrating existing use case patterns. Leverage packaged integration content from SAP or SAP partners.		●	●
SAP Process Orchestration	To be avoided	Don't use for new integration use cases, as product is set for end of maintenance by 2027/2030. Gradually move existing use case patterns to SAP Integration Suite.	●		
SAP Application Interface Framework	General recommendation	Use whenever you want business users to monitor interfaces and to do error handling with SAP S/4HANA.	●	●	
Third Party	Possible exception	Use only for very specific scenarios (e.g., country specific scenarios for business to government use case patterns) and if it can't be built on SAP BTP with reasonable effort.			●

Figure 5.16 Decision Table for Process Integration

PumpTech decided it had a strong preference for SAP Integration Suite for both new and transitioning use cases, and it advised against SAP Process Orchestration for new integrations due to the scheduled end of maintenance. SAP Application Interface Framework received a general recommendation for business user monitoring of interfaces. Lastly, PumpTech considered third-party services as a possible exception in specific scenarios that SAP BTP can't accommodate with reasonable effort.

Defining Integration Best Practices

The third phase of SAP Integration Solution Advisory Methodology involves governing the implementation of integrations and achieving transparency in integration design and implementation. However, in this chapter, due to the scope limitations, we'll bypass a detailed exploration of this phase. Instead, we'll guide you to the publicly available SAP Integration Solution Advisory Methodology documentation for architecture blueprints, decision criteria, and best practices. To gain a comprehensive understanding of these practices, we encourage you to consult the SAP Integration Solution Advisory Methodology content directly in SAP's resources at *http://s-prs.co/v586305*.

Enabling a Practice of Empowerment

The fourth and final phase of SAP Integration Solution Advisory Methodology is about elevating integration to a strategic level within the organization. This phase aims to accomplish the following:

- Recognize integration as a key discipline
- Position integration as a strategic differentiator
- Provide a holistic view of the organization's integration landscape
- Democratize the process of integration delivery
- Foster a culture of knowledge sharing regarding best practices
- Enhance communication between teams and system integrators

As an intermediate result of the final phase, PumpTech is introducing a specific integration architect role to lead and centralize its integration efforts. This is a shift from the previous practice, in which application teams handled integration tasks. The integration architect will be responsible for crafting company-wide integration guidelines, ensuring that all integration practices align with organizational strategies and objectives. Additionally, PumpTech is establishing an Integration Center of Excellence, which will be charged with overseeing best practices, standardizing integration processes, and fostering a knowledge-sharing culture to streamline and optimize how the organization's systems interconnect.

5.4.8 Extensibility

SAP customers often face complex challenges when extending their software solutions to meet specific business needs. These challenges include maintaining a clean core—which involves keeping the core SAP system standardized without custom modifications while extending functionality, ensuring cloud readiness to leverage cloud technologies, and adopting future-proof strategies to accommodate new technologies and methodologies. These complexities necessitate a structured approach to extending SAP solutions effectively without compromising the integrity and upgradability of the core system.

SAP Application Extension Methodology offers a structured, technology-agnostic framework designed to support enterprise architects, developers, and project teams in extending SAP applications. This methodology facilitates the assessment of extension use cases, the selection of appropriate technologies, and the definition of target solutions that align with business needs and IT strategies. By standardizing the extension process, the methodology helps to ensure consistency and efficiency, thereby reducing errors and misalignments between business requirements and technical implementations.

To ensure a systematic approach from the initial assessment to the final implementation of the extension solution, SAP Application Extension Methodology is methodically structured into these three sequential phases:

1. **Assess extension use case**
   This foundational phase focuses on understanding the business and system context. Here, the team documents the specific business needs and technical requirements, establishing a solid base for the extension project.
2. **Assess extension technology**
   This phase involves defining and categorizing extension styles and tasks. It bridges the gap between business requirements from phase 1 and the technical capabilities within SAP's ecosystem, facilitating a clear mapping of business needs to technological solutions.
3. **Define extension target solution**
   This final phase is about selecting suitable technical building blocks and crafting a detailed implementation plan. It culminates in defining the target solution that incorporates the chosen technologies and fulfills the outlined business objectives.

In the process of extending SAP applications, it's crucial for enterprise architects to focus on defining an overarching extensibility strategy at the outset, rather than delving into a detailed analysis of each application. By establishing a comprehensive extensibility framework early on, architects can set clear guidelines and principles that govern the extension process for various applications and projects.

SAP Application Extension Methodology offers guidance by initially recommending the definition of extension styles and tasks in various tiers such as user interface, application, and data. An illustration of this can be found in Table 5.3.

ID	Extension Style	Extension Task
A01	Business logic extension	Add custom field to UI service.
A02	Business logic extension	Add custom field to an API.
A03	Business logic extension	Adapt a standard business process with custom logic (e.g., prefill/validate field within logical unit of work).
A04	Business logic extension	Create application logic.
A05	Business logic extension	Create an API for UI.
A06	Business logic extension	Create an API for integration.
A07	Business logic extension	Consume an API.

Table 5.3 Example of Extensions Styles and Tasks for Application Tier

The application tier is crucial because it's responsible for enhancing and customizing business logic to meet specific business requirements. This is where the core functionalities of a business application are extended or modified. By implementing additional business logic within this tier, organizations can realize their unique business needs. This could involve adding new process steps to an existing workflow, exchanging current steps for more efficient ones, or completely reconfiguring the sequence of operations to streamline the processes.

Extension task guidance like in Figure 5.17 is a set of directives and best practices designed to assist in the development and implementation of technical extensions. The goal of extension task guidance is to provide a structured approach for enterprise and domain architects to follow, ensuring that extensions are developed in a way that aligns with the overall system architecture and business objectives.

ID	Extension Style	Extension Task	Technical Extension Building Block	Guidance	Color Code
A03	Business logic extension	Adapt standard business process with custom logic	Modification	No modification of the standard allowed	
			Classic extensibility (ABAP)	Only if other options are not possible with reasonable effort	
			Key user extensibility	Should always be first option to be evaluated	
			On-stack developer extensibility	For tightly coupled and more complex extensions	
			SAP BTP side-by-side extension	For loosely coupled extensions	

Legend: Allowed Only in exceptional cases Prohibited

Figure 5.17 Example of Guidance for One Extension Task

The example gives the extension style business logic and maps various extension technologies to the **Adapt standard business process with custom logic** extension task. For each of the extension technology building blocks, it gives guidance for when to use what technology in the given context. The color coding highlights the general guidance of allowed and prohibited technologies and technologies that should only be applied in exceptional cases.

In addition, PumpTech can leverage a *Solution Standardization Board* (SSB) as a governance body to oversee the application of SAP standards and ensure that extensions are built only when standard SAP functionalities can't accommodate the company's specific business processes. This board would consist of key stakeholders—such as enterprise architects, SAP solution experts, and business process owners—who collectively possess a deep understanding of both SAP system capabilities and the company's unique requirements.

The SSB would be responsible for the following tasks:

1. **Reviewing extension requests**
   Assessing proposed extensions against existing SAP standard solutions to determine if the standard functionality can meet the business needs without customization
2. **Guarding best practices**
   Ensuring that any proposed extension aligns with SAP best practices and PumpTech's IT strategy to avoid unnecessary complexity and future maintenance challenges
3. **Maintaining system integrity**
   Making decisions that protect the integrity of the SAP system by preventing redundant or conflicting extensions and promoting reusable solutions
4. **Evaluating business impact**
   Considering the potential business impact and ROI of an extension versus utilizing standard SAP functionality
5. **Approving extensions**
   Approving the development of custom extensions only when it's clear that the SAP standard can't be adjusted to meet the necessary business requirements

By establishing such a board, PumpTech can maintain a disciplined approach to extending its SAP system, ensuring that extensions are justified and strategically sound and add genuine value to the business. This framework helps to minimize unnecessary customizations, streamline operations, and preserve the upgradeability of the SAP system.

Custom Code Analysis

Although conducting a detailed analysis of custom code usage within an SAP environment usually extends beyond the standard role of an enterprise architect, it's nonetheless advisable to incorporate this task into the broader transformation strategy. The enterprise architect is responsible for overseeing the assessment of existing custom code within the legacy system to determine its compatibility with and necessity in the new environment.

During the transition to SAP S/4HANA, SAP aids organizations in managing custom code through a suite of tools and services designed to simplify the process. Custom code analysis within the SAP Readiness Check framework allows businesses to identify and evaluate existing custom code that may not be compatible with the new environment. SAP then provides recommendations for adapting this code, leveraging tools like the ABAP Test Cockpit (ATC) for seamless migration and optimization. Within SAP services, AI is utilized to intelligently cluster custom code, simplifying the process of pinpointing which custom code objects one must keep, and which can be decommissioned.

The goal is to streamline the custom code, ensuring it aligns with the new, simplified data model of SAP S/4HANA and utilizes the advanced processing capabilities of the SAP HANA database. This approach reduces the footprint of custom developments, enhances performance, and aligns custom code with best practices. This mitigates risks associated with the transition and leverages the full potential of the SAP S/4HANA platform.

5.4.9 Initiatives and Roadmap

The *initiatives catalog* is an essential precursor to crafting the overall enterprise architecture roadmap. It provides a detailed and organized framework of actions required to fulfill strategic goals, and it lays the groundwork by identifying and prioritizing specific, actionable initiatives. It thereby outlines the steps necessary for the transformation journey. This preplanning step ensures that each initiative is aligned with broader business objectives, resources are effectively allocated, and potential risks are assessed. In essence, it establishes a clear and structured pathway that guides the subsequent development of a coherent and comprehensive roadmap for enterprise-wide change.

By having an initiatives catalog, enterprise architects can effectively prioritize and evaluate initiatives based on their strategic alignment, feasibility, and potential impact. This allows enterprise architects to make informed decisions about which initiatives to pursue and how to sequence them, ultimately enabling the creation of a roadmap that outlines the timeline and dependencies of each initiative. Without an initiatives catalogue, enterprise architecture planning would lack the necessary visibility and structure to effectively manage and coordinate the complex interplay of initiatives. This would in turn hinder the organization's ability to achieve its strategic objectives.

We'll explore PumpTech's roadmap in the following sections, starting with the next few years and then looking toward the horizon and the eventual final state.

Initial Roadmap

In the provided excerpt from PumpTech's overall initiatives catalog in Figure 5.18, we can see a structured outline of planned initiatives designed to transition the organization's IT infrastructure and business processes. Each initiative is listed with key components such as a concise description, the alignment with business or IT goals, the enhancement of specific business capabilities, the applications involved, dependencies on other projects, and a projected timeline.

No	Initiative	Description	Business/IT Goal	Business Capabilities	Applications	Dependencies	Timeline
1	Conversion and cloud migration	Brownfield conversion of existing SAP ERP system	Cloud transition	All	SAP S/4HANA Cloud Private Edition	None	2026
2	Cloud migration	Move all SAP NetWeaver-based systems to private cloud environment	Cloud transition	All	SAP S/4HANA Cloud Private Edition, SAP EWM, SAP TM, SAP Advanced Planning and Optimization	Initiative 1	2026
3	Central services	Establishment of central finance, central procurement, group reporting, and master data	Centralized services	Finance, procurement, master data	Central finance, group reporting, central procurement, SAP Ariba, SAP Master Data Governance	Initiative 1	2026
4	HR	Cloud transition of all HR processes	Cloud transition	HR	SAP SuccessFactors	Initiative 1	2027
5	Supply chain execution	Replatform from sidecar deployment to embedded solutions	Landscape consolidation	Warehouse management, transportation management, production planning	SAP S/4HANA Cloud Private Edition (embedded TM, EWM, PP/DS)	Initiative 1, 2	2029
6	Clean core and business AI	Enablement of latest business AI capabilities and clean core adaption for all processes	Innovation, agility	All	All	Initiative 1, 2, 3, 4, 5	2029

Figure 5.18 Excerpt from PumpTech's Initiatives Catalog

As an example, the second initiative in the catalog excerpt is titled **Cloud migration**. The description indicates that it involves a move to a private cloud environment, which aligns with the broad IT goal of cloud transition, which is in turn part of PumpTech's strategy to modernize its infrastructure. It impacts all business capabilities and involves applications like SAP S/4HANA in a private cloud, with no dependencies and a set timeline for 2025.

As a further illustration, initiative number 5, **Supply chain execution**, is dedicated to the transition of existing SAP EWM, SAP TM, and SAP Advanced Planning and Optimization PP/DS solutions to embedded components within SAP S/4HANA Cloud Private Edition. The target year for this initiative is 2029, marking it as one of the final, more complex stages of the transformation. This is due to the significant enhancements PumpTech has made to these systems, recognizing their critical importance and the complexity involved in their migration. Consequently, the success of this initiative depends upon the completion of initiatives 1 and 2, making it a critical final step in the company's overall transformation journey.

Drawing from the comprehensive initiatives catalogue, PumpTech has subsequently developed a detailed transformation roadmap, which is illustrated in Figure 5.19.

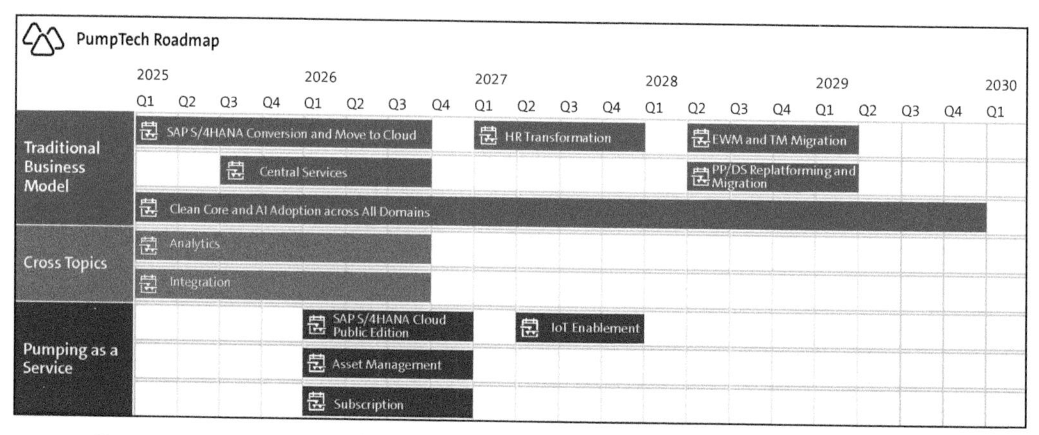

Figure 5.19 Roadmap for PumpTech

The initial cluster of PumpTech's roadmap is centered on modernizing its traditional business operations. This begins with the transition of the SAP ERP system to SAP S/4HANA in the private cloud through a concurrent conversion and cloud migration project. Simultaneously, other SAP NetWeaver-based products like SAP EWM and SAP TM are also moved to the private cloud via a lift-and-shift approach. The activation of central services concurs with the SAP S/4HANA go-live event, and following this, as a secondary measure, the HR function will transition to the cloud with the implementation of SAP SuccessFactors. Finally, as previously noted, supply chain execution systems—SAP EWM, SAP TM, and SAP Advanced Planning and Optimization—will be transferred onto the SAP S/4HANA platform as embedded capabilities.

The roadmap's second cluster addresses overarching themes such as analytics and integration. Here, PumpTech has opted for fully cloud-based solutions from the outset to hasten the reporting and integration processes associated with the adoption of the new business model. This strategy significantly enhances business agility by merging both traditional and new business operations within a singular, unified system.

In the roadmap's third cluster, dubbed **Pumping as a Service**, PumpTech activates its new business model, beginning with the deployment of SAP S/4HANA on a public cloud. This is complemented by asset management solutions on SAP BTP and the establishment of subscription functionalities on SAP BTP. Subsequent to this, PumpTech implements IoT capabilities. This tactic ensures that the principal integration necessities bridging the traditional and the new business models are met, enabling PumpTech to operate both business models simultaneously within an integrated landscape, albeit at varying velocities.

At this point, the roadmap has helped lay the foundation for the formulation of transition architectures. These architectures will serve as incremental blueprints, guiding the organization through its digital transformation journey. For a structured and measured progression, PumpTech has designated three distinct time horizons, which we cover in the following sections.

2025–2026+ Horizon

Figure 5.20 summarizes the horizon of the years 2025 and 2026. At this stage, the conversion to SAP S/4HANA and the migration to the private cloud for all SAP NetWeaver systems has taken place. In addition, central services like Central Finance and central procurement are enabled and analytics and integration capabilities are running on SAP BTP.

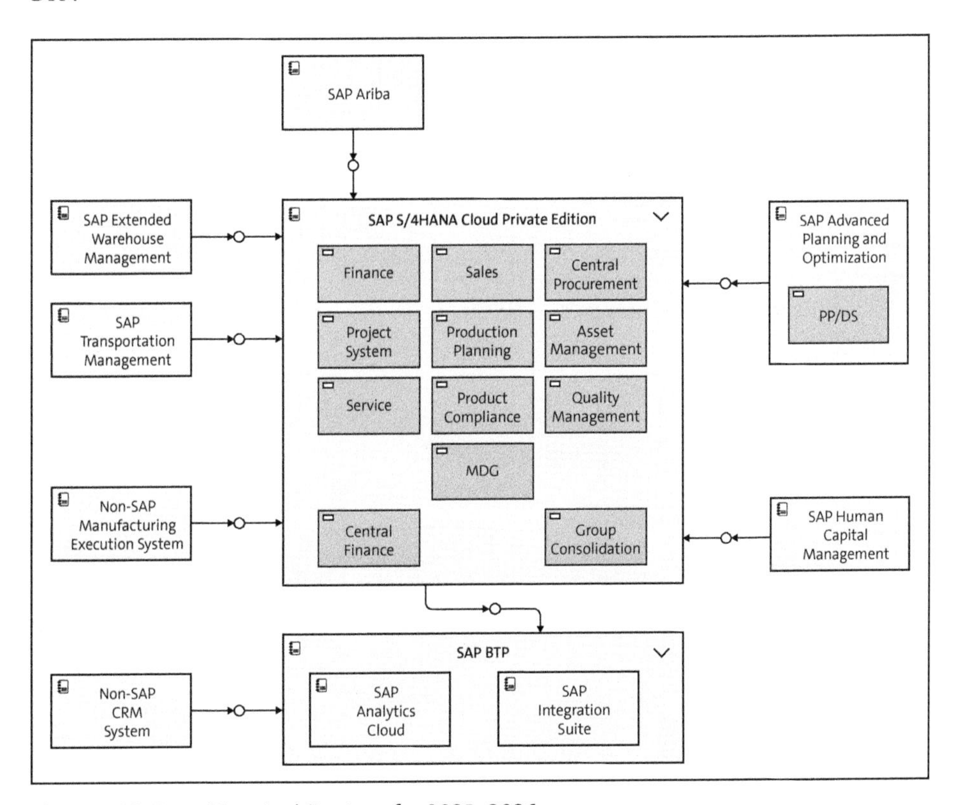

Figure 5.20 Transition Architecture for 2025–2026+

2026–2027+ Horizon

In the next stage, as depicted in Figure 5.21, the new business model is implemented with SAP S/4HANA Cloud Public Edition. This includes supporting applications for asset management and subscription billing on SAP BTP.

2027–2028 Horizon

Figure 5.22 presents the subsequent stage, in which the critical HR cloud transformation is achieved with HR capabilities now running on SAP SuccessFactors in the cloud.

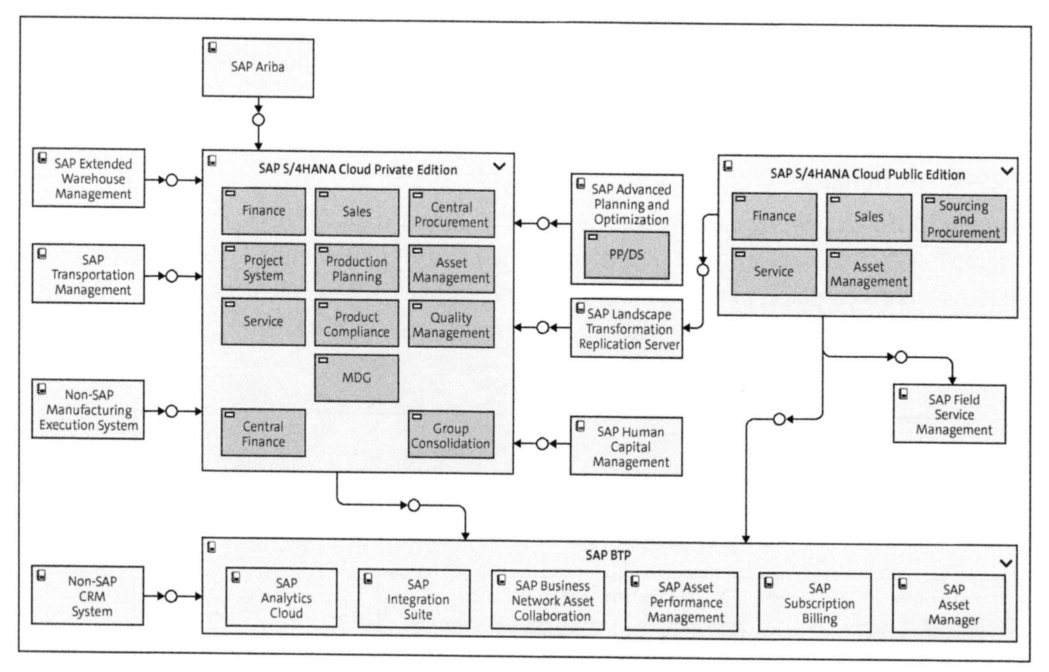

Figure 5.21 Transition Architecture for 2026–2027+

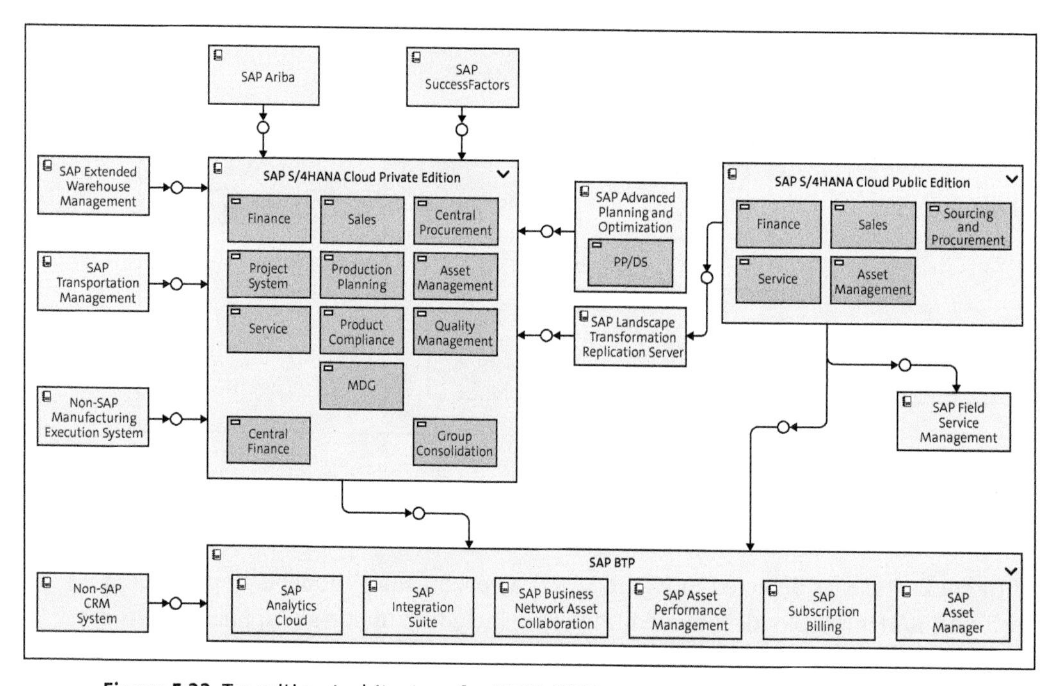

Figure 5.22 Transition Architecture for 2027–2028

Final State and Beyond

The final state of the architecture was detailed in the Section 5.4.4 on application architecture. Although the roadmap concludes with the comprehensive transformation, it doesn't specify initiatives for the period thereafter. However, some prospective initiatives could include the following:

- **Sustainability initiatives**
  Utilizing cloud-based tools to monitor and manage the company's environmental impact and energy consumption and to streamline resource usage
- **Customer experience enhancement**
  Developing more personalized and responsive customer service platforms with cloud-based AI, including chatbots and customized service portals
- **Global expansion support**
  Using the scalability of the cloud to support expansion into new markets with minimal physical infrastructure investment
- **Blockchain for supply chain**
  Exploring blockchain technology to improve transparency, traceability, and efficiency within the supply chain

5.4.10 Transition Scenario Evaluation

PumpTech has an essential decision to make in choosing a method to transform its current SAP ERP system.

The company must carefully evaluate the following approaches, which are depicted in Figure 5.23:

- **Brownfield**
  This approach would upgrade the existing system directly to SAP S/4HANA, maintaining current processes and ensuring business continuity with minimal changes.
- **Selective data transition**
  This method offers a more customized transformation by selectively migrating data and processes, allowing for gradual adoption of new features and process improvements.
- **Greenfield**
  A fresh start, this option involves building a new system from the ground up with SAP S/4HANA, adopting the latest best practices, and enabling a complete reengineering of business processes.

The transition to SAP S/4HANA can take several pathways, each with distinct strategies and implications. A system conversion maintains a low level of process reengineering, offering full historical data migration and typically adopting a big-bang approach to rollout, which is ideal for businesses seeking minimal disruption. In contrast, shell conversion involves medium to low process reengineering and allows for selective historical data migration. It also offers more flexibility in rollout, which can be either phased or big bang. Mix and match, as part of selective data transition, requires a medium to high reengineering effort and partly migrates data. This is suitable for companies wanting to combine process redesign with data continuity. Finally, a new implementation, also known as a *greenfield approach*, demands high process reengineering with no historical data migration. This provides an opportunity for a complete overhaul and innovation with the flexibility of a phased or big-bang rollout. Each option allows customization to the organization's needs, balancing the needs for innovation, data preservation, and system disruption.

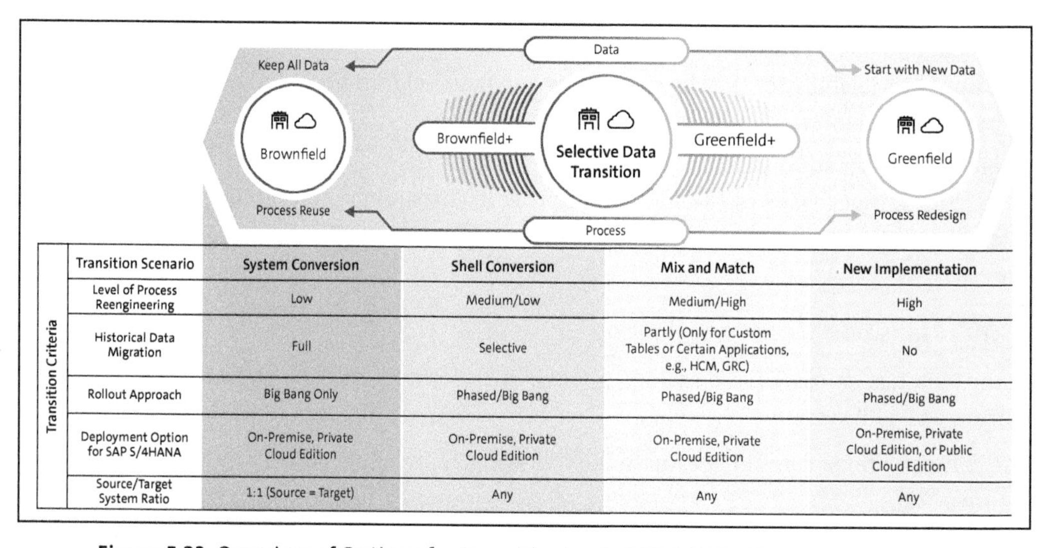

Transition Scenario	System Conversion	Shell Conversion	Mix and Match	New Implementation
Level of Process Reengineering	Low	Medium/Low	Medium/High	High
Historical Data Migration	Full	Selective	Partly (Only for Custom Tables or Certain Applications, e.g., HCM, GRC)	No
Rollout Approach	Big Bang Only	Phased/Big Bang	Phased/Big Bang	Phased/Big Bang
Deployment Option for SAP S/4HANA	On-Premise, Private Cloud Edition	On-Premise, Private Cloud Edition	On-Premise, Private Cloud Edition	On-Premise, Private Cloud Edition, or Public Cloud Edition
Source/Target System Ratio	1:1 (Source = Target)	Any	Any	Any

Figure 5.23 Overview of Options for Transitioning to SAP S/4HANA

PumpTech is utilizing a methodical evaluation and scoring approach, as shown in Figure 5.24, to ascertain the most suitable option for transition to SAP S/4HANA. This process begins with the identification of criteria, in which PumpTech starts with a set of template criteria and then modifies it after discussions to cater to its specific needs. Following this, the weighting of criteria stage allows PumpTech to prioritize these criteria by assigning them significance values on a scale of 1 to 5, with 1 being the least important and 5 being essential.

Once the criteria are established and weighted, the valuation of selected criteria phase involves scoring each criterion based on its performance, from 1 (poor) to 5 (excellent), with documentation to support the reasoning behind each score. This step ensures that all considerations are evaluated according to their impact on the transition process.

Criteria	System Conversion	Selective Data Transition		New Implementation
		Shell Conversion	Mix and Match	
Process reengineering	Full process reuse	Predominantly process reuse	Predominantly process redesign	Full process redesign
Rollout strategy	Big bang	Big bang or phased	Typically phased by company codes	Typically phased by company codes
System consolidation or split	Not possible, only 1 to 1 replacement possible	Possible, carve-out by application in system split scenario is also possible	Possible, carve-out by application in system split scenario is also possible	Possible
Historical data migration	All data including full history	All data or selective history	Selective history	Only master data and open items, no history
Deployment	Private cloud, on-premise	Private cloud, on-premise	Private cloud, on-premise	Public cloud, private cloud, on-premise
Data transformation	Data model changes without field and value mapping	Data model changes with field and value mapping	Data model changes with field and value mapping	New data construction, possible field and value mapping
Typical project cost (relatively)	Low	Medium	High	High
Typical project duration	6-9 months	7-10 months	10-36 months	10-36 months

Figure 5.24 Comparison of Different Transition Options

PumpTech has used the following decision criteria:

- **Business impact and downtime**
  The impact on business operations and the acceptable amount of downtime during transition
- **Cost**
  The total cost of ownership, up-front migration costs, and potential savings from improved efficiency
- **Custom code adaptation**
  The effort required to adapt or retire existing custom code within the SAP ERP environment
- **Data migration complexity**
  The volume of data, the complexity of historical data migration, and the need for data cleansing or archiving
- **Degree of process reengineering needed**
  The degree to which the company can reuse existing processes effectively or will need to reengineer them significantly to leverage SAP S/4HANA capabilities
- **Integration with other systems**
  The transition's potential effects on existing integrations with third-party and legacy systems

- **User training and change management**
  The extent of training needed for users to adapt to the new system and the readiness of the organization for change
- **Future-readiness and scalability**
  The ways in which each option positions PumpTech for future growth and technology adoption
- **Timeline for implementation**
  The speed at which each transition option can be implemented within the desired timeline

Finally, the analysis of results is automated through Microsoft Excel, which calculates the scores and indicates the winner among the transition options. This step takes into account the closeness of scores, noting that if two options have scores that are not significantly different, they are both considered equally viable. This comprehensive approach enables PumpTech to make a data-driven decision on the transition method that best aligns with its organizational priorities and readiness for the change to SAP S/4HANA.

After an evaluation using the scoring approach, PumpTech concluded that a two-phased transition strategy for the SAP ERP system would serve their objectives best, as depicted in Figure 5.25.

To support their innovative pumping-as-a-service business model, PumpTech has chosen to harness the capabilities of SAP S/4HANA Cloud Public Edition. Consequently, a greenfield implementation is the sole transition scenario for deploying in the public cloud.

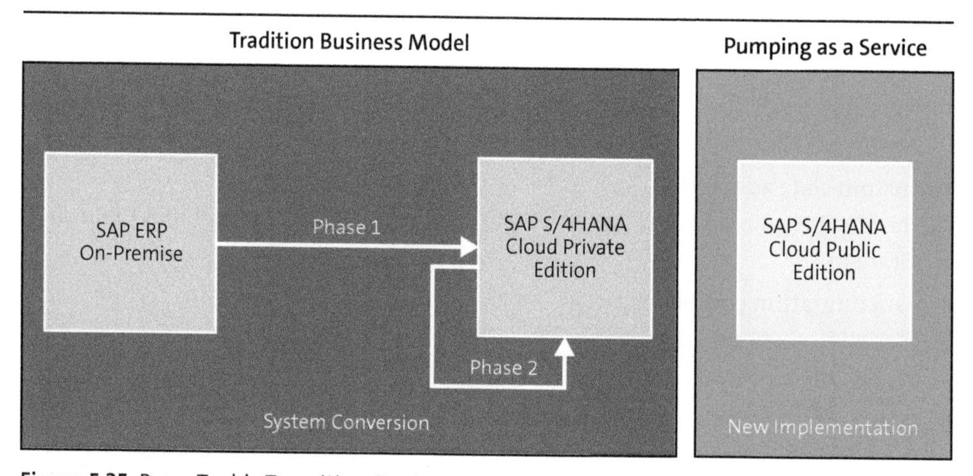

Figure 5.25 PumpTech's Transition Strategy

In the first phase, system conversion emerged as the optimal approach. This decision likely hinged on several key factors:

- **Business continuity**
  PumpTech may have determined that maintaining its existing processes during the initial transition would minimize operational disruptions. System conversion enables this by upgrading the existing system to SAP S/4HANA while keeping the core processes intact.
- **Cost-effectiveness**
  A direct conversion has been evaluated as more cost-effective than a full-scale greenfield implementation, especially when considering the need to preserve extensive historical data.
- **Private cloud migration**
  The decision to migrate to a private cloud is aligned with PumpTech's overall cloud strategy.
- **Skillset and readiness**
  PumpTech has assessed their internal team's familiarity with the existing systems, deeming that the team's skill set and readiness level is more suited for a system conversion, which requires less retraining compared with a new implementation.

For the second phase, the back-to-standard approach implies that PumpTech plans to align more closely with SAP's standard functionality after the system conversion. This strategy was likely chosen based on criteria such as the following:

- **Process optimization**
  After stabilizing the S/4HANA environment, PumpTech seeks to optimize processes, capitalizing on the new system's best practices without the immediate pressure of a system overhaul.
- **Innovation adoption**
  The second phase allows PumpTech to adopt new innovations and best practices at a controlled pace, further improving efficiency and performance.
- **Cost and complexity reduction**
  By moving toward standardization, PumpTech aims to reduce the complexity and cost of future upgrades and customizations, simplifying maintenance and support.
- **Change management**
  A phased approach gives more time for thorough change management, ensuring that the organization and its stakeholders are ready for the change and minimizing resistance and disruption.

Through this staged approach, PumpTech could leverage the immediate benefits of S/4HANA while setting the stage for ongoing improvements and alignment with SAP's standard practices. This balanced method appears to provide a pragmatic path that mitigates risk, manages cost, and aligns with the company's strategic direction for digital transformation. Adopting this strategy, PumpTech can execute phase 1 with limited effort, simultaneously facilitating the rollout of a new business model in the public cloud.

5.5 Summary

In this chapter, we explored the details of PumpTech's SAP cloud transformation journey, utilizing SAP EA Framework as our navigational tool. We described the use case through a structured approach to architecture transformation in the cloud, taking into account critical considerations such as a clean core, two-tier architecture, and organizational readiness. We explored how the cloud has transformed business operations, focusing on its impact on agility, scalability, and innovation.

The heart of the chapter detailed the practical steps in the transition from traditional on-premise infrastructures to a flexible and efficient cloud architecture. The roadmap we laid out is both aspirational and grounded, aligning with the company's overarching strategy and the specific steps needed to actualize it. We saw how PumpTech's established SAP systems, once symbols of operational stability, have been transformed to meet the demands of a rapidly evolving digital economy. The shift to cloud architecture represents an opportunity to escape the limitations of traditional systems and embrace a new paradigm of operational flexibility and strategic decision-making.

At the operational level, we witnessed how PumpTech navigated the migration of its SAP-centric landscape to a future-ready cloud infrastructure. This migration wasn't merely a technical upgrade but a strategic pivot that unlocked new avenues for growth and innovation. By successfully executing this cloud transformation, PumpTech positioned itself to leverage enhanced operational efficiency, foster innovation, and deliver exceptional experiences to customers and employees. However, it's clear that such a transformation demands careful planning, structured execution, and ongoing management to fulfill its potential.

With the roadmap now defined, PumpTech is set to draft transition architectures that will provide tangible stepping stones toward achieving its cloud objectives. The company has outlined three time horizons for this phased approach: the immediate 2025–2026+ horizon, the intermediate 2026–2027+ horizon, and the longer-term 2027–2028+ horizon. Each horizon brings its own set of initiatives and milestones, shaping a clear trajectory for technological and business process advancements.

As we approach the conclusion of PumpTech's narrative in this chapter, it's apparent that the company's cloud transformation is more than an IT project—it's a business imperative that touches every aspect of their organization. It reflects a concerted effort to align IT capabilities directly with business objectives, fostering a resilient and competitive organization in the digital age. The journey of cloud transformation, as illustrated by PumpTech, is ongoing and iterative, requiring leaders to maintain a dual focus on today's operational excellence and tomorrow's strategic success.

Now, let's move on to our next use case: business transformation.

Chapter 6
Business Transformation

In today's fast-paced, ever-evolving world, business transformation is no longer an option—it's a necessity. In this chapter, you'll discover how business transformation is not just a buzzword but is a strategic necessity for most SAP customers, and you'll navigate the complexities of change with SAP enterprise architecture planning.

Business transformation often involves a fundamental shift in how a company operates, usually aimed at coping with a significant change in the company's market or competitive environment. However, some companies fail in this endeavor, often due to a lack of vision, resistance to change, or an inability to effectively manage the transformation process. One notable example is Blockbuster, a video rental company that failed to adapt to the digital age. Blockbuster was once a staple in American entertainment, but it failed to foresee and adapt to the rise of streaming services and the shift in consumer preferences toward digital content. It had the opportunity to purchase Netflix early on but passed on the deal, viewing the internet as a niche market. This failure to innovate and adapt its business model ultimately led to Blockbuster's bankruptcy in 2010. On the other hand, companies that have embraced digital transformation, like Netflix, have thrived.

In this chapter, we'll delve into the utilization of SAP EA Framework to facilitate business model transformations, such as transitioning from a product sales approach to a pay-per-use model. The case study will demonstrate how SAP EA Framework and the principles of enterprise architecture transformation can deliver a comprehensive perspective on the organization's existing situation, establish the sought-after future state, and help the organization devise a strategic roadmap to reach its defined objectives. Furthermore, the case study will illustrate how the alignment of business and IT requirements can influence the determination of the target architecture and the transformation roadmap, driven by both business and IT imperatives.

6.1 Business and IT Context

Business transformation is a broad term that encompasses the profound changes an organization undertakes to align its operational structure, culture, and core objectives with its evolving business strategy. It's a holistic process that can involve changes in the organization's business model, processes, people, and technology. The goal of business transformation is to improve operational efficiency, foster innovation, and position the organization for sustained growth in a constantly evolving market landscape.

As depicted Figure 6.1, business transformation is a comprehensive approach to evolving a company in multiple dimensions to achieve long-term success and adaptability. It encompasses *strategic transformation*, which redefines business models and market strategies; *process transformation*, which focuses on optimizing and reengineering core operations for efficiency; *organizational transformation*, which restructures the company to better align it with strategic goals; *cultural transformation*, which is aimed at fostering a supportive and innovative organizational culture; and *digital transformation*, which integrates advanced technologies to enhance all aspects of business operations and customer interactions. Together, these transformations drive holistic improvement and position the business to thrive in a rapidly changing environment.

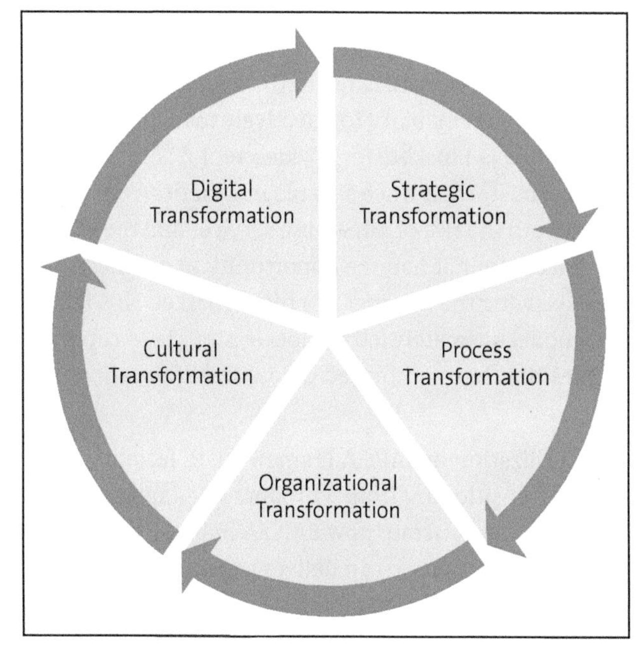

Figure 6.1 Elements of Business Transformation

We'll establish the business and IT contexts for business transformation in detail in the following sections, to set the stage for our use case.

6.1.1 Business Context

At its heart, business transformation is a response to a dynamic business environment. It may be necessitated by a variety of factors, including disruptive technologies, evolving customer expectations, regulatory changes, competitive pressures, and mergers and acquisitions (see Chapter 8). It's about an organization's ability to adapt, innovate, and remain agile amidst these challenges.

The global business landscape is undergoing a seismic shift, with the boundaries between industries becoming increasingly blurred. Traditional ways of doing business are being challenged and supplanted by innovative, digitally driven models. In this context, business transformation isn't just a desirable option; it's a strategic imperative. This is especially pertinent for SAP customers who must navigate the complexities of this evolving landscape and embrace new business models, even in classic industries such as the automotive sector.

Traditionally, car manufacturers like Ford, GM, and Toyota focused on vehicle design, manufacturing, and sales. However, with the rise of electric vehicles, autonomous driving, and connected car technologies, the traditional automobile industry is increasingly overlapping with the technology, chemical, and energy sectors. Therefore, innovation in the automobile industry is no longer limited to the research and development (R&D) teams responsible for creating the next generation of vehicles. It's crucial for every department and discipline to embrace innovation and actively contribute to the transition from best practices to industry-leading practices, as outlined for the automotive industry in Figure 6.2.This approach extends to the very forefront of the industry, allowing cross-functional teams to explore novel methods of delivering exceptional value to customers. As a result, this generates improvements in revenue, profitability, and sustainability.

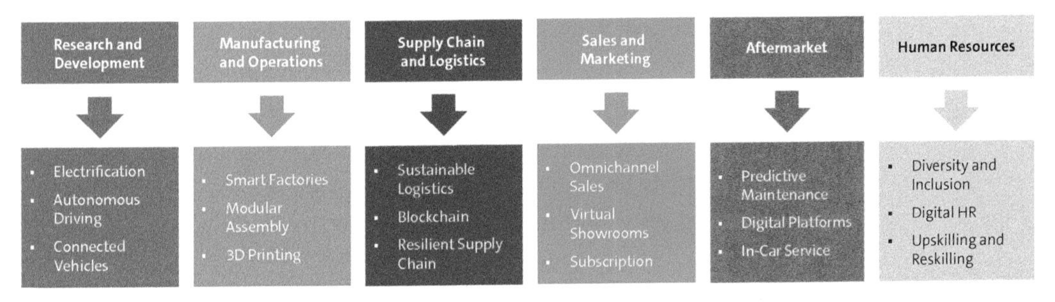

Figure 6.2 Transformation in the Automotive Industry

The majority of the world's leading companies are clients of SAP and are thus either directly or indirectly impacted by the aforementioned shifts in business models and the blurring of industry boundaries. These companies are currently transitioning from legacy SAP ERP systems or other third-party ERP systems to SAP S/4HANA Cloud. This transition involves undertaking substantial transformation programs, and consequently, the initial phase or early stages of transformation present ideal opportunities to reassess changes in business models and the necessary business transformations. Such reassessment should be incorporated into the planning process for this transformation.

6.1.2 IT Context

With the advent of digital technologies, the role of IT has undergone a significant transformation. Traditionally, IT has been viewed as a support function, responsible for managing and maintaining the organization's technology infrastructure. However, in recent years, IT has emerged as a critical enabler of business transformation, playing a pivotal role in driving innovation, efficiency, and growth.

Historically, IT departments were primarily focused on providing technical support, managing hardware and software systems, and ensuring data security. Their main responsibility was to keep the lights on and ensure that the organization's technology infrastructure was operational. While this role was essential for the smooth functioning of day-to-day operations, it often limited the involvement of IT in strategic decision-making and business transformation initiatives.

In today's digital era, however, the role of IT has expanded beyond its traditional boundaries. IT is now seen as a strategic partner that is actively involved in shaping the organization's overall business strategy and driving transformational change. IT has the potential to unlock new opportunities, improve operational efficiency, enhance customer experiences, and create new revenue streams.

Contrary to the perception that IT may hinder business transformation, it's increasingly recognized as a key accelerator of transformational initiatives. IT can provide the necessary tools, technologies, and capabilities to enable organizations to adapt to changing market dynamics and seize new opportunities. By leveraging emerging technologies such as artificial intelligence (AI), machine learning, cloud computing, and big data analytics, IT can help organizations gain a competitive edge and drive innovation.

While the new role of IT in business transformation offers immense potential, it also presents challenges that need to be addressed. These challenges include aligning IT initiatives with business goals, managing the complexity of technology integration, ensuring data privacy and security, and developing the necessary skills and capabilities within the IT workforce.

Aligning the business with IT is essential for driving business transformation. It ensures that technology investments are aligned with business objectives and enables effective collaboration between business and technology teams. By aligning business and IT strategies, organizations can leverage technology to streamline operations, enhance efficiency, and deliver innovative solutions that drive business growth and success in today's digital era.

In the same way the role of IT has changed in the context of business transformations, the role of an enterprise architect has significantly evolved in recent years. Traditionally, enterprise architects focused on designing and managing the technical aspects of an organization's IT infrastructure. However, with the increasing complexity and interdependence of business and technology, the role of an enterprise architect has expanded to encompass a broader set of responsibilities.

This change also requires enterprise architects to have or obtain a new set of capabilities so that they can effectively lead business transformations. To lead successful business transformations, enterprise architects must now possess strong strategic thinking skills, the ability to align technology with business objectives, and a deep understanding of organizational dynamics. This shift in the architect's role is necessary because business transformations involve not only technological changes but also changes in business models, processes, and culture. By embracing this change, enterprise architects can effectively guide organizations through complex transformations, ensuring that technology initiatives are aligned with business goals and driving sustainable growth.

Main Tasks of the Enterprise Architect in Business Transformations

The classic role of the enterprise architect is evolving in the context of business transformation. Enterprise architects must now have a broader focus on strategic alignment, innovation, and change management. The following list highlights key tasks enterprise architects now undertake in driving successful business transformations:

1. **Aligning business and IT strategy**
   The enterprise architect plays a crucial role in aligning the business strategy with the IT strategy. This involves understanding the goals and objectives of the organization and identifying how technology can support and enable those goals. The enterprise architect works closely with business stakeholders and IT leaders to ensure that the IT strategy is aligned with the overall business strategy, helping to drive innovation, efficiency, and competitive advantage.
2. **Developing business and IT architecture**
   Another key task of the enterprise architect is to develop and maintain the business and IT architecture. This involves understanding the current state of the organization's business capabilities and processes, information systems, and technology infrastructure, and it also involves designing a future state that aligns with the business strategy. The enterprise architect creates blueprints and models that capture the organization's business capabilities, processes, data, applications, and technology components, ensuring that they are well documented and communicated to stakeholders.
3. **Developing transformation roadmap**
   The enterprise architect is responsible for developing a transformation roadmap that outlines the steps and initiatives needed to achieve the future state architecture. This involves identifying the key projects, initiatives, and investments required to close the gaps between the current and future state. The enterprise architect collaborates with business and IT stakeholders to prioritize and sequence these initiatives, considering factors such as business value, technical feasibility, and resource constraints.
4. **Establishing a governance model**
   Enterprise architects play a critical role in establishing a governance model for

business transformations. This involves defining the processes, roles, and responsibilities for decision-making, oversight, and control throughout the transformation journey. The enterprise architect ensures that there's a clear framework in place for guiding and monitoring the implementation of the architecture, ensuring that it remains aligned with the business strategy and objectives. This includes establishing governance boards, defining decision-making criteria, and implementing review processes to ensure compliance with architectural standards and principles.

5. **Aligning with adjacent functions**
   Enterprise architects collaborate and align with adjacent functions such as value management, organizational change management, risk management, and other relevant disciplines. They work closely with these functions to ensure that the business transformation initiatives are effectively planned, communicated, and executed. The enterprise architect provides architectural guidance and support to these functions, helping them understand the impact of the transformation on their respective areas and ensuring that their activities are aligned with the overall architecture and strategy. This collaboration helps to ensure a holistic and integrated approach to business transformations.

6.2 Problem Statement

In the rapidly evolving landscape of the automotive industry, Fusion Motors, a venerable (and fictitious) automotive original equipment manufacturer (OEM) known for its reliable combustion engine vehicles, stands at a critical juncture. Faced with increasing pressure from environmental regulations, shifting consumer preferences toward electric vehicles, and the disruptive impact of technological advancements, Fusion Motors must undergo a profound transformation to stay competitive. This chapter delves into the strategic overhaul of Fusion Motors as it transitions from its traditional manufacturing roots to embrace a new, sustainable business model centered on electric mobility, autonomous driving technologies, and digital services. By exploring the challenges, strategies, and outcomes of this pivotal shift, we gain insights into the complexities and opportunities inherent in redefining a legacy automotive company for the future.

6.2.1 Company Profile

Fusion Motors traditionally excelled in producing internal combustion engine cars and selling them through dealership networks, but now, it seeks to transition toward electric vehicles, direct selling, and subscription-based models. We'll describe the envisioned transition from the traditional business model to the new business model in this section.

Fusion Motors Traditional Business Model

Fusion Motors is a leading automotive company that follows a typical production process for cars with internal combustion engines. The process begins with design and engineering, which determine the car's specifications and features. Next, the company procures raw materials and components from various suppliers. These materials are then assembled on an assembly line, where specialized workers and automated machinery work in tandem to build the car. The assembly process involves tasks such as welding, painting, and installing components like the engine, transmission, and electrical systems. Quality control checks are conducted at multiple stages to ensure that the car meets the company's standards, and finally, the completed cars are shipped to dealerships for sale to customers.

The traditional business model of Fusion Motors involves selling cars through a network of dealerships to end customers. Fusion Motors designs, manufactures, and markets the vehicles, while the dealerships act as intermediaries between Fusion Motors and the customers. Fusion Motors establishes agreements with dealerships, granting them the right to sell their vehicles in specific territories. Dealerships purchase the vehicles from Fusion Motors at wholesale prices and then sell them to customers at retail prices, earning a profit margin. Fusion Motors provides support to dealerships in terms of marketing, training, and after-sales service. This model allows Fusion Motors to focus on manufacturing, while dealerships handle the sales and distribution aspects, creating a symbiotic relationship between the two entities.

The aftersales business model of Fusion Motors typically operates by providing a range of services and products to customers after they have purchased their vehicles. This includes regular maintenance and servicing, repairs, warranty support, and the sale of spare parts and accessories. Fusion Motors has also established a network of authorized service centers and dealerships to ensure convenient access to these services for their customers. Additionally, Fusion Motors may offer extended warranty packages, roadside assistance programs, and customer support helplines to further enhance the ownership experience. The aftersales business model aims to foster customer loyalty, satisfaction, and retention by providing reliable and efficient support throughout the lifespan of the vehicle.

Fusion Motors' New Business Strategy

The traditional business model of Fusion Motors is undergoing significant changes due to several factors. Firstly, there's a growing customer demand for sustainability, with consumers increasingly seeking environmentally friendly transportation options. Secondly, changing global legal frameworks are pushing for a shift toward electric vehicles, with governments implementing stricter regulations and incentives to promote these vehicles' adoption. Additionally, the market is witnessing increasing competition, not only from traditional automakers but also from new players in the industry, such as tech companies and startups. Lastly, new mobility concepts, such as ride-sharing and autonomous vehicles, are reshaping the way people perceive and use cars, requiring car manufacturers to adapt their business models to cater to these evolving trends. As a result, car manufacturers are compelled to embrace sustainability, invest in electric vehicle technology, explore new business models, and collaborate with other stakeholders to stay competitive in this rapidly changing landscape.

For all these reasons, Fusion Motors has decided to implement a new business strategy called AutoVision, which focuses on the following strategic priorities:

- **Obtaining new customers through a shift to a direct-selling model**
  - Establishing a direct-to-consumer sales channel to eliminate middlemen and enhance the customer experience
  - Cutting out dealership markups to offer competitive pricing and increased affordability
  - Implementing online platforms and mobile apps to facilitate seamless purchasing and customization experiences
  - Offering virtual showrooms and test drives to provide customers with convenient and personalized buying options
- **Driving new mobility by introducing a subscription model for ultimate flexibility**
  - Offering subscription-based ownership models to cater to changing consumer preferences
  - Providing customers with the option to switch between different vehicle models, based on their needs and preferences
  - Including maintenance, insurance, and other services in the subscription package for a hassle-free ownership experience
  - Allowing customers to easily upgrade to newer models as they become available, ensuring they always have access to the latest technology and features
- **Embracing sustainability with a shift in focus to battery electric vehicles**
  - Investing in research and development to produce a diverse range of high-quality and affordable battery electric vehicles (BEVs)
  - Establishing a comprehensive charging infrastructure to alleviate range anxiety and encourage widespread adoption of BEVs
  - Collaborating with renewable energy providers to ensure the use of clean energy sources for charging BEVs
  - Offering attractive incentives and benefits, such as tax credits or reduced charging costs, to customers who choose BEVs
- **Putting software at the heart of Fusion Motors' business model**
  - Developing advanced software systems to enhance vehicle performance, safety, and connectivity

- Introducing over-the-air updates to continuously improve and add new features to vehicles
- Leveraging AI and machine learning to provide personalized driving experiences and predictive maintenance
- Partnering with technology companies to integrate smart home and IoT capabilities with vehicles for seamless connectivity

6.2.2 IT Landscape

Fusion Motors operates three SAP ERP systems in three different regions: AMERICAS, EMEA, and APJ. These SAP ERP systems, as depicted in Figure 6.3, are based on an outdated template that was built over fifteen years ago and is heavily customized. The template covers all necessary processes, including source-to-pay, plan-to-fulfill, lead-to-cash, and finance. The recruit-to-retire process has been implemented in a non-SAP solution, and an internally developed solution is used for acquire-to-decommission. Additionally, various other SAP and non-SAP solutions are utilized as shown in Figure 6.3.

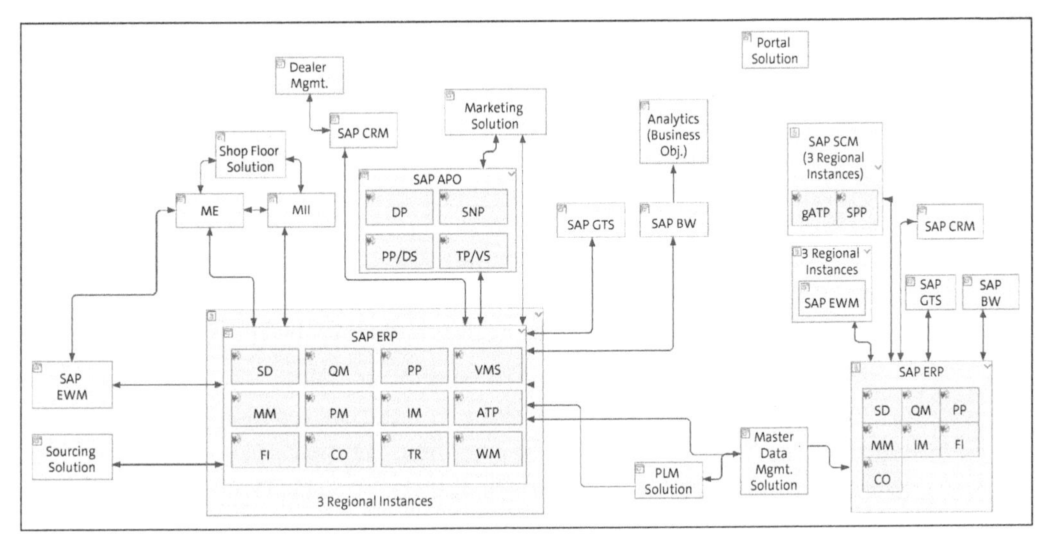

Figure 6.3 As-Is Landscape of Fusion Motors

In addition to these regional SAP ERP systems, there's another dedicated SAP ERP system for aftersales business, and it's primarily focused on the lead-to-cash process for the automotive aftermarket. Other SAP and non-SAP solutions—such as SAP Customer Relationship Management (SAP CRM), SAP Supply Chain Management (SAP SCM) (with global available-to-promise [gATP] and service parts planning [SPP]), and three regional instances of SAP Extended Warehouse Management (SAP EWM) for warehousing and distribution centers—are also important for the aftermarket processes.

One of the main IT challenges currently faced by Fusion Motors is the extensive customization, which has made it difficult for the company to adopt innovations and carry out smooth upgrade projects. Another challenge is integrating SAP and non-SAP solutions, particularly in terms of performance and the lack of standardized interface technologies and monitoring capabilities. From a business perspective, implementing new requirements has been extremely challenging in the past, and the high level of customization in the system has made it nearly impossible to leverage standardized best practice processes from SAP. As a result, new business requirements have had to be implemented mostly through custom development, leading to low agility in responding to changing requirements or legally mandated changes.

Fusion Motors has previously explored options for transforming to SAP S/4HANA. Due to budget constraints, the company took a brownfield approach, which involved converting the SAP ERP systems to SAP S/4HANA and only replacing and replatforming applications that are now embedded in the SAP S/4HANA core (e.g., SAP EWM). However, when the new CEO took over at Fusion Motors and introduced the AutoVision strategy, it became clear that the legacy landscape with all its pain points could not support the new strategy using a brownfield approach. Nonetheless, cost savings remain an important aspect of Fusion Motors' strategy, so the company must reach a compromise that supports the significant shift in the business model and resulting business requirements while also maximizing cost savings during the transformation process.

6.3 Applying the Enterprise Architecture Transformation Approach

In Chapter 2, we discussed SAP EA Framework, which consists of five pillars: methodology, reference content, tooling, practice, and services. Now, we'll showcase the practical application of the SAP EA Framework using the aforementioned use case.

The initial stage requires us to pinpoint the specific modules within enterprise architecture planning that we'll implement, along with the corresponding deliverables for each module. Following that, we need to determine the appropriate sequence for iterating over these various modules. In the final stage, we must establish a comprehensive plan for the enterprise architecture tasks to be undertaken, including identifying any additional roles that we need to incorporate into the enterprise architecture planning process.

Fusion Motors' primary objective is to comprehend the influence of the new business strategy on the enterprise architecture and the overall planning of the transition to SAP S/4HANA. The cost-saving aspects are explored in Chapter 9, specifically in the total cost of ownership (TCO) reduction use case. Consequently, we direct our attention in this chapter toward the aspect of business transformation.

When considering the modular approach of enterprise architecture planning, we must prioritize the architecture vision and business architecture domains (see Figure 6.4).

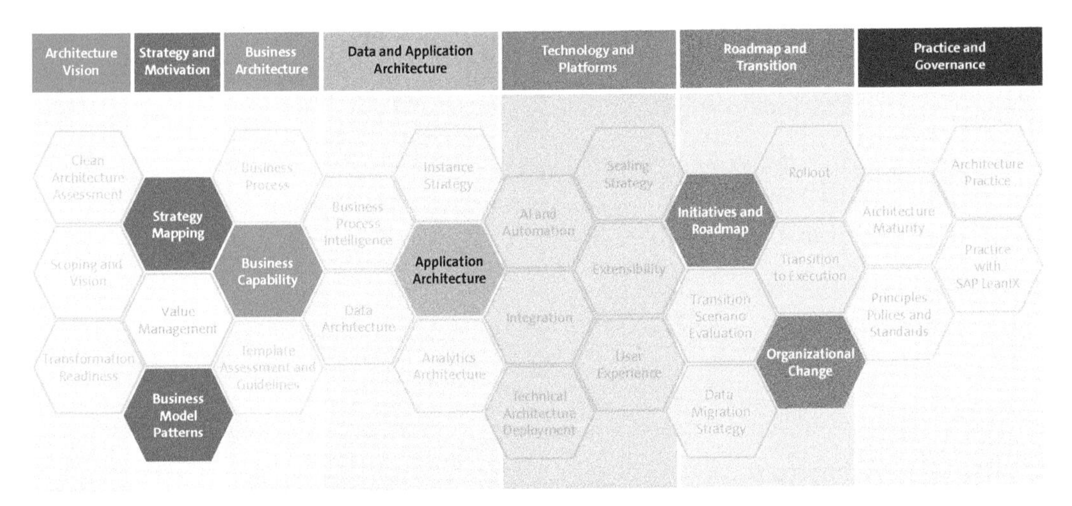

Figure 6.4 Modules in Scope for Fusion Motors

Let's apply these modules to our Fusion Motors use case:

- **Business model patterns and strategy mapping**
  Fusion Motors is on the brink of a significant business model transformation that will have far-reaching effects on its people, technology, and processes. To begin, we must incorporate the business strategy and the changes to the business model into our enterprise architecture planning.
- **Business capability**
  From a business architecture standpoint, considering the strategic changes, we need to examine the necessary business capabilities. Initially, we will not focus on the modules for business process and template assessment and guidelines, as we can address them in a later iteration of our enterprise architecture planning.
- **Application architecture**
  However, we must discuss the impact on the target architecture based on the selected business capabilities. While other modules of the application and data architecture are important, they're not required in the first iteration. The same applies to the technology and platforms modules.
- **Initiatives and roadmap**
  Once we've developed an initial target architecture that incorporates the required changes aligned with the new business strategy, we can begin drafting the first roadmap for Fusion Motors. We can consider other modules from the same domain in subsequent iterations.
- **Organizational change management**
  Since Fusion Motors already has a robust enterprise architecture practice in place, we won't include governance modules related to the enterprise architecture practice in our planning. However, we'll select *organizational change* as a crucial aspect

because the transformation will bring about significant changes to the operating model of Fusion Motors, and we need to integrate organizational change management (OCM) into the transformation approach from the outset.

Additional iterations could then put focus on additional important aspects of the application architecture and data and analytics. For example, iteration 2 would be an architecture deep dive, with the following modules in focus:

- Instance strategy
- Integration
- Scaling strategy

Iteration 3 would be about data and analytics, with the following modules in focus:

- Data architecture
- Analytics architecture
- Data migration strategy

We should plan other iterations over additional modules in an agile way after we've completed the first iterations. In addition, in each iteration, we may need to look again at already delivered modules and enhance them based on the results of the current iteration.

In addition to determining the modules and iterations, we must determine the roles necessary to execute the modules. Specifically, for the strategy and business domain, it's crucial to have input from various business stakeholders because IT can't handle this responsibility by itself. These roles typically include functional domain leads, business process owners, and project and program managers. On the IT side, we would involve IT functional domain leads, IT architects, and IT managers. Furthermore, for specific topics, it may be necessary for us to consult additional experts with expertise in the relevant functional or solution areas.

6.4 Key Deliverables and Outcomes

In this section, we'll outline the main deliverables and results from the modules that we discussed in the preceding section. These deliverables are in fact produced through a series of workshops involving various participants and roles from the organization, as previously mentioned. We'll provide a brief explanation of each deliverable's creation process and the necessary prerequisites, and we'll also showcase the final outcome or a sample of the ultimate deliverable.

6.4.1 Strategy Mapping

In this section, we'll delve into the *strategy mapping* approach, starting with an analysis of the organizational map to gain a clearer understanding of Fusion Motors' structure. Next, we'll examine the business context in which Fusion Motors operates, identifying key stakeholders and actors within its ecosystem. Finally, we'll develop a strategy map that connects strategic priorities, business goals, and value drivers.

Organizational Structure

Understanding the *organizational structure* of a company is crucial in enterprise architecture planning. The organizational structure provides the blueprint of the internal framework of a company, indicating how different departments and teams are interlinked. This insight is essential for developing an enterprise architecture that aligns with the existing organizational hierarchy and workflows. By acknowledging this structure, we can design the architecture to streamline processes, enhance communication, and foster efficient decision-making processes, thereby improving overall operational efficiency.

In addition, understanding the organizational structure allows enterprise architects to identify the key stakeholders involved in various processes. These stakeholders play a significant role in the acceptance and successful implementation of the designed architecture. Hence, understanding their roles, responsibilities, and pain points can enable architects to design solutions that cater to their specific needs, improving the likelihood of buy-in and successful adoption of the architecture.

Finally, the organizational structure can reveal the strategic direction and goals of the company. The structure often mirrors the company's priorities, whether they are focused on customer service, innovation, operational efficiency, or other objectives. By understanding these priorities, enterprise architects can ensure that the architecture supports these strategic goals and thus facilitates their achievement. This alignment between the enterprise architecture and the company's strategic direction is critical for driving business value and ensuring the long-term success of the architecture.

In practice, almost every organization has a kind of diagram of its organizational structure. It's typically therefore not required, as part of the enterprise architecture planning process, to create such a diagram from scratch. However, it's sometimes difficult to find the right level of abstraction in the available diagrams. A structure that goes into all the details and depicts several levels of hierarchies and roles isn't required for the enterprise architecture planning processes.

Hints for the Organizational Structure

The following points offer key insights into how to derive the organizational structure during enterprise architecture planning:

- **Structure**
  Organizations are generally structured in one of two ways: divisionally or functionally (for more information, refer to Chapter 2). When you're planning for enterprise architecture, we suggest that you start with a divisional structure. This approach will provide you with an initial understanding of the primary business divisions and their respective requirements, which will be crucial as the process progresses.
- **Scope**
  Even though the scope of your current activities is only specific to one business unit and doesn't cover the entire organization, we recommend starting with the full organizational structure. This will enable you later on to depict preliminary dependencies and overlaps with other organizational units (e.g., group level business processes in conjunction with divisional specific processes) in the realm of the business architecture.
- **Iterations**
  You should begin with a broad portrayal of a divisional structure, understanding that it doesn't need to be perfect in the first iteration. As the enterprise architecture planning develops, the organizational structure diagram will naturally progress as well. Initially, concentrate on the components of your current iteration, and during each subsequent iteration, reassess the diagram and supplement it with additional structuring elements as needed.

In our use case, Fusion Motors has a global organizational structure that is divided into various divisions to cater to different aspects of its business. This structure enables the company to effectively manage its operations in different regions and product categories, and it underscores the company's commitment to delivering high-quality products and services to its customers worldwide.

We can depict the current organizational structure of Fusion Motors in a simplified way as shown in Figure 6.5.

<table>
<tr><td colspan="10">Group</td></tr>
<tr><td colspan="7">Passenger Cars</td><td>Luxury Cars</td><td>Battery Production</td><td>Software Delivery</td></tr>
<tr><td colspan="3">EMEA</td><td colspan="2">APJ</td><td colspan="2">AM</td><td>Global</td><td>Global</td><td>Global</td></tr>
<tr><td>Germany</td><td>South Africa</td><td>England</td><td>China</td><td>India</td><td>Brazil</td><td>Mexico</td><td>Italy</td><td>France</td><td>Germany</td></tr>
</table>

Figure 6.5 Fusion Motors' Organizational Structure and Main Locations

The company's divisional structure includes a passenger cars division that is further divided into three regions: EMEA (Europe, the Middle East, and Africa), APJ (Asia-Pacific and Japan), and AM (Brazil and Mexico). This regional division allows the company to cater to the specific needs and preferences of different markets, taking into account cultural, economic, and regulatory differences. It also enables the company to manage its operations more efficiently and effectively, as each regional division can focus on its own set of challenges and opportunities.

In addition to the passenger cars division, Fusion Motors has a battery production division and a software delivery division, both of which operate globally. The battery production division is responsible for the production of batteries for the company's electric vehicles and for ensuring the quality and safety of the products. The software delivery division, on the other hand, is tasked with developing and maintaining the software used in the company's vehicles, including software for navigation, entertainment, and vehicle control systems. These divisions highlight the company's commitment to technological innovation and sustainability.

Business Context Model

A *business context model diagram* is an essential tool in enterprise architecture planning. It's designed to present a high-level snapshot of an organization's interactions with the external environment, and it illustrates the various entities that the organization interacts with and the flow of information and materials between these entities and the organization. In situations where the architectural scope doesn't encompass the entirety of the business, the model proves useful in defining the scope and delineating the interactions between entities within and outside this scope in the business's ecosystem.

The business context model diagram is relevant in the following areas of enterprise architecture planning:

1. **Strategic overview**
   The diagram is instrumental in providing strategic management leaders with a comprehensive view of interactions, roles, and relationships within the value system. It assists them in gaining a holistic understanding of the organizational landscape.
2. **Operational insight**
   Those responsible for operational development can leverage the diagram to help them identify and comprehend underlying processes, roles, and transactions, giving them a deeper insight into operational intricacies.
3. **External engagement perspective**
   The diagram offers a broad perspective on how the enterprise interacts with the external environment. It emphasizes the exchanges with business partners, including the flow of products, data, financial transactions, and contractual obligations.

4. **Interface considerations**
   The diagram highlights specific organizational interfaces that require scrutiny during the design and deployment phases, thus assisting in risk mitigation.
5. **Integration with business model**
   The diagram can be incorporated into a broader business model, providing clarity on the architecture of the value system and thereby enhancing the understanding of the business's functional dynamics.

Hints for the Business Context Model

It is essential that you follow these guidelines for developing a business context model in the context of enterprise architecture planning:

1. Begin by identifying the primary actors in the business context, then progressively add the interdependencies and linkages among them. However, avoid getting too detailed in the initial stages. For purposes of enterprise architecture planning, a high-level representation is usually sufficient.
2. Don't concentrate solely on the current business context; take into account potential future changes to the model. Depending on the business strategy and evolving business models, entirely new actors may enter the business context.
3. Highlight changes in the business context that are known today because this will be valuable input for the later architecture planning activities in the business architecture domain.
4. Utilize generative AI, which can provide an initial understanding of the key players in the industry that you should consider during the business context modeling process. While generative AI may not be able to create an exact model, it can offer valuable insights, especially when there's no existing business context model for the organization.

Fusion Motors is embedded into the business context as depicted in Figure 6.6. The light green boxes represent relationships that are likely to change as a result of Fusion Motors' change in business model.

Currently, the main customers of Fusion Motors are corporate car fleets and dealerships. Private customers purchase their cars from Fusion Motors through dealerships, but with the company's intention to implement a direct-selling business model, this relationship will change. Private customers will also become direct customers of Fusion Motors; they will place orders for cars or subscription-based services. This change in the business context will also require a new relationship with the dealerships.

Fusion Motors currently relies on various component suppliers, but significant changes are expected in this area as well. A paradigm shift is anticipated in the existing relationship with the electronics supplier because Fusion Motors plans to develop its own software, with the supplier potentially only providing electronic components without software. Additionally, Fusion Motors will need to establish a completely new supply chain for battery production. The main supplier in this case will provide battery cells, which Fusion Motors will then assemble into batteries at its production facility in Germany.

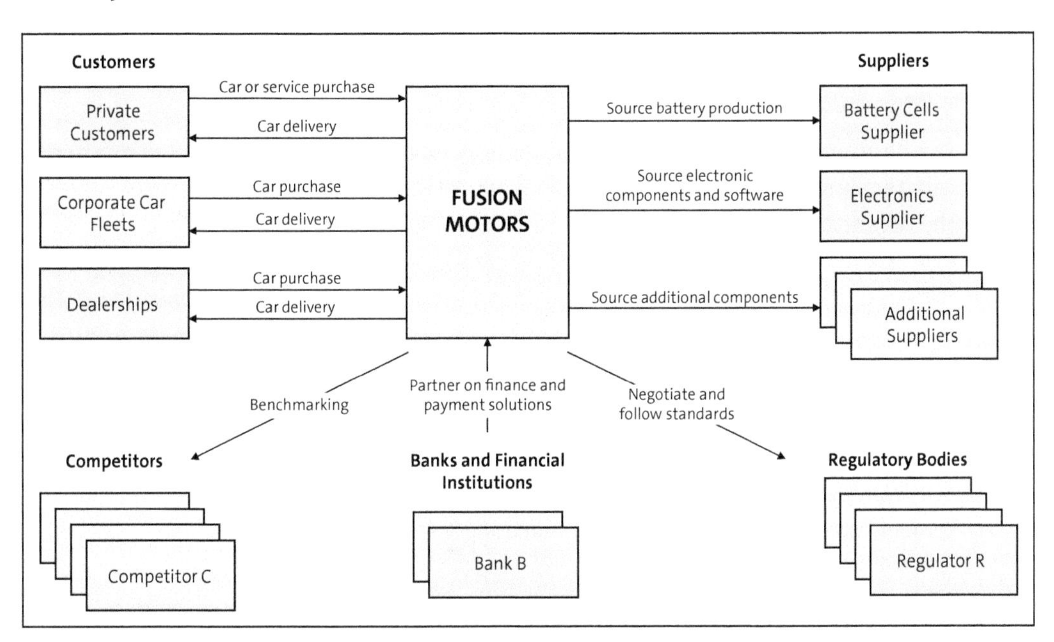

Figure 6.6 Business Context of Fusion Motors
(Light Green = Change Expected; Blue = No Change Expected)

Regulatory bodies also play a crucial role in the business context. This is particularly true for large regions such as the European Union, where it is expected that only BEVs will be available for sale by 2035. Fusion Motors must maintain close contact with regulatory bodies to comply with standards and regulations while also actively participating in shaping those standards and regulations as much as possible. New regulations may lead to changes in the business model or impact roadmap items.

The company's relationship with financial institutions and banks may also change as a result of the shift toward direct selling and subscription-based services. Fusion Motors may form partnerships with these institutions to offer customized payment solutions.

Lastly, Fusion Motors' relationship with competitors is expected to remain unchanged. Fusion Motors will continue to compare its own cars with those of competitors in terms of features and price. Currently, joint ventures are not being considered by Fusion Motors management.

Strategy Map

The *strategy map* is an important artifact of enterprise architecture planning as it provides a visual representation of the organization's strategic priorities and how they are linked to business goals. By clearly outlining the strategic priorities, the strategy map helps align the efforts of different departments and teams toward a common objective. It allows stakeholders to understand the overall direction of the organization and how their individual efforts contribute to larger goals. This alignment is crucial for effective enterprise architecture planning as it ensures that the architecture decisions and investments are in line with the organization's strategic priorities.

Furthermore, the strategy map helps identify and prioritize value drivers that are critical to achieving business goals. Value drivers are the key objectives that create value for the organization and its stakeholders. By mapping these value drivers to strategic priorities, the strategy map enables the organization to focus its resources and efforts on the areas that have the greatest impact on achieving the organization's goals. This helps in making informed decisions about resource allocation, investment priorities, and technology adoption within the enterprise architecture planning process.

Hints for Creating a Strategy Map

A strategy map is a crucial tool for aligning business strategies with architecture planning. The following are key tasks for creating a strategy map that you should keep in mind:

1. Determine the extent of the strategy map exercise. You can develop a strategy map for different levels within the organization, such as the corporate level, regional level, business unit level, or line of business level. Depending on the overall scope of the enterprise architecture planning, you can establish the strategy map accordingly with the same scope.
2. Examine the organization's existing strategy documents. Some information—such as corporate reports, websites, and interviews with top-level executives—may be publicly accessible, and internal strategy departments may provide additional information.
3. Utilizing the shared and available information, create an initial draft of the strategy map. Then, conduct stakeholder interviews to validate the strategy map and incorporate the input gathered from the stakeholders.
4. Increase awareness among business departments, such as the strategy department, about the significance of the strategy map and its connection to enterprise architecture planning. It's crucial for the organization to understand that enterprise architecture planning isn't solely a technical task carried out by IT. By utilizing the strategy map, you can ensure that the company's strategic priorities and goals are incorporated into comprehensive transformation planning.

5. Note that the strategy mapping exercise doesn't involve defining or reviewing an organization's strategy. Its purpose is to help the enterprise architect and all stakeholders involved in planning to understand the strategy so that it can be mapped to future enterprise architecture planning activities. As a result, creating a strategy map usually doesn't require a significant amount of effort, and you can typically create it after conducting a few brief stakeholder interviews.

Fusion Motors recently published a new strategy document called AutoVision, which outlines the strategic priorities in Section 6.2. At the corporate level, Fusion Motors has a dedicated strategy department led by a chief strategy officer (CSO), who is responsible for shaping the company's strategic direction. After carefully reviewing the AutoVision strategy, we approached the strategy department to introduce the scope of the enterprise architecture planning cycle. Department staff agreed that the strategy map should initially focus on the corporate level of Fusion Motors and should consider the overall strategic priorities rather than specific priorities for individual business units or lines of business (LoBs). The strategy department hasn't yet determined how the defined strategic priorities will be broken down for business units, regions, and LoBs, but this can be addressed in a later iteration of the strategy map.

As a first step, we map the strategic priorities and business goals for Fusion Motors and put them into the strategy map, as shown in Figure 6.7.

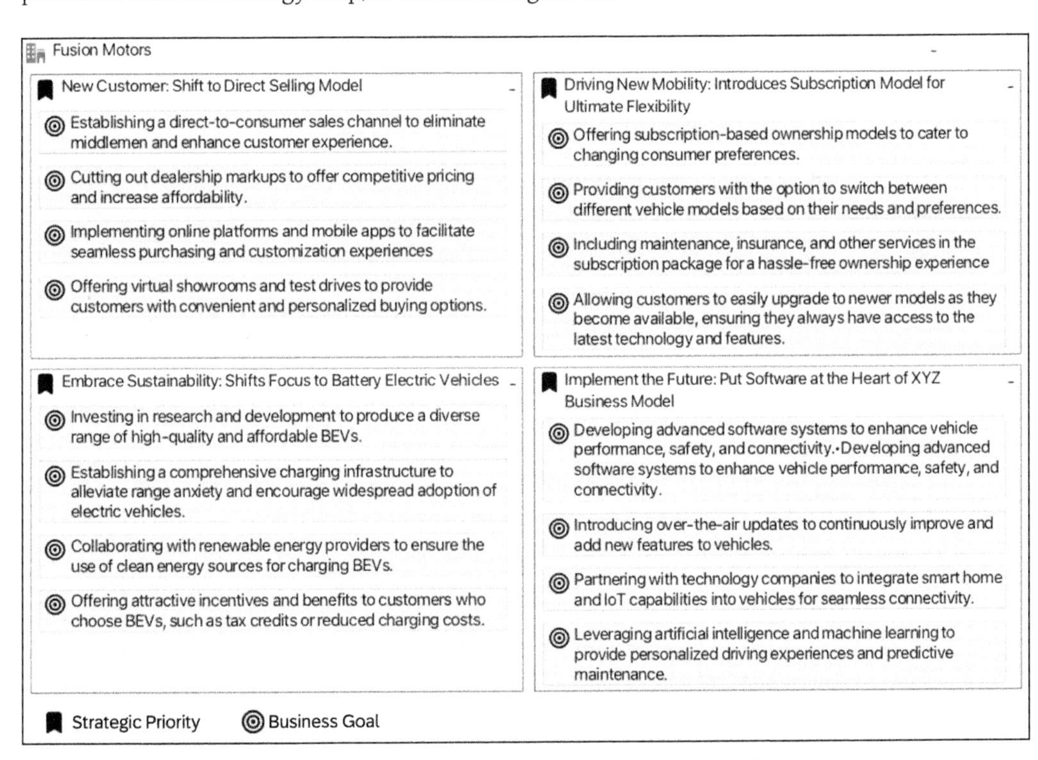

Figure 6.7 Strategy Map Levels 1 and 2: Strategic Priorities and Business Goals

In certain cases, and especially if you do the first iteration of the strategy map yourself, this level 1 and level 2 hierarchy of the strategy map is sufficient. However, we recommend that for most (but not all) important business goals, you should also discuss value drivers and create the strategy map up to level 3. In Figure 6.8, we demonstrate the mapping from one particular strategic priority and also the mapping of value drivers to business goals.

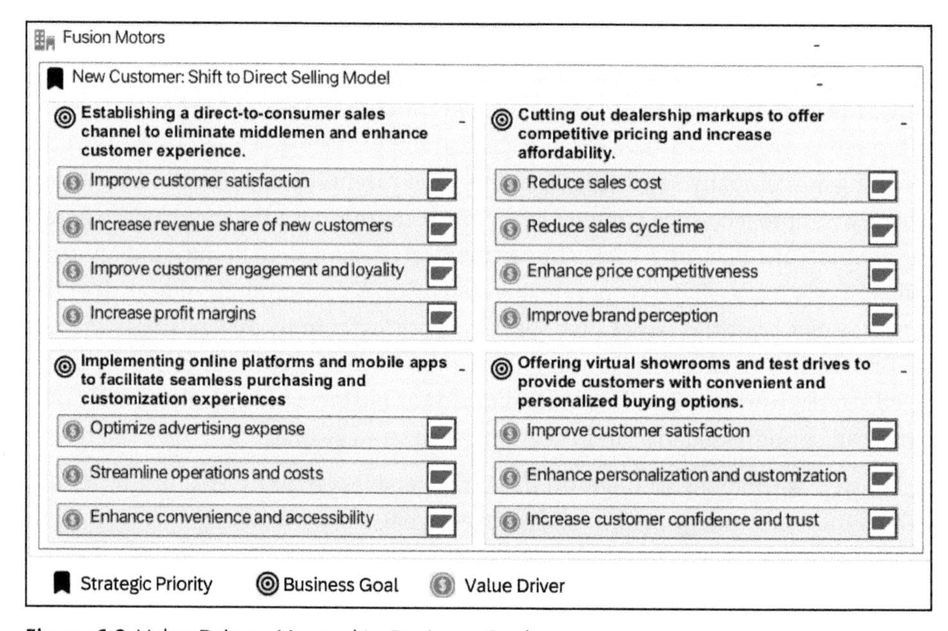

Figure 6.8 Value Drivers Mapped to Business Goals

SAP has developed reference content for value drivers that you can utilize for mapping exercises. The benefit of using SAP reference content is that it automatically maps value drivers to business capabilities in the subsequent section of the business domain. In practice, you can employ both approaches: selecting value drivers from the SAP reference content and using value drivers you derive from the organization's strategy document.

Each value driver can potentially be linked to specific KPIs, each of which has a defined timeframe and includes a specific achievable target. For instance, the value driver **Increase revenue share of new customers** can be linked to a KPI such as **Achieve a 20% increase in revenue share of new customers by 2025**. This mapping of KPIs to value drivers is particularly crucial for defining the value proposition of the transformation later in the process.

6.4.2 Business Model Patterns

Business model patterns are standardized templates or frameworks that describe the key elements and relationships of a business model. They provide a structured approach to understanding and designing business models, allowing organizations to leverage proven strategies and best practices. These patterns help identify the core components of a business model, such as value propositions, customer segments, revenue streams, and cost structure.

The concept of business model patterns has been developed and refined by various researchers and practitioners in the field of business strategy and innovation. While it's difficult to attribute the invention of business model patterns to a single individual or organization, scholars like Alexander Osterwalder and Yves Pigneur have made significant contributions to the field. Their work on the business model canvas (BMC), a popular tool for visualizing and designing business models, has been widely adopted by entrepreneurs and managers worldwide.

Business model patterns are crucial for enterprise architecture planning because they provide a systematic approach to analyzing and aligning the various components of a business model. By using predefined patterns, organizations can avoid reinventing the wheel and leverage successful strategies that have been proven to work in similar contexts. These patterns help organizations identify potential gaps or weaknesses in their existing business models and enable them to make informed decisions about how to improve or innovate their offerings.

The relationship between business model patterns and a strategy map lies in their complementary nature. While business model patterns focus on the structure and components of a business model, a strategy map provides a visual representation of an organization's strategic priorities and the cause-and-effect relationships between them. By using business model patterns in conjunction with a strategy map, organizations can ensure that their business model supports and aligns with their overall strategic priorities.

Hints for Scoping Business Model Patterns

Business model patterns assist in breaking down a business model into standardized components. The following points are important to consider:

1. To help you identify the patterns of business models within the scope, we recommend that you follow the sequence of deliverables outlined in this chapter. You can start with the organizational structure, then consider the business context, and finally examine the strategy map. By following this approach, you'll typically gain enough information to identify the appropriate business model patterns.
2. You should utilize the business model canvas as a tool to map the business model patterns. The business model canvas is a well-known and widely understood tool for representing an organization's business model. While our focus isn't on defining or reviewing the organization's business model, we aim to understand the current business model and any anticipated changes.

3. As with the strategy map, the scope of the business model pattern exercise can vary from the corporate level to specific regional or business unit models. The organizational map that has already been created can assist you in mapping the identified business model patterns to the relevant elements of the organization.
4. You can select a list of business model patterns from the following website: *https://businessmodelnavigator.com/explore*.

Based on the current information and completed deliverables, we've selected the business model patterns for Fusion Motors depicted in Figure 6.9, and we've carefully organized and mapped them according to the structure of the business model canvas. From the list of business model patterns (see the text box for details), we selected the most suitable ones based on our understanding of Fusion Motors' current and future business models. Then, we mapped those patterns to the appropriate sections of the BMC. Integrating the patterns into the canvas helps clarify how they align with and support Fusion Motors' overall business strategy.

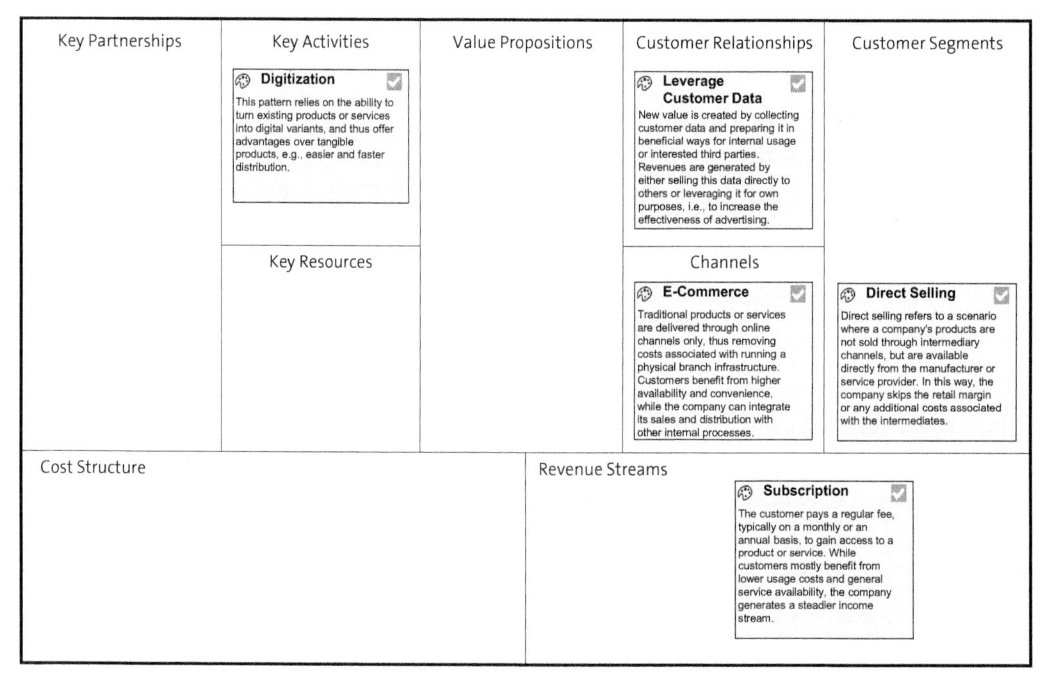

Figure 6.9 Identified Main Business Model Patterns Mapped to the BMC

However, not all business model patterns are applicable to all business units within Fusion Motors. The luxury car division, in particular, has unique customer relationship requirements and will therefore continue with its existing business model of direct sales to customers. However, for passenger cars, a new business model of direct selling will be introduced alongside subscription-based services and the option for customers to purchase or subscribe to cars directly through Fusion Motors' e-commerce shops.

Nonetheless, the business model patterns **Digitization** and **Leverage Customer Data** will be relevant for all units and must be integrated into the core of all business models and processes at Fusion Motors. As a result, we can also align the business model patterns with the organization structure as shown in Figure 6.10.

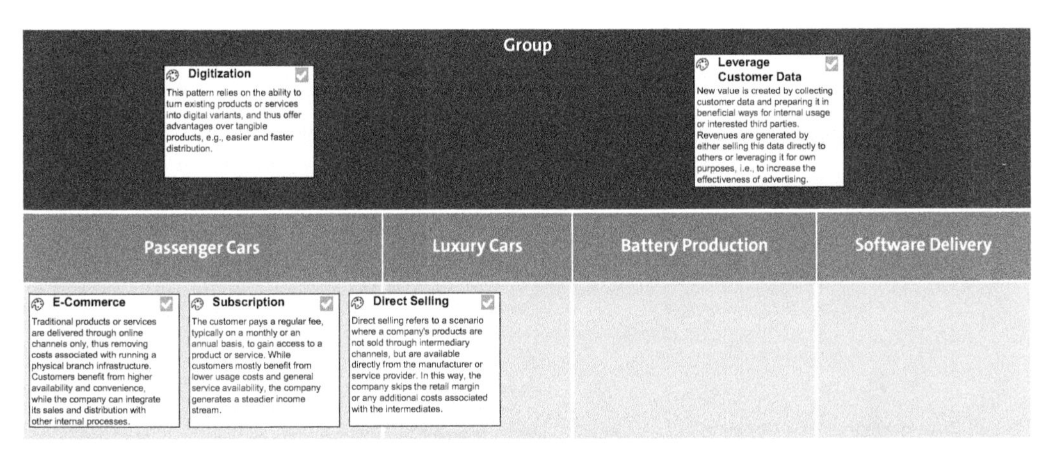

Figure 6.10 Business Model Patterns Mapped to the Organizational Structure of Fusion Motors

6.4.3 Business Capability Map

A *business capability map* provides a high-level view of what a business does and can do, irrespective of how it's done or who does it. This abstraction helps you focus on the strategic aspects of the business, rather than getting involved in operational details. Starting enterprise architecture planning with business capability mapping allows organizations to identify gaps, redundancies, and opportunities for innovation in their abilities, and it allows them to align their IT investments with business strategy. In contrast, focusing on business processes too early can limit this strategic view, as processes are often specific, detailed, and tied to current operational practices. Therefore, a capability-driven approach provides a more strategic, flexible, and future-focused foundation for enterprise architecture planning.

While business process scoping and mapping are crucial and required components of every enterprise architecture planning cycle, our preliminary focus for this iteration is on the business transformation arising from Fusion Motors' corporate strategy. Thus, we've initially opted to concentrate on business capabilities. Nonetheless, it's essential that we address business processes in the subsequent iteration, ensuring that we accurately map them to relevant aspects of both the business strategy and the model.

Next, we'll trace the business capabilities derived from the strategy map and the business model patterns, and we'll take a closer look at assessing business capabilities.

Business Capabilities Derived from the Business Strategy Map

We can begin creating the business capability map by using our strategic deliverables: specifically, the strategy map and the business model patterns. As mentioned earlier, SAP provides reference content that establishes the connection between value drivers in the strategy map and business capabilities. Therefore, once we've created the strategy map and identified the value drivers, we can immediately determine the business capabilities that support these value drivers, as well as the business goals and strategic priorities of the organization, as shown in Figure 6.11.

Figure 6.11 Business Capabilities Derived from Strategy Map

Business Capabilities Derived from Business Model Patterns

We can apply the same logic to business model patterns. SAP has also mapped the available business model patterns to its reference content of business capabilities, and each specific business model pattern is assigned to the business capabilities at level 3, based on a contribution score. This score indicates the extent to which a particular business capability is required to support the selected business model pattern, on a scale from 1 (low) to 4 (high). The score is then aggregated to higher levels of the business capability map to identify the most relevant business areas for the business model pattern. Figure 6.12 shows the business capabilities filtered only for highest score of 4.

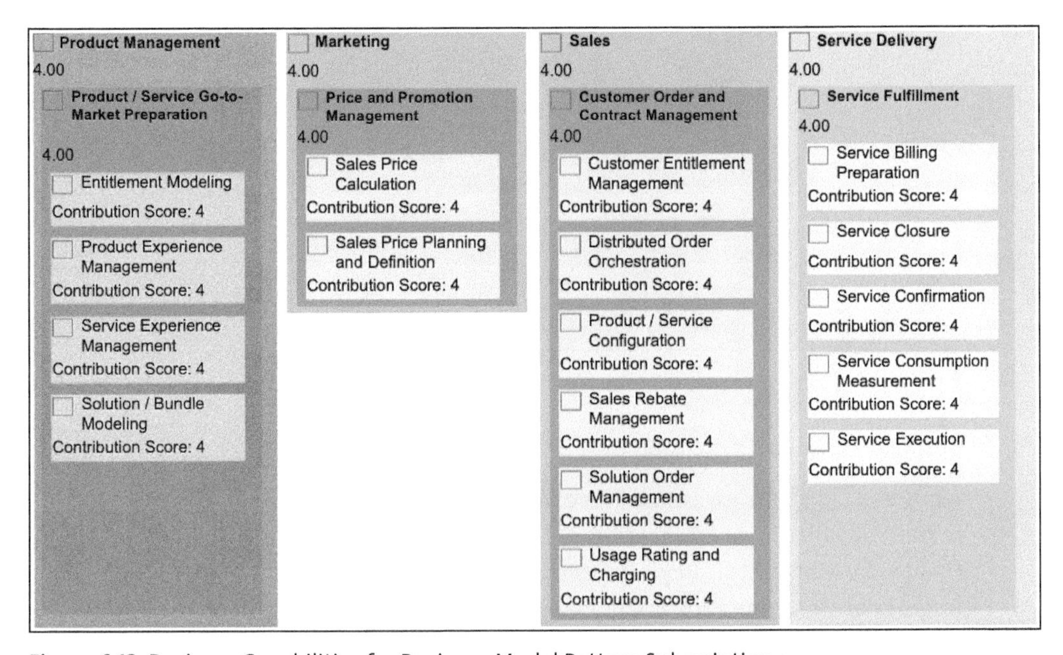

Figure 6.12 Business Capabilities for Business Model Pattern Subscription

Based on the strategy map and business model patterns, we've derived an initial set of business capabilities, but we haven't completed the overall business capability mapping. With this approach, we've only selected the capabilities that are of the highest strategic importance to the organization. To fully complete the business capability mapping exercise, we also need to consider nondifferentiating business capabilities. However, since our focus here is on business transformation, we'll only concentrate on the strategic business capabilities in this case study. By integrating the methodologies of business model patterns and strategy maps, we can construct a comprehensive business capability map that is derived from Fusion Motors' strategy, as shown in Figure 6.13.

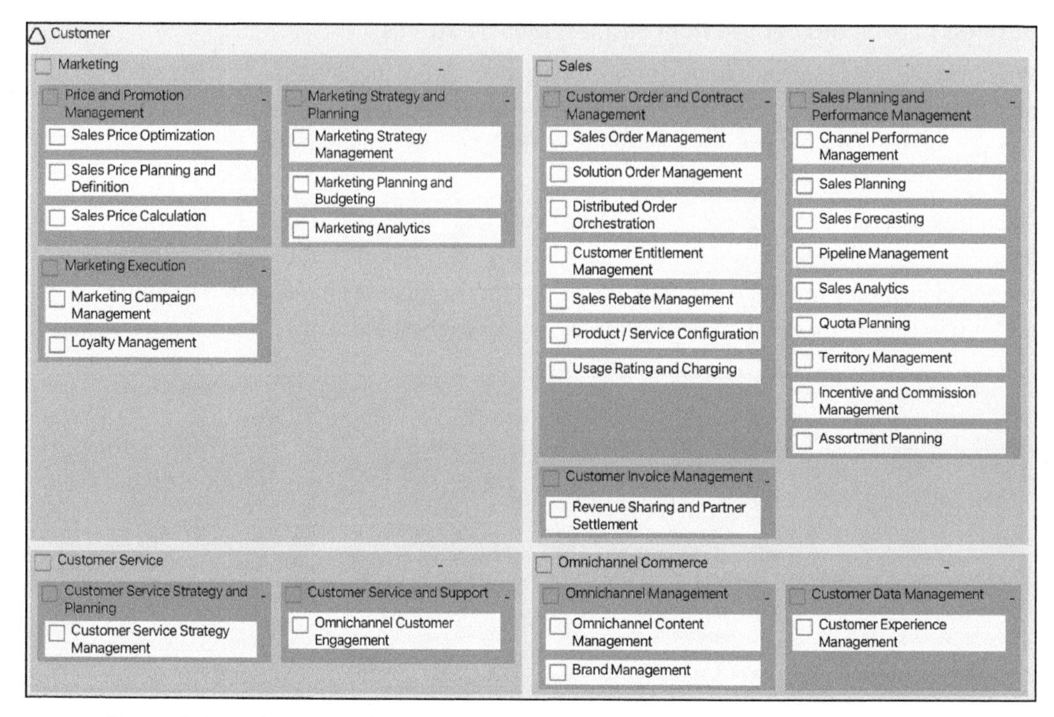

Figure 6.13 Business Capability Map for Customer Enterprise Domain

Business Capability Assessment

A *business capability assessment* involves analyzing the identified business capabilities in more detail, using certain criteria. This assessment helps prioritize the capabilities that have the highest strategic impact, which we can then use as input for designing the target architecture and roadmap.

The assessment also directs enterprise architects to create the target architecture and roadmap in the following ways:

- **Target architecture**
  They should clearly connect prioritized capabilities to the designed target architecture.
- **Initiatives**
  When planning initiatives for customers, they should take into account the prioritization of capabilities and align them with the prioritization of initiatives.
- **Roadmap**
  When creating the roadmap, they should implement the prioritized business capabilities in a timely manner.

Business Capability Assessment Use Cases

For strategic decisions, make your assessment based on the following questions:

- What are our strongest and weakest capabilities?
- What capabilities provide strategic differentiation?
- What capabilities can be leveraged in new markets?
- Where should we invest our resources?
- Are we investing in the right areas (where we have challenges and where we see strategic impact)?
- Are we aligned with all stakeholders on the strategic priorities of the organization?
- Are we investing according to our strategic goals?
- Where can technology add more strategic value?
- Where can we use technology to lower costs?

For operational efficiencies, make your assessment based on the following questions:

- Where do we have multiple processes and technologies supporting one capability?
- Are we investing to resolve current pain points?
- Which capabilities are costing too much to support?
- Which capabilities should we outsource?
- Where do we need more employee education?
- Where do we need process improvement?

Using the initial business capability map we created from the strategy and business model patterns, we conduct a business capability assessment for Fusion Motors. This assessment evaluates the readiness and benefits of the business, and we utilize a specific assessment grid for this purpose. During a collaborative discussion with key stakeholders in the company, we place the identified business capabilities on the assessment grid, as shown in Figure 6.14.

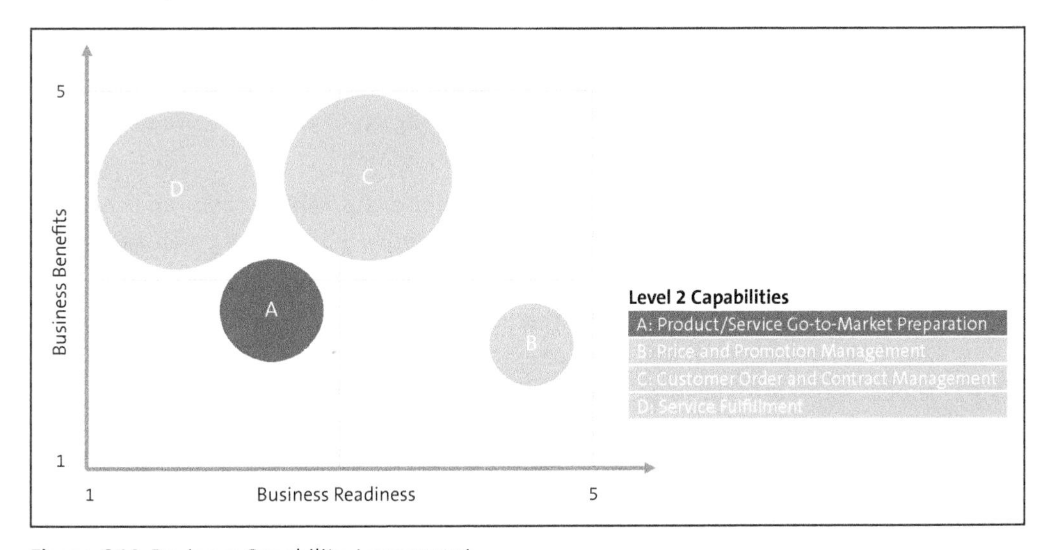

Figure 6.14 Business Capability Assessment

During the assessment, we utilize the level-2 capabilities of the business capability map, which we derived from the scoping of the business model pattern. We plot business readiness on the *x* axis and business benefits on the *y* axis to evaluate the capabilities. The size of the circle represented the geographic impact of each capability. From the perspective of the assessment grid, the critical business capabilities are located in the upper-left and upper-right corners. For Fusion Motors, we identified service fulfillment and customer order and contract management as critical capabilities due to their high business benefits and contributions, despite their lower level of business readiness. Additionally, these two capabilities have a significant geographic impact, so we should give careful consideration to these capabilities during the roadmap planning process.

6.4.4 Application Architecture

In the preceding sections, we've developed the business architecture for Fusion Motors by connecting the company's corporate strategy to the strategy map and business model patterns, thus identifying the company's business capabilities. In this section, we'll illustrate how we can now derive the desired application architecture based on the business architecture. This will allow us to identify the key components of the target application architecture that will assist Fusion Motors in implementing its strategy from an IT standpoint.

To begin with, we align the applications with the business capabilities identified in Figure 6.13. In the SAP reference content, SAP has already performed this alignment using SAP products. However, it's clear that in practice, companies will use both SAP and non-SAP solutions together. Nonetheless, the SAP reference content serves as a helpful tool for creating the application map, particularly in an SAP-centric architecture. Non-SAP solutions can also be aligned with the business capabilities. In Figure 6.15, we see an extract of the application map.

The applications are directly linked to the business capabilities and color coded based on their lifecycle stage, such as retire, keep, investigate, or introduce. We explain the stages and color coding on the application map as follows:

- Fusion Motors has already decided to implement SAP S/4HANA Cloud and SAP Analytics Cloud, so those applications are in the *introduce* stage and color coded dark green.
- Fusion Motors hasn't yet decided whether to adopt some other solutions, but it has identified them as potential candidates to meet the business needs. Therefore, those applications are in the *investigate* stage and color coded light green.
- Fusion Motors has chosen not to use SAP solutions for certain business capabilities like **Marketing Strategy Management** and **Marketing Planning and Budgeting**, and it instead has opted for third-party applications. Therefore, those applications are in the *keep* stage and color coded blue.

- Fusion motors has decided to retire other applications, which means they'll no longer be maintained in the future. They are therefore in the *retire* stage and color coded red.

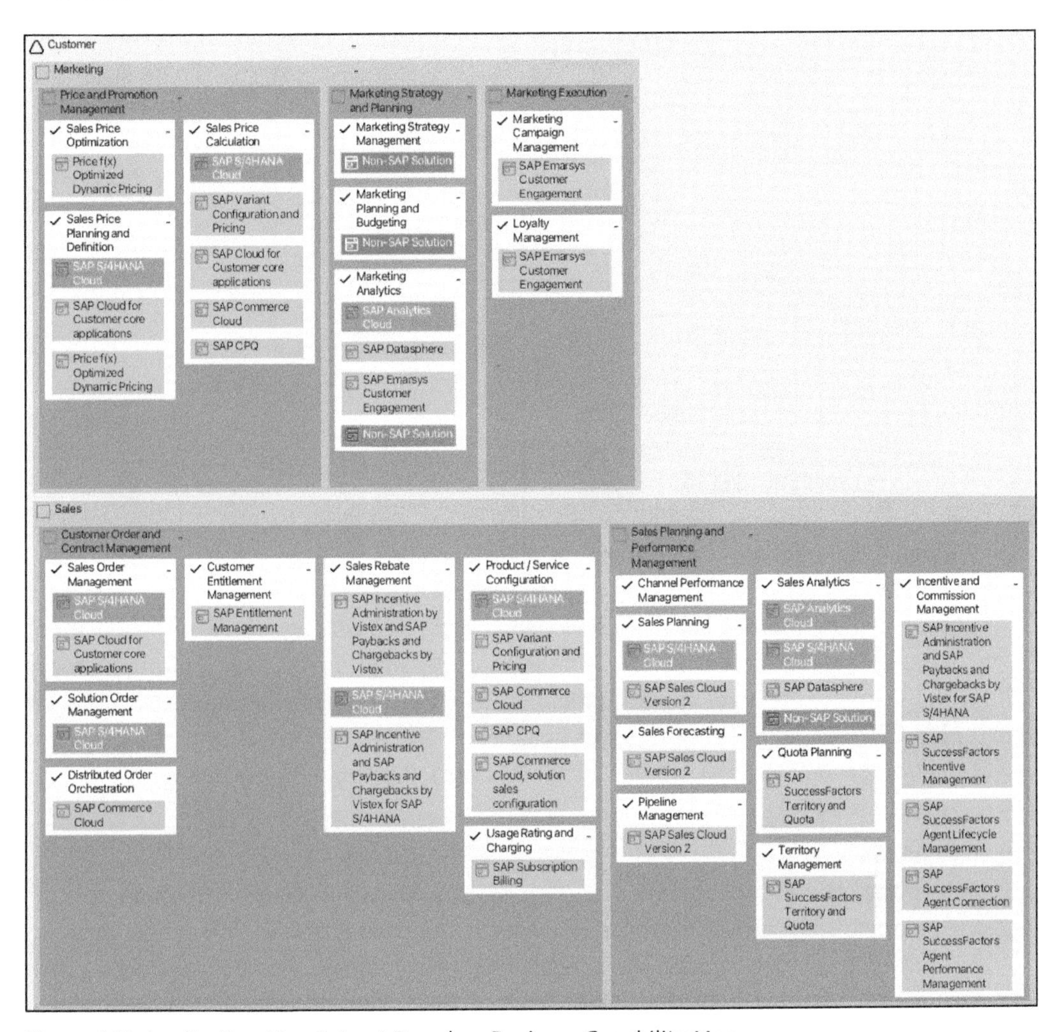

Figure 6.15 Application Map Extract Based on Business Capability Map
(Light Green = Investigate; Dark Green = Introduce; Blue = Keep; Red = Retire)

Also note that in some cases, such as with the **Channel Performance Management** business capability, no suitable application has been identified yet. Therefore, the mapping for that capability should remain empty, and further investigations are needed to find a suitable solution.

Using the application map as a basis, we can now develop an initial overview of the target application architecture, as shown in Figure 6.16. Typically, we would have constructed a thorough business capability map and application map, but for the sake of simplicity in this particular scenario, our focus has been on the customer enterprise domain. As a result, the application architecture will only provide a partial representation of the entire architecture.

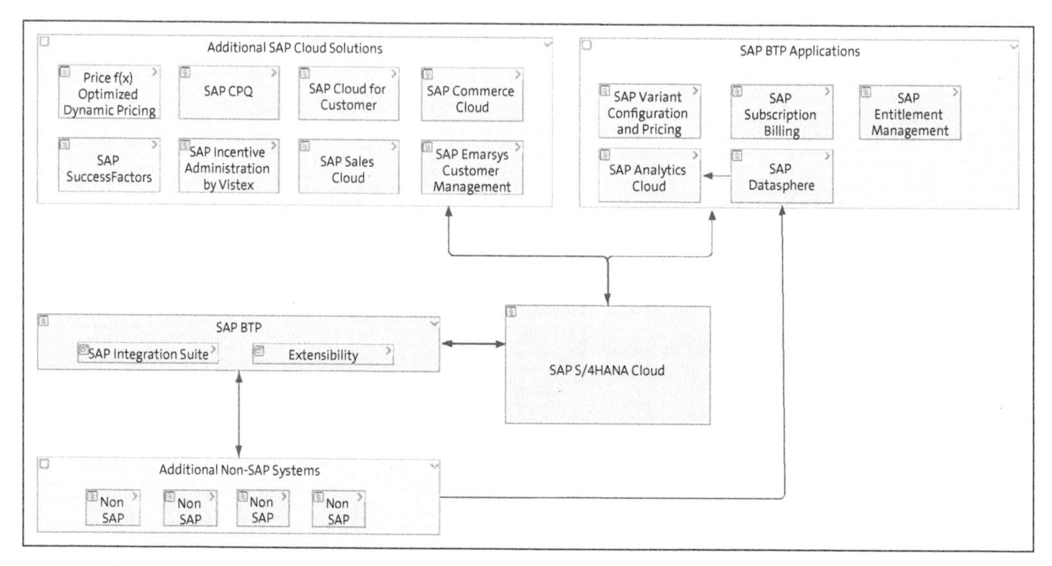

Figure 6.16 Target Application Architecture for Marketing and Sales

It's worth mentioning that we previously determined that we won't fully implement the target application architecture module at this stage. Instead, we'll primarily focus on the high-level target application architecture, without considering specific integration flows, deployment options, or instance strategy decisions. However, we suggest that in future iterations, we should address the following modules at a minimum, for the following reasons:

- **Instance strategy**
  In the current architecture of Fusion Motors (shown in Figure 6.3), there are currently three regional instances. However, in our desired application architecture, we haven't yet determined the overall instance strategy. We are considering two options: having a single global instance for Fusion Motors or having a global instance for most regions and local instances only for countries with strong data privacy requirements or geopolitical issues.
- **Deployment strategy**
  Fusion Motors has decided to adopt SAP S/4HANA Cloud Public Edition in the future. However, we haven't thoroughly analyzed whether this is the best option based on the requirements we've derived from the business capabilities. Therefore, in the next iteration, the deployment strategy module should validate this decision or propose alternative options such as SAP S/4HANA Cloud Private Edition or a two-tier ERP scenario.
- **Integration**
  Fusion Motors will have various integration scenarios involving SAP-to-SAP and

SAP-to-non-SAP connections. The target application architecture will include several additional applications in the landscape, such as SAP Integrated Business Planning for Supply Chain (SAP IBP) replacing parts of SAP Advanced Planning and Optimization. Therefore, it's crucial to carefully evaluate and establish the appropriate integration architecture.

- **Analytics and data strategy**
  In the proposed target application architecture, we've included SAP Datasphere and SAP Analytics Cloud. These solutions will need to connect with multiple SAP and non-SAP solutions to gather data for analytical and planning purposes. This necessitates the development of a comprehensive data and analytics strategy.

6.4.5 Initiatives and Roadmap

Now that we've completed the main architectural components, including a directional target architecture, we can begin planning the transition. Our initial focus for this iteration was on the expected changes to the business architecture and target application architecture, in light of Fusion Motors' new corporate strategy and business models. Therefore, our roadmap development will primarily address these aspects and will not yet include other elements that we will discuss in future iterations, such as the transition roadmap from SAP ERP to SAP S/4HANA. In this initial iteration, we need to coordinate with Fusion Motors' business stakeholders to determine the overall timeline for implementing the necessary strategic capabilities and business models. We'll use this information in subsequent iterations, where we'll focus on the landscape transformation aspects.

Now, let's dive into the initiatives catalog and roadmap for Fusion Motors.

Initiatives Catalog

An initiatives catalog, as we introduced in Chapter 2, Section 2.2.6, is a crucial artifact in enterprise architecture planning because it provides a comprehensive overview of all the proposed initiatives within an organization. It serves as a centralized repository that captures and categorizes various initiatives—such as technology upgrades, process improvements, and organizational changes—that are aligned with the enterprise architecture strategy.

Hints for Creating the Initiatives Catalog

You should consider following these steps when creating an initiatives catalog:

1. **Determine the list of initiatives**
   You can derive these initiatives directly from the target architecture and through collaboration with other stakeholders. It's also important to consider any existing projects within the organization and gain an understanding of each initiative, including its rationale, objectives, planning, and resource requirements.

2. **Align the initiatives with the organization's strategy**
   Utilize the strategic priorities and business goals identified in the strategy mapping exercise to assess the extent to which each initiative contributes to a strategic priority and/or business goal.
3. **Evaluate each initiative in terms of value and feasibility**
   Use a value/effort matrix to assess and categorize each initiative based on its level of value and effort required, ranging from low to high for each dimension.
4. **Establish a general timeline for each initiative**
   While it may not be possible to define exact timelines at this stage, we recommend that you provide an indicative timeline based on the expected effort required for each initiative.
5. **Identify and note any dependencies between initiatives**
   Note any dependencies between initiatives so that you can properly sequence and prioritize them.

Figure 6.17 provides the initiatives catalog for Fusion Motors with a focus on the direct sales strategic priority.

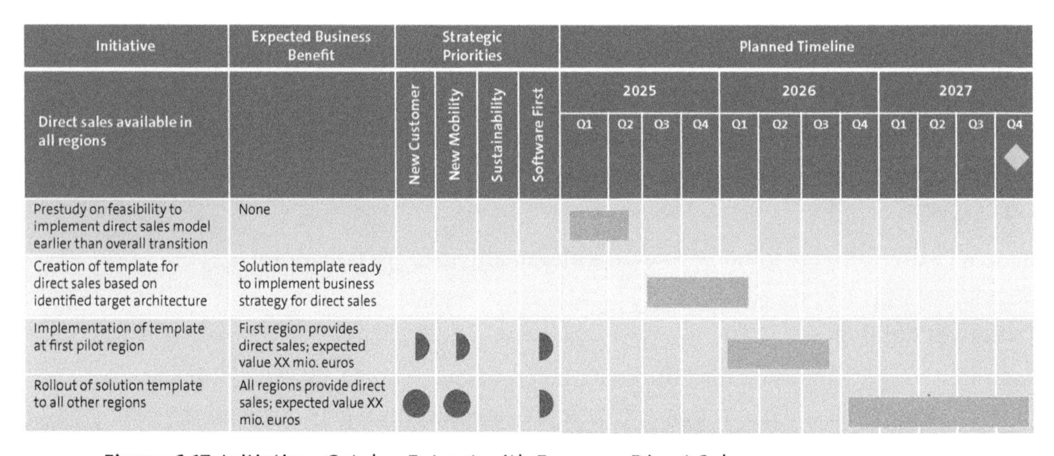

Figure 6.17 Initiatives Catalog Extract with Focus on Direct Sales

Fusion Motors has set a significant goal to make direct sales available in all regions by the end of 2027. The first step toward achieving this goal is to conduct a feasibility study on implementing a direct sales-focused architecture alongside the overall transition. Fusion Motors acknowledges that the transition to SAP S/4HANA won't be completed by the time the first regions need to enable direct sales, but it also doesn't make sense to implement the new business model in the existing outdated system. Therefore, the new business model needs to be implemented in the target architecture earlier, which requires a parallel landscape.

In addition to this, Fusion Motors plans to create a streamlined and flexible template approach using cloud solutions and then implement this template in the first region. Once the template is successfully implemented, it can be rolled out to all regions. These initiatives have a significant impact on Fusion Motors' strategic priorities, such as attracting new customers and embracing new mobility solutions, as outlined in Fusion Motors' corporate strategy.

Roadmap

With the availability of the initiatives catalog, we can now create the initial version of the roadmap. Since we haven't included all the necessary modules in our architecture planning, the focus of the roadmap is currently on the business transformation aspects of the overall transition planning (see Figure 6.18). As a result, the roadmap will need to evolve into a comprehensive plan as we plan additional aspects in future iterations (such as the SAP S/4HANA transition strategy or instance strategy).

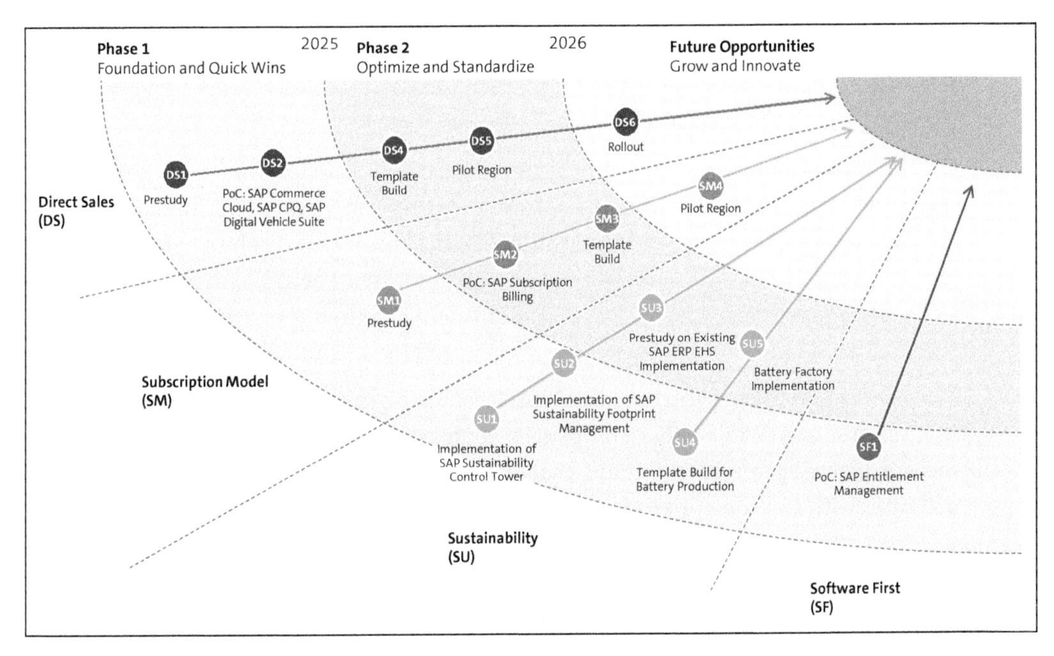

Figure 6.18 Roadmap for Business Transformation

The previous section discussed the direct sales initiatives that have been organized and included in the roadmap based on the initiatives catalog. Other main strategic priorities have been identified in the same manner. Fusion Motors faces similar challenges in all other areas as it undergoes a significant transformation, which includes both a business transformation (as described in this use case) and a landscape transformation from the existing SAP ERP system to the future target architecture on SAP S/4HANA.

The landscape transformation will require careful planning and will take several years to complete. However, Fusion Motors can't afford to wait until the landscape transformation is finished to implement its new business models, such as direct sales and battery production. As a result, Fusion Motors has decided to simultaneously implement certain new business models as *speedboats* in a new cloud-based model, alongside the landscape transformation. This decision will have implications for the integration architecture, as the new business models in the cloud will need to be integrated with the existing legacy architecture.

6.4.6 Organizational Change Management

SAP's project experience shows that transformation projects frequently stall or fail to achieve their objectives because of related nontechnical problems.

Over the past ten years, data shows that most transformation projects fail to meet their objectives due to nontechnical factors. A significant 75% of these failures are linked to challenges such as resistance to change, inadequate skills, poor communication, and resource constraints. These issues emphasize the critical role of human and organizational factors in the success of complex projects.

On the other hand, only 25% of project failures are caused by technical difficulties, including challenges in mapping requirements, system sizing, and integration with legacy systems. This highlights that while technical problems can be a barrier, they are less common than issues related to people and processes. Focusing on improving team skills, communication, and stakeholder alignment can therefore significantly increase the likelihood of project success.

Hence, we suggest that you take OCM into account right from the start of the transformation program. While OCM isn't usually seen as a central aspect of enterprise architecture planning, we recommend that you integrate the overall OCM concept and activities with the enterprise architecture planning process.

Since Fusion Motors has only completed the initial phase of enterprise architecture planning, there's no need for detailed OCM activities at this point. However, we propose creating an initial plan for future OCM-related activities, as shown in Figure 6.19.

As one of the initial steps in the enterprise architecture planning process, we suggest that Fusion Motors conducts a stakeholder identification analysis. This analysis will help identify the key stakeholders who can influence the direction and outcome of the transformation. Additionally, we propose conducting an initial workshop on OCM risk identification and mitigation planning with the identified stakeholders. During these workshops, we'll focus on understanding any risks related to people, such as competing transformation initiatives and planned organizational restructuring. We'll then define specific actions to mitigate these risks, assigning clear responsibilities, due dates, and completion criteria.

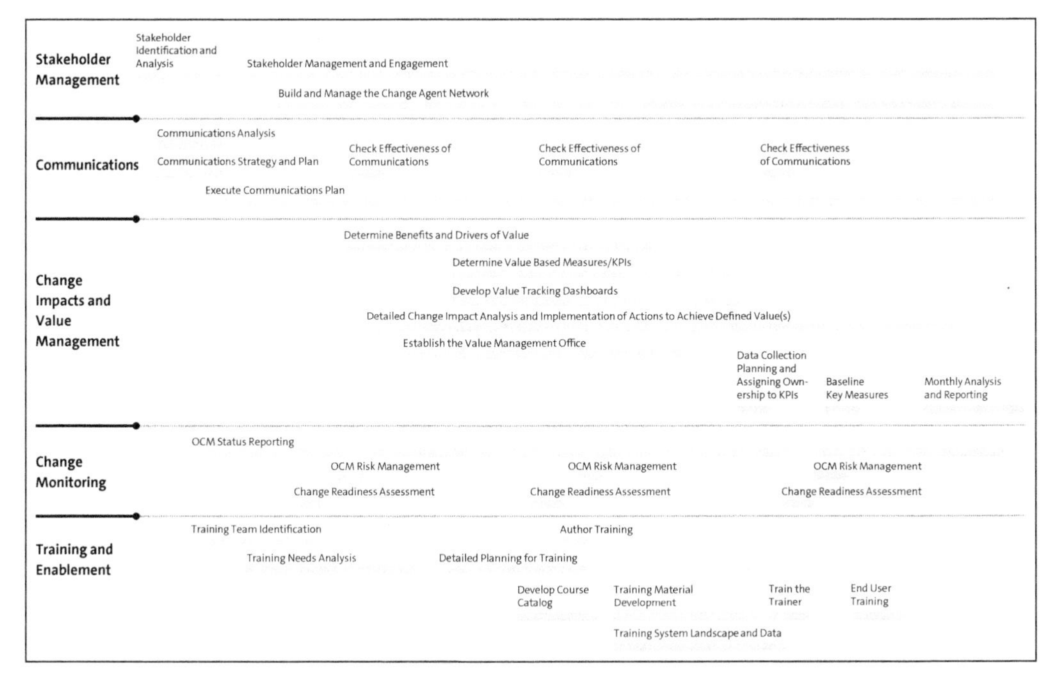

Figure 6.19 OCM Planning Activities

6.5 Summary

In this chapter, we've utilized the SAP EA Framework to analyze a business transformation scenario. By considering a fictional case of an organization that undergoes a change in its business model and embraces a new business strategy, we've observed the crucial role of enterprise architecture planning in assisting the organization during the initial planning phases of the transformation. It has become evident that enterprise architecture encompasses more than just dealing with IT systems; it's instrumental in comprehending and translating a business strategy into a practical architecture and roadmap.

Therefore, we can summarize the following key takeaways:

- **Business transformation is omnipresent**
  Organizations are constantly faced with the need to adapt and innovate in order to remain competitive. The blurring of industry boundaries has further intensified the need for transformation, as traditional sectors are being disrupted by new players and business models.
- **Alignment of business and IT**
  Enterprise architecture planning helps align business strategy with IT strategy. It ensures that IT applications and infrastructure are designed and implemented in a way that supports overall business goals and expected business model changes. This

alignment is crucial for successful business transformation as it ensures that investments are targeted toward achieving the desired business outcomes.

- **Importance of iterations**
  Enterprise architecture planning requires managing complex landscapes and interdependencies. To better handle this complexity, the planning process is divided into iterations. Each iteration can concentrate on specific enterprise architecture planning modules or delve into different levels of depth for each module.
- **Planning is about people**
  Effective organizational change management is an essential component of enterprise architecture planning and should be integrated right from the start. By incorporating OCM, an organization can ensure that it can fully utilize the new architecture to successfully accomplish its strategic priorities and facilitate long-lasting business transformation.

In the twenty-first century, amid the global challenges of maintaining a livable planet for current and future generations, no business transformation can succeed without incorporating a thorough sustainability strategy. In the next chapter, we'll explore another use case demonstrating the connection between sustainability and enterprise architecture.

Chapter 7
Sustainability

Sustainability is the next big priority for digital transformation, and enterprise architecture will play an important role in making an organization a sustainable enterprise. From an IT perspective, there are many important architecture decisions your company needs to make to reach the goal of being a sustainable enterprise.

Sustainability is a rapidly growing priority across all industries, and every C-level manager is discussing it. This critical topic has multiple dimensions, so large and strategic SAP customers need guidance to help them define their target application architecture and roadmap. This guidance helps them evolve into intelligent and sustainable enterprises, so the role of the enterprise architect is more crucial than ever in helping companies achieve their sustainability goals.

Although sustainability is not a new concept, its urgency has intensified. The 2030 Agenda for Sustainable Development, which was adopted by all United Nations member states in 2015, aims to transform the world by providing a shared blueprint for peace and prosperity for people and the planet. Central to this agenda are the seventeen Sustainable Development Goals (SDGs), which form an urgent call for global action by all member countries.

Driven by the need to implement these SDGs, organizations committed to sustainability are investing significant time and resources in identifying priorities and selecting appropriate technological solutions to address global challenges. Environment, health, and safety (EHS) and overall sustainability topics have evolved from being mere compliance and cost drivers to becoming competitive advantages and differentiators.

Enterprise architecture plays a pivotal role in achieving these SDG goals by aligning prioritized business capabilities with the best-fit applications, deriving the transformation roadmap, and defining the future state architecture.

SAP's goal is to move us toward a world that has no emissions, no waste, and no inequality. SAP's approach to corporate sustainability is focused on making a positive impact on the economy, society, and the environment within the limits of the planet. SAP places special emphasis on comprehensive management and reporting of environmental, social, and governance (ESG) factors, taking action on climate change, promoting a circular economy, and fulfilling our social responsibilities.

This use case explores how a company's sustainability strategy can be integrated into its enterprise architecture to improve operational efficiency, reduce waste and emissions, and promote social and environmental responsibility. You'll learn how enterprise architecture transformation supports a company's sustainability strategy and helps a company reach its ESG goals, with sustainability as one of the key strategic priorities both for SAP and for many of SAP's customers.

7.1 Business and IT Context

As with our previous use cases, you can look at sustainability from two angles: business and IT. Both the business and IT contexts for sustainability emphasize the interconnectedness among corporate strategy, technological innovation, and social responsibility. Companies that integrate sustainable practices into their core business strategies while leveraging appropriate IT solutions are better positioned to thrive in a rapidly changing global landscape. Businesses face risks related to regulatory noncompliance, supply chain disruptions due to climate change, and reputational damage from unsustainable practices, and IT plays a crucial role in collecting, managing, and analyzing vast amounts of data related to sustainability metrics. This involves implementing systems and technologies to track energy consumption, emissions, waste management, etc.

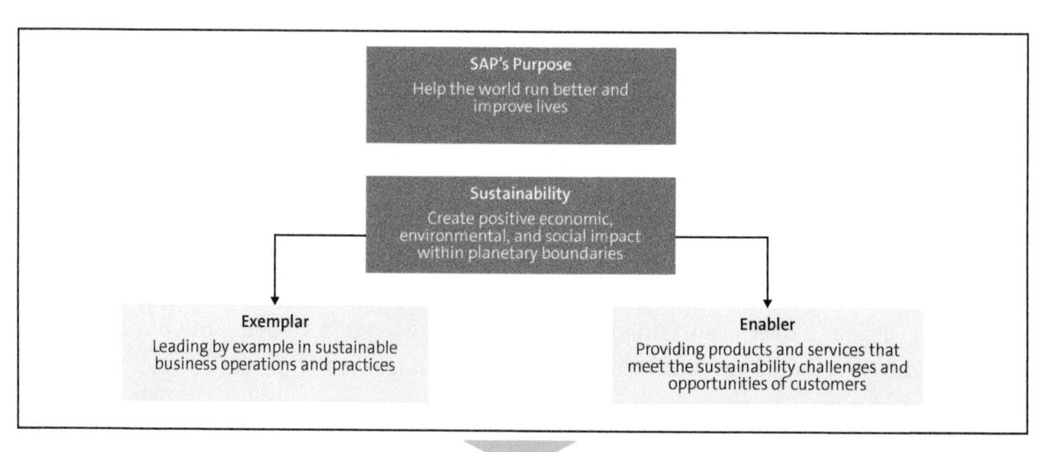

Figure 7.1 SAP's Approach to Sustainable Impact

SAP's focus on sustainability aligns with the SDG goals. SAP aims to achieve "zero" in sustainability metrics and integrates sustainability data into business processes in all value chains to drive performance, innovation, and market growth. SAP's approach to sustainable impact is illustrated Figure 7.1.

Let's look at the business and IT context from a sustainability perspective and examine sustainability's impact on SAP applications.

7.1.1 Business Context

Sustainability is the next big strategic business priority for digital transformation. Sustainability is quickly gaining momentum as the foremost strategic topic for any organization, primarily due to the following reasons:

- Companies need to comply with new reporting standards from regulatory bodies like the International Sustainability Standards Board (ISSB), the Securities and Exchange Commission (SEC), and the European Financial Reporting Advisory Group (EFRAG). They also need to ensure audit-proof accounting for greenhouse gas quantities.
- The European Union Deforestation Regulation (EUDR) is a new mandatory regulation designed to ensure that products sold in the European Union don't contribute to global deforestation or forest degradation. SAP Green Token—an SaaS cloud solution for end-to-end traceability and transparency of sustainability information—helps to address the EUDR requirements for your business.
- The Food Safety Modernization Act (FSMA) is a new law set to go into effect in January 2026. It mandates that all companies that operate in the United States and handle specified food categories—primarily fresh and fresh-frozen items such as vegetables, fruits, cheeses, and meats—must maintain and be able to provide product-tracing information if contaminated food is found in the supply chain. This means companies must be able to trace the origin of contaminated food and how it has moved through the supply chain. Violating this requirement carries hefty fines. Additionally, the law extends the product-tracing requirement to include foreign supply chain partners. SAP Green Token and SAP Global Batch Traceability are applications that help to address the FSMA requirement.
- Companies need to achieve ESG transparency along the value chain (i.e., scope 1, 2, and 3 emissions transparency) on the company and product levels. This business context is linked to the **Zero Emissions with Climate Action** sustainability priority in Figure 7.1.
- Data on and insights into material flow, traceability, recovery, and reuse strive toward zero electronic waste and phasing out single-use plastics. This business context is linked to the **Zero Waste with Circular Economy** sustainability priority in Figure 7.1.

- Extended producer responsibility (EPR) schemes place responsibility for the environmental impact of products on producers, who pay fees based on the volume of plastic packaging of their products in the market (e.g., a fee per ton of plastic).
- Plastic taxes are imposed on the production, use, or disposal of plastic products. The purpose of a plastic tax is to discourage the use of single-use plastics and to incentivize the use of more sustainable, biodegradable, or recyclable materials.
- Tracking and gaining insights into worker safety data are linked to the **Zero Inequality with Social Responsibility** sustainability priority in Figure 7.1.

Greenhouse Gas Emissions Definitions: Scope 1, 2, and 3

Scope 1, 2, and 3 emissions are categories used to classify different sources of greenhouse gas emissions within an organization or related to a specific activity. The details on them are as follows:

- **Scope 1 emissions**
  Scope 1 includes direct greenhouse gas emissions from sources that are owned or controlled by the organization. This includes emissions from sources like company-owned vehicles, emissions from on-site fuel combustion in boilers or furnaces, and process emissions from manufacturing.
- **Scope 2 emissions**
  Scope 2 includes indirect emissions from the generation of purchased electricity, heat, or steam that an organization consumes. These emissions occur outside of a company's direct control but are a result of its energy consumption. For instance, if a company uses electricity generated from a coal-fired power plant, the emissions from the plant would fall under scope 2 for that company.
- **Scope 3 emissions**
  These are indirect emissions that occur because of the organization's activities but are not directly owned or controlled by the organization. Scope 3 emissions typically encompass a broader range of sources, including emissions from the supply chain, business travel, employee commuting, waste disposal, and product life cycles. These emissions can be the most challenging to measure and manage as they often involve numerous stakeholders beyond the immediate control of the organization. Collecting accurate data makes measuring and assessing these emissions difficult, and with different suppliers involved in the supply chain, there's no clarity on who exactly is accountable for reducing scope 3 emissions.

7.1.2 IT Context

Typically, in an organization, there are two groups who are responsible for strategic priorities related to sustainability: one in operations and the other in corporate. The operations team's priorities are managing regulatory requirements and managing sustainable operations, and the team is primarily responsible for managing product safety and compliance throughout the product lifecycle. On the other hand, corporate's sustainability priorities are sustainability performance management, environment footprint management, the circular economy, and sustainability collaboration.

Based on our experience, there are many architecture-related questions that the company needs to answer from an IT perspective so that it can reach the goal of becoming a sustainable enterprise. Some of the key architecture questions are as follows:

1. How do we determine the priorities for sustainability to define the future-state IT architecture?
2. What are the target applications for the traditional EHS areas such as product compliance and EHS currently running in SAP ERP?
3. What are the architecture options and the product roadmap for next-generation SAP sustainability applications such as SAP Sustainability Footprint Management, SAP Responsible Design and Production, and SAP Sustainability Control Tower?
4. What are the target architecture options and the roadmap for the following solutions?
   - The incident management solution
   - The environment management solution
   - The waste management solution
   - Substance volume tracking, hazardous substances management, and dangerous goods management

These are some of the key questions to be answered during the architecture sessions, and this is by no means an exhaustive list. For example, today's customers have a choice between implementing classic or new capabilities of SAP S/4HANA for product compliance. *Classic capabilities* are all features of SAP ERP-based product safety and product compliance solutions, graphical user interface (GUI) and Web Dynpro transactions, the specification database, and the workbench approach. New capabilities mean new process models, SAP Fiori apps, data models, core data services (CDS), and integration into the product master, sales, delivery, procurement, and transportation. Full scope is not yet available and is in the innovation roadmap. Of course, both options have pros and cons, and once new capabilities support the full scope for SAP S/4HANA, existing classic capabilities will no longer be available.

Using enterprise architecture methodology, content, and tools (described in Chapter 2 and Chapter 4) is an excellent way to address these questions, which will help us to define the business capability map for sustainability priorities, the baseline target application architecture, and the roadmap for SAP sustainability applications. We will then revisit, revalidate, and incrementally update this based on the change events at the customer strategic plan and SAP product roadmap planning levels. Figure 7.2 shows the steps we follow during the transformation planning phase from strategy to roadmap for SAP sustainability solutions.

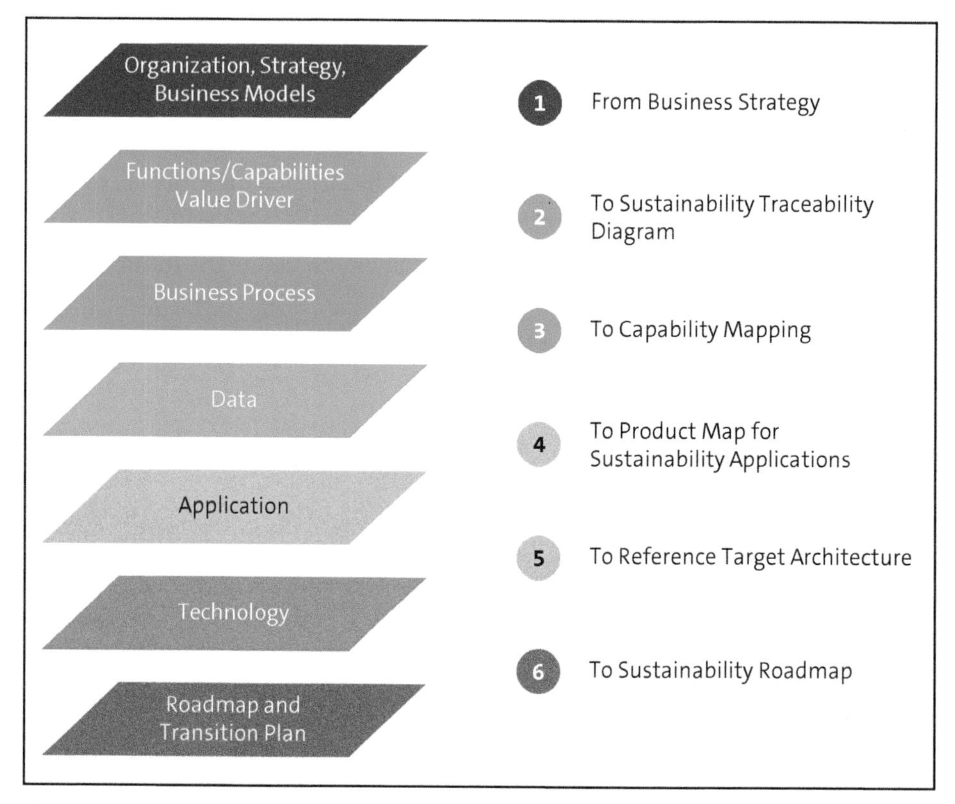

Figure 7.2 Strategy to Roadmap for Sustainability

The enterprise architect plays a pivotal role in the SAP sustainability transformation initiative. The enterprise architect helps to align the company's technology infrastructure and applications with sustainability goals. Their responsibilities typically include the following:

- **Mapping sustainability goals to IT strategy**
  This includes understanding the organization's sustainability objectives and translating them into a comprehensive IT strategy. It involves identifying how technology can support and enable sustainability initiatives across the enterprise.
- **Designing sustainable IT solutions**
  This includes working closely with stakeholders to design and implement IT solutions that support sustainability efforts. It might also include developing systems to track and manage environmental data, energy consumption, the carbon footprint, and other sustainability metrics.
- **Integration of sustainability into applications**
  This includes integrating sustainability considerations into existing SAP applications or implementing new modules that facilitate the tracking, measurement, and reporting of sustainability-related data. This integration might involve SAP S/4HANA applications and SAP sustainability solutions.

- **Data management and analytics**
  This includes overseeing the collection, management, and analysis of sustainability-related data. This involves ensuring data accuracy, integrity, and accessibility to support decision-making and reporting on sustainability performance.
- **Collaboration and stakeholder management**
  This includes collaborating with various teams—including sustainability, IT, operations, and business units—to ensure alignment of strategies, requirements, and implementation plans. Managing stakeholders and ensuring buy-in for sustainability initiatives is crucial.
- **Adoption and change management**
  This includes driving the adoption of sustainable practices through change management initiatives, and it also includes training employees on new systems, processes, and the importance of sustainability in their roles.
- **Compliance and regulatory alignment**
  This includes ensuring that IT systems and processes comply with relevant sustainability regulations, standards, and reporting requirements. It might involve keeping abreast of changing regulations and adjusting systems accordingly.
- **Continuous improvement**
  This includes facilitating continuous improvement by monitoring and evaluating the effectiveness of sustainability initiatives and IT solutions. It also includes identifying areas for optimization and innovation to further enhance sustainability efforts.

Overall, the enterprise architect acts as a bridge between sustainability objectives and IT capabilities, ensuring that technology is leveraged effectively to support, measure, and drive the organization's sustainability transformation.

7.2 Problem Statement

In this section, we'll demonstrate how to utilize the enterprise architecture transformation approach, using a case study example. Specifically, we'll examine how to apply the transformation approach in the context of a multinational company in the consumer products industry.

GreenCon Inc. is a fictitious multinational company with business units in multiple industries, including consumer products, chemicals, and life sciences. It is considering a move toward multi-instance target architecture for SAP S/4HANA, with global SAP S/4HANA as one instance and another separate SAP S/4HANA instance for the company's China business. China business rules state that either data can't leave the country or that local copies of data must be maintained within the country's borders.

7.2.1 Company Profile

One of the strategic priorities for GreenCon Inc. in this business transformation initiative is to reach its ESG goals with sustainability. GreenCon Inc. wants to build a comprehensive business capability map and target application architecture diagram as well as a sustainability roadmap. This will provide the company with a clear framework for future prioritization on sustainability.

The problem statement is about GreenCon Inc., which has an ambitious and strategic ESG goal of measuring and automating its greenhouse gas emissions from its end-to-end supply chain. Currently, its emissions are measured manually, and hence, the goal is to automate the metrics in the next three years. GreenCon Inc. has reached out to SAP for help with defining the target architecture and roadmap for the applications that drive the sustainability capabilities and processes.

In addition, EHS (operations stakeholders) and sustainability (corporate stakeholders) groups within GreenCon Inc. are looking to address some of the following priorities and issues:

- Achieving end-to-end visibility in value chain product compliance.
- Improving operational efficiency and compliance.
- Supporting real-time visibility, decision-making, and action using automated collection of metrics and dashboards.
- Creating a roadmap for environment compliance and waste management solutions in SAP S/4HANA. Both of these solutions need to undergo significant improvements and data model changes in SAP S/4HANA so that the company can retire and phase out non-SAP incident management, risk management, and management of change and convert to SAP solutions.
- Evaluating SAP Sustainability Control Tower for overall sustainability reporting via SAP Analytics Cloud solutions.

7.2.2 IT Landscape

Currently, GreenCon Inc. has one global SAP ERP instance for all its business units (as depicted in Table 7.1). Many heterogenous non-SAP and SAP peripheral systems are deployed, and these include treasury, financial supply chain management (FSCM), SAP Global Trade Services (SAP GTS), SAP Extended Warehouse Management (SAP EWM), SAP Transportation Management (SAP TM), SAP Advanced Planning and Optimization, SAP Supply Network Collaboration, and SAP HANA Enterprise Edition.

Business Unit	Industry Segments	Sales Revenue
BU 1	Consumer products	60% of total revenue
BU 2	Hi-tech products	15% of total revenue

Table 7.1 GreenCon Inc. Business Segments and Divisions

Business Unit	Industry Segments	Sales Revenue
BU 3	Pharmaceutical products	15% of total revenue
BU 4	Industrial, machinery, and components industry	10% of total revenue

Table 7.1 GreenCon Inc. Business Segments and Divisions (Cont.)

The as-is application architecture diagram we've received as an input from GreenCon Inc. is heat mapped with lifecycle attributes (as shown in Figure 7.3). SAP Supply Chain Management (SAP SCM) and SAP Advanced Planning and Optimization applications are flagged as **Retire** (red), whereas the core ERP application is in **Phaseout** mode (orange) and the value case (including a business case) is being prepared to move to SAP S/4HANA. Many of the existing applications will be maintained even in the target architecture. Lifecycling of applications helps to eliminate systems you don't need and replatform the applications, cutting costs and energy use. For example, SAP Hybris Commerce is an old solution tagged as **Keep**; it will be replatformed to SAP Commerce Cloud as the target application. This also helps to reduce your company's carbon footprint and make the planet greener.

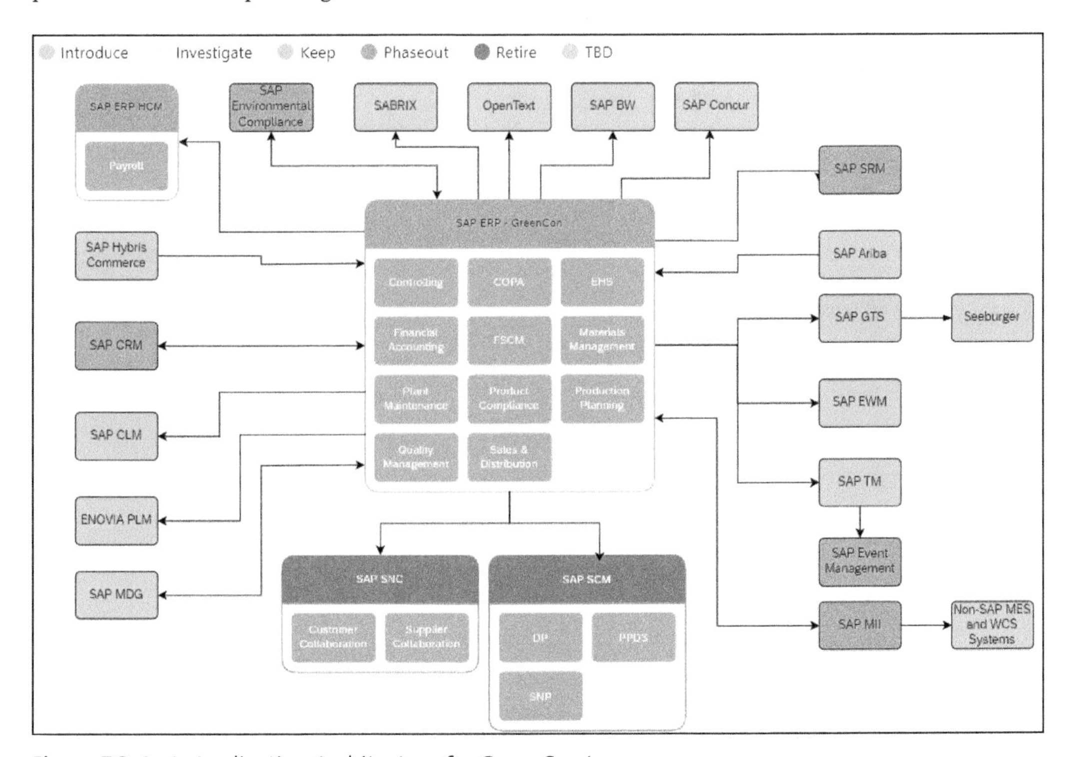

Figure 7.3 As-Is Application Architecture for GreenCon Inc.

7.3 Applying the Enterprise Architecture Transformation Approach

In Chapter 2, you learned about SAP EA Framework, which consists of five pillars: methodology, reference content, tooling, practice, and services. Now, let's apply the enterprise architecture transformation approach to this use case to define the target architecture and roadmap for sustainability capabilities for GreenCon Inc. to address the strategic priorities.

Based on our understanding of the customer's situation and requests, the key architecture modules we've identified in scope for the North Star service are as follows:

- **Strategy mapping**
  This module assesses the sustainability objectives of GreenCon Inc. and how they relate to the goals and drivers of the organization.
- **Business capability**
  This module focuses on heatmapping the business capabilities that are relevant to sustainability in scope and out of scope for the SAP S/4HANA transformation. It also focuses on pain points and opportunities to move to new SAP applications.
- **Application architecture**
  This module focuses on designing the target application architecture for SAP sustainability applications, along with the high-level solution components map. This will provide GreenCon Inc. with visibility into the business capabilities mapping to solution capabilities of SAP applications.
- **Initiatives and roadmap**
  This module helps to determine a set of transformation initiatives for sustainability; lists the major roadmap clusters, phases, and dependencies; and then sequences them on a relative timescale to create the sustainability roadmap for GreenCon Inc.

Figure 7.4 shows the modular end-to-end enterprise architecture planning, which helps in enabling key architecture decisions for sustainability transformation.

We will perform the following steps to derive a target architecture and roadmap based on the business capability-driven approach:

1. Scope architecture modules, business domain, business area, and business capabilities, based on SAP Reference Business Architecture.
2. Identify required solution capabilities and solution components based on the business capability scoping.
3. Heatmap business capabilities to identify the opportunities, pain points, and requirements of GreenCon Inc.
4. Identify deployment types (public cloud, private cloud, and on-premise).
5. Review and validate as-is application architecture and categorize it as per the lifecycle of solution components, in terms of which applications would be retired, phased out, kept, introduced, investigated, and decided upon later.

6. Identify application architecture impacts (risks and assumptions) and provide recommendations.
7. Define and design target application architecture by business domain, including high-level information flows.
8. Define an application architecture roadmap aligned with the SAP product roadmap.

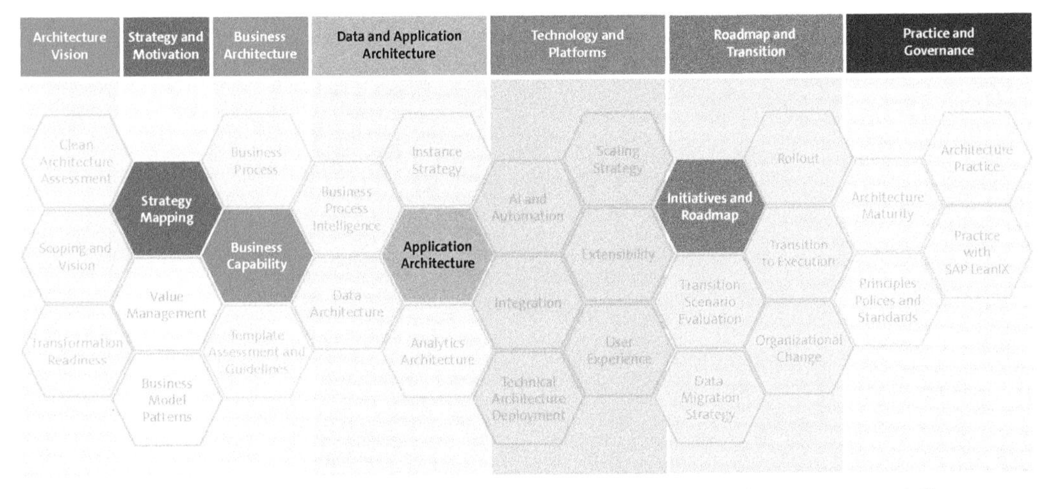

Figure 7.4 Modules in Scope for Typical North Star Service Architecture for SAP Sustainability

---

Multiple Approaches to Designing the Target Architecture

Many roads lead to the same place, and we can also derive the target architecture and roadmap by using the business process-driven approach or by using business model patterns.

In Chapter 4, you learned about the different tools for enterprise architecture planning with SAP. Now, we'll use an internal SAP tool to define the architecture scope for GreenCon Inc. for the sustainability transformation initiative. As you can see from Figure 7.5, the level 1 business domain (**R&D/Engineering** and **Sustainability Management**) is scoped, and then, the following level 2 business areas are scoped to be relevant for GreenCon Inc.'s sustainability priorities:

- **R&D/Engineering**
  - **Product Compliance Management**
- **Sustainability Management**
  - **Circular Economy**
  - **Sustainability Performance Management**
  - **Sustainable Operations**
  - **Environmental Footprint Management**
  - **Sustainability Collaboration**

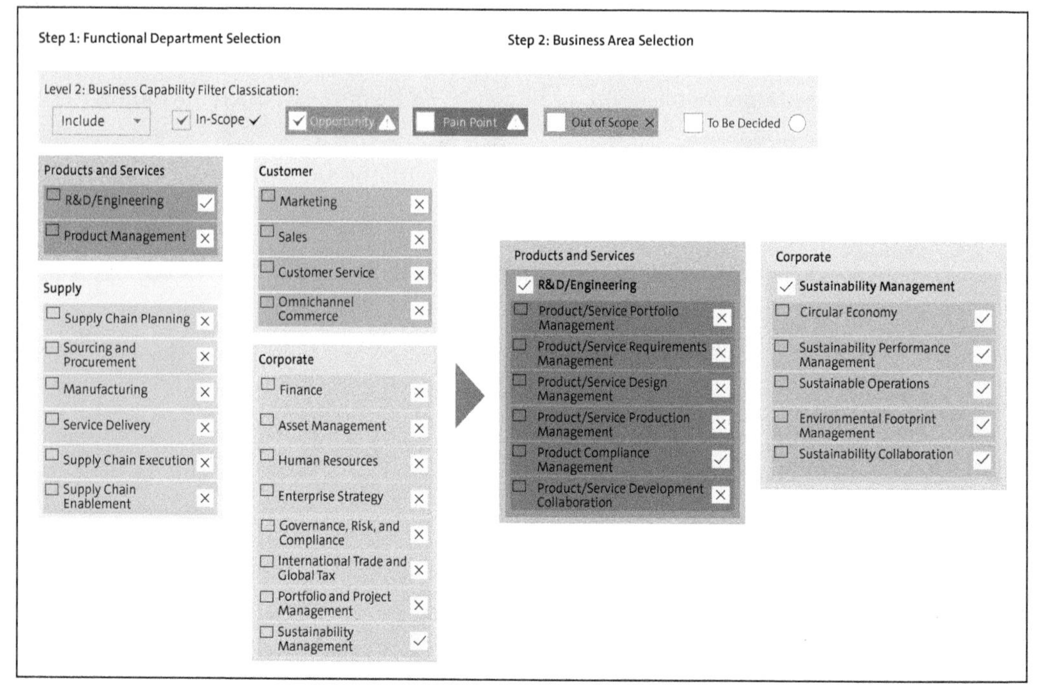

Figure 7.5 Scoping for Business Domain and Business Area

We follow this by scoping the business capabilities that are needed to create a sustainable enterprise, as shown in Figure 7.6. Key pain points identified during the scoping session are as follows:

- Sustainability steering and reporting
  - Compliance with relevant ESG key performance indicators (KPIs) from different (and changing) standards
  - Challenges in reporting and communication with stakeholders, also in relation to GreenCon's self-imposed targets
- Waste management
  - Calculating and disclosing EPR fees
  - Discharging responsibilities toward plastic and other material taxes
- Sustainability data collection
  - Gathering data from source systems across the company as well as from external sources, which often requires manual work
  - Integrating and harmonizing financial and nonfinancial data, which is a challenge
  - Need for accuracy, completeness, consistency, granularity, and timeliness of data

- Hazardous substance management
  - Identification of all relevant regulations, standards, and directives on the type of product at hand and its application
  - Lack of compliance visibility and easy access to compliance information for decision-making and customer inquiries

We've also identified opportunities to move to new SAP applications. Some of the key opportunities are as follows:

- **Environment footprint calculation**
  Assess and record environmental footprints.
- **Environmental footprint analysis**
  Evaluate, interpret, and analyze environmental footprints across the end-to-end value chain.
- **Sustainability information collaboration**
  Exchange sustainability data and KPIs with business partners and collaborate across business networks to increase data accuracy on scope 1, 2, and 3 footprints and to enable benchmarking.

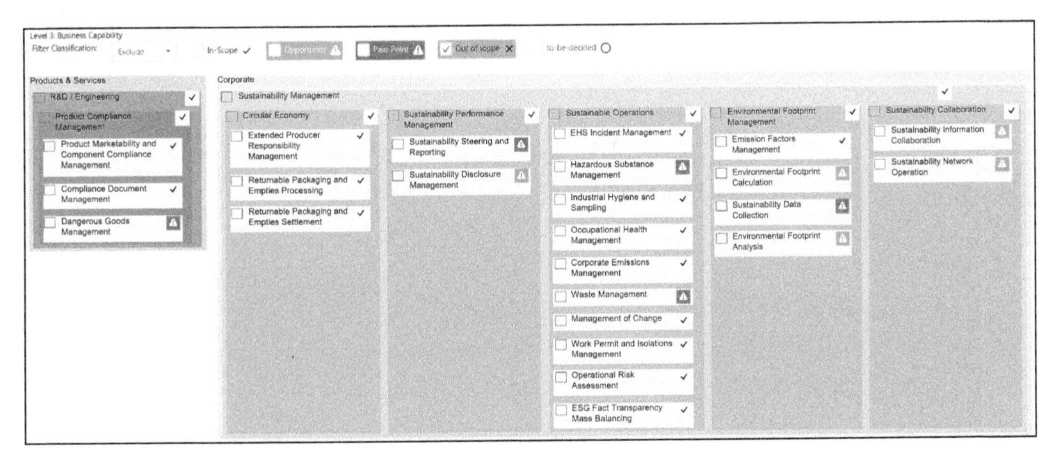

Figure 7.6 Scoping for Business Capability and Heatmapping with Opportunities and Pain Points

Typical Architecture Roadmap Patterns for Sustainability

While we've already done the scoping and we understand customers' need for architecture guidance on sustainability priorities, it's important that we also understand there's a typical roadmap pattern by which customers can transform from the as-is SAP ERP architecture to the target SAP S/4HANA and new sustainability applications from SAP. One example of such a pattern is shown in Figure 7.7.

This is one of the roadmap patterns, and you need to keep this mind while also being flexible enough to create a roadmap unique to your customer's situation.

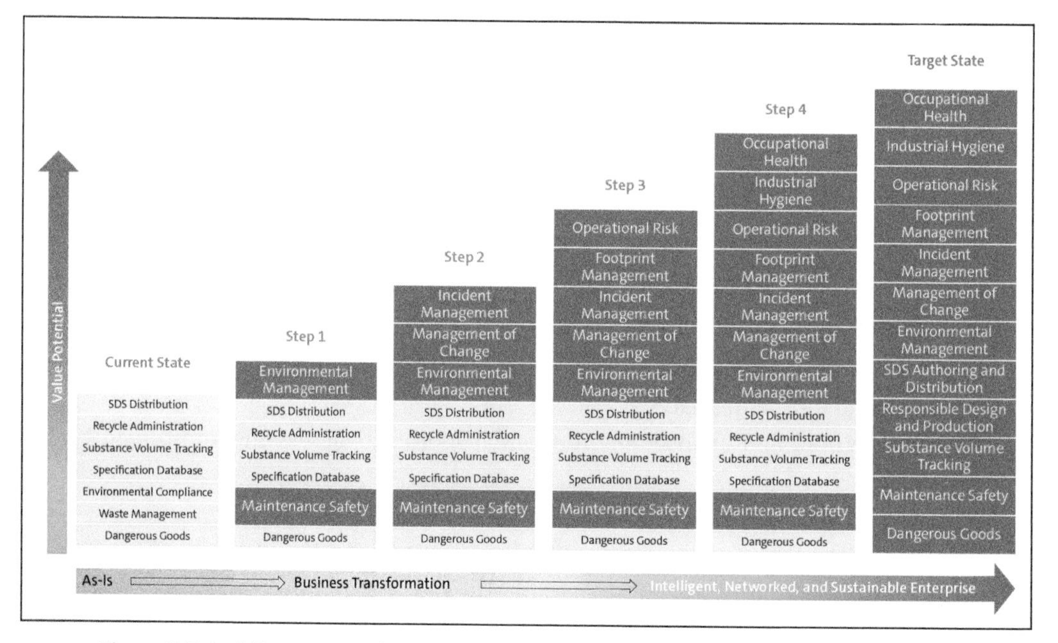

Figure 7.7 Architecture Roadmap Pattern for SAP Sustainability

7.4 Key Deliverables and Outcomes

In this section, we'll outline the main deliverables and results from the modules that we discussed in the preceding section. We'll provide a brief explanation of each deliverable's creation process and the necessary prerequisites. Furthermore, we'll showcase the final outcome or a sample of the ultimate deliverable.

After we conduct the scoping and collect inputs from the customer, GreenCon Inc., we need to execute the following steps to create the desired outcome and deliverables for the customer:

1. Strategy mapping
2. Creating as-is application architecture review that includes the following:
   - IT and business, as-is and to-be strategy, and change drivers
   - Reviewed and validated as-is application architecture, which is categorized as per the following lifecycle: introduce, investigate, phase out, and retire
3. Business capability heatmapping that includes the following:
   - Reviewing business capabilities according to SAP EA Framework (SAP Reference Business Archiitecture and SAP Reference Solution Architecture content) and mapping to customer business objectives and pains and gains
   - Collecting requirements (functional and nonfunctional)

4. Creating a high-level solution component map
5. Designing target application architecture by completing the following steps:
   - Conducting workshop sessions as per business domain
   - Sketching target application architecture
   - Focusing on key sustainability business areas
6. Creating a multiyear, strategic roadmap that focuses on sustainability

7.4.1 Strategy Mapping

The purpose of strategy mapping is essentially to help you understand GreenCon Inc.'s strategic objectives related to creating a sustainable future for the company and how these objectives impact the target architecture and roadmap for SAP sustainability solutions. You can document the objectives as shown in Figure 7.8. GreenCon has four strategic drivers: top-line growth, bottom line, green line, and strategic and transformation drivers. The green line strategic driver (in dark blue) is aimed at the sustainability goal of optimized environmental impact (in light blue), which is linked to five sustainability objectives (in dark green). Light-green objectives are not measured under sustainability.

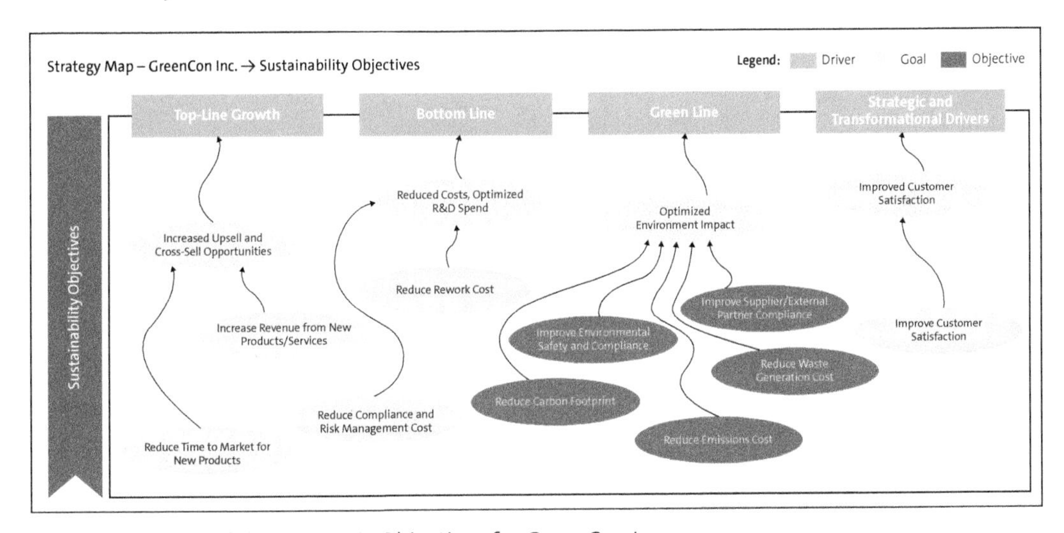

Figure 7.8 Sustainability Strategic Objectives for GreenCon Inc.

You'll find that the key value drivers that are relevant for GreenCon Inc. are the sustainability objectives (indicated by the dark green items in Figure 7.8), which are as follows:

1. Reduce emissions costs by improving overall actionable insights into the product and the corporate footprint, which yield smarter, more effective footprint-related business decisions.

2. Reduce waste generation costs by allowing improved decisions and resource management based on responsible design and production insights.
3. Reduce the carbon footprint by using returnable packaging instead of single-use packaging and having full inventory transparency to make supply chains more sustainable in the future.
4. Improve environmental safety and compliance by providing checks to support product tracking.
5. Improve supplier and external partner compliance by enabling suppliers to monitor the statuses of sourcing projects and submit their quotes.

7.4.2 Business Capability Heatmap

You perform the business capability heatmapping by focusing on pain points and opportunities (as identified during the scoping phase), and then you map it with the sustainability priorities of zero emissions, zero waste, and zero inequality (as shown in Figure 7.9).

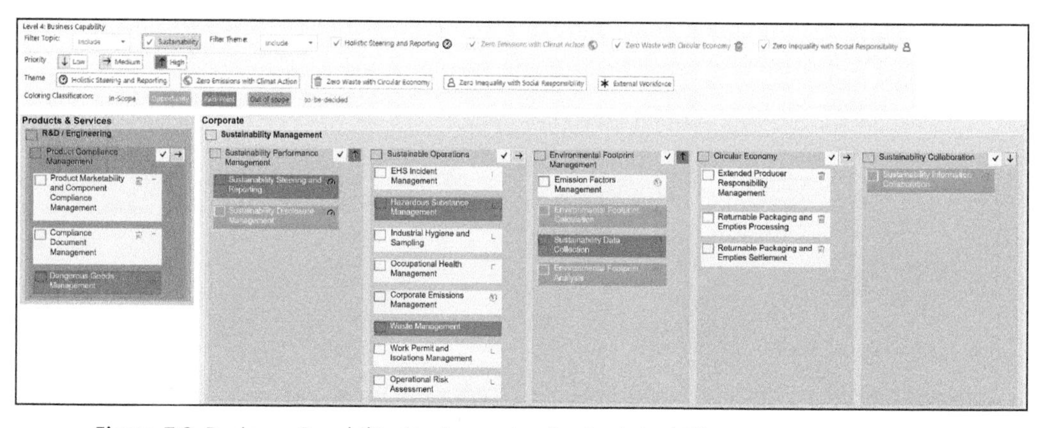

Figure 7.9 Business Capability Heatmapping for Sustainability

As an example, you can see from the business capability heatmap that there are four key pain points (color coded in red) in the capabilities: **Sustainability Steering and Reporting**, **Hazardous Substance Management**, **Waste Management**, and **Sustainability Data Collection**. However, GreenCon Inc. has also identified four opportunities (color coded in green) to move toward the target architecture and roadmap: **Sustainability Disclosure Management**, **Environment Footprint Calculation**, **Environment Footprint Analysis**, and **Sustainability Information Collaboration**.

Heatmapping provides you with a solid decision-making foundation for future investment planning and priorities. Classifying the business capabilities in terms of their priority, from **High** (color coded in red) to **Low** (color coded in light green), has the highest strategic impact that you can leverage as an input for the target architecture and the roadmap for sustainability (as shown in Figure 7.10).

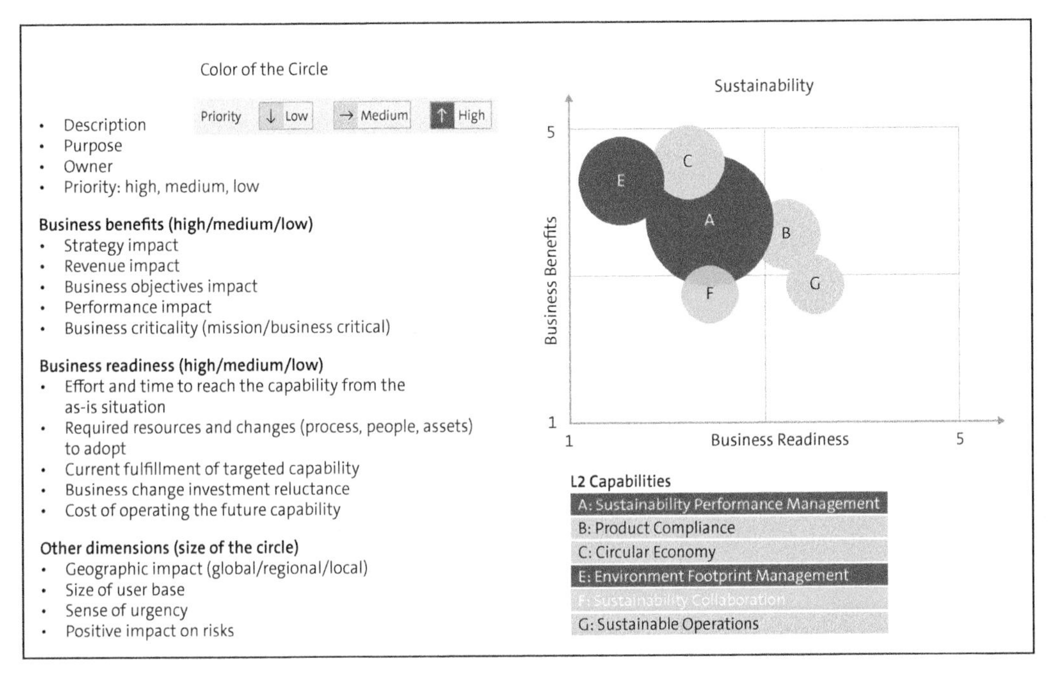

Figure 7.10 Business Capability Classification for Sustainability

As an example, let's look at **E: Environment Footprint Management**. The current fulfillment of this capability is low, but the business benefit is high. Currently, this is managed "manually" (outside of the system), and managing it in the system is a huge opportunity in the target architecture, with SAP S/4HANA as the core.

Prioritization has an impact on the architectural planning when you create the target architecture and roadmap (i.e., you can clearly link the prioritized capabilities to the designed target architecture). Also, you can realize the prioritized business capabilities in a timely manner during roadmap creation.

7.4.3 High-Level Solution Component Map

You can derive the high-level *solution component map* (also known as a product map) for GreenCon Inc. by mapping the business capabilities to the solution capabilities of SAP applications, as shown in Figure 7.11.

As an example, SAP Sustainability Footprint Management is a public cloud solution on SAP Business Technology Platform (SAP BTP), along with the digital core of SAP S/4HANA Cloud Private Edition. You need to choose a deployment option for each of the applications: whether it will be in the public cloud, the private cloud, or an on-premise platform. GreenCon Inc. has decided to go for SAP S/4HANA Cloud Private Edition (via RISE with SAP) due to its cloud-first strategy, and it has a long-term vision for cloud architecture to reduce its TCO and realize a higher return on investment (ROI). These goals tie in with our cloud- and TCO-focused use cases in Chapter 5 and Chapter 9, respectively.

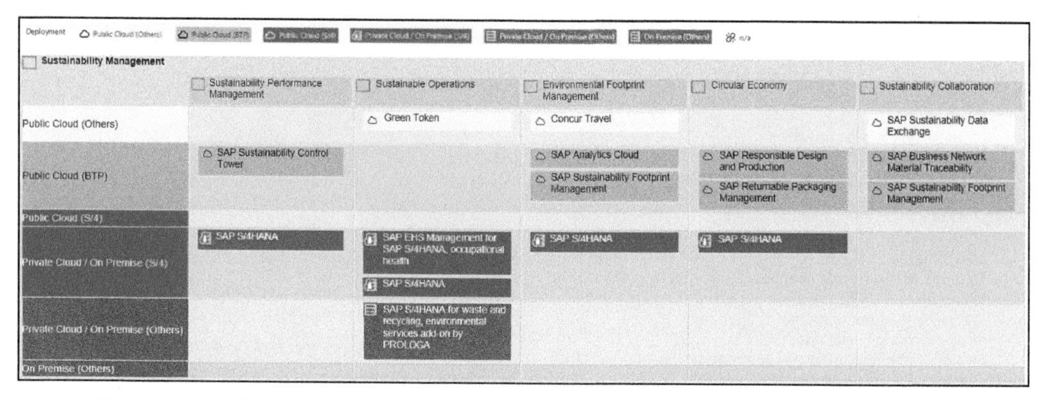

Figure 7.11 High-Level Solution Component Map for Sustainability Management

7.4.4 Application Architecture

Designing the target architecture for sustainability applications for GreenCon Inc. is not an easy task since the capabilities and processes are embedded within all the business domains, such as R&D and engineering, sourcing and procurement, the supply chain, and corporate functions.

Key architecture findings for GreenCon Inc. are as follows:

- Most of the product compliance areas implemented at GreenCon Inc. are available in classic SAP S/4HANA for product compliance. As of today, other than waste management, everything can be converted to SAP S/4HANA (until the specification database is completely taken out). The specification database is a centralized repository and database that includes specifications such as chemical composition, physical properties, safety data, regulatory compliance information, and sustainability metrics for each item.
- Waste management is deprecated in SAP S/4HANA, and GreenCon Inc. is looking to implement greenfield new waste management. New waste management has new data models in SAP S/4HANA, and there are three applications to be evaluated by GreenCon for its waste management use cases: waste and recycling in SAP S/4HANA

Cloud Private Edition; an SAP S/4HANA Cloud Private Edition waste and recycling environmental services add-on by PROLOGA; and an SAP Waste and Recycling automated route planning option by PROLOGA.

- All the customizations done in substance volume tracking, hazardous substance management, and dangerous goods solutions will continue to work in the classic mode in SAP S/4HANA. Later, when the old data model needs to be converted to a chemical compliance info object (CCI), this needs to be reevaluated and analyzed to determine whether any standard functionality is available in SAP S/4HANA.
- All waste management WRICEFs need to be investigated when the new waste management module is implemented in SAP S/4HANA.
- The target application for recycling administration is SAP Responsible Design and Production, which has been generally available since 2021.
- The company should investigate SAP EHS Management, environment management (which is a successor to SAP Environment Compliance 3.0) for potential use in China as a greenfield implementation.
- In the end state with SAP S/4HANA, all the chemicals used in risk management, incident management, environment management, and chemical data management will be replaced with CCI (the replacement for the specification database on the product compliance side).

Product Compliance: Classic versus New Capabilities

Today, SAP customers have a choice. They can implement new product compliance in parallel with classic product compliance, with both CCI and a specification database coexisting for marketability assessment functionality with integration into product master, sales, delivery, procurement, and transportation. However, we recommended that customers choose either classic or new capabilities, depending on their needs, since there may be no synchronization between classic and new data models. It's important to note that classic capabilities will only be usable through SAP S/4HANA release 2029. Afterward, customers will need to switch to new capabilities. Classic capabilities will be supported until 2036, given seven years' maintenance for each SAP S/4HANA release.

Refer to SAP Note 3015382 for more information.

The sketch for the target application architecture for SAP sustainability solutions for GreenCon Inc. is illustrated in Figure 7.12, which also highlights sustainability priorities.

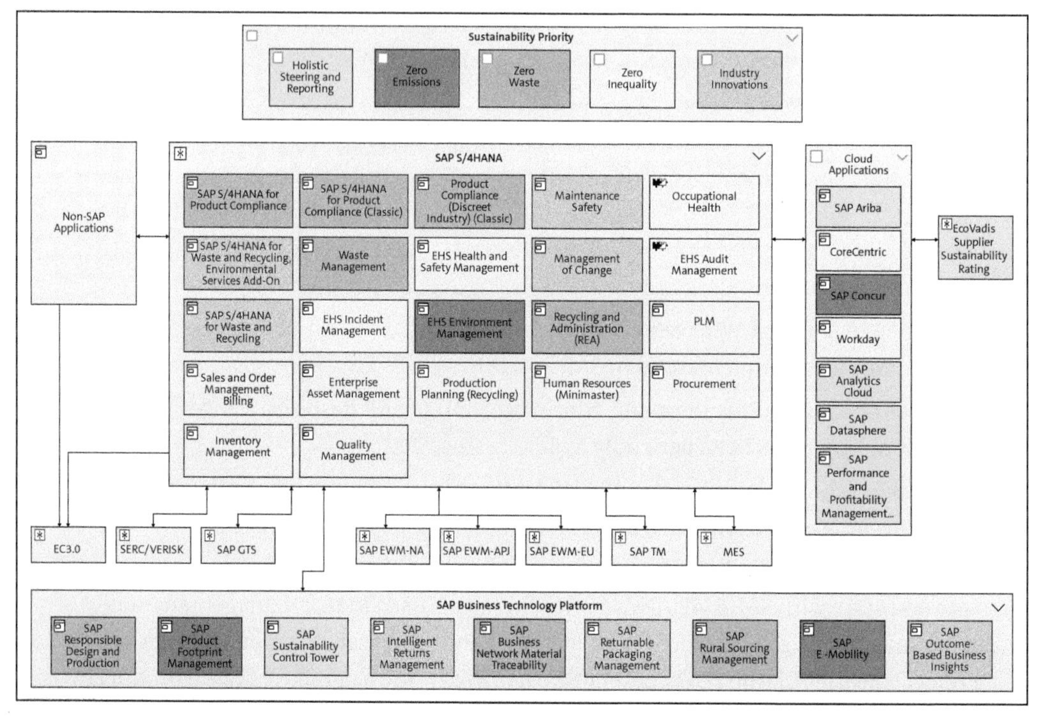

Figure 7.12 Target Application Architecture for SAP Sustainability Solutions

To meet its sustainability challenges, GreenCon Inc. needs to choose the right SAP solutions from the SAP Cloud for Sustainable Enterprises portfolio. Based on our architecture sessions and aligned with the company's sustainability priorities, we recommend the following key SAP sustainability applications as target applications (see Figure 7.13):

- **SAP Sustainability Control Tower**
  This application helps make sustainability data accessible, auditable, and forward thinking. It helps unlock the power of data to help GreenCon Inc. record, report, and act on its sustainability goals and comply with regulatory reporting.
- **SAP Sustainability Footprint Management**
  This application will help calculate GreenCon Inc's carbon footprint from cradle to gate and will consider materials, transportation, and production.
- **SAP EHS Management, environment management**
  This application will help GreenCon Inc. to measure, calculate, and track greenhouse gas emission inventories at the enterprise level to deliver a single primary source of data. This will be a huge benefit since it won't only collect and integrate emissions data but also combine direct (scope 1) and indirect (scope 2 and 3) emissions data that includes both plant-owned assets and services and operations.
- **SAP Responsible Design and Production**
  This solution will help GreenCon Inc. manage its EPR obligations and plastic taxes so it can control and eliminate costs related to the downstream waste system and make

design changes to eliminate waste. This will help ensure compliance and better decision-making.

- **Waste management component of SAP EHS Management, environment management**
  This application enables waste generators to reduce the environmental impact of their waste by supporting the handling of waste disposal processes in compliance with regulations and laws. GreenCon decided to implement the SAP S/4HANA Cloud Private Edition, waste and recycling, environmental services add-on by PROLOGA in phase 1 of the transformation roadmap.
- **SAP S/4HANA for product compliance**
  SAP S/4HANA for product compliance offers GreenCon Inc. complete end-to-end support across all relevant processes: from product development, purchasing, and production to sales and distribution. Deploying automated compliance assessment and reporting, GreenCon can mitigate reputational risk while shortening time to market.

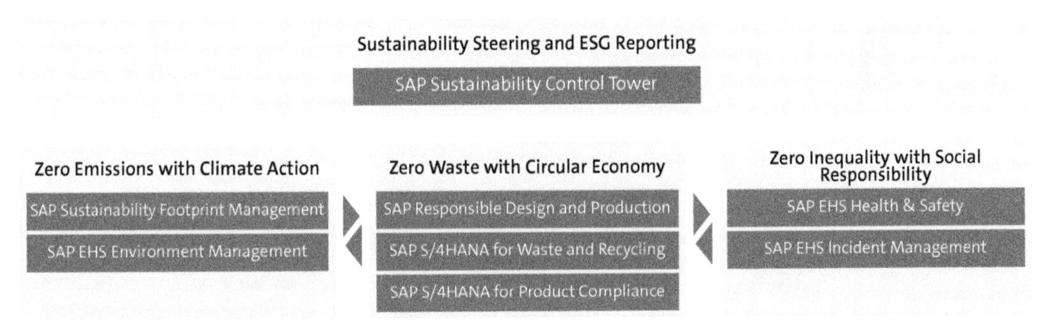

Figure 7.13 Recommended SAP Sustainability Target Applications for GreenCon Inc.

SAP EHS Management will provide a holistic approach to worker safety that systematically embeds operational risk management across the entirety of GreenCon Inc.'s operations. This will help the company respond proactively to hazards and withstand future health and safety challenges.

SAP Green Ledger

SAP is pioneering SAP Green Ledger (a ledger-based accounting system for carbon), and the vision is to move from averages to actuals and implement transactional accounting (i.e., recording, reporting, and acting on carbon in sync with financial data for every business transaction). Managing sustainability data alongside financial data allows the company to do the following:

- Flexibly drill down on revenue, cost, and margin in relation to emissions along organizational structures and value chains.

- Set carbon budgets at the cost center level.
- Embed carbon into business decision-making at the point where a decision is made.

7.4.5 Sustainability Roadmap

Based on our extensive architecture workshop sessions, we've developed a comprehensive roadmap for GreenCon Inc. (illustrated in Figure 7.14).

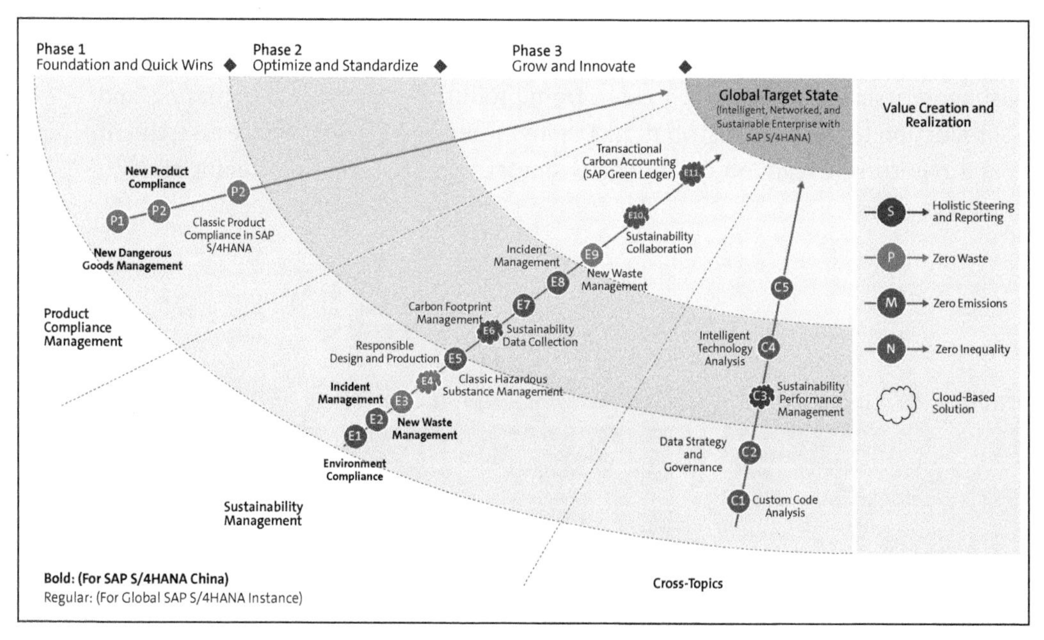

Figure 7.14 Sustainability Roadmap for GreenCon Inc.

The key clusters are as follows:

- **Product Compliance Management**
  GreenCon has a choice of whether to use new capabilities or let capabilities remain in classic form for some time. GreenCon's China business plans to use new capabilities since the transition scenario is greenfield implementation, but GreenCon as a parent company will retain its classic capabilities due to its system conversion transition approach.
- **Sustainability Management**
  This is the big piece of the roadmap that can manage the environmental footprint of a company, manage and steer sustainability performance, ensure sustainable operations, and enable a circular economy. GreenCon has decided to implement the latest solution portfolio (i.e., new waste management).

Remember!

This roadmap is adaptable and can be tailored to suit the specific needs and priorities of any organization. SAP typically works closely with its customers to develop a roadmap that aligns with the specific starting point and customers' unique business objectives and sustainability priorities.

7.5 Summary

This chapter covered the case study for GreenCon Inc., in which we defined the target architecture and roadmap for SAP sustainability applications by using a step-by-step approach from SAP EA Framework. To summarize, EHS and sustainability are core elements of the SAP S/4HANA application and will become available immediately when SAP S/4HANA system is deployed. There are two main scenarios for the EHS and sustainability transformation:

1. GreenCon Inc. will deploy SAP S/4HANA new waste management, new environment footprint management, and new product compliance for marketability assessments as part of its SAP S/4HANA deployment in China. This will support the replacement of SAP ERP waste management, which is deprecated in the SAP S/4HANA application.
2. GreenCon Inc. will deploy SAP Sustainability Control Tower. This will enable end-to-end visibility of compliance through to the production equipment on the factory floor.

Most of the EHS functionalities that GreenCon uses will continue to be used in the new SAP S/4HANA environment in the classic mode. We suggest that GreenCon plan to use new waste management functionality in the China instance of SAP S/4HANA because the old SAP ERP waste management functionality is deprecated in SAP S/4HANA. GreenCon will have to discuss the requirements for China implementation and plan accordingly if a separate SAP S/4HANA environment is to be planned. We also recommend that the company implements new product compliance, new waste management, incident management, environment management, SAP Responsible Design and Production, management of change, incident management, and risk assessment, specifically since all these can be implemented as greenfield implementations.

This completes our sustainability use case. Now, we'll shift gears to discuss mergers, acquisitions, and divestitures in the next chapter.

Chapter 8
Mergers, Acquisitions, and Divestitures

Mergers, acquisitions, and divestitures constitute one of the strategic priorities for business transformation. Mergers and acquisitions are a daunting task from an enterprise architecture perspective: every system and process used by both companies must be accounted for, analyzed, and harmonized to achieve synergies, and hence, SAP EA Framework plays a key role.

This use case explores how a company can integrate its mergers, acquisitions, and divestitures (MAD) strategy into its enterprise architecture to provide strategic guidance in every phase of the deal in both the buyer's and the seller's organization. Enterprise architects are vital to *mergers* and *acquisitions* (M&A) success, and MAD is one of the key strategic business priorities for many SAP customers in many industries. M&A constitutes a growth strategy that helps to grow the company inorganically by enabling it to acquire strategic markets, build complementary portfolios, expand its niche, differentiate its capabilities, and carve out strategic flexibility, agility, and focus. In contrast, *divestiture* is a scaling-down decision to grow organically by focusing on differentiation and core capabilities, raising capital, reducing debt, and complying with regulatory requirements.

There are two more terms associated with divestiture: carve-out and spin-off. A *carve-out* involves a company creating a new, separate entity out of a part of its existing business. This new entity often becomes an independent company, and the parent company may sell a minority stake in the new company through an initial public offering (IPO) or to a strategic investor. A *spin-off* involves a company creating a new, independent company by distributing shares of the new company to its existing shareholders. Unlike a carve-out, a spin-off typically results in the parent company fully separating from the new entity.

The key benefits of using enterprise architecture methodology, content, tools, and services in MAD are as follows:

- Providing a single source of truth that all stakeholders can understand in the predeal, during deal execution, and post-deal phases of the MAD cycle

- Assisting with heatmapping and clustering the entire business capability, the business process, and the business scenarios to bring standardization and harmonization to the two entities
- Reducing IT risk by providing transparency into the IT landscape by removing redundancies and achieving synergies
- Accelerating IT synergies and integration with the transition service agreement (TSA) guidelines and timeframe
- Modeling North Star service architectures to ensure that the objectives of MAD are fulfilled between the two entities

This chapter will dive into the utilization of SAP EA Framework to facilitate MAD, and it will include a case study that demonstrates how enterprise architects can scope both a divestiture and an acquisition. In addition, we'll examine how the case studies achieve their transition strategy or target architecture and roadmap.

8.1 Business and IT Context

You must know that MAD for an organization has an important business context, and enterprise architecture can be an excellent enabler to provide a holistic view. Let's look at the business and IT context of MAD as a strategic priority.

8.1.1 Business Context

MAD is a strategic move for an organization, and it impacts the organization's structure, stakeholders, business capabilities, and business processes. MAD strategies are used by companies for various reasons, including expanding market share, gaining access to new technologies or markets, achieving cost synergies, and focusing on core competencies. Each strategy has its own implications for the companies involved, their employees, their shareholders, and the overall market.

Two real-life examples and reasons for M&A are as follows:

- Adding niche and innovative capabilities to a product portfolio. For example, Bristol-Myers Squibb acquired Celgene to get into the cell therapy business.
- Expansion into one or many product categories in high-priority and high-growth markets. For example, Mondelez acquired Ricolino to expand route-to-market capabilities.

Two real-life examples and reasons for divestitures are as follows:

- Strategic flexibility and compelling financial profile. For example, Johnson & Johnson may spin-off a consumer health business for strategic flexibility and to provide each company with a compelling financial profile that more accurately reflects the strengths and opportunities of the business.

- Enhanced agility and focus. For example, 3M may spin-off a health care business to better position it for long-term success.

Types of Mergers and Acquisitions

As an enterprise architect, you often encounter various strategic transactions that reshape companies' ownership, structure, and operations. These include mergers, acquisitions, and divestitures, each of which has unique characteristics and implications. While there are different types of M&A, they all present similar IT challenges, such as system integration, data migration, and aligning business processes. Some different types of M&A are as follows:

- **Horizontal M&A**
  Horizontal mergers involve companies that offer similar products or services, and they aim to reduce competition and consolidate the companies' market position. For example, Disney and Pixar merged to combine creative forces and dominate the animated film industry.
- **Vertical M&A**
  Vertical mergers occur between companies at different stages of the same supply chain, and they aim to increase efficiency and control over the production process. A notable example is Tesla acquiring SolarCity, which allowed Tesla to integrate its electric vehicle business with solar energy solutions.
- **Market extension M&A**
  In a market extension merger, companies that offer the same products or services in different markets join forces to expand their customer base. For instance, T-Mobile and Sprint merged to enhance their reach in the US telecom market.
- **Product Extension M&A**
  Product-extension mergers happen when companies producing related products in the same industry merge to broaden their product offerings and access more customers. An example is Google acquiring Fitbit, which enabled Google to expand its wearable technology and health data portfolio.
- **Conglomerate M&A**
  Conglomerate mergers occur between companies in completely unrelated industries, often to diversify business interests and reduce risk. For example, Berkshire Hathaway acquiring Precision Castparts brought together a financial services giant and a manufacturer of aerospace components.
- **Concentric M&A**
  Also known as a congeneric merger, a concentric merger occurs between companies in the same industry that offer different products or services that complement each other. An example is Microsoft acquiring LinkedIn, which combined Microsoft's software expertise with LinkedIn's professional networking platform, creating synergies in the digital workspace.

8.1.2 IT Context

From an IT perspective within the framework of enterprise architecture, MAD presents several key contexts and challenges that need to be addressed. Some critical IT use cases are as follows:

- **System landscape optimization**
  Mergers and acquisitions often result in redundant IT infrastructure, including servers, data centers, and networking equipment. Enterprise architecture helps in assessing the current infrastructure landscape of both organizations and developing a rationalization plan to consolidate resources, optimize costs, and streamline operations.
- **Application integration and rationalization**
  Organizations on either side of a merger or acquisition typically have disparate sets of applications supporting various business functions. Enterprise architecture facilitates the integration and rationalization of these applications by identifying overlaps, defining integration patterns, and creating a roadmap for consolidating or retiring redundant applications. Examples include SAP-to-SAP rationalization, non-SAP-to-SAP rationalization, and legal entity rationalization.
- **Data integration and management**
  Organizations on either side of a merger or acquisition may have different data structures, formats, and governance policies. Enterprise architecture plays a crucial role in designing data integration solutions, establishing data standards, and implementing data governance frameworks to ensure consistency, accuracy, and compliance across the integrated organization. This may include designing transition strategies such as selective data transition, for example.
- **Technology alignment**
  M&A can result in technology misalignment between the organizations, leading to compatibility issues and integration challenges. Enterprise architecture helps in aligning the technology stacks, evaluating compatibility between systems and platforms, and developing a roadmap for technology standardization and modernization.

Enterprise architects perform critical activities in each of the MAD phases, both from the seller side and the buyer side, by providing strategic guidance and technical expertise to ensure a smooth integration or separation of systems, processes, and technologies between the involved entities (see Figure 8.1).

The enterprise architect activities involved in each of the phases are as follows:

- **Preparation**
  This is the assessment and analysis phase that occurs on a continuous basis to look for potential MAD opportunities. The enterprise architect assesses the business capabilities, existing IT infrastructure, applications, data systems, and architectures

of both the acquiring company and acquired and divested companies. This evaluation helps identify redundancies, gaps, and opportunities for integration or separation to help the organization stay relevant in the future.

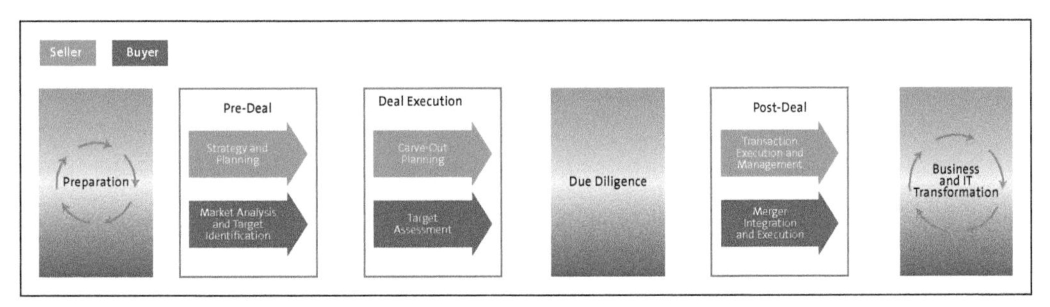

Figure 8.1 Phases of MAD

- **Pre-deal**
  This phase focuses on strategic planning, either by M&A (on the buy side) or the divestment (on the sell side) of their branches or business areas:
  - *MAD strategy and planning (sell side)*: In this phase, enterprise architects evaluate the overall business strategy assessment and review the business capabilities. Most large enterprises have grown organically or inorganically and accumulated a lot of redundancies in capabilities that need to be identified for the target architecture and prepared for the transformation. This is a highly confidential phase, and the organization needs a trusted advisor to oversee it without compromising the deal.
  - *Market analysis and target assessment (buy side)*: Integration planning is the key in this phase. Enterprise architects develop a comprehensive integration plan that aligns with the strategic goals of the MAD. They create a roadmap for combining or separating systems, ensuring interoperability, data migration, and seamless operation during and after the transition.
- **Deal execution**
  In this phase, the organizations focus on analyzing and planning their transformation according to their corporate strategy, either by M&A activities in the growth phase or by divestitures in the restructuring phase, as follows:
  - *Carve-out planning (sell side)*: This is the stage when the organization makes the decision to organically grow its own business capability by divestiture. It plans a carve-out or spin-off (i.e., it develops a detailed plan for the divestiture process, including timelines, key milestones, resource allocation, and IT system separations) to enhance its agility and focus.
  - *Target assessment (buy side)*: This is the stage when the organization makes the decision to inorganically grow its own business capabilities via M&A. Enterprise architects can pick up the outputs generated in the earlier phase of preparation

and the pre-deal phase, and they can iterate to the next level of detail to validate all the assumptions and recommend the first baseline version of the target architecture and roadmap.

- **Due diligence**
  In this phase, the organization has decided to move ahead with MAD. Enterprise architects need to work in close collaboration with business and IT stakeholders on the following key aspects:
  - *Stakeholder management*: Enterprise architects collaborate with various stakeholders, including IT teams, business leaders, and external consultants, to communicate the technical aspects, progress, and impacts of the integration or separation. Clear communication is vital to managing expectations and ensuring a smooth transition.
  - *Technology alignment*: Enterprise architects ensure that the IT strategies, technologies, and architectures of both entities align with each other and with the overall business objectives post-MAD. This alignment is crucial for optimizing resources and achieving synergy.
  - *Risk management*: Enterprise architects identify potential risks associated with the integration or separation of systems, including data security, compatibility issues, and operational disruptions. They develop mitigation strategies to address these risks and ensure business continuity.
  - *Governance and compliance*: Enterprise architects establish governance frameworks and ensure compliance with regulatory requirements throughout the MAD process. They oversee adherence to industry standards and best practices, and they define architecture principles.
- **Post-deal**
  This phase focuses on IT synergies due to post-merger optimization, transaction execution and management, merger integration, and execution. This phase is usually conducted under time pressure because it needs to adhere to TSA commitments. Enterprise architects must do the following:
  - *Post-merger optimization*: After the integration or separation is complete, enterprise architects continue to monitor and optimize the combined or separated IT landscape. They refine architectures, streamline processes, and identify opportunities for further enhancement and innovation.
  - *Transaction execution and management (sell side)*: This is the phase when the carve-out planning done earlier is further detailed to be made ready for execution.
  - *Merger integration and execution (buy side)*: This is the phase when the target architecture and roadmap are designed to guide the execution team and the architects develop a plan for integrating IT systems and infrastructure, including data migration, application integration, and technology standardization.

- **Business and IT transformation**
  This is the phase when the enterprise architect's role is multifaceted, combining technical expertise with strategic insight to facilitate a successful transition during MAD activities while minimizing disruptions and maximizing the benefits of the combined or separated entities. This is conducted in a continuous improvement mode to enhance and enrich the architecture at regular intervals, based on business priorities and pain points.

Enterprise Architect Involvement in Mergers and Acquisitions

Almost 90% of enterprise architects are involved in M&A activities to support post-merger integration efforts, according to the January 2021 LeanIX Mergers & Acquisitions Survey (see *http://s-prs.co/v586306*). IT transparency, rationalizing applications, developing a target IT landscape, building a common business capability map, and developing scenarios during post-merger integration are among the top uses cases enterprise architects believe they contribute to the most.

8.2 Problem Statement

Consider a multinational pharmaceutical products company, PharmCo Inc., that is divesting its consumer products business and acquiring a cell and gene business called NewCo Inc. Figure 8.2 and the following list illustrate the definition of MAD, using these fictitious companies as examples:

- **Merger**
  This occurs when PharmCo and NewCo combine to form a new entity called MergeCo. It's a strategic move in which both companies agree to pool their resources, operations, and personnel to create a stronger, more competitive entity. A real-world example occurred when Exxon and Mobil merged to form ExxonMobil, creating one of the largest oil companies in the world. The merger was valued at around $81 billion.
- **Acquisition**
  In an acquisition, PharmCo takes over NewCo by purchasing a controlling stake or ownership in it. The acquired company may retain its name and identity (as shown in Figure 8.2) or be absorbed into the acquiring company. A real-world example occurred when Facebook acquired Instagram for approximately $1 billion in cash and stock. This acquisition helped Facebook expand its social media reach and integrate Instagram's popular photo-sharing app.
- **Divestiture**
  Divestiture is the process of selling off assets, subsidiaries, or divisions of a company. It's often done to streamline operations, focus on core competencies, or raise capital. Companies might divest noncore businesses that are underperforming or

not aligned with their strategic objectives. For example, in our case study, business unit 3 (BU3) is sold to another company called SpinCo (see Figure 8.2). A real-world occurred when Kraft Foods split into two companies: Kraft Foods Group (focusing on North American grocery products) and Mondelez International (focusing on global snacks). This divestiture aimed to create more focused and competitive companies.

- **Carve-outs and spin-offs**
  Divestiture is like selling to another company, and a spin-off means the part that is spun off is run independently. In both cases, you "carve out" a part of the business from the mother company. In this case, PharmCo is a carve-out, and SpinCo is a spin-off since it legally becomes another company. A real-world example of a carve-out is DowDuPont, which was formed from the merger of Dow Chemical and DuPont, which then executed a series of carve-outs, resulting in the creation of three independent companies: Dow Inc., DuPont Inc., and Corteva Agriscience. A real-world example of a spin-off occurred when Abbott Laboratories spun off its research-based pharmaceuticals business into a new company called AbbVie. This spin-off allowed AbbVie to focus on its drug development and commercialization efforts.

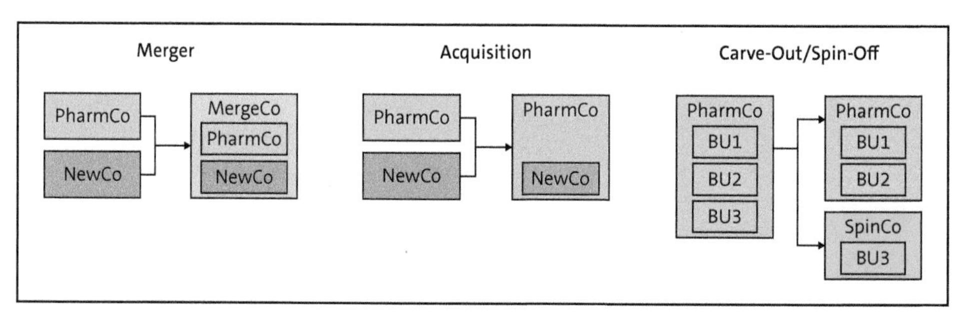

Figure 8.2 Definition of Mergers, Acquisitions, and Divestitures Using Fictitious Companies

Now, let's dive deeper into our two use cases along with their problem statements.

8.2.1 Divestiture Use Case: Company Profile and IT Landscape

In this first use case, enterprise architects get involved in the deal execution phase for the carve-out planning. BU3, which is dedicated to consumer products, is being divested and carved out as SpinCo from the existing IT architecture so that PharmCo can achieve strategic flexibility and enhanced agility in order to put more focus on its pharmaceutical products.

PharmCo has two SAP ERP production systems, one for US business units and another in Canada. The Canadian SAP ERP system has comingled data for pharmaceutical and consumer products, and the US SAP ERP system has larger representation of global processes.

Table 8.1 shows that BU1, which represents pharmaceutical products, accounts for 60% of PharmCo's total revenue and is their core business. Similarly, BU2 and BU3, which represent medical devices products and consumer products (respectively), account for 15% and 25% (respectively) of total revenue for PharmCo. The board of directors of PharmCo decided to spin-off BU3 to focus only on its pharmaceutical business along with its related medical devices business.

Business Unit	Industry Segments	Sales Revenue
1	Pharmaceutical products	60% of total revenue
2	Medical devices products	15% of total revenue
3 (SpinCo)	**Consumer products**	**25% of total revenue**

Table 8.1 PharmCo's Divestiture of BU3 (SpinCo)

SpinCo has comingled pharmaceutical data along with consumer products data in the SAP ERP system in Canada. SpinCo company codes need to be carved out from PharmCo's ERP system along with all the pharmaceutical data (including the historical data, depending on the divestiture and TSA agreement). Usually, historical data is kept under the parent company, for legal and statutory reasons.

SpinCo has approached the SAP services team (i.e., SAP Transformation Hub) to provide advisory services on its transition strategy to move to SAP S/4HANA and carving out from the existing SAP ERP system. This is from the seller's perspective on how to separate IT systems.

8.2.2 Acquisition Use Case: Company Profile and IT Landscape

In this second use case, enterprise architects get involved in the post-deal phase for merger integration and execution. PharmCo is acquiring NewCo to enhance its product portfolio with a niche capability, and at the same time, it's entering a new market with its cell and gene products.

Table 8.2 shows that PharmCo decided to acquire a new and niche innovative business of cell and gene products to augment its pharmaceutical business model as per the growth strategy.

Business Unit	Industry Segments	Sales Revenue
1	Pharmaceutical products	60% of total revenue
2	Medical devices products	15% of total revenue
4 (NewCo)	**Cell and gene products**	**TBD**

Table 8.2 PharmCo's Acquisition of BU4 (NewCo)

NewCo has a unique end-to-end cell and gene process for manufacturing, the supply chain, and the sales business domain. This is a make-to-order (MTO) process, while PharmCo mainly has a make-to-stock (MTS) process set up in its SAP S/4HANA system.

PharmCo has approached the SAP services team (i.e., SAP Transformation Hub) to design the target architecture and roadmap that will consider the new capabilities and applications integrations due to its acquisition of NewCo. This is from the buyer's perspective on how to integrate IT systems from two different companies with two different ERP systems—one being SAP S/4HANA and the other being a non-SAP system. The objective is to design the target architecture using the best of both worlds.

8.3 Applying the Enterprise Architecture Transformation Approach

In Chapter 2, you learned about SAP EA Framework, which consists of five pillars: methodology, reference content, tools, practice, and services. Now, let's apply the SAP EA Methodology and transformation approach to these two use cases to evaluate the transition strategy for SpinCo. and to design the target architecture and roadmap for PharmCo.

8.3.1 Divestiture Use Case

Based on our understanding of the customer's situation and requests, in Figure 8.3, we've depicted the key architecture modules identified in scope for the divestiture use case. The modules are as follows:

- **Strategy mapping**
  This module will focus on understanding the strategic importance of the TSA in the context of a divestiture of Spin Co., enabling enterprise architects to effectively align the business strategy with IT requirements and project timelines.
- **Business capability**
  This module focuses on identifying the business capabilities relevant to the divestiture of the consumer products division (BU3) and heatmapping them to determine what falls within and outside the scope of the business and IT transformation.
- **Application architecture**
  This module focuses on reviewing the target application architecture to document the impacts of the divestiture on the scope of business capabilities. While other modules of the application and data architecture are important, they are not required in the first iteration. The same applies to the technology and platforms modules. However, data architecture must be reviewed in a subsequent iteration of the architecture review.
- **Transition scenario evaluation**
  This is the most important module scoped since SpinCo's primary goal is to evaluate

the transition scenarios of the spin-off of its consumer products business units to the target application and roadmap. This is also to ensure how much historical data is needed to set up the target.

The modular end-to-end enterprise architecture planning helps in enabling key architecture decisions for MAD. PharmCo, in its divestiture of BU3 (i.e., SpinCo), decided to perform transition scenario evaluation with the help of the SAP enterprise architect. In a typical delivery of this module, SAP customers evaluate the transition strategy options: greenfield (a.k.a. new implementation), brownfield (a.k.a. system conversion), and selective data transition. But here, the IT arm of the divested company wants to dig deep and get SAP's recommendation on the different options for the selective data transition approach. The Data Management and Landscape Transformation (DMLT) group also played a key role in deciding and recommending the transition options, especially with regard to data-related needs with data quality and data integration services. SAP's DMLT team has several tools that can assist with the planning and management of the divestiture.

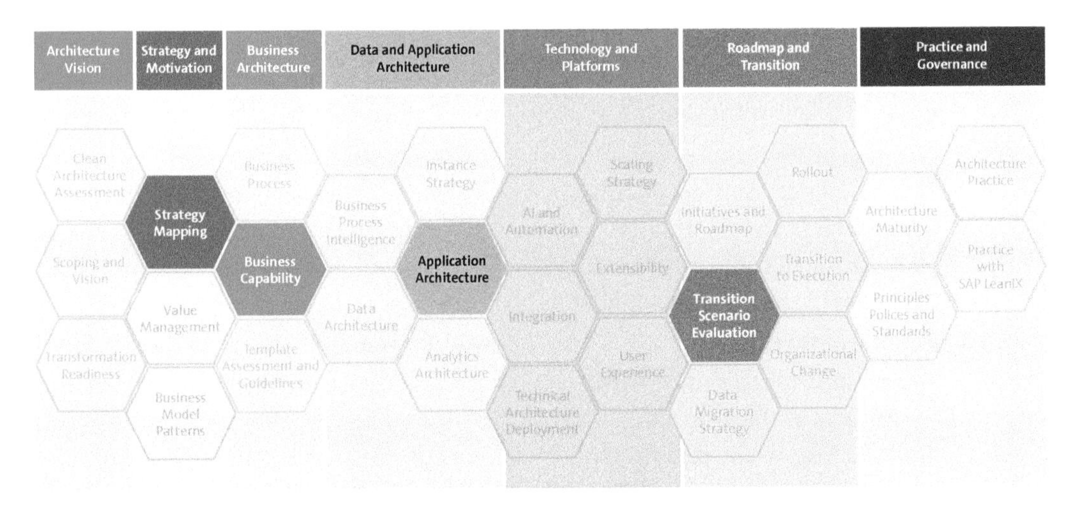

Figure 8.3 Use Case 1: Transition Strategy to Move to SAP S/4HANA for SpinCo

8.3.2 Acquisition Use Case

Based on our understanding of the customer's situation and requests, the key architecture modules we've identified in scope for the acquisition use case are as follows (see Figure 8.4):

- **Strategy mapping**
  This module will focus on understanding the strategic importance of the TSA in the context of the acquisition of NewCo, enabling enterprise architects to align the business strategy with IT requirements and project timelines effectively.

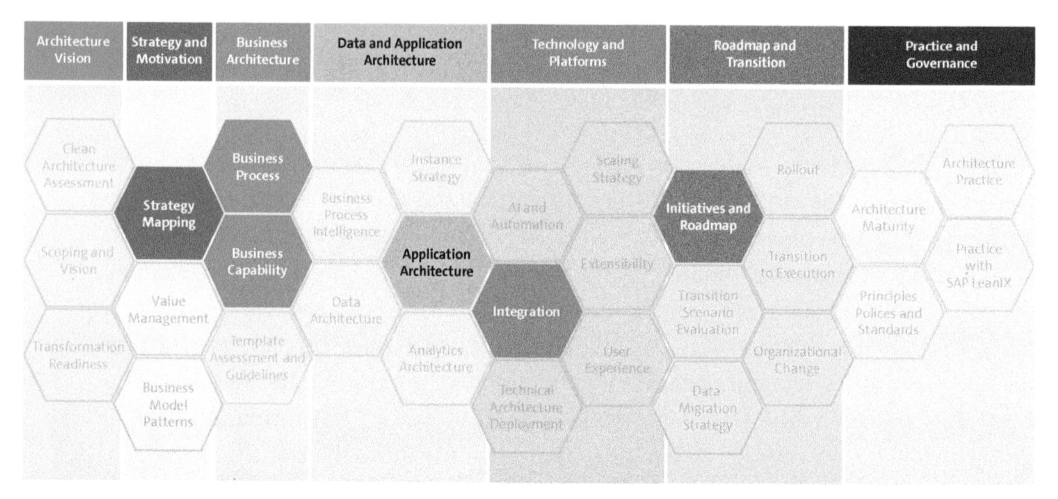

Figure 8.4 Target Architecture and Roadmap Design for PharmCo with Integration of NewCo

- **Business capability**
  This module focuses on identifying the business capabilities relevant to the acquisition of the cell and gene products division (BU4) and heatmapping them to determine what falls within and outside the scope of the business and IT transformation.
- **Business process**
  This module focuses on what business processes are in scope for the architecture design. This is an important module for evaluating the end-to-end business process of the acquired company (NewCo) to assess its fit with existing processes and applications of the parent company (PharmCo) to decide on the best of both worlds.
- **Application architecture**
  This module focuses on reviewing the target application architecture to document the impacts of the acquisition on the scope of business capabilities as well as business process. While other modules of the application and data architecture are important, they are not required in the first iteration. The same applies to the technology and platforms modules. However, data architecture must be reviewed in a subsequent iteration of the architecture review as this is important from the M&A perspective.
- **Integration**
  This module focuses on integration strategy and the critical integration list derived from the application architecture module. This module includes several options to integrate SAP and non-SAP cloud and on-premise applications after the acquisition, impacting the two separate IT landscapes of PharmCo and NewCo.

- **Initiatives and roadmap**
  Once we've developed an initial target architecture that incorporates the required changes aligned with the acquisition strategy, we can begin drafting the first roadmap for NewCo. Other modules from the same domain can be considered in subsequent iterations.

PharmCo, in its acquisition of BU4 (i.e., NewCo), wants to perform target architecture and roadmap design and get SAP's recommendation on integration of business capabilities, business processes, and applications.

8.4 Key Deliverables and Outcomes

In this section, we'll outline the main deliverables and results from the modules that we discussed in the preceding section. We'll provide a brief explanation of each deliverable's creation process and the necessary prerequisites. Furthermore, we'll showcase the outcome or a sample of the ultimate deliverable.

After you conduct the scoping and collect inputs from the customer, you need to execute the following steps to create the desired outcome and deliverables for the customer in each of the following use cases.

8.4.1 Divestiture Use Case

Enterprise architects for SpinCo performed these steps to derive their transition strategy to move to SAP S/4HANA using a selective data transition approach:

1. Understand midterm and long-term strategic priorities. In this case, SpinCo's midterm priorities are to move its current SAP ERP into SAP S/4HANA within the TSA timeframe of two years. Otherwise, they'll need to pay a huge penalty and fine to PharmCo.
2. Analyze PharmCo's overall business capability map and identify business capabilities that are relevant to the carve-out for the selective data transition scenario evaluation.
3. Define the scope of the data transition by identifying the specific business units, processes, and data sets that need to be migrated to SAP S/4HANA.
4. Analyze the current data landscape to understand data dependencies, interdependencies, and any potential data quality issues.
5. Describe the transition path options for selective data transition and explain their impacts (high-level pros and cons). The options identified during the scoping are as follows:

- *Option 1*: Split Canada, create a global template with pilot implementation in Canada (CA), and then do the North American (NA) rollout.
- *Option 2*: Create a shell template using the NA system, harmonize and then bring Canadian data and implement it, and then do the NA rollout.
- *Option 3*: Conduct a system merge of NA and CA, split Canada, harmonize data and create one S/4HANA template, and then do a big-bang rollout.

6. Jointly define all the criteria for the selective data transition. Typical criteria used for this evaluation are as follows:
   - *Financial risk*: The primary financial risk for the divested company is that if they don't meet the TSA, they'll have to pay a heavy penalty. The timeframe is usually 18 to 22 months to carve out the system landscape and start operating independently.
   - *Deployment speed and duration*: Deployment speed and duration are the key criteria for adhering to the TSA commitments. They refer to the time taken to carve out the business units (i.e., the company codes and the associated data) from PharmCo to SpinCo.
   - *Business fit and flexibility*: These criteria answer the question of whether the solution fulfills all the future business requirements (e.g., global processes, harmonization, standardization). Harmonizing disparate business processes is crucial for maximizing operational efficiency and realizing synergies.
   - *Risk of transition*: This criterion evaluates the inherent risks associated with selective data transition, including data loss, system downtime, security vulnerabilities, and regulatory compliance issues. It involves conducting risk assessments and developing mitigation strategies to address potential challenges and uncertainties. Assessing the risk of transition helps in identifying and prioritizing mitigation efforts to ensure a successful data transition process.
   - *Dual maintenance*: Typically, you would need an N + 2 landscape strategy to integrate the NewCo capabilities into the PharmCo entities. In an N + 2 strategy, PharmCo maintains the existing SAP landscape (N) along with two additional instances: one for the acquiring or merged entity (NewCo) and another for the divested entity (SpinCo). This approach allows for a smoother transition period when both the existing entity and the new entity can operate independently on their respective systems. It provides more flexibility and time for data migration, testing, and training before they fully transition to the new landscape.

N + 1 and N +2 Landscape Strategies

In SAP terminology, the *N + 1* and *N + 2* landscape strategies refer to approaches for managing system landscapes in terms of system upgrades, patches, and overall maintenance. *N* is the production environment where actual business operations are conducted (i.e., business-as-usual systems). These strategies help organizations ensure stability, support development, and test environments effectively while minimizing disruptions to production systems. Let's take a closer look at each:

- **N + 1 landscape strategy**
  The N + 1 landscape strategy involves maintaining one additional environment beyond the production environment. This additional environment is typically used for testing purposes. N + 1 is the testing environment, a separate system for testing upgrades, patches, and new developments before they are deployed to the production environment. This environment often mirrors the production environment as closely as possible.
- **N + 2 landscape strategy**
  The N + 2 landscape strategy involves maintaining two additional environments beyond the production environment. These additional environments are typically used for development and quality assurance and testing purposes. N and N + 1 are the same as described previously, while N + 2 is the development environment (i.e., a system where new developments, configurations, and customizations are first created and tested by developers).
  - *TCI and effort*: Total cost of implementation (TCI) and effort encompass the overall cost and effort required to execute the selective data transition successfully. They include expenses related to data migration tools, workforce training, system customization, implementation, and post-transition support. Evaluating TCI and effort helps in budgeting and resource planning, ensuring that the transition process remains cost effective and efficient.
  - *Change management*: Effective change management is indispensable for navigating the cultural and organizational shifts accompanying M&A and divestiture activities. Enterprise architects representing PharmCo and SpinCo collaborate with all the stakeholders to communicate the vision, manage resistance, and foster a culture of collaboration and innovation.

7. Add weights for all the criteria, depending on the experiences of and perspectives from the group.
8. Jointly evaluate and perform scoring to choose the best-fit scenario.
9. Based on the transition net, recommend the relevant selective data transition scenario applicable for SpinCo.

Table 8.3 shows the outcome of the transition scenario evaluation, which is derived from the architecture workshop session with SpinCo stakeholders (refer back to step 5).

Criteria	Description	Option 1: Split, Pilot Implementation, and NA Rollout	Option 2: Shell Template, Harmonize, and NA Rollout	Option 3: System Merge, Split, Harmonize, and Big Bang
Financial risk	TSA costs and cost of not meeting the deadline of 18 to 22 months	Low	Low to medium	High
Business fit and flexibility	Consumer business capability as well as process coverage, impact on continuous business operations and business commitment	Low	Medium	High
Deployment speed and duration	Speed to reach the defined target architecture (including TSA commitments), time to implement, and time to value and create single-instance SAP S/4HANA consumer template	First go live will take 9 to 12 months. Deployments will take 9 to 12 months per deployment.	First go live will take 12 to 15 months. Deployments will take 6 to 9 months per deployment.	First go live will take 15 to 18 months. No deployments since it's a big-bang rollout.
Dual maintenance	N + 2 or N + 3 maintenance, retrofit of custom code	N + 3 (High)	N + 3 (High)	N + 1
Risk of transition	Number and effort of the extraordinary change events, expected cost for development (including custom code remediation), testing, training	First go live is low. Deployments are high.	First go live is medium. Deployments are medium.	First go live is medium.
TCI and effort	Transition and deployment effort, TCI	X (first go live) + 3X (deployments)	1.5X (1st go live) + 2X (deployments)	2.5X (big bang)
Change management	Business commitment and stakeholder commitment, business change management	First go live is low. Deployments are high.	First go live is medium. Deployments are low.	Medium to high

Table 8.3 SpinCo's Transition Scenario Evaluation

We recommended option 2 based on this evaluation to create a shell template with US data, harmonize for the Canadian business unit, and then bring Canadian data to start the first pilot implementation. The North American rollout will follow. The primary reason for this recommendation is to get into SAP S/4HANA as a quick win with a smaller business but with a larger coverage of business processes from the template of North America. We also offered quantitative analysis of scoring with weighting of the criteria and then doing a "what-if" analysis based on cost, benefit, and risk, as depicted in Figure 8.5.

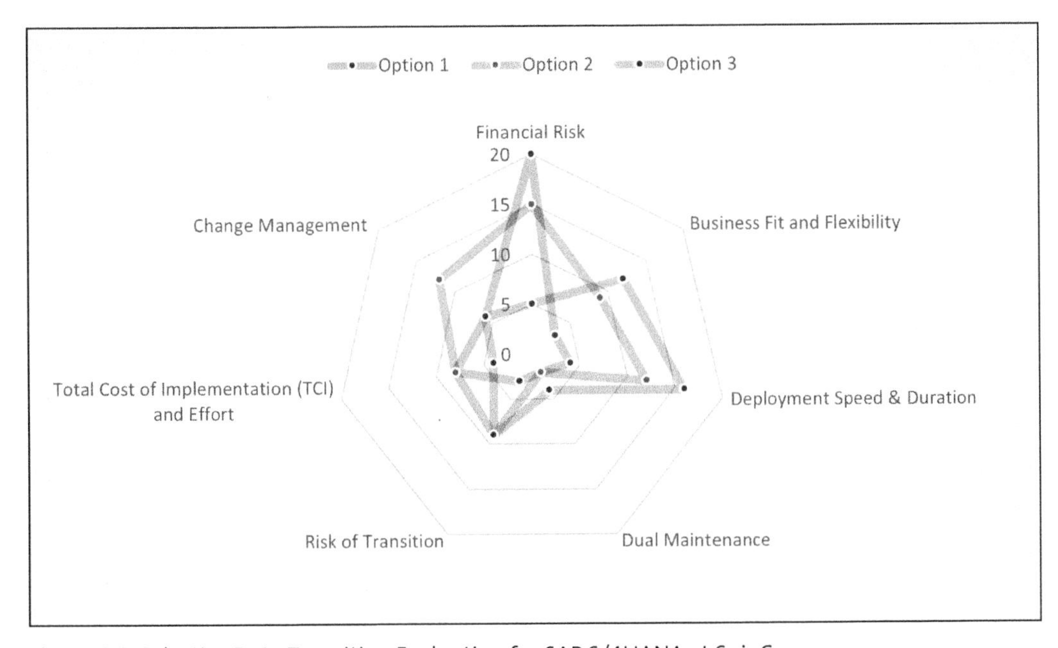

Figure 8.5 Selective Data Transition Evaluation for SAP S/4HANA at SpinCo

8.4.2 Acquisitions Use Case

Enterprise architects for NewCo performed the following steps to derive the target architecture and roadmap, based on the business capability-driven approach but also deep diving for critical business processes:

1. Understand the strategic priorities for integrating NewCo capabilities into PharmCo. The key priority is to implement cell and gene therapy business capabilities and processes into PharmCo's SAP S/4HANA system. You need to understand the pain and gain and architecture requirements. The business footprint diagram for PharmCo is depicted in Figure 8.6. Here, as an example, you can see the links among core capabilities such as operational procurement, production execution, quality management, and logistics execution performed in SAP S/4HANA.

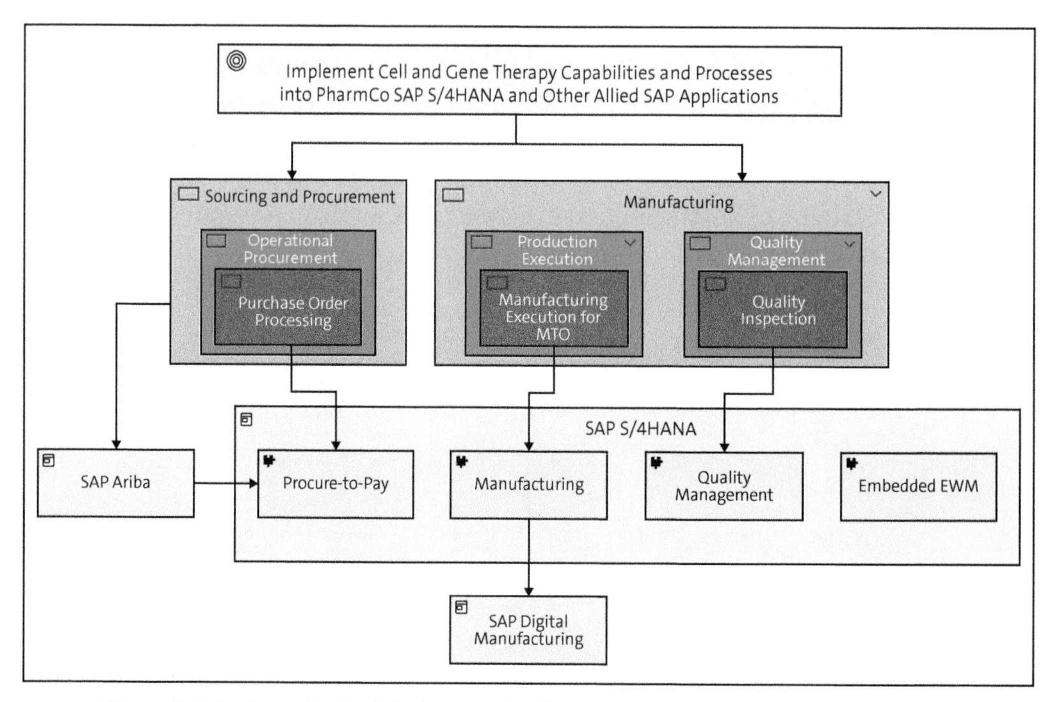

Figure 8.6 Business Footprint Diagram for PharmCo

Business Footprint Diagram

As per SAP EA Framework, a *business footprint diagram* describes the links between business goals, organizational units, business functions, and services, and it maps these functions to the technical components delivering the required capability. A business footprint diagram provides clear traceability between a technical component and the business goal that it satisfies while also demonstrating ownership of the services identified. This is also explained in Chapter 2.

A business footprint diagram demonstrates only the key facts linking organization unit functions to delivery services, and it is utilized as a communication platform for senior-level (C-suite) stakeholders.

2. Assess the cell and gene therapy business capabilities that will support requirements. Perform heatmapping opportunities and determine pain points, fit, and gaps with respect to PharmCo's existing capabilities in the system. Identify the critical requirements that are not possible to configure in the standard SAP system and will need large enhancements using custom applications. (These are also known as *big rocks* for solution architecture.) The business capability heatmap is shown as an example in Figure 8.7.

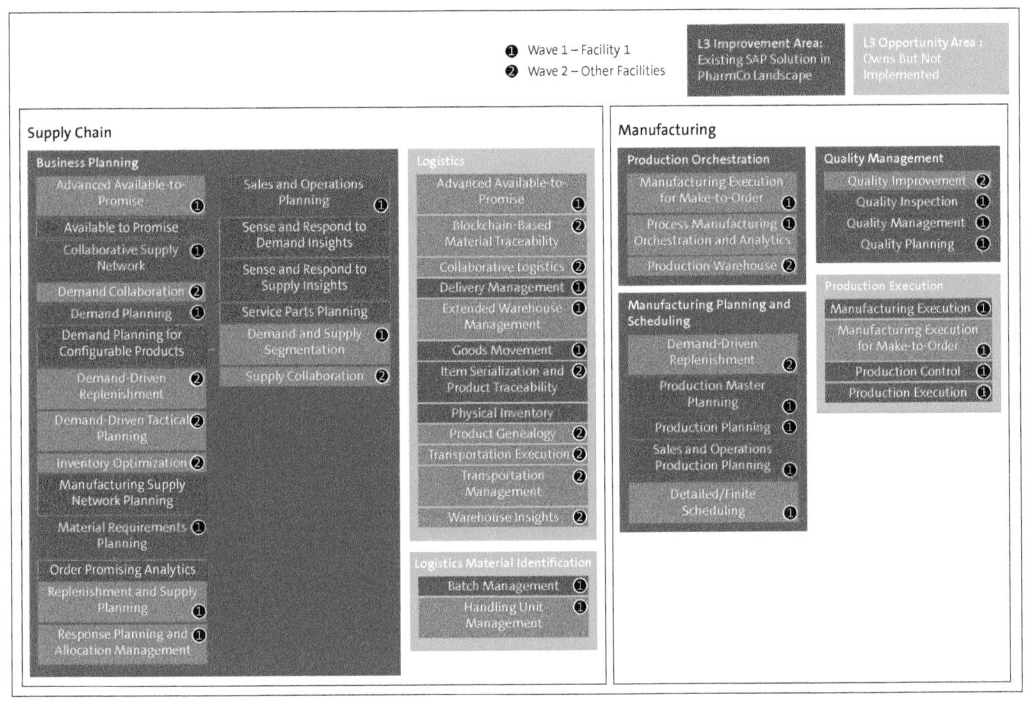

Figure 8.7 Business Capability Heatmap for PharmCo

3. Perform deeper analysis into cell and gene therapy requirements and conduct high-level architecture design options for big rocks. Some examples of identified big rocks are as follows:
   - Capture and track a tag ID in each transaction. The chain of identity needs to be ensured, and a unique patient identification (a tag ID) must be verified and checked at every stage of the supply chain and manufacturing process.
   - Capture shelf life in hours and minutes.
   - Design an MTO (batch size 1) cell and gene therapy process.
4. Review and integrate NewCo's end-to-end process flow with PharmCo's, as shown in Figure 8.8. For example, the **Patient Scheduling**, **Patient Drug Infusion**, and **Manage/Monitor** processes are mostly managed outside SAP applications and hence shown in a different color.
5. Review and validate as-is application architecture and planning target application architecture and categorize them as per the lifecycle of solution components, in terms of which applications would be placed in the *retire*, *phase out*, *keep*, *introduce*, *investigate*, and *to be decided* categories for the target application architecture. In the aftermath of M&A, redundant applications and systems of PharmCo present significant challenges. Through comprehensive portfolio assessments using lifecycle

attributes, enterprise architects identify overlapping functionalities, rationalize applications, and prioritize integration efforts to streamline operations and reduce technical debt.

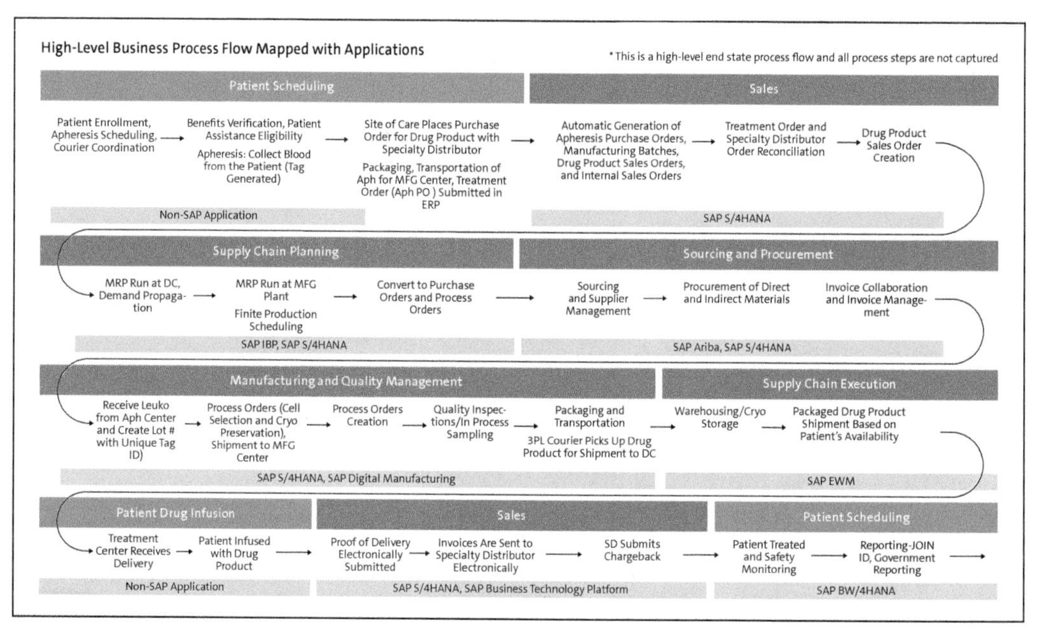

Figure 8.8 High-Level, End-to-End Business Process Flow with Mapped Applications in PharmCo

6. Identify application architecture impacts (along risks and assumptions) and provide recommendations.
7. Develop interim and application architecture and assess consequences of choices.
8. Define and design target application architecture by critical business domain for manufacturing, supply chain, and sales, including high-level information flows.
9. Identify a critical integration list. There are many existing interfaces that need to be changed, and new (and interim) interfaces must be designed for the NewCo integration project. A few of the examples are listed in Table 8.4.

Application 1	Application 2	Integration Domain	Interfaces
SAP S/4HANA	Patient scheduling system	Private cloud to public cloud	Sales order confirmation, shipment schedule confirmation
SAP S/4HANA	SAP Digital Manufacturing	Private cloud to public cloud	Stock creation, inventory adjustment, process order, order status

Table 8.4 Critical Integration List for NewCo Integration Project

Application 1	Application 2	Integration Domain	Interfaces
SAP S/4HANA	SAP Ariba	Private cloud to public cloud	Suppliers, purchase requisitions, invoices
SAP S/4HANA	SAP IBP	Private cloud to public cloud	Demand planning, sales and operations planning
SAP S/4HANA	Non-SAP application	Business to government	Chargeback payment, contract pricing request
SAP S/4HANA	Specialty distributor	Business to business	Purchase order, chargeback request, invoice, price authorization, price adjustment response

Table 8.4 Critical Integration List for NewCo Integration Project (Cont.)

10. Prioritize the current initiatives and planned initiatives with respect to ease of implementation versus value potential.
11. Evaluate the phase-wise capability and application architecture adoption.
12. Define an application architecture roadmap aligned with the SAP product roadmap, as shown in Figure 8.9. Treat this only as an example where the roadmap sequencing is done based on the engineering runs for the first facility, followed by clinical and commercial trial operations and then rollout to other facilities.

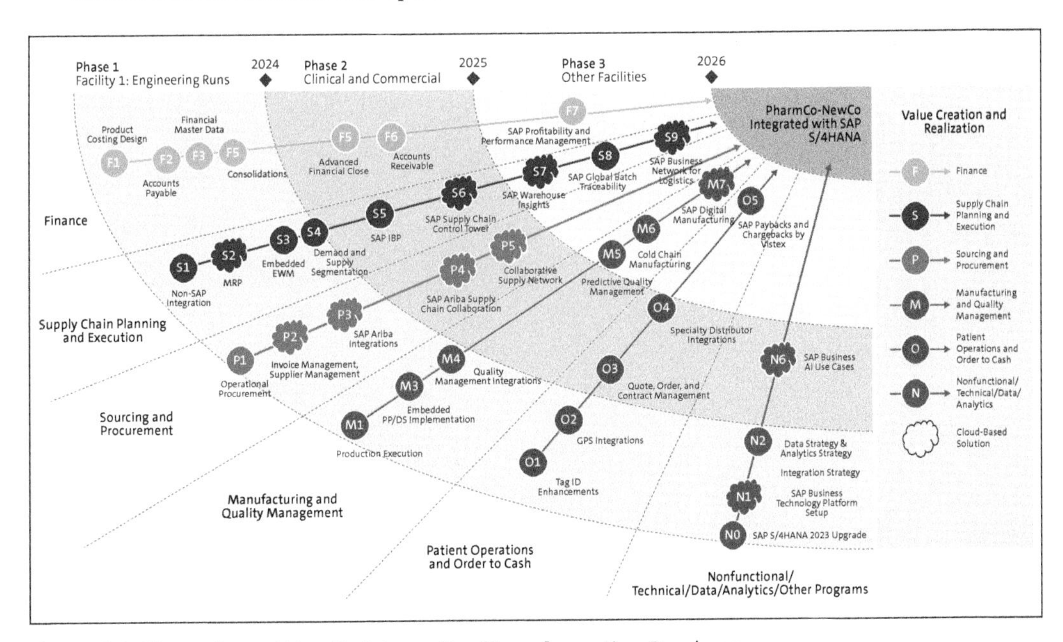

Figure 8.9 PharmCo and NewCo Integration Transformation Roadmap

8.5 Summary

In this chapter, we delved into the critical role of the enterprise architect in facilitating a successful M&A and divestiture for PharmCo, which is divesting its consumer products business to SpinCo and acquiring a cell and gene business called NewCo to enrich and enhance its strategic pharmaceutical portfolio. In the dynamic landscape of PharmCo's modern business, MAD has become an integral strategy for growth, consolidation, and restructuring. However, these activities pose unique challenges to enterprise architecture, demanding a strategic approach to ensure seamless integration, optimization of resources, and preservation of business value.

This completes our MAD use case. In the next chapter, we'll take a detailed look at reducing total cost of ownership in a large transformation initiative.

Chapter 9
Reducing Total Cost of Ownership

Understanding total cost of ownership helps you, as an enterprise architect, make informed architecture decisions about the selection, design, and management of SAP solutions to ensure they provide value while minimizing costs over their lifetime.

This use case will provide details on how the total cost of ownership (TCO) is one of the key criteria when deciding the best enterprise resource planning (ERP) solution for the business. Reducing TCO is one of the key strategic priorities during the planning phase of a digital transformation journey, and it feeds into the value case for the overall business case for ERP transformation to SAP S/4HANA. Besides, TCO analysis is one of the critical criteria in making the build versus buy or best-of-breed versus standard integrated ERP product selection decision.

In this chapter, we'll explore the utilization of SAP EA Framework to facilitate a reduction in TCO. We'll provide a case study that involves planning a cloud transformation (reminiscent of Chapter 5), but with the specific purpose of reducing TCO, which is our focus here. We'll see how enterprise architects can scope their transformation approach to deliver on a target architecture, transition strategy, and roadmap. But first, let's introduce some key concepts for TCO to lay the groundwork for the rest of the chapter.

9.1 Understanding Total Cost of Ownership

Let's start with understanding what is meant by *TCO* and its ingredients. The first step in calculating the TCO of an ERP system is to identify and estimate the different types of costs involved. Another key consideration is capital expenditure (CapEx) versus operational expenditure (OpEx). We'll discuss these topics in the following sections.

9.1.1 Types of Costs

Figure 9.1 shows the typical components of TCO and how these are visible to the stakeholders. During the product selection phase, most of the negotiations take place with respect to software license costs and hardware and infrastructure costs. These two are the visible costs, as shown in the Figure 9.1, which is the tip of the iceberg. But there are many other costs below the surface, which are visible only when you start the implementation of the software and running the applications for the end users.

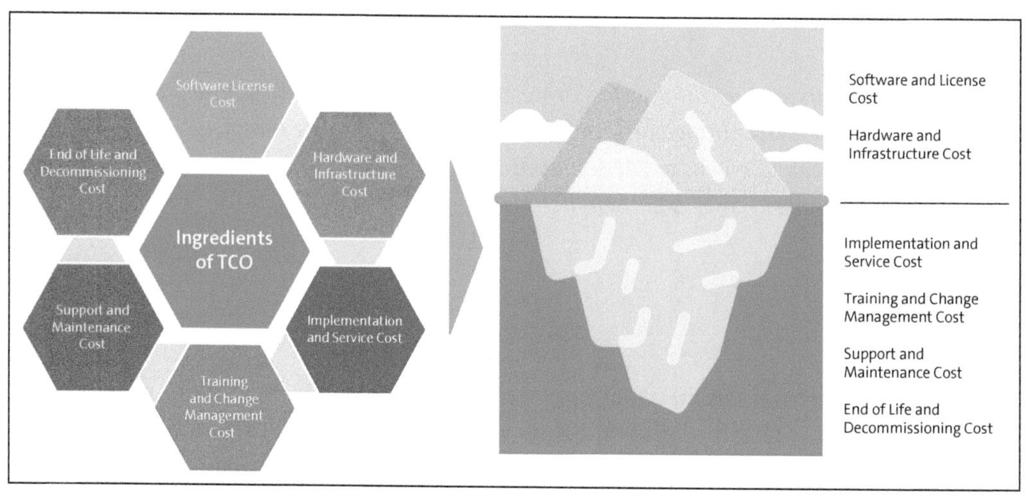

Figure 9.1 Ingredients of TCO and Its Visibility to Stakeholders

The key ingredients of TCO for purchasing ERP applications are as follows:

- **Software license cost**
  This is the acquisition cost that is the initial cost for purchasing software. Software costs include the license fees, subscription fees, and cloud fees for the ERP software and any additional modules or integrations. License costs are associated with purchasing ERP applications, which is based on a variety of metrics such as the number of users, transactions, etc. The annual subscription cost for an application is the recurring cost in exchange for products or services. For example, pricing for SAP software depends on the SAP deployment options. During the cloud transformation, there's a ramp-down of perpetual license costs and a ramp-up of the cloud subscription cost.
- **Hardware and infrastructure costs**
  These include the costs of the servers, storage, network, and devices required to run the ERP system. Data center costs are associated with data center facilities, including power, cooling, and physical security if managed on-premise. The main cost driver for TCO is usually the infrastructure cost, which can be owned by SAP or hyperscalers.
- **Implementation and service costs**
  These are the costs incurred in deploying the applications. These are also called *consulting fees* and *professional services fees*, which are for external consultants and system integrators who assist with implementation, configuration, customization, and integration of SAP S/4HANA. Some examples of these costs, assuming SAP S/4HANA is the core cloud ERP, are as follows:

- Internal resources costs: These are costs associated with internal staff dedicated to the project, including project managers, IT staff, and business analysts.
- Data migration costs: These are the costs of efforts and tools required for data extraction, cleansing, transformation, and loading from legacy systems to SAP S/4HANA.
- Customization and development costs: These are the costs of developing custom code and applications or modifying standard SAP functionalities to meet specific business requirements.
- Integration costs: These are the costs associated with integrating SAP S/4HANA with other enterprise systems, such as SAP ERP, SAP Customer Relationship Management (SAP CRM), SAP Supply Chain Management (SAP SCM), and third-party applications.
- Testing costs: These are costs associated with testing and test automation tools used during the SAP Activate implementation phases. Test management tools (e.g., artificial intelligence [AI]-driven test automation tools from Tricentis) can help in effectively allocating resources for testing activities, minimizing the need for additional staff, reducing labor costs, and reducing overall project timelines.
- Cloud services costs: These are costs related to cloud infrastructure, including IaaS and PaaS, for hosting SAP applications.

- **Training and change management costs**
  These are the costs associated with moving to a new ERP application. The costs incurred in educating employees on how to use the new ERP application (e.g., costs for training end users, IT staff, and administrators) are called *training costs*. The costs involved in managing the transition to the new ERP application, ensuring smooth adoption, and minimizing disruption (e.g., costs related to managing organizational change, including communication, workshops, and support materials) are called *change management costs*.
- **Support and maintenance costs**
  These are the ongoing costs for maintenance and support services, including application management services, bug fixes, and enhancements. Management of applications also includes system management, server management (e.g., patching and server setup), and incident management.
- **Operational costs**
  These include staff salaries, overhead, costs of utilities, and other expenses related to running the ERP system. Some examples of operational costs are as follows:
  - System monitoring costs: These are the costs of tools and resources for monitoring system performance, availability, and security.
  - Risk and compliance management costs, which include the following:
    - Security costs: These are the costs of investments in security measures to protect sensitive data and ensure compliance with regulations. There are also

costs for implementing security measures, including firewalls, encryption, and intrusion detection systems.

- Compliance costs: These are the costs associated with meeting industry-specific regulatory requirements, such as audits, certifications, and reporting. There are also expenses related to ensuring compliance with industry regulations and standards, such as the European Union's General Data Protection Regulation (GDPR), the Health Insurance Portability and Accountability Act of 1996 (HIPAA), and the Sarbanes-Oxley Act (SOX).

- Upgrade and innovation costs: These are the costs of periodic upgrades to the latest SAP S/4HANA releases and enhancements for the purpose of leveraging new features and functionalities.
- Innovation project costs: These are the costs of investments in innovation projects, such as adopting new SAP technologies (e.g., SAP Business Technology Platform [SAP BTP], SAP Fiori, SAP Business AI) or integrating with Internet of Things (IoT), AI, and machine learning.
- Business process reengineering costs, which include the following:
  - Process redesign costs: These are costs associated with reengineering business processes to align with best practices enabled by SAP S/4HANA.
  - Change impact costs: These are costs related to the impact of process changes (including temporary productivity losses during the transition period) on business operations.
- Test management and test automation costs, which include the following:
  - Lower maintenance effort costs: These include the costs of automated tests that can be reused across multiple releases to reduce the effort needed for regression testing and ongoing support.
  - Efficient defect detection and resolution costs: These are the costs of test management efforts to ensure that all aspects of the software are thoroughly tested, and any defects are identified and resolved before deployment. This reduces the number of defects reported during the support and maintenance phase and thus lowers the overall cost of defect resolution.
- Scalability and flexibility costs, which include the following:
  - Scalability costs: These are costs related to scaling the SAP S/4HANA environment to accommodate business growth efforts, such as adding users, expanding storage, and enhancing processing power.
  - Flexibility costs: These are the costs of investments in ensuring the system's flexibility to adapt to future business needs and technological advancements.

- **End-of-life and decommissioning costs**
  These are the costs associated with decommissioning an application. When the ERP application reaches end of life, the costs incurred to retire it include costs of data extraction, system shutdown, archiving, and disposal of hardware. These costs

encompass various activities and processes required to transition away safely and efficiently from the old ERP system or any other legacy applications. For example, there are costs incurred in running parallel applications and temporary dual maintenance.

To summarize, the TCO is the sum of all costs involved in the purchase, implementation, and support of SAP software during its lifetime. The TCO approach is all about considering and doing analysis of all types of direct and indirect costs to mitigate risks.

9.1.2 Capital versus Operational Expenditures

During the on-premise era, when database servers and application servers were hosted on premise at customer data centers, the purchase price of an SAP software package used to be seen as a CapEx, whereas the implementation, service, and support costs were considered OpEx. However, in the cloud era, all these costs are moving towards being OpEx, which will reduce the TCO in the long run. By carefully analyzing and managing these TCO drivers, organizations can optimize their SAP S/4HANA implementation projects, ensuring a balance between costs and the value derived from the new system. Also note that in the context of ERP transformation, CapEx and OpEx have distinct definitions, and their dynamics often change with the shift toward cloud-based solutions.

CapEx are the funds a company uses to acquire, upgrade, and maintain physical assets such as property, industrial buildings, and equipment. In the context of ERP transformation, this includes the following:

- Purchasing hardware, such as servers, storage devices, and networking equipment
- Purchasing software licenses, such as perpetual software licenses for ERP solutions
- Purchases involved in infrastructure setup

OpEx are the ongoing costs of running a product, business, or system. These are the costs required for the day-to-day functioning of the ERP system, and they include maintenance, support, and operational costs. In the context of ERP transformation, these includes the following:

- Subscription fees for cloud services
- Costs for software maintenance and support contracts
- Salaries for IT staff involved in the day-to-day management of the ERP system
- Utilities and operational costs for running data centers
- Costs of regular updates, patches, and minor upgrades

With the shift to cloud-based ERP solutions, the balance between CapEx and OpEx often changes, primarily due to the nature of cloud service models (such as software as a service [SaaS], infrastructure as a service [IaaS], and platform as a service [PaaS]).

Cloud transformation has the following impacts on CapEx:

- **Reduced CapEx**
  Cloud transformation typically reduces the need for significant up-front investment in physical infrastructure and perpetual software licenses. This is because companies no longer need to purchase and maintain their own hardware or invest heavily in data center facilities.
- **Shift to subscription-based models**
  Instead of paying a large sum up front, companies pay recurring subscription, which are considered operational expenses.

Cloud transformation has the following impacts on OpEx:

- **Increased OpEx**
  As companies move to cloud-based models, OpEx increases because they pay recurring subscription fees for SaaS, IaaS, and PaaS.
- **Predictable OpEx**
  OpEx becomes more predictable with cloud services due to regular subscription fees. This facilitates better budgeting and financial planning.
- **Reduced maintenance and support costs**
  Since the cloud service provider handles most of the maintenance, updates, and support, internal operational costs can be reduced.
- **Scalability and flexibility**
  Cloud services offer scalable solutions where companies can adjust their usage and costs based on their needs, avoiding overprovisioning and underutilization of resources.

By transitioning to cloud-based ERP systems, companies can achieve a more flexible and financially predictable model, shifting the financial burden from high CapEx to manageable and scalable OpEx.

9.2 Business and IT Context

Let's look at the business and IT context of how to reduce TCO for business and the IT transformation of your company from an enterprise architecture perspective. Understanding the business and IT context helps you identify strategies that can effectively minimize costs while achieving the desired outcomes of the transformation.

9.2.1 Business Context

There are different drivers to reduce TCO from a business perspective. Business drivers to reduce TCO focus on aligning IT investments with the broader strategic goals of the organization. These drivers help ensure that the ERP transformation not only delivers technological benefits but also supports business objectives. The main drivers of reducing TCO for a large ERP transformation program are as follows:

- **Cloud transformation**
  The simple act of embarking on the journey of cloud transformation is a key driver of reducing TCO. This is because it puts the company on the path to becoming an intelligent and sustainable enterprise. Cloud transformation also has a high impact on reducing TCO elements like software license costs, hardware and infrastructure costs, implementation and service costs, training and change management costs, and support and maintenance costs (i.e., moving from on-premise to IaaS to PaaS to SaaS has direct, positive impact on all the cost elements in reducing TCO). Some specific examples of how cloud transformation can reduce TCO are as follows:
  - Lower initial costs: Cloud ERP solutions eliminate the need for substantial upfront investments in hardware and infrastructure cost, converting CapEx to predictable OpEx.
  - Pay-as-you-go model: Organizations only pay for the resources they use, allowing for better cost management and budgeting.
  - Faster time to market: Cloud ERP systems can be deployed more quickly than on-premise solutions, enabling faster realization of benefits and return on investment (ROI).
  - Continuous innovation: SAP S/4HANA Cloud continuously innovates and offers new features, allowing organizations to leverage the latest technologies without additional investment.
- **Process standardization**
  Standardizing processes is a key success factor in cloud transformation. When we refer to *standard processes*, we mean those provided out of the box in the ERP product. However, over time, extensive customization to meet specific business needs can make the system difficult to manage and maintain. Therefore, implementing business processes as close to the standard as possible, with minimal customization, can reduce complexity and maintenance costs and in the long run help reduce TCO. For example, implementing industry best practices to ensure efficient and effective use of ERP systems reduces the need for costly customization and rework.

  Process standardization has a high impact on reducing TCO elements like implementation and service costs, training and change management costs, and support and maintenance costs. In other words, the more processes are standardized and follow best practices, the greater chance the company has to decrease its implementation and service costs due to greater reusability and reduced complexity.
- **Process harmonization**
  A major challenge for many organizations is the difficulty in maintaining consistent processes among different divisions and regions. For example, a make-to-order process for heavy equipment manufactured in the United States is different from the

process for heavy equipment manufactured in Europe. During transformation planning, there's a choice between *process reuse* and *process reengineering* depending on which transition strategy is chosen. For example, there are costs associated with reengineering business processes to align with best practices enabled by SAP S/4HANA. Also, there are expenses related to the impact of process changes on business operations, including temporary productivity losses during the transition period. So, process harmonization during an ERP transformation can lead to substantial cost savings by streamlining operations, reducing complexity, improving data quality, and optimizing IT infrastructure. These benefits collectively contribute to a lower total cost of ownership, making the transformation more cost-effective and sustainable in the long term. Best practice is to harmonize as much as possible.

Process harmonization is an important factor in reducing costs in the business and in IT. It's important to understand that process harmonization doesn't automatically mean a single instance (in the sense of a single production system). Process harmonization can also be achieved with multiple production systems, and it depends rather on the setup of the nonproductive systems (also called development systems or the development track).

Process harmonization has a high impact on reducing the TCO elements like implementation and service costs, training and change management cost, and support and maintenance costs. In other words, increasing the number of process variants will greatly increase these three cost elements, thereby increasing the TCO.

- **Process automation**
  Once the process is standardized and harmonized, there's an immense potential to automate it. Process automation reduces the TCO for an ERP transformation by increasing efficiency, lowering labor and maintenance costs, enhancing accuracy, and improving scalability. Automation eliminates repetitive and time-consuming manual tasks, thereby freeing users to focus on high-value tasks. Besides, automation reduces the need for manual intervention, decreasing labor costs associated with routine tasks. Automated processes can be scaled up or down more easily than manual processes, accommodating business growth without significant additional costs. For example, SAP Build Process Automation is the simpler, faster way to enhance business efficiency and agility with confidence.

  Process automation has a high impact on reducing the TCO elements like implementation and service costs, training and change management costs, and support and maintenance costs. In other words, automating the processes has the long-term benefits of increasing efficiency, accuracy, and scalability.

- **Data harmonization**
  Many large organizations don't have their data harmonized. Customer data may reside in one system, material master in another system, supplier data in another, and financial data in another. In addition, the processes of create, read, update, and

delete (CRUD) are different and diverse in these four master data domains. Data harmonization reduces the TCO for a large transformational program by improving data quality, streamlining data management, and enhancing operational efficiency. For example, harmonizing data removes duplicates and inconsistencies, ensuring a single source of truth. This reduces errors and the need for corrective actions, leading to cost savings. Also, having consistent and high-quality data improves decision-making, reducing the costs associated with poor decisions based on inaccurate or incomplete information.

Data harmonization has a high impact on reducing the TCO elements like implementation and service costs. In other words, harmonizing data can reduce the time and effort required for integrating and migrating data from different sources, leading to lower implementation costs.

9.2.2 IT Context

Now, let's focus on IT context. The main drivers of reducing TCO from an IT perspective within the framework of enterprise architecture are as follows:

- **Cloud architecture**
  The target state of cloud transformation is to get to scalable, flexible, and agile cloud architecture. Cloud services in the cloud architecture offer the ability to dynamically scale resources up or down based on demand, ensuring that you only pay for what you use and avoiding the costs associated with overprovisioning on-premise hardware.
- **Landscape consolidation**
  This is one of the key IT drivers of reduced TCO. The following are some of the cost elements that have the highest impact on TCO reduction:
  - Reduced hardware and infrastructure costs: Consolidating multiple systems onto fewer, more efficient servers reduces hardware costs, data center space, power consumption, and cooling requirements.
  - Lower support and maintenance costs: Managing a consolidated landscape is simpler and requires fewer IT resources, reducing the cost of system administration, monitoring, and support. Besides, fewer systems mean fewer maintenance contracts, updates, patches, and support agreements, lowering overall maintenance cost.
- **Application rationalization**
  This involves strategically identifying and evaluating an organization's application inventory to determine what should be harvested, invested in, retired, or consolidated. This process helps reduce the TCO for a large ERP transformation program in several ways, such as lowering licensing, maintenance, and support costs. By rationalizing and standardizing based on a common set of applications and platforms,

management becomes simpler and the need for diverse skill sets is reduced. Additionally, a rationalized application portfolio enables quicker deployment of new technologies and innovations, reducing time-to-market and associated costs.

- **Data center centralization**
  By leveraging the centralized data centers of hyperscalers, organizations can significantly reduce their TCO for large and complex ERP transformations, achieving a more efficient, scalable, and cost-effective IT environment.
- **Clean architecture**
  Achieving a clean architecture involves minimizing extensions; ensuring data is clean, complete, and accurate; using standard application programming interfaces (APIs) where possible; and implementing SAP best practices for noncore business capabilities. The rationale behind SAP's *clean core* paradigm is to allow customers to extend their SAP S/4HANA software while making updates seamless and enabling agile business and innovation adoption. Clean architecture facilitates faster software deployment and easier adoption of both SAP and business innovations, as well as regulatory changes. For your business, this approach reduces TCO while maintaining the flexibility needed to create value for the organization and your customers.

The role of an enterprise architect in reducing TCO for a large transformation initiative involving systems like SAP S/4HANA, SAP Integrated Business Planning for Supply Chain (SAP IBP), SAP Extended Warehouse Management (SAP EWM), SAP Transportation Management (SAP TM), SAP Global Trade Services (SAP GTS), and SAP Ariba, as well as for the retirement of mainframe legacy systems, is pivotal. Here's how an enterprise architect can contribute to TCO reduction:

- **Strategic planning**
  Enterprise architects play a key role in defining the strategic direction of the transformation initiative. They assess current systems, identify pain points, and define the North Star service architecture. By aligning the transformation roadmap with strategic priorities and business goals, enterprise architects ensure that investments are optimized to reduce long-term TCO.
- **Technology selection**
  Enterprise architects evaluate various technology options for the transformation, considering factors such as scalability, interoperability, and vendor lock-in. They recommend solutions that offer the best value proposition in terms of solution capability, cost-effectiveness, and TCO reduction over the entire lifecycle.
- **Standardization and rationalization**
  Enterprise architects promote standardization and rationalization of systems and processes across the enterprise. By reducing complexity and redundancy, they streamline operations and drive efficiencies that result in lower TCO. This may involve consolidating duplicate solution capabilities, retiring obsolete systems, and enforcing standardized platforms.

- **Optimized infrastructure**
  Enterprise architects design infrastructure solutions that balance performance, reliability, and cost-efficiency. Whether it's on-premise data centers or private cloud or public cloud services, they optimize infrastructure configurations to support the needs of SAP systems and minimize operational costs over time.
- **Governance and compliance**
  Enterprise architects establish governance frameworks and compliance standards to ensure that the transformation initiative adheres to regulatory requirements and industry best practices. By proactively addressing risk and compliance issues, they mitigate potential financial penalties, legal liabilities, and reputational damage that could inflate TCO.

In summary, enterprise architects play a multifaceted role in reducing TCO for large transformation initiatives by driving strategic alignment, optimizing technology investments, optimizing infrastructure, ensuring governance and compliance, and fostering a culture of continuous improvement.

Total Cost of Ownership is Also Known as Total Cost of Operations

The cost of the cloud is also known as the total cost of operations. This TCO includes the cost of operating and supporting the application. It includes the following cost components:

- **Cloud infrastructure costs**
  These includes hardware-related expenses such as compute, storage, backup, and the network and its operations. Infrastructure includes everything from the data center to the operating system in the cloud value chain, and infrastructure costs are the costs of consumption of IT services to operate a cloud product. You can also distinguish between infrastructure provided by SAP infrastructure from various suppliers such as Amazon Web Services (AWS), Microsoft Azure, and Google Cloud Platform (which we call hyperscaler infrastructure).
- **Application management costs**
  These includes costs for operating the solutions. They include system management (e.g., system backup), database management, server management (e.g., patching, server setup), and incident management costs.
- **Cloud support costs**
  These include all expenses related to support activities based on the standard support agreement. An example is costs for delivering product support (i.e., answering how-to questions from customers).
- **Other costs**
  These include royalties for third-party solutions.

The main cost driver for TCO is usually infrastructure cost. For example, for one of the large customers, 40% of the overall cost of the cloud is the infrastructure cost, followed by 25% for application management costs, 20% for other costs, and 15% for cloud support costs.

9.2.3 Total Cost of Ownership Analysis

There are different ways in which you can do a TCO analysis for your ERP transformation. Two primary ways to do this are as follows:

- **Standalone TCO analysis**
  You do a standalone TCO analysis with an as-is versus to-be comparison of cost elements. You can do this by discussing different deployment scenarios and their TCO implications with the customer. Typical steps for this analysis are depicted in Figure 9.2.

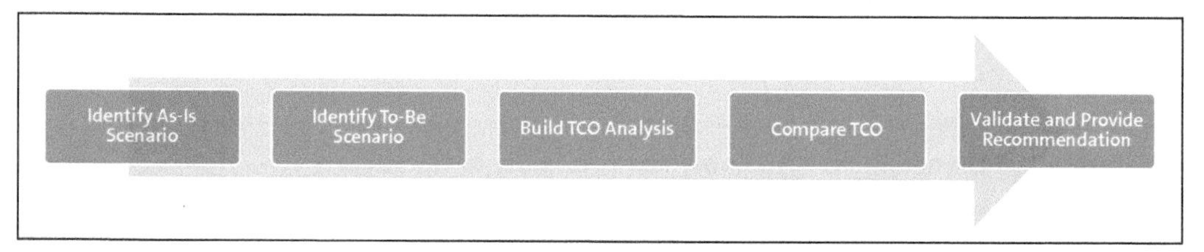

Figure 9.2 Standalone TCO Analysis

You can select as-is and to-be scenarios based on Table 9.1.

As-Is State	To-Be State
SAP ERP baseline	■ SAP ERP lift and shift ■ SAP S/4HANA (on-premise) ■ SAP S/4HANA Cloud Private Edition ■ SAP S/4HANA Cloud Public Edition
On-premise SAP S/4HANA baseline	■ SAP S/4HANA Cloud Private Edition ■ SAP S/4HANA Cloud Public Edition
Non-SAP ERP baseline	■ SAP S/4HANA (on-premise) ■ SAP S/4HANA Cloud Private Edition ■ SAP S/4HANA Cloud Public Edition

Table 9.1 As-Is and To-Be Scenarios

Based on the identified as-is and to-be scenarios for the customer, you can build a TCO analysis using different cost elements as described in Section 9.1. You can then do a TCO analysis by comparing different scenarios, identifying the best scenario, and making a recommendation.

- **TCO analysis within a business case**
  You can do a TCO analysis within a business case without making a comparison with as-is costs. First, prepare the business case, and then, use it to integrate TCO within a business case and ROI calculation. Typical steps for this analysis are depicted in Figure 9.3.

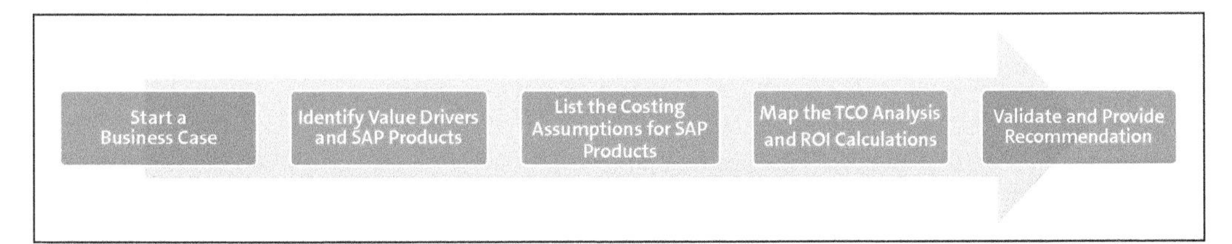

Figure 9.3 TCO Analysis within a Business Case

In this approach, the first step is to create a new business case for ERP transformation based on identified value drivers and target SAP products. There are cost assumptions for each of the target SAP products. You need to perform TCO analysis and then map it with the ROI calculations. Make the analysis, identify the correct approach, and make recommendations.

You'll typically use standalone TCO analysis when comparing the costs of different options or solutions without considering the specific needs and objectives of a business. This analysis is useful for identifying the direct and indirect costs associated with owning and operating a particular product or service, and it can help you make informed purchasing decisions. For example, the following are a few situations where you can use standalone TCO analysis:

- Cost comparison of different applications or technologies
- Budgeting and financial planning
- Making a procurement decision related to buying a new application

On the other hand, you'll use TCO analysis along with the business case when a more comprehensive assessment is required. This approach incorporates the TCO analysis into the broader strategic planning and decision-making process of the organization. It takes into consideration the specific needs, goals, and objectives of the business, as well as the potential benefits and risks associated with the investment. This helps in evaluating the costs in the context of the overall business impact and helps you make a more informed decision. For example, the following are a few situations where you can use standalone TCO analysis:

- Making strategic investments where understanding the full financial impact and expected business benefits are critical
- Working on new initiatives that need evaluation of both costs and benefits

- Working on transformation programs, such as digital transformation, ERP transformation, and mergers and acquisitions (M&A)

9.3 Problem Statement

Let's study an example of a fictitious global conglomerate called Digital Aerospace Technologies Inc. (DAT Inc.), which is preparing a business case for its transformation to move to the cloud. One of the key objectives of this case for the cloud is to reduce TCO by removing 30% of redundant applications and IT components within one year and save $1.2 million in application costs per year for the next five years. The use case we share here is a real-life mix of multiple customer experiences during their phase 0 transformation planning, especially from the perspective of enterprise architecture. This is also a case of a mix of multiple industries such as aerospace and defense, chemicals, and consumer products. DAT Inc. wants to perform the TCO analysis within a business case.

In the following sections, we'll briefly look at the company's profile and its IT landscape to understand the pain points.

9.3.1 Company Profile

DAT Inc. designs, manufactures, and services aircraft engines for commercial airlines and defense contractors, and it has operation in three regions: North America (NA; the United States and Canada), Latin America (LA), and Europe. This company has a growth strategy and has recently acquired another company, TechD Inc., which focuses on exclusive technology solutions for the defense industry and chemical products to strengthen its forward and backward integration processes and capabilities. However, the company is facing numerous challenges to achieving this growth strategy due to fierce competition in the market and huge supply chain issues that make it hard to meet customer demand. In addition, the company's large, complex, monolithic, and antiquated IT landscape (especially in the manufacturing business domain) is not agile and is too inefficient to meet the growing business demand. This increases TCO.

The business divisions of both companies are shown in Table 9.2.

DAT Inc.			TechD Inc.	
Commercial Aerospace	Defense	Services	Technology Solutions and Services	Chemical

Table 9.2 Business Divisions of DAT Inc. and TechD Inc.

DAT Inc. is embarking on a digital transformation journey in response to macroeconomic factors and emerging digital trends. The combined entity of DAT Inc. and TechD Inc. has two main reasons for initiating this transformation:

- **Customer expectations and supply chain efficiency**
  DAT Inc. faces increased customer expectations for product innovation. Additionally, the supplier base is shrinking, and many suppliers must adjust to demand fluctuations across their customer base. Aerospace and defense companies need to efficiently manage their global supply chains and provide better products at lower costs.
- **Global sourcing and supply chain management**
  Contractual relationships, low manufacturing costs, new capabilities, and a lack of capacity require DAT Inc. and TechD Inc. to source products globally. The need to reduce overall delivery time and risk drives them to track products in the supply chain more closely or move production closer to customers.

9.3.2 IT Landscape

DAT Inc. is modernizing its mainframe applications and planning a transformation program to adopt SAP S/4HANA as its digital core cloud ERP system. Each of the three regions currently has its own ERP landscape, alongside legacy mainframe ERP applications. The company aims to consolidate its ERP landscape through application rationalization, process harmonization, and standardization. The key strategic priority is to reduce the total cost of ownership and achieve maximum ROI. The landscape summary is shown in Figure 9.4.

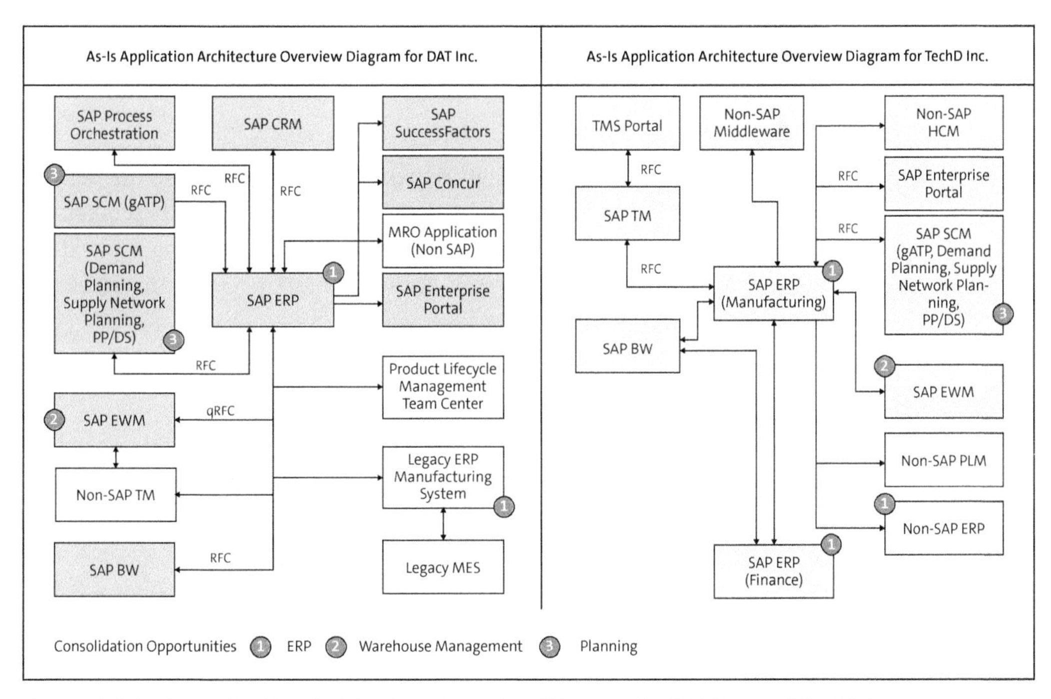

Figure 9.4 As-Is Application Architecture Overview Diagram for DAT Inc. and TechD Inc.

The combined entity of DAT Inc. and TechD Inc. is performing a TCO analysis within a business case. It wants to evaluate all the TCO ingredients, starting with software license costs, hardware and infrastructure costs, implementation and service costs, training and change management costs, and support and maintenance costs. It has identified the two key transformation drivers (i.e., one business driver and one IT driver) for its phase 0 transformation to meet the objective of reducing the TCO. These two drivers are as follows:

- **Business process harmonization and standardization**
  DAT's intent is to identify the process areas to standardize and harmonize across all its business units and regions to reduce complexity and pain points. Some of the business pain points are as follows:
  - There is a lack of process harmonization among business units and in the three regions, especially in sales, supply chain, manufacturing, and finance business processes in both DAT Inc. and TechD Inc.
  - There are broken process steps, and there's no end-to-end visibility of the business process due to manufacturing and finance running in two different ERP systems for TechD Inc.
  - There's a lack of visibility of stock and hence incorrect commitment of delivery dates to customers, with no backorder process for legacy orders.
  - Customer service representatives don't have visibility into processes running in legacy ERP applications in DAT Inc.
  - There's duplication of master data and residing in multiple locations with minimum governance processes.
- **Landscape consolidation**
  DAT Inc. aims to not only to consolidate its existing ERP landscape and the peripheral applications but also to transition to cloud architecture following its acquisition of TechD Inc. This move is driven by the goals of reducing TCO, achieving application rationalization, and harmonizing processes. Some of the IT pain points are as follows:
  - No common designs align with various ERP systems (i.e., SAP and non-SAP systems in two companies spread across business units and regions).
  - All existing SAP systems have a huge amount of custom code. There's a strong desire to go back to standard and utilize SAP Best Practices.
  - There's a significant number of classified systems for individual programs with extremely limited information.
  - There are inconsistent and point-to-point integrations, and there's a strong need to move to open and event-driven architecture.

Based on the pain points, we've identified three landscape consolidation opportunities, as depicted in Figure 9.4:

❶ **ERP consolidation**
There are two ERPs in DAT Inc.'s landscape, one SAP and one legacy ERP. There are also three ERPs in TechD Inc.'s landscape, two SAP and one non-SAP ERP system. This means there's an opportunity to reduce the number of ERP applications.

❷ **Supply chain execution (i.e., warehouse management consolidation)**
This is an opportunity, but considering the nature of the business, the volume of the transactions, and the complexity of the production warehouses and distribution centers, we need to make a detailed assessment to provide a recommendation.

❸ **Supply chain planning consolidation**
Having one planning solution is the goal, and there's an opportunity to harmonize and then consolidate into one global supply chain planning solution.

There are other opportunities for consolidation (e.g., analytics, transportation management) in DAT Inc.'s IT landscape, but we won't cover these in this chapter.

9.4 Applying the Enterprise Architecture Transformation Approach

Let's now do the scoping for the problem statement of DAT Inc. by applying SAP EA Framework.

The enterprise architect is onboarded and applies SAP EA Methodology and transformation approach to scope the modules that will provide guidance to reduce the TCO for DAT Inc. and the acquired company, TechD Inc. During the scoping session, the enterprise architect identifies the following typical use cases:

- Target processes harmonization toward a global template valid for all business segments
- Instance strategy recommendations
- Transition strategy design during creation of the consolidation roadmap
- Complex interim landscapes and a parallel run of legacy and new solutions

TCO analysis is crucial for DAT Inc. since this is a large-scale transformation initiative involving multiple systems like SAP ERP, legacy ERP, SAP SCM, and SAP EWM (as shown in Figure 9.4), as well as the retirement of mainframe legacy systems. Table 9.3 shows some scenarios and use cases that the enterprise architect can identify for TCO analysis for DAT Inc.

TCO Elements	Scenario	Use Case	Evaluation Requirements
System Consolidation and Retirement	DAT Inc. plans to retire multiple mainframe legacy systems and consolidate their functionalities into SAP S/4HANA, SAP IBP, SAP EWM, and SAP BW/4HANA.	Calculate the TCO of maintaining and operating the legacy systems versus migrating their functionalities to the SAP landscape. Consider factors such as hardware maintenance, software licensing, support contracts, and personnel costs.	The modules to be evaluated are business capability and instance strategy.
License and Subscription Costs	DAT Inc. needs to assess the licensing and subscription costs associated with SAP S/4HANA, SAP IBP, SAP EWM, and SAP BW/4HANA, including any additional modules or users required for the transformation.	Compare the TCO of licensing and subscription fees for SAP systems to the costs of maintaining licenses for the mainframe legacy systems. Consider long-term contracts, volume discounts, and potential negotiation opportunities with SAP.	The SAP accounts team along with the SAP value advisory team uses the SAP Value Lifecycle Manager tool to evaluate and compare the TCO with different cost elements.
Infrastructure and Hosting Costs	Determine the infrastructure and hosting requirements for SAP S/4HANA, SAP IBP, SAP EWM, and SAP BW/4HANA, including hardware, servers, storage, and cloud services.	Evaluate the TCO of on-premise hosting versus cloud-based solutions for SAP systems. Consider factors such as initial setup costs, ongoing maintenance, scalability, and data security.	The RISE with SAP team evaluates the infrastructure and the hosting requirements for SAP S/4HANA and other applications with the help of an SAP enterprise architect and a RISE with SAP cloud architect.
Integration and Customization Expenses	Assess the integration and customization efforts required to connect SAP S/4HANA, SAP IBP, SAP EWM, and SAP BW/4HANA with other internal and external systems.	Calculate the TCO of developing and maintaining integration interfaces and customizations for SAP systems. Consider factors such as development resources, third-party tools, middleware licenses, and ongoing support costs.	Modules to be evaluated are business process and application architecture to identify the high-level interfaces connecting different applications in scope. This is to be followed by evaluating integration modules as the next iteration.

Table 9.3 Matrix of TCO Elements, Scenarios, and Use Cases

TCO Elements	Scenario	Use Case	Evaluation Requirements
Training and Change Management	Plan for training and change management activities to support the transition to SAP S/4HANA, SAP IBP, SAP EWM, and SAP BW/4HANA.	Estimate the TCO of training programs, documentation, user support, and organizational change management initiatives. Consider factors such as the number of users, training materials, instructor costs, and productivity loss during the transition period.	Module to be evaluated is organizational change; this will be scoped in subsequent iterations of the architecture review.
Operational Support and Maintenance	Determine the ongoing operational support and maintenance requirements for SAP S/4HANA, SAP IBP, SAP EWM, and SAP BW/4HANA.	Calculate the TCO of in-house support teams, third-party maintenance contracts, software updates, patches, and system monitoring tools. Consider factors such as SLAs, downtime costs, and the complexity of system configurations.	Module to be evaluated is transition to execution; this will be scoped in subsequent iterations of the architecture review.
Risk and Compliance Costs	Identify potential risks and compliance requirements associated with the transformation initiative, including data security, regulatory compliance, and business continuity.	Assess the TCO of implementing risk mitigation measures, security controls, compliance audits, and insurance coverage. Consider factors such as legal fees, fines, reputational damage, and the cost of noncompliance.	Module to be evaluated is transition to execution.

Table 9.3 Matrix of TCO Elements, Scenarios, and Use Cases (Cont.)

By conducting comprehensive TCO analysis of these scenarios and use cases, DAT Inc. can make informed decisions about the large transformation initiative involving SAP systems and legacy system retirement. Cost calculations and recommendations are not covered as part of this book, but we do cover the impact of cost elements (which we described earlier as the ingredients) on the TCO.

Based on our understanding of DAT Inc. scenarios and the ask, in Figure 9.5, we show the key architecture modules we identified in scope for the use of reducing TCO.

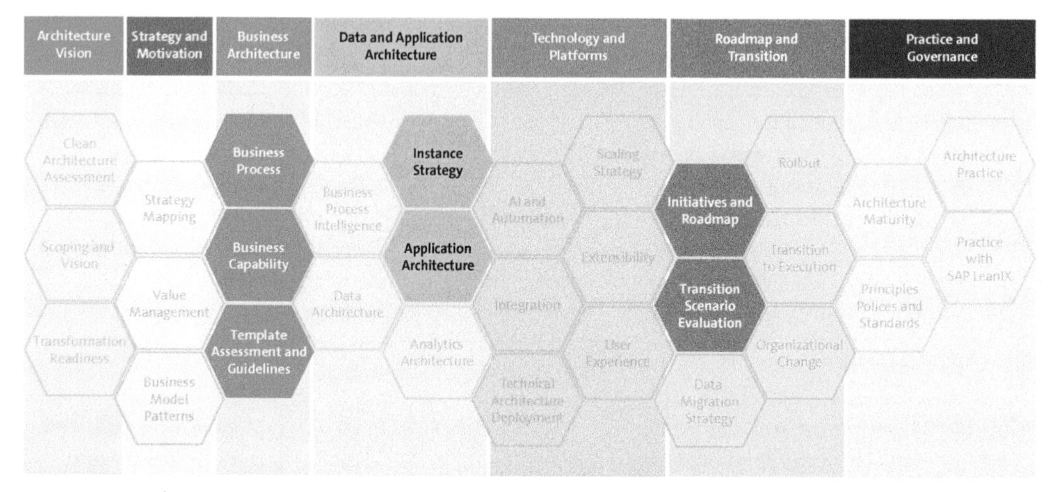

Figure 9.5 Modules in Scope for DAT Inc.

Let's walk through how to apply each module to our DAT Inc. and TechD Inc. use case:

- **Business process**
  This module focuses on the review of the as-is process map and heatmapping with pain points, plus the identification of dependencies among business processes, overlaps, process variants, and harmonization and standardization potential. This module will also focus on the evaluation of reuse versus renewed processes for DAT Inc. and TechD Inc. and will make the best of both worlds.
- **Business capability**
  This module focuses on deriving the top prioritized business capabilities from the business process that has the high harmonization potential.
- **Template assessment and guidelines**
  This module focuses on the review of the overall analysis and design template approach (e.g., global, local, partial template). This module also helps in reviewing organizational units (e.g., business divisions, regional units) and the main markets they operate in, plus the scope of their operations (e.g., sales, manufacturing, finance), including level of process independency, level of master data independency, and organizational unit independency. Key requirements for the template design are identification of major localization requirements of all the three regions (NA, LA, and Europe) in scope and evaluation of the template's initial high-level scope.
- **Instance strategy**
  This module focuses on the landscape consolidation pattern. There are multiple ERPs in DAT Inc.'s IT landscape, and this module helps in making decisions about single instances versus regional instances.

- **Application architecture**
  This module focuses on the impact of consolidation on the target application architecture, including deployment strategy analysis and replatforming options of components in scope. This module will assess the impact of TCO based on the cost elements described in the introduction section. It will also evaluate the three rating options (the number of process variants, the number of SAP systems, and the number of data centers) scoped by the enterprise architect based on the cost elements.
- **Transition scenario evaluation**
  This module focuses on the following:
  - Determining transition options for the landscape consolidation
  - Identifying and evaluating relevant scenarios based on a predefined set of criteria
  - Determining the best-fit option, considering the sequences of consolidation
- **Initiatives and roadmap**
  This module includes a review of ongoing and planned initiatives and their impact on the parallel run. This will help identify new required initiatives with mapping to overall dependent initiatives, and it will identify their impact on rollout. This will also help build an initiatives catalog for DAT Inc. based on the ongoing, planned, and new required initiatives and their time horizons. Last but not least, this module will build a strategic roadmap for the next five years, based on the initiatives catalog.

9.5 Key Deliverables and Outcomes

In this section, we'll outline the main deliverables and results from the modules that we discussed in the preceding section. We'll provide a brief explanation of each deliverable's creation process and the necessary prerequisites. Furthermore, we'll showcase the outcome or a sample of the ultimate deliverable. Our focus will be on the outcome, which has an impact on reducing the TCO.

9.5.1 Business Process

In our analysis of DAT Inc.'s business processes as part of the enterprise architecture transformation, we identified the pain points and harmonization potential as illustrated in Figure 9.6.

The major pain points are in the **Make to Inspect**, **Plan to Optimize Fulfilment**, **Order to Fulfill**, and **Invoice to Cash** process areas. These are also the process areas where the harmonization potential is high or medium. If these processes are harmonized throughout their business divisions and regions, there's a high probability that the TCO will be reduced in the target architecture. Let's take a closer look at each process:

- The **Make to Inspect** process covers production planning, production operations, quality management, and production performance management for DAT as well as

TechD. After production, finished products are received into the manufacturing warehouse. The manufacturing process is scattered over two different ERP systems: SAP and legacy ERP.

- The **Plan to Optimize Fulfillment** process covers the definition of supply chain, manufacturing, and service-fulfillment strategies, followed by demand, inventory, supply planning, and alignment of the different plans through sales and operations planning and performance measurement.
- The **Order to Fulfill** process covers customer order and contract management, followed by different fulfillment processes for physical products, services, and subscriptions.
- The **Invoice to Cash** process covers the processing of customer invoices and the subsequent management of accounts receivable, as well as the collection of customer payments.

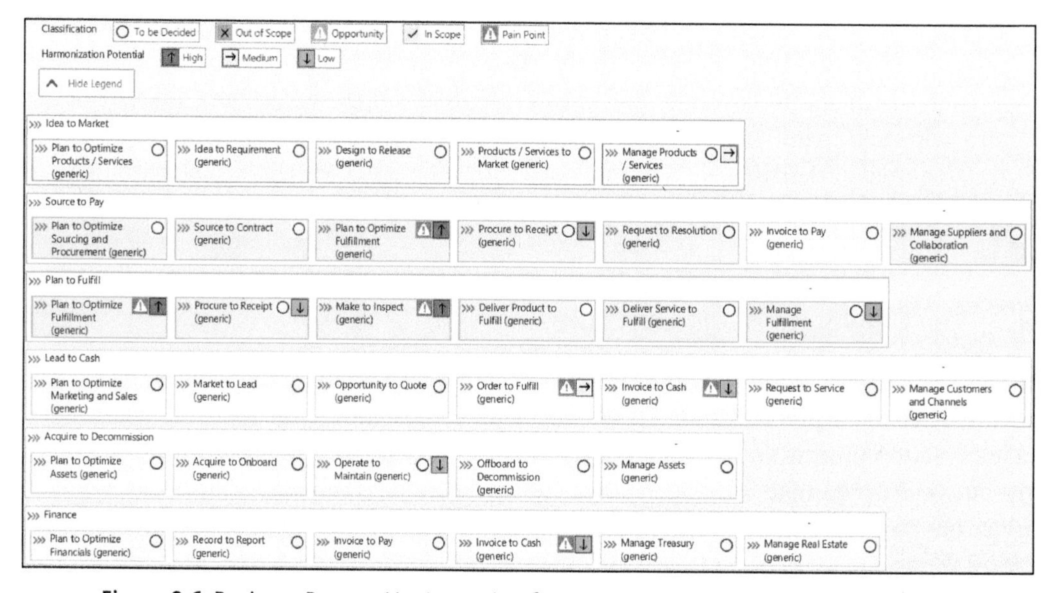

Figure 9.6 Business Process Heatmapping for DAT Inc.

9.5.2 Business Capability

After we identify the business process heatmapping with pain points and harmonization potential, we derive the business capability, along with the heatmapping of the pain points and opportunities, as shown in Figure 9.7.

Heatmapping of business capabilities has identified additional pain points specific to aerospace business and defense business units of DAT Inc. These are maintenance, repair, and overhaul (MRO) operations and enterprise portfolio and project management (EPPM), along with pain points in other areas such as supply chain, sales, and sustainability (which are marked in red). For example, one of the major pain points in the

MRO area is the processing of repairs that involve assemblies. In most cases, users at DAT Inc. need to disassemble and test the parts individually before sending them to the respective suppliers for repair. This individual processing is very time consuming because it doesn't recognize the individual boards when they're received as an assembly. As a result, users were forced to create a work order for each part of the assembly, which adds to the time and effort required for processing. Similarly, in the EPPM business area, the project manufacturing management and optimization (PMMO) capability has a pain point of managing different stock ownerships for materials within the bill of materials structure. Stock ownership could be at different levels, such as individual stock, grouped stock, and plant stock.

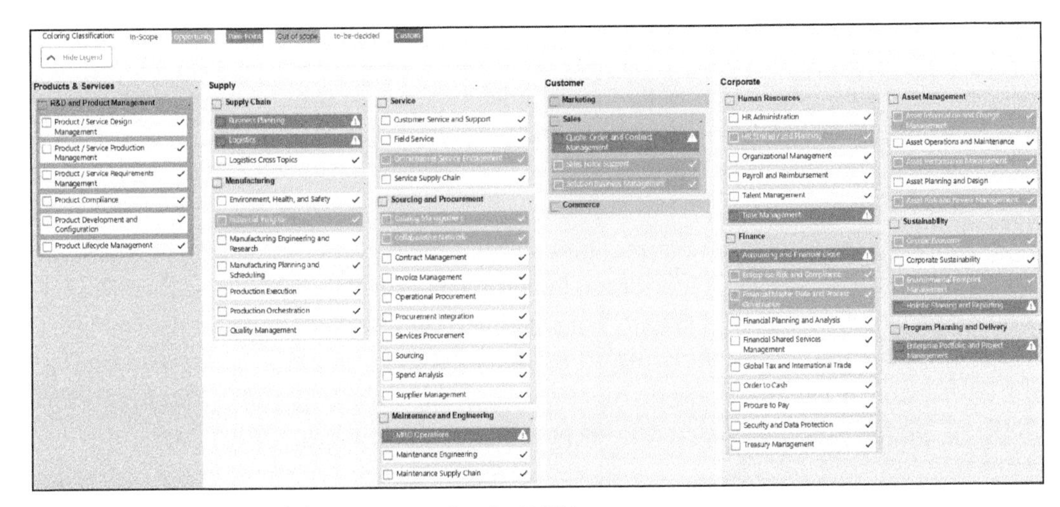

Figure 9.7 Business Capability Heatmapping for DAT Inc.

9.5.3 Template Assessment and Guidelines

The enterprise architect has done a deep analysis of the harmonization potential, as shown in Figure 9.8. They did this to assess the commonalities of processes between DAT Inc. and TechD Inc. for the global template. The enterprise architect uses the template assessment and guidelines module to design the global template strategy for DAT Inc., including the acquired company, TechD Inc. They use these to determine the needs and benefits that will drive such a strategy and identify the key building blocks that can be used to operationalize the strategy. There are three fundamental patterns that the enterprise architect identifies during transformation planning for DAT Inc. and TechD Inc., considering the business divisions and the business domains of the two entities. These three patterns will also become the three flavors of the global template design. They are as follows:

- **Innovation**
  There are business divisions that need a business-driven redefinition of their IT architectures, in support of a business model transition. These areas fundamentally

redefine key qualities of the enterprise architecture, such as data-centric architecture and modular cloud architecture. For example, the warehouse management department of the commercial aerospace business unit has fully mature and automated warehouses with advanced capabilities such as automated storage and retrieval systems, pick by voice, and pick by light. These warehouses can adopt SAP Business AI capabilities to further enhance their business model and achieve further growth.

- **Optimization**
  There are business divisions that can reap the benefits of IT architectures and foster automation and the use of AI. This pattern will foster timely adoption of cloud innovations, and it will also ensure the flexible extension of capabilities on the business platform. For example, the sales department of the defense business unit has lot of scope for automation as its foundational capabilities like order management are already mature. It can leverage embedded AI capabilities within SAP S/4HANA to go for further optimization.
- **Renovation**
  Certain business divisions require a bottom-up renovation of their IT architectures. These units need this renovation because their growth has paralleled the historical expansion of the ERP system. For example, the manufacturing process is not only diverse throughout the two business units (commercial aerospace and defense) but also complex among the divisions as well as the two organizations. This means that the same discrete manufacturing process is running differently in two places.

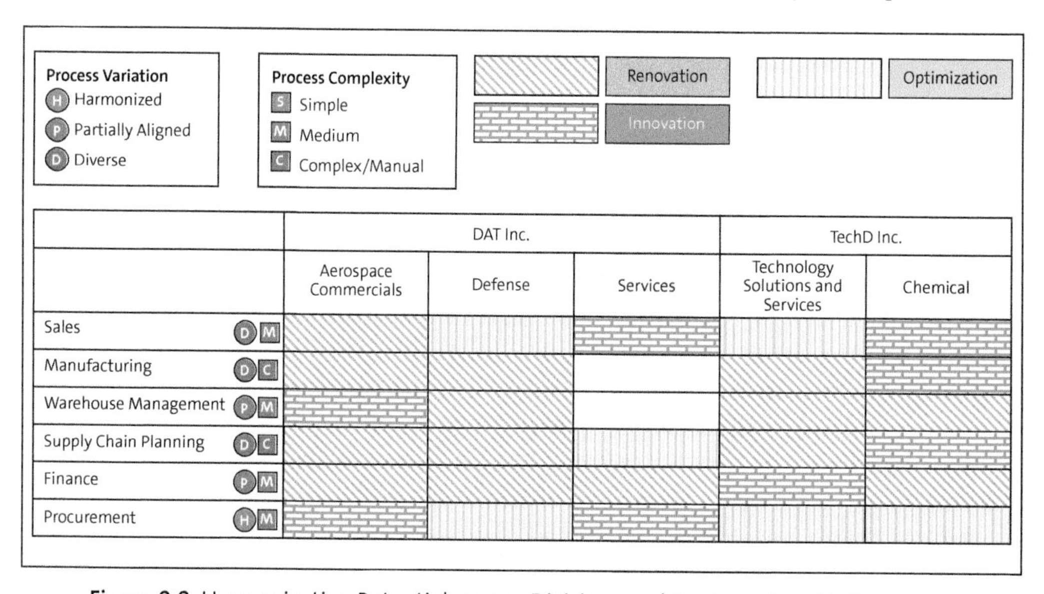

Figure 9.8 Harmonization Potential among Divisions and Business Capabilities

Based on the matrix of **Process Variation** versus **Process Complexity** among business domains and divisions, the enterprise architect creates a global template by identifying the common and SAP Best Practice processes for sales, manufacturing, warehouse management, supply chain planning, finance, and procurement. The three flavors of the template (i.e., renovation, optimization, and innovation) are built to roll out to respective business divisions and business domain.

9.5.4 Instance Strategy

The instance strategy module has guided DAT Inc. in making decisions regarding the consolidation of its SAP landscape architecture. It has enabled DAT Inc. to determine the most suitable landscape option by considering its unique business strategy, operational needs, and technological infrastructure. This well-proven and robust methodology evaluates the pros and cons of various instance options both qualitatively and quantitatively, with a focus on SAP Best Practices criteria.

With the exception of additional data centers used for high availability or disaster recovery, DAT Inc. operates primarily out of a single data center for its core SAP ERP application and doesn't use a hyperscaler's location for all its SAP on-premise software. At most, it maintains primary data centers in the key geographic regions of NA, LA, and Europe. Additionally, it commonly adopts SaaS solutions such as SAP Concur and SAP SuccessFactors for nondifferentiating processes. For most SAP business applications in DAT Inc., network stability, bandwidth, and latency are rarely significant constraints, although exceptions may occur for applications managing production plants or warehouse processes. We illustrate the instance strategy options we've evaluated in Figure 9.9.

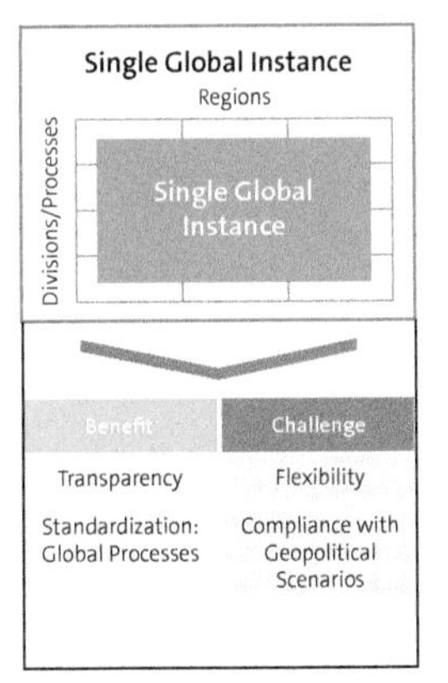

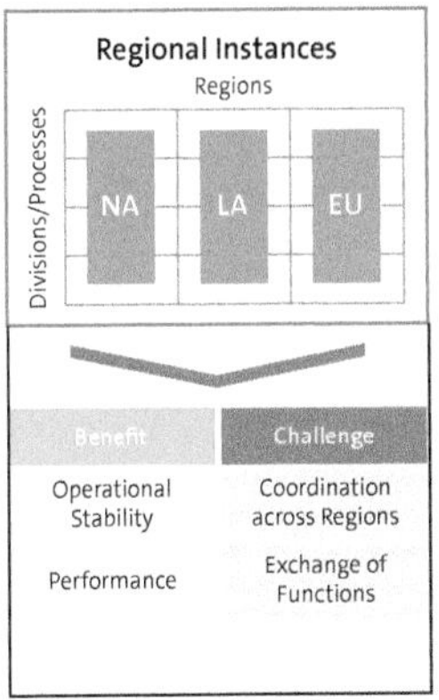

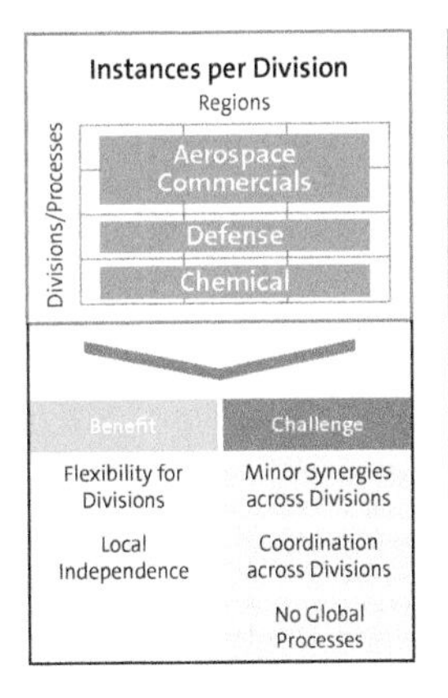

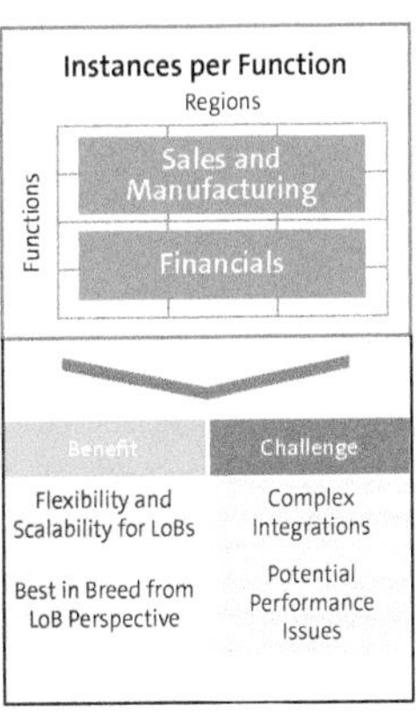

Figure 9.9 Instance Strategy Options for DAT Inc.

The approach to and methodology for evaluating these options, using both qualitative and quantitative analysis, are consistent with those outlined in Chapter 5 on cloud transformation. The enterprise architect assigned to DAT Inc. has applied this same methodology and recommended a single global instance as the preferred option. This recommendation is based on the two key business drivers identified for DAT's phase 0 transformation: business process standardization and harmonization, plus landscape consolidation, which are crucial for achieving the objective of reducing TCO.

Process harmonization is crucial for reducing costs both in business and in IT. However, it's important to note that process harmonization doesn't necessarily equate to having a single instance or a single production system. A single system alone doesn't ensure process harmonization; without proper governance, multiple process variants can still emerge within the same system.

Process harmonization can also be achieved with multiple production systems, depending largely on how nonproductive systems (such as development systems) are set up. This means that DAT Inc. can achieve business process harmonization with a virtual single instance, as process harmonization doesn't rely solely on having a single production system. DAT Inc. has taken this advice into consideration and has planned a deeper exploration of the architecture pattern outlined in Figure 9.10 for the next iteration. You can see the landscape strategy, which we explore in detail in the instance strategy module for nonproduction systems in the landscape. Many SAP customers have implemented option 2 to achieve process harmonization using the global template approach. This means one development system, multiple quality systems, and multiple production systems. This is one of the ways to achieve TCO.

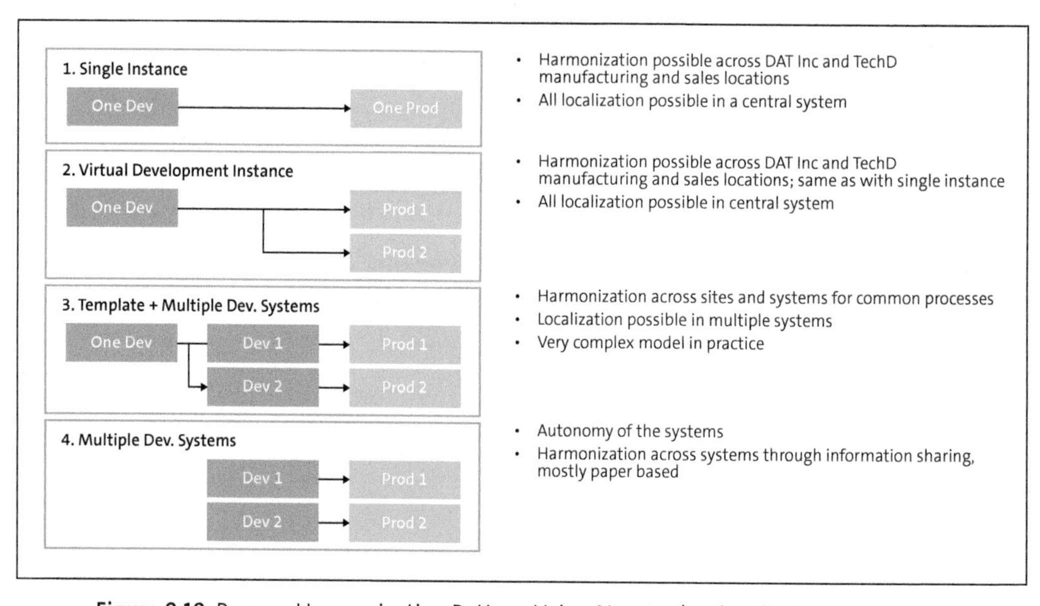

Figure 9.10 Process Harmonization Pattern Using Nonproduction System Strategy

System consolidation without process harmonization is suboptimal for reducing TCO. This means that even if DAT Inc. consolidates all SAP and non-SAP applications according to the three scenarios depicted in Figure 9.4 without harmonizing processes for sales, manufacturing, finance, warehouse management, and planning, the TCO won't be reduced.

In summary, after this assessment, DAT has decided that the first iteration of its instance strategy will be a single global instance with a single development system and a single production system.

9.5.5 Application Architecture

The target application architecture as illustrated in Figure 9.11 is the recommended framework from the first iteration of the workshop session. The enterprise architecture team offered guidance on exploring the capabilities of SAP S/4HANA Cloud Private Edition, and some target applications, marked as "Introduce" have already been decided upon. However, other target applications remain in the "Investigate" phase. This indicates that follow-up actions, such as providing presales demonstrations to showcase the applications' capabilities, are needed to help DAT Inc. make informed decisions.

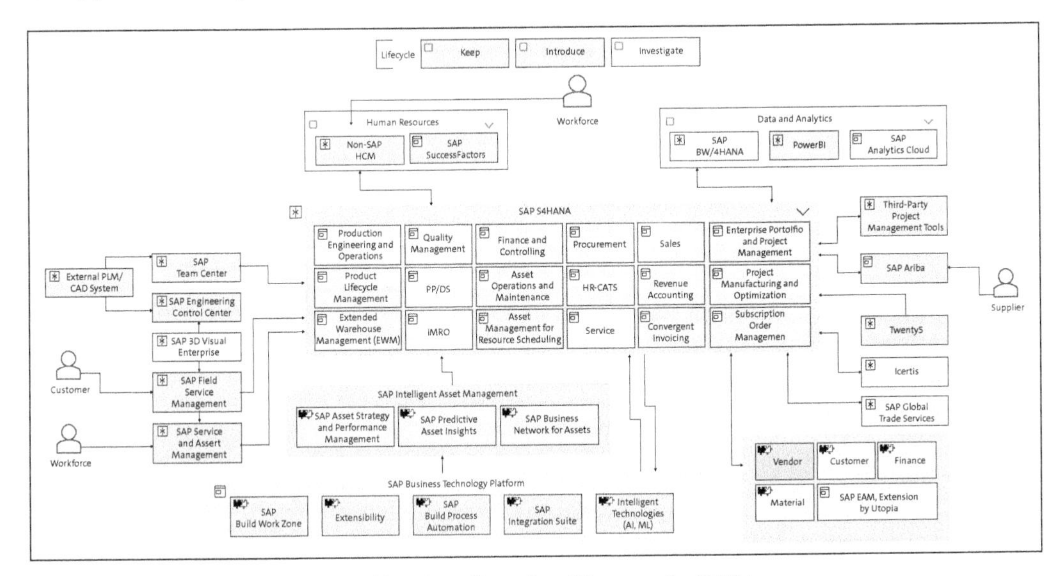

Figure 9.11 Target Application Architecture Overview Diagram for DAT Inc.

We won't cover all the target applications, but following are some of the key applications introduced in the target application architecture for DAT Inc. to address its current pain points:

- **SAP S/4HANA for manufacturing production engineering and operations**
  This component will integrate and connect engineering and manufacturing processes to meet the production requirements of DAT Inc. and TechD Inc.
- **PMMO**
  This component provides project-based integration of engineering, manufacturing, and supply chain processes for execution on a single platform. This is a delight for business users in DAT Inc.

The consolidation and harmonization of all the applications of both companies have an impact on all the cost elements of TCO, as shown in Table 9.4.

TCO Ingredients	Description of DAT Inc.	Number of Process Variants	Number of SAP Systems	Number of Data Centers
Software license cost	Cost incurred in purchasing new license for SAP S/4HANA	Depends on license models	Depends on license models	Medium
Hardware and infrastructure cost	Cost incurred in setting up the RISE with SAP infrastructure, including the hyperscaler infrastructure	No impact	High	High
Implementation and service cost	Cost for implementation and cost of cloud services, along with cost for resources onboarded by partner	High	Medium	Low
Training and change management cost	Cost to enable users to learn and use SAP S/4HANA	High	Medium	Low
Support and maintenance cost	Cost for maintenance and support needed for SAP S/4HANA	High	Low	High
End-of-life and decommissioning cost	Needs to be calculated and predicted for future	Low	Medium	TBD

Table 9.4 TCO Decision Matrix Used for DAT Inc.

Increasing the number of process variants can have significant impacts on various cost elements of the TCO. Here's how it can affect different cost components:

- **Implementation and service costs**
  - Complexity and customization: More process variants require additional customization and configuration during the implementation phase, leading to higher consulting and development costs.
  - Longer deployment times: Increased process variants can extend the implementation timeline, resulting in higher project management and labor costs.
- **Maintenance and support costs**
  - Increased maintenance effort:

  - More process variants lead to a more complex system, which requires additional effort to maintain and support, thus increasing labor and operational costs.
  - Higher support costs: The complexity of managing multiple process variants can lead to more frequent issues and higher support requirements, increasing the cost of the help desk and technical support services.
- **Training and change management costs**
  Managing changes among multiple process variants can be more challenging and costly, requiring additional resources for communication, training, and implementation.

Next, increasing the number of SAP systems has a high impact on various cost elements of the TCO. Here's how it can affect different cost components:

- **Hardware and infrastructure costs**
  - Increased resource requirements: Each SAP system requires its own IT resources (e.g., servers, storage), and that increases infrastructure costs.
  - Higher monitoring and management costs: More systems mean increased complexity in system monitoring and management, leading to higher costs for IT operations.
- **Implementation and service costs**
  - Higher initial setup costs: Each additional SAP system requires its own setup, configuration, and customization, leading to higher initial implementation costs.
  - Consulting and development costs: Implementing multiple systems often necessitates more consulting and development resources, further increasing costs.
  - Integration costs: Integrating multiple SAP and non-SAP systems (such as SAP S/4HANA, SAP EWM, SAP TM, and SAP BW/4HANA) with each other can be more complex and costly, requiring additional integration development and maintenance effort.
- **Maintenance and support costs**
  - Increased maintenance effort: More SAP systems mean more maintenance tasks, such as applying patches, updates, and bug fixes, which increase labor and operational costs.
  - Higher support requirements: Multiple systems require more extensive support resources, including help desk and technical support, which result in higher support costs.
- **Training and documentation costs**
  - Additional training needs: Employees must be trained on each system, increasing the overall training expenses.
  - Complex documentation: Creating and maintaining comprehensive documentation for multiple systems is more resource-intensive and thus increases associated costs.

Increasing the number of data centers can also have a high impact on various cost elements of TCO. Here are the cost components it can affect and how it can affect them:

- **CapEx**
  - Infrastructure costs: Each additional data center requires substantial investment in physical infrastructure, including servers, storage, networking equipment, and facilities (power, cooling, and physical security).
  - Building and construction costs: Setting up new data centers involves high construction or leasing costs for the physical space.
- **OpEx**
  - Maintenance and utilities: Operating multiple data centers increases ongoing costs for maintenance, power, cooling, and facilities management.
  - Staffing costs: More data centers necessitate additional personnel for IT support, facilities management, security, and administration, thus increasing labor costs.
  - Energy costs: Each data center consumes significant energy, contributing to higher utility bills.

There are other layers such as labor and IT resources that impact the TCO. For example, labor and IT resources are cheaper in Southeast Asia than in the United States. In summary, increasing the number of process variants, SAP systems, and data centers typically leads to higher costs among various costs elements of TCO, including implementation, maintenance, support, training, integration, compliance, operational efficiency, IT infrastructure, licensing, innovation, and change management. Simplifying and harmonizing processes can help reduce these costs and improve overall efficiency.

9.5.6 Transition Scenario and Evaluation

DAT Inc.'s intent is to consolidate the landscape into one SAP S/4HANA instance. It wanted to conduct a preliminary assessment (without conducting a comprehensive assessment) of which of the two major choices on the table—a greenfield (new implementation) scenario or a brownfield (system conversion) scenario—is the best transition strategy for moving to SAP S/4HANA. The enterprise architect recommended having a workshop with key stakeholders of DAT Inc. and TechD Inc. to discuss the seven guiding questions that can influence the choice of the transition scenario to SAP S/4HANA.

The results are illustrated in Figure 9.12. This shows that most key stakeholders from DAT Inc. lean toward the brownish right side, which means that they would prefer to have the transition from SAP ERP to SAP S/4HANA be a brownfield implementation (i.e., a system conversion approach). It also shows that most key stakeholders from TechD Inc. lean toward the greenish left side, which means that they would prefer to have the transition from SAP ERP to SAP S/4HANA be a greenfield implementation (i.e., a new implementation approach).

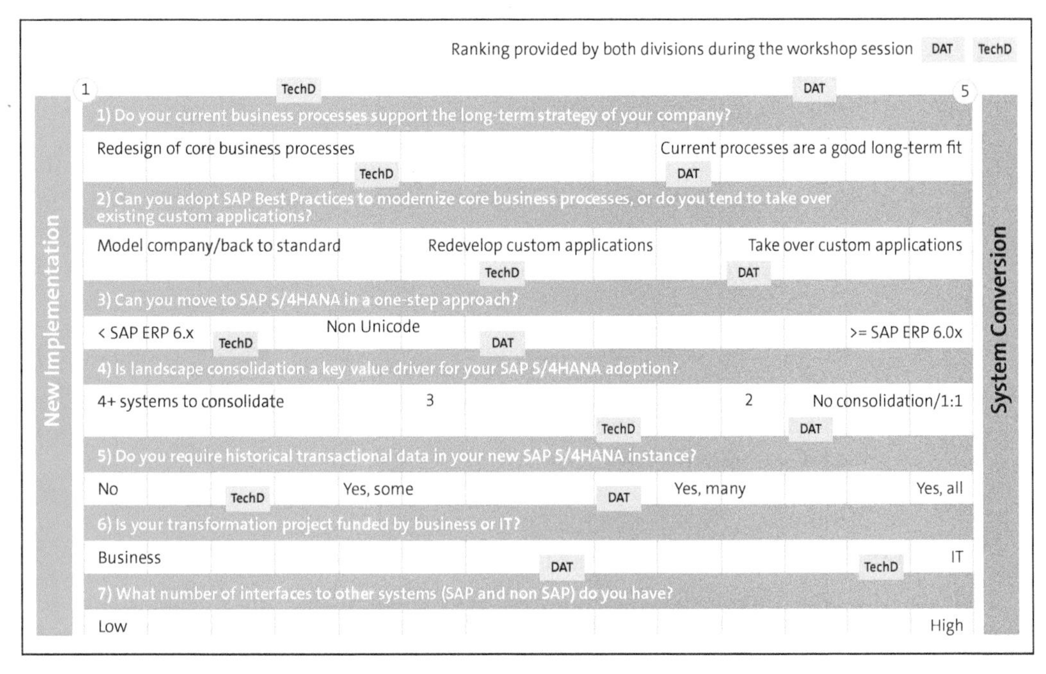

Figure 9.12 Transition Scenarios for DAT Inc. and TechD Inc.

After the discussion of the transition scenarios, we conducted a poll on all the seven guiding questions, and the results clearly showed that DAT Inc. prefers the system conversion approach to transitioning to SAP S/4HANA. This approach allows DAT Inc. to leverage its existing investment in SAP ERP systems to help in lowering the TCO. The primary challenge will be integrating the acquired company, TechD Inc., either using the overall system consolidation strategy or through a greenfield implementation.

As a next iterative approach as part of the North Star service, the transition strategy will be further delved into using the following detailed criteria and guiding questions:

- **Process reengineering**
  This is based on the question of whether DAT Inc. wants to fully reuse the existing processes set up in SAP ERP or wants to fully redesign the process based on the SAP Best Practices approach in SAP S/4HANA. The question to ask when evaluating this criterion is "What is the grade of required process transformation to the future target system?"
- **Data scope**
  This criterion is extremely important and directly influences the system's sizing strategy. It doesn't make sense to collect complete historical data in SAP S/4HANA without a strong business or legal justification. The question to ask when evaluating this criterion is "How much data do you need?"

- **Deployment option**
  This is the decision between RISE with SAP and the public cloud as the only two strategic options in DAT Inc.'s deployment strategy in the transformation program to move to SAP S/4HANA. The question to ask when evaluating this criterion is "How will your future systems be deployed?"
- **Source-target ratio**
  The priority is the consolidation of the systems in all regions and organizations. This is the starting hypothesis unless the detailed assessment provides a different recommendation. The question to ask when evaluating this criterion is "What is the ratio of your current as-is systems (ERP and legacy) to your to-be systems (SAP S/4HANA)?"
- **Rollout approach**
  This is an important criterion to use when deciding whether to have a big-bang rollout in all divisions and countries at once or to have a phased rollout by application, country, or division. The question to ask when evaluating this criterion is "How will your future systems be rolled out?"

9.5.7 Initiatives and Roadmap

We've seen the initiatives catalog at work in the previous case study chapters, but as a refresher, let's review the fact that the initiatives catalog is a list of initiatives derived from the strategy map, capability map, and target architecture that may help in realizing the goals in DAT Inc.'s strategy map. Initiatives lead to outcomes through implementation of a set of business capabilities in the capability map. The capabilities, in turn, are enabled through the solutions that can be found in the target architecture designed for DAT Inc. Figure 9.13 shows the list of initiatives derived as part of this phase 0 planning.

(Rating Scale: 1 to 5) | (High: 5; Low: 1)

#	Initiative	Category	Solution Cluster	Org. Unit	Subinitiatives (Projects)	Status	Approach	Ease of Implement	Value Potential
1	SAP S/4HANA Implementation	Core ERP	E1 SAP S/4HANA	DAT Inc.	System Conversion of SAP S/4HANA	New	Brownfield	4	5
2	SAP S/4HANA Implementation	Core ERP	E2 SAP S/4HANA	TechD Inc.	Greenfield Implementation of SAP S/4HANA (Manufacturing)	New	Greenfield	3	4
3	SAP S/4HANA Implementation	Core ERP	E3 SAP S/4HANA	TechD Inc.	Greenfield Implementation of SAP S/4HANA (Finance)	New	Greenfield	2	5
4	SAP S/4HANA Implementation	Core ERP	E4 Legacy ERP	DAT Inc.	Greenfield Implementation of Legacy ERP (Manufacturing) System	New	Greenfield	1	3
5	SAP S/4HANA Implementation	Core ERP	E5 SAP SCM	DAT Inc.	Greenfield Implementation of SAP SCM gATP to SAP S/4HANA Advanced ATP	New	Greenfield	3	4
6	SAP EWM Implementation	Supply Chain Execution	S1 SAP EWM	DAT Inc.	SAP EWM 9.5 to SAP S/4HANA EWM Migration	In-Progress Migration	Migration	3	4
7	SAP EWM Implementation	Supply Chain Execution	S2 SAP EWM	TechD Inc.	Greenfield Implementation of SAP EWM Warehouses to SAP S/4HANA EWM	Planned	Greenfield	2	3
8	SAP IBP Implementation	Supply Chain Planning	I1 SAP SCM	DAT Inc.	Global Template Design, Build, and Implementation of SAP IBP	New	Greenfield	3	5
9	SAP IBP Implementation	Supply Chain Planning	I2 SAP SCM	TechD Inc.	SAP SCM Move to SAP IBP Global Template	New	Greenfield	3	4

Figure 9.13 Initiatives Catalog for Consolidation

These initiatives are prioritized in a value and efforts matrix as illustrated in Figure 9.14, based on perceived relative business value and intensity of effort involved in the implementation phase. As an example, the first initiative is the system conversion project for DAT Inc.'s SAP ERP application to transition to SAP S/4HANA. The system conversion approach will bring your unchanged business processes to the new technical platform. This is a complete technical in-place conversion of an existing ERP software system in SAP Business Suite to SAP S/4HANA, and it will also ensure selective adoption of the most relevant innovations at the own speed decided by DAT Inc. Per Figure 9.14, this initiative E1 (system conversion of SAP ERP) is the ideal candidate where the ease of implementation and the value are both high.

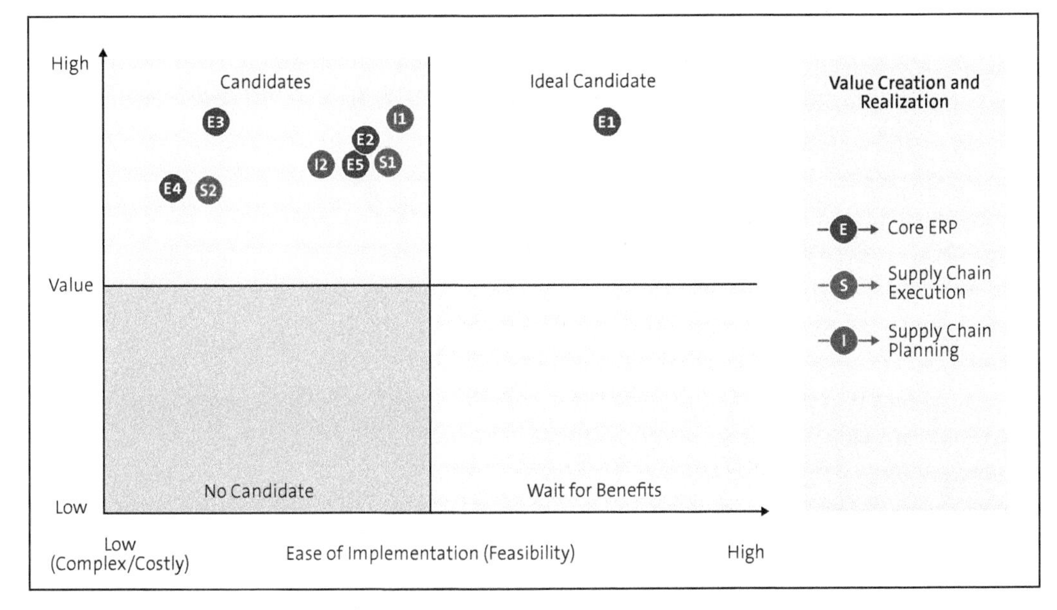

Figure 9.14 Initiatives Prioritization

Based on the prioritization of the initiatives, the transformation roadmap for consolidation of DAT Inc. and TechD Inc. systems is planned as shown in Figure 9.15. The key information on the transition roadmap along with the sequence of the target milestones (the diamond symbols in the green sections) can be summarized as follows:

- Before the start of the ERP consolidation, DAT Inc. wants to derisk the warehouse management systems by migrating from SAP EWM to SAP S/4HANA EWM (initiative S1). There are multiple rollouts planned per group of warehouses, as shown in Figure 9.15. The rollout for the group of TechD Inc. warehouses is planned to be implemented as the last milestone (initiative S2).
- The next milestone (initiative I1) is the supply chain planning global template, with pilot greenfield implementation of SAP IBP, replacing the existing SAP SCM demand planning (DP), supply network planning (SNP) solution for TechD Inc. Replacement

of the SAP SCM (DP, SNP) solution for DAT Inc. (initiative I2) is planned to occur toward the end but well before the end of the maintenance period for SAP.

- The core ERP consolidation initiative will start from the system conversion of SAP ERP (initiativeE1) plus the greenfield implementation of advanced available-to-promise (ATP) (initiative E5). This program will also include a project that involves replacing the TechD Inc. manufacturing system (initiative E2). Interim architecture will be complex as the finance system for TechD Inc. will still be in legacy SAP ERP at the end of this milestone.
- The next initiative (E3) is planned to move the full finance capabilities from the SAP ERP (Finance) instance to the consolidated SAP S/4HANA instance in a greenfield implementation approach.
- The last initiative (E4) is planned to move all the business units of DAT Inc. running in legacy ERP to the consolidated instance of SAP S/4HANA instance in a greenfield implementation approach.

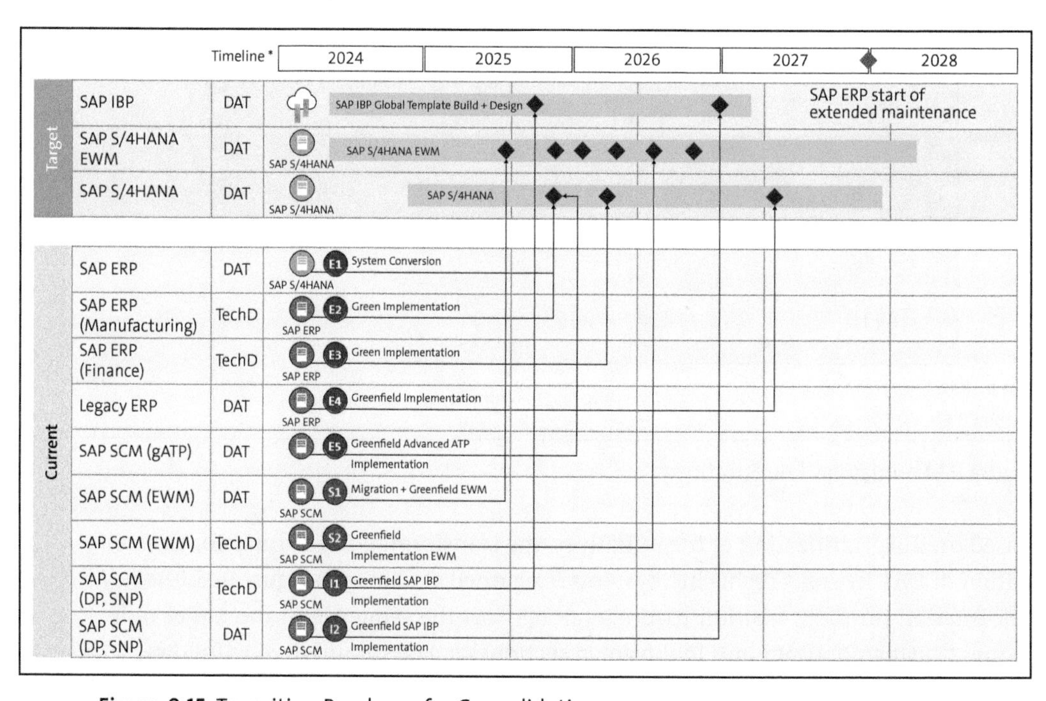

Figure 9.15 Transition Roadmap for Consolidation

9.6 Summary

In this chapter, we explored a case study that applied the elements of TCO and assessed how to provide recommendations for reducing TCO during transformation planning. In many organizations, IT doesn't decide on the purchase of applications; instead, the business units choose specific applications. Often, these applications are managed with a very poor governance structure, which can significantly impact costs and, consequently, the TCO for a large transformation program.

In short, reducing TCO depends largely on two critical factors:

1. Harmonization of business processes, in which best practice is to harmonize as much as possible
2. Centralization of data centers for SAP solutions, in which best practice is to centralize to regional data centers (or even to a single global data center)

This completes our use case of reducing TCO. In the next chapter, we'll look in detail at patterns for additional use cases in a large transformation initiative.

Chapter 10
Patterns for Additional Use Cases

Identifying a pattern to solve a specific use case for customers is one of the highly effective habits of a successful enterprise architect. We'll discuss some key patterns for additional use cases in this chapter, which is relevant to the transformation planning phase.

In the last five chapters, we dove into specific use cases for cloud transformation; business transformation; sustainability; mergers, acquisitions, and divestitures (MAD); and reducing total cost of ownership (TCO) as use cases for enterprise architecture. This chapter will briefly explore additional use cases that are relevant to the topic of enterprise architecture with SAP regarding planning, management, and transformation. The most relevant patterns for additional use cases based on our experience with enterprise architecture transformation for SAP customers are as follows:

- **Rollout strategy assessment**
  The *rollout strategy* (also known as the *deployment strategy* in the partner ecosystem) is the plan for implementing SAP solutions (typically by line of business, region, country, or another logical grouping unit) so that the business can consume the solutions in a sustained manner with maximum adoption and minimum risk of disruption.
- **Two-tier ERP architecture**
  Two-tier ERP architecture enables customers to maximize their existing investment in their ERP system while allowing for the flexibility and agility they need to run their business. This use case will show the transition from the central model to a federated model, powered by two-tier ERP deployment.
- **Split application architecture**
  This use case is about *split application architecture*, in which different application components are divided into separate interconnected systems (e.g., finance in one SAP system and logistics in another SAP system). The split architecture is an antipattern; in general, SAP never recommends this as the pattern to be realized as it makes the architecture complex and difficult to maintain in the long run. However, we'll still cover it in this chapter because many businesses find themselves in this situation.

All these additional patterns solve very specific business and IT use cases. Let's explore them further.

10.1 Rollout Strategy Assessment

Rollout strategy assessment is usually for customers who have defined a multiyear transformation roadmap that includes target architecture. The typical use cases are as follows:

- The customer is concerned about whether the defined solution and planned rollout follow SAP Best Practices.
- The customer is facing challenges related to the timeline of the planned rollout and needs to speed up deployment.
- The customer has complex interim landscapes and is running legacy and new solutions in parallel.
- The customer wants to evaluate potential alternatives to the initial plan.

The following sections will explain the business context and the IT context, and they will provide a mini case study for a rollout strategy assessment. We'll then apply SAP EA Framework and review the key deliverables.

10.1.1 Business Context

The rollout strategy is typically applicable to either conglomerates or global enterprises undergoing transformation, and it can involve multiple solutions, business divisions, regions, and business units (e.g., parts, vehicles, services/warranty). Correctly designing the rollout strategy is very important for the business since the rollout can either minimize disruption (if you start it in a region or business unit that contributes less revenue) or maximize disruption and risk (if you start it in a region or business unit that contributes major revenue to the company).

10.1.2 IT Context

In the context of IT, the rollout strategy for application architecture involves outlining the sequence of solution implementations, typically organized around lines of business (LoBs), regions, or applications as rollout clusters, over a broad timeline. As the rollout progresses, the interim IT architecture can become complex, potentially requiring the creation of temporary interfaces that may be discarded once the final architecture is in place. The rollout approach is all about what portion of the global solution to roll out per wave from source architecture (e.g., SAP ERP) to target architecture (e.g., SAP S/4HANA) within the most appropriate transition path.

10.1.3 Problem Statement

This use case is about a large fictitious retail company, RETKA Inc., that has a presence in multiple countries and has a global rollout project that is at risk. The issues the company has highlighted are as follows:

- There are recurring delays in cluster rollouts.
- Clusters have gone live but are yet to be stabilized.
- The global template is not in a state to enable a quick and efficient industrial rollout.
- Test phases have repeatedly taken longer than planned, especially due to delays in delivery of interface developments.

RETKA Inc. has reached out to SAP to conduct a rollout strategy assessment and come up with a new rollout approach. An SAP enterprise architect has been assigned, has started analyzing the problem statement using SAP EA Framework, and will provide detailed recommendations on these issues.

10.1.4 Applying the Enterprise Architecture Transformation Approach

Based on the understanding of RETKA Inc's ask and requirements, the modules scoped for this rollout strategy assessment are depicted in Figure 10.1.

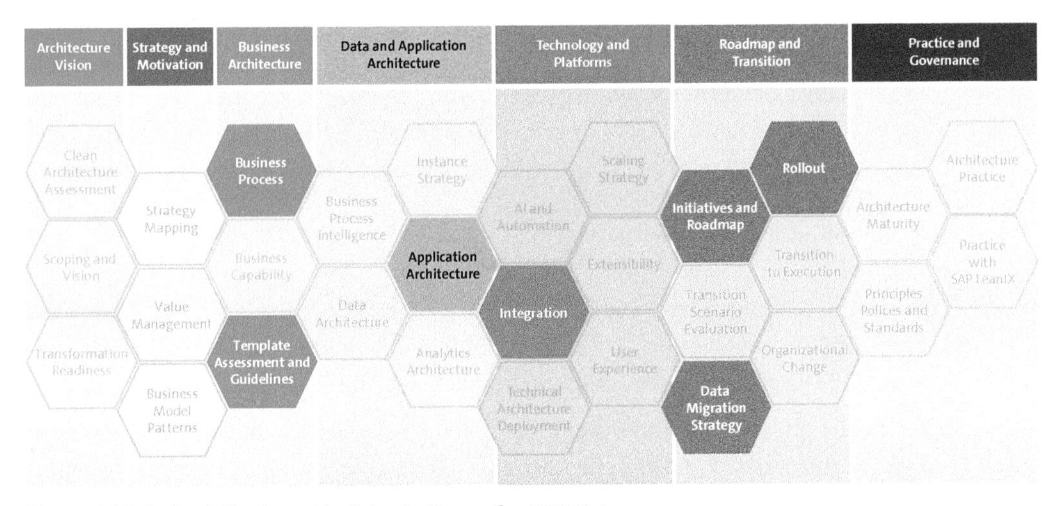

Figure 10.1 Rollout Strategy Modules in Scope for RETKA Inc.

Scoping details for each of these modules are as follows:

- **Business process**
  The scope is to review the end-to-end process for order to cash, source to pay, and record to report (finance), and to identify dependencies among the processes that are critical for the rollout strategy.

- **Template assessment and guidelines**
  In the case of RETKA Inc., the global template is already defined, so the scope is only to get the inputs to conduct the rollout strategy assessment (especially the rollout dimensions and clusters) and recommend SAP's approach.
- **Application architecture and integration**
  We scope the end-to-end solution architecture and integration architecture, assessing them to determine the solution complexity reduction potential.
- **Data migration strategy**
  We review the data migration strategy according to the rollout plan, based on functional domains, systems, data objects, potential data merges, mapping requirements, fit/gap, data transformation requirements, custom objects, data harmonization, and cleansing potentials.
- **Initiatives and roadmap**
  In this module, we review ongoing and planned initiatives and assess their impact on the parallel run. We then identify any new required initiatives, map them to overall dependent initiatives, and evaluate their impact on the rollout. This process will culminate in the creation of an initiatives catalog and the development of a strategic roadmap for the next five years.
- **Rollout**
  This is a *focus module*, which identifies and short-lists rollout options and different potential rollout sequences. It defines assessment criteria for the rollout options and sequences, and then it assesses options and sequences according to the defined criteria. It also reviews as-is nonproductive landscape and software change management practices and required nonproductive landscape supporting the parallel run.

10.1.5 Key Deliverables and Outcomes

Based on the modules we scoped in the previous section, we present several deliverables to RETKA Inc., but we don't cover all of them in this chapter. One key deliverable we do cover is SAP's recommended rollout approach, which is illustrated in Figure 10.2:

❶ We define the rollout dimensions and clusters for RETKA Inc. as per the scope dimension matrix. We can explain the initiative by breaking it down into "who," "what," and "where" components as follows:
  - *Who* consists of the business units that are in scope for the rollout.
  - *What* is the end-to-end process and the initiative such as order to cash, source to pay, and record to report (finance).
  - *Where* consists of the geographical entities containing the countries, regions, or company codes where the first pilot implementation will occur, followed by a phased or big-bang rollout there. The Americas and Europe are the rollout regions for RETKA Inc, and the first pilot is in the Canada region.

Each small block represents a potential rollout cluster that can be validated and aggregated, if necessary. For example, the SAP S/4HANA Sales solution can be deployed in all food retail stores in the Pacific region at one go, so it constitutes a single-block rollout cluster.

❷ We conduct a rollout options evaluation, considering the criteria listed in Figure 10.2. After we determine the building blocks for the implementation, we need to put them into a sequence and decide which steps to perform in parallel. Typical recurring patterns are big bang, successive implementation, parallel implementation, and pilot implementation.

❸ The next step is rollout sequencing and evaluating parallelization considerations, such as business priorities and value, SAP product roadmap, technical restrictions, business continuity risk, etc. Based on our recommendation, RETKA Inc. decides to roll out to the Canada region first as the pilot, followed by the United States and Europe.

❹ After determining what to roll out and where to deploy it, RETKA Inc. wants to accelerate the rollout by doing a few parallel rollouts to achieve time to value. The adequate level of parallelization of the rollout steps is mainly driven by the following factors:

- The availability of internal and external resources
- The duration of the hypercare and stabilization support
- The need to avoid critical project phases (e.g., final integration test, final preparation, cutover)

10

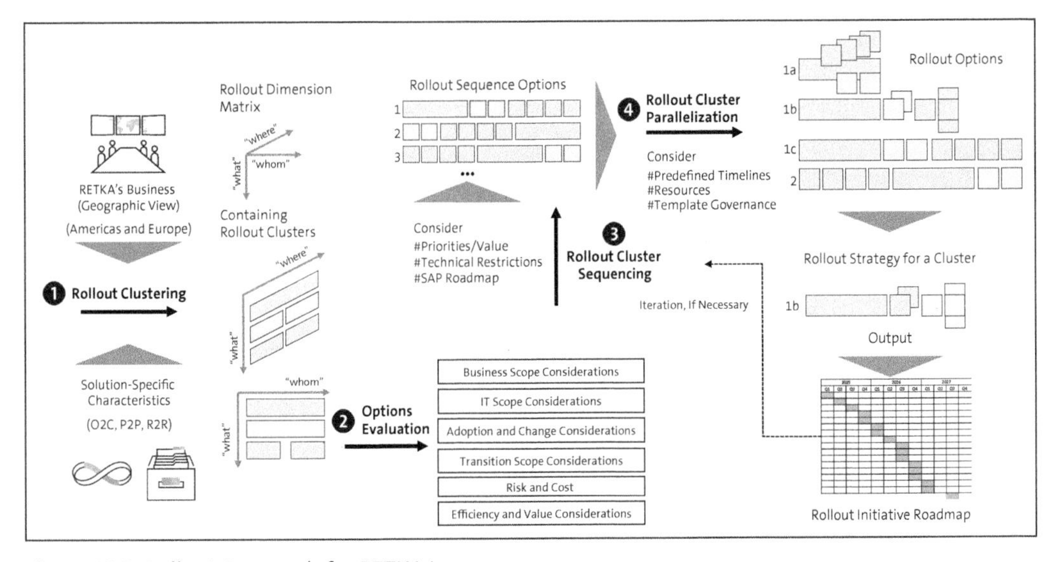

Figure 10.2 Rollout Approach for RETKA Inc.

10.2 Two-Tier Architecture

SAP's *two-tier ERP strategy* with SAP S/4HANA Cloud is a decentralized and federated deployment approach. It enables business and IT to collaborate in building sustainable, scalable solutions. This strategy provides the benefits of cloud deployment, such as speed, standardization, and low TCO, while also maximizing existing investments. For example, the two tiers can be the organization's headquarters and its subsidiaries. By choosing a subsidiary ERP system built on a software-as-a-service (SaaS) paradigm, customers can significantly reduce up-front costs, including hardware, software, data center, and implementation expenses. This significantly lowers their cost of ownership and allows them to deploy a best-in-class system quickly while staying within budget. But as you've seen by now, having a rich SaaS-based ERP system that meets a subsidiary's functional and budgetary requirements is not enough. It's also critical for the selected subsidiary system to support deep information integration with the headquarters' ERP system.

In the following sections, we'll explain the business and IT contexts and provide a mini case study of a two-tier architecture. We'll then apply SAP EA Framework and review the key deliverables.

10.2.1 Business Context

In reality, there are multiple use cases for this decentralized and federated model, but the typical use cases for two-tier architecture from the business context are illustrated in Figure 10.3. We explain them further here:

- **Scenario 1**
  This is the headquarters and subsidiaries model, where headquarters is running an on-premise ERP and subsidiaries run cloud ERP. In this scenario, the organization can reap the following benefits:
  - Seamless integration of headquarters and subsidiaries
  - Flexibility that allows subsidiaries to be independent or dependent
  - The ability to free up subsidiaries' scope and allow them to innovate
  - Facilitation of fast onboarding of subsidiaries

  Subsidiaries could be sales, manufacturing, distribution, shared services, consulting, service delivery, field service, and repair.
- **Scenario 2**
  The organization can think of spinning off complete LoBs and running them on cloud ERP. For example, it could turn finance as shared services into a separate legal entity and have the rest of the business running on an on-premise ERP system. In this scenario, organizations can reap the following benefits:
  - Rapid innovation for large corporations with different LoBs

  - Ease of introducing new innovations to the business without interfering with on-premise installation

- **Scenario 3**
  Organizations can bring their subcontractors or dealers onto cloud ERP while continuing to run their on-premise solution so as to have a network linking headquarters, dealers, and subcontractors. In this scenario, organizations can reap the following benefits:
  - The integration of large corporations with their vendors, dealers, and subcontractors to form an entire supply chain network
  - High visibility across the supply chain
  - Process automation that results in the reduction of manual intervention

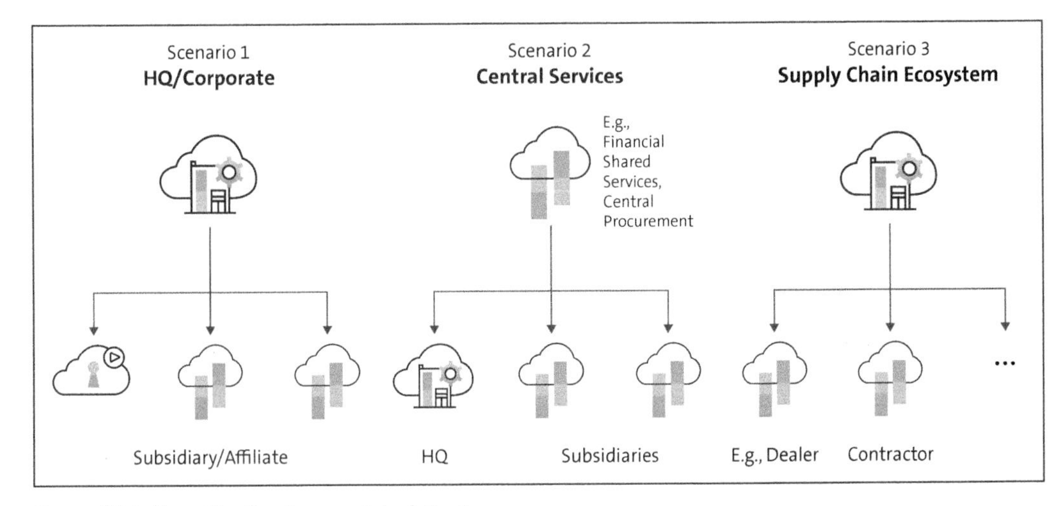

Figure 10.3 Two-Tier Deployment Architecture

10.2.2 IT Context

Many large enterprises struggle with monolithic architecture and want to transform it into agile and scalable architecture. They can start cloud transformation today with the deployment of cloud software in a specific area of the business. Refer to Chapter 5 on cloud transformation for further details. In the IT context, one way to adopt early two-tier architecture is to adopt predelivered integration into SAP S/4HANA Cloud Public Edition.

A few examples of the data flow between tier 1 and tier 2 are illustrated in Figure 10.4:

- As a first example, a production order and a planned order will be replicated from the manufacturing execution system to SAP S/4HANA.
- As a second example, tier 2 is equipped to run the procurement and manufacturing operations with the ability to manage local inventory. Tier 1 would prefer that the

purchasing operations be independently be carried out by tier 2 for purposes of operational ease and cost effectiveness.

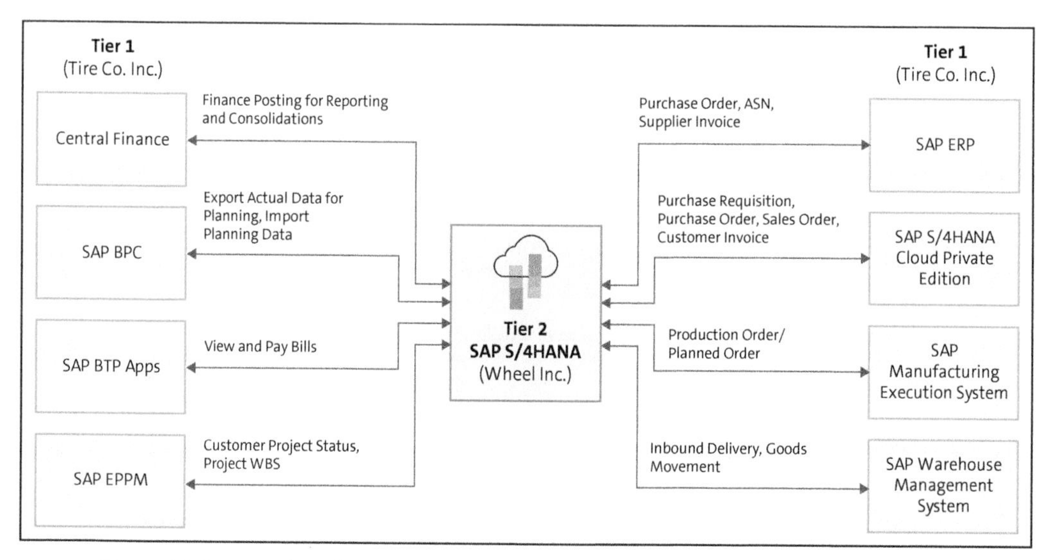

Figure 10.4 Examples of Integrations between Tier 1 and Tier 2

10.2.3 Problem Statement

Let's consider a fictitious company, Tire Co Inc., which is in heavy manufacturing and distribution and has acquired another company, Wheel Inc., as its sales subsidiary. Tire Co's strategic priority is to take over the customer base and market share of Wheel Inc. Tire Co is using SAP ERP, whereas Wheel Inc. has decided to deploy SAP S/4HANA Cloud. This is known as a *headquarters and subsidiaries deployment model*, where the headquarters runs on an on-premise or private cloud ERP and the subsidiaries run a public cloud ERP. One of the key problem statements is to map the best-practice end-to-end process of sales from central stock and drop shipments with advanced shipment notification (ASN), due to the following key reasons:

- **Inventory visibility**
  Lack of real-time visibility into central stock levels is leading to out-of-stock and overstock situations for Tire Co.
- **Order processing and coordination**
  Coordinating between central stock and the supplier for drop shipments is complex and time-consuming.
- **ASN**
  Managing discrepancies between ASN and actual deliveries leads to potential delays and disputes.

10.2.4 Applying the Enterprise Architecture Transformation Approach

Based on our understanding of Tire Co Inc's requirements, we scope the following modules for this pattern of two-tier architecture, as depicted in Figure 10.5:

- **Business process**
  We scope the best-practice process of sales from central stock and drop shipments with ASN as something to be explained in detail to customers using the two-tier architecture deployment model.
- **Application architecture**
  The primary scope of the discussion is to address the integration between SAP ERP and SAP S/4HANA Cloud using SAP Business Technology Platform (SAP BTP).
- **Integration**
  We scope this module to focus on understanding the capabilities and potential usage of two-tier document replication using SAP Event Mesh for Tire Co Inc. *SAP Event Mesh* is a fully managed cloud service that enables applications to communicate through asynchronous events. Using it will help Tire Co Inc. create responsive applications that operate independently and participate in event-driven business processes across the business ecosystem, enhancing agility and scalability.
- **Extensibility**
  We scope this module to introduce Tire Co. to SAP's clean core paradigm, which provides an overview of all relevant extensibility options, identifies how to use them in specific customer situations, provides input into the platform design, and creates activities for the implementation of the required capabilities.

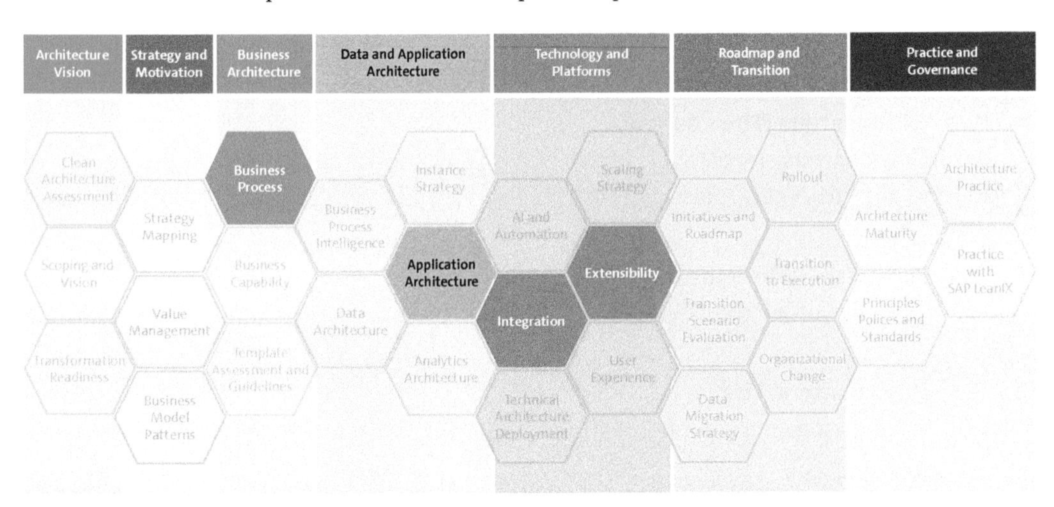

Figure 10.5 Architecture Modules Scoped for Tire Co. Inc.

10.2.5 Key Deliverables and Outcomes

The key outcome we deliver after the architecture workshop session is the end-to-end sales process flow across the two-tier architecture, as illustrated in Figure 10.6.

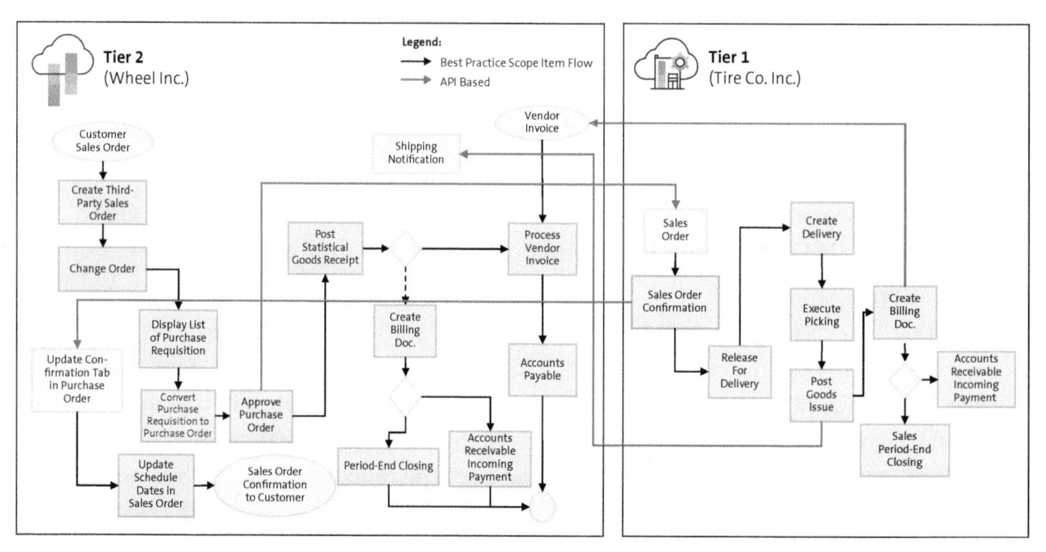

Figure 10.6 Sales from Central Stock and Drop Shipment with ASN

Let's take a closer look at this process flow:

- **Subsidiary (Wheel Inc.)**
  A customer order at the local sales office creates follow-on documents (a purchase order, sales order, and order confirmation), and an API (the red line in Figure 10.6) sends the purchase order information for sales order creation at tier 1.
- **Headquarters (Tire Co.)**
  Headquarters creates a sales order with the sold-to as the sales office and the ship-to as the customer. Headquarters also confirms the delivery date to the tier 2 sales office.
- **Subsidiary (Wheel Inc.)**
  The sales office provides an order confirmation to the customer with confirmed dates.
- **Headquarters (Tire Co.)**
  Headquarters posts goods for customer delivery, with ASN to tier 2 and billing to tier 2.
- **Subsidiary (Wheel Inc.)**
  The subsidiary posts a statistical goods receipt for tier 1 drop shipment, plus customer billing.

Please note that all the mapping in SAP BTP is decided based on project-based requirements, and the flow can be automated as per the business requirements.

10.3 Split Architecture

Split architecture is architecture that's designed based on splitting the application components into two different SAP systems. Having split application architecture goes against the idea of having an integrated core, and hence, this is an antipattern. But split architecture still exists in many SAP customer landscapes. Many large automotive and consumer industry customers have logistics in one SAP ERP system and finance in another SAP ERP system, and SAP ERP isn't built for such a split. Hence, many custom enhancements and modifications are needed to make it work, and that comes with high effort and some severe restrictions. Many times, a functional split is considered a better choice to bend the software to accommodate the organization's silos than to bend the silos to accommodate a single global instance.

On the other hand, *modular architecture* is architecture that's designed as a collection of discrete modules or application components that can be independently developed, upgraded, and retired. A simplified view of split versus modular architecture is shown in Figure 10.7. Logistics application components such as sales and distribution, materials management, and production planning are in one SAP S/4HANA instance, whereas finance application components such as the general ledger, accounts receivable, and accounts payable are in another SAP S/4HANA instance. This is an example of split architecture since the core ERP system is split.

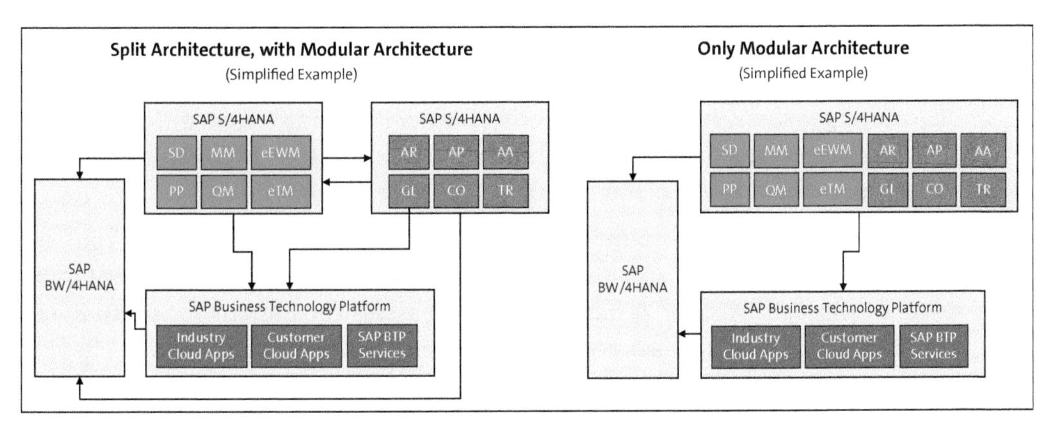

Figure 10.7 Split versus Modular Architecture

The following sections will explain the business context and the IT context, and they will provide a mini case study of a split architecture. We'll then apply SAP EA Framework and review the key deliverables.

10.3.1 Business Context

Some organizations aim to adopt split architecture to enhance their agility, flexibility, and scalability. Businesses also seek to accelerate innovation and iteration, and breaking down applications into smaller, more manageable components can facilitate that. Essentially, what is needed is a modular architecture, rather than a split architecture. If you decide on split architecture, then you need to give up on process harmonization as a goal.

10.3.2 IT Context

From an IT perspective, a split architecture is a system that is divided into separate components that are responsible for specific tasks. Splitting logistics and finance components is a common example from the past. However, this should generally not be the target architecture in the SAP S/4HANA world because there are major risks related to the need to replicate every logistics and finance transaction in both systems to maintain the integrity of the entire ERP system.

10.3.3 Problem Statement

Let's look at a scenario for a large multinational company, MFG. Co. Ltd., which currently has a single global instance of SAP ERP connected to SAP Extended Warehouse Management (SAP EWM), SAP Transportation Management (SAP TM), SAP Advanced Planning and Optimization, and SAP Business Warehouse (SAP BW) on SAP HANA. In addition to these, the company has multiple legacy applications for its manufacturing and finance processes. MFG. Co. Ltd. has decided to halt the deployment of its legacy manufacturing plants to SAP ERP, as it feels it's no longer strategic to continue investing in legacy systems. Instead, it's exploring a new SAP S/4HANA strategy to deploy the remaining legacy manufacturing plants. This will lead to a split architecture situation, with some manufacturing plants in SAP ERP and others in SAP S/4HANA.

The company's main constraint is that it can't change its existing customer communications and channels— specifically, its centralized sales order management capabilities—through SAP ERP. It has asked SAP to assess the options for splitting the architecture for manufacturing from the architecture for other capabilities. The specific request is for SAP to conduct a detailed assessment to determine whether MFG. Co. Ltd. can treat the plant in SAP S/4HANA as just a manufacturing location and pass it confirmed process orders. This is the minimum requirement in SAP S/4HANA for manufacturers.

10.3.4 Applying the Enterprise Architecture Transformation Approach

Based on our understanding of MFG. Co. Ltd.'s requirements to build a futuristic SAP S/4HANA platform as an interim state architecture for manufacturing plants, we've scoped the following modules for this pattern of two-tier architecture, as depicted in Figure 10.8:

- **Business process**
  We've broken down the sales order process and the manufacturing process into steps to analyze which steps will stay in SAP ERP and which steps will move to SAP S/4HANA. The customer has make-to-stock (MTS) as well as make-to-order (MTO) scenarios.
- **Application architecture**
  We've defined the interim state application architecture, considering that manufacturing capabilities and processes will be in SAP S/4HANA while sales and order management capabilities and processes will be in SAP ERP.
- **Integration**
  We've assessed different integration domains and patterns to review which is the best integration approach to connecting SAP ERP and SAP S/4HANA in the various order and manufacturing scenarios.

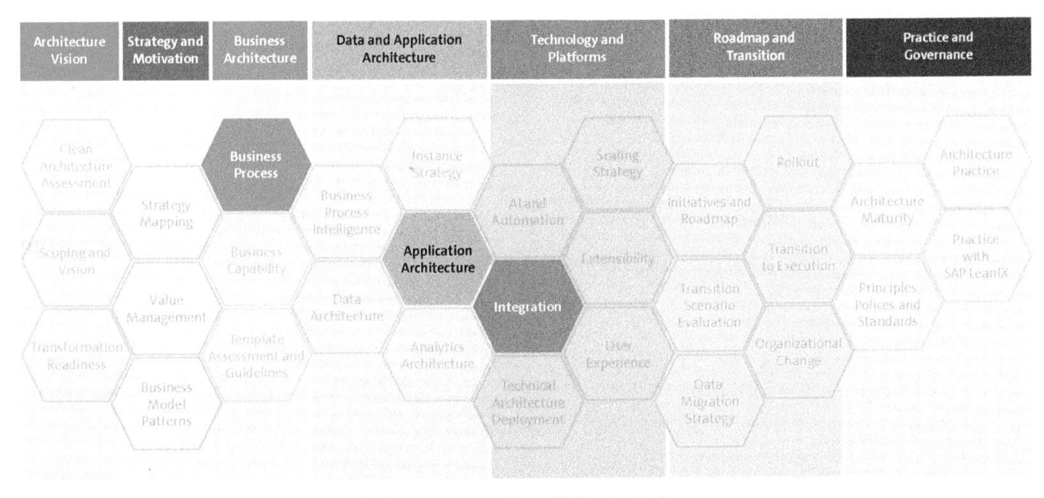

Figure 10.8 Split Architecture Modules in Scope for MFG. Co. Ltd.

10.3.5 Key Deliverables and Outcomes

After the scoping and workshop sessions, we evaluated many options. The best option is shown in Figure 10.9, and it involves SAP ERP for sales order management and SAP S/4HANA for subsales orders, planning, procurement, manufacturing, and logistics connected with peripheral systems like SAP Advanced Planning and Optimization, SAP EWM, and SAP TM. We consider this option to be the best fit due to its flexibility and coverage of all manufacturing scenarios. This makes it the interim architecture that brings planning, manufacturing, and logistics data into a simplified landscape for faster reaction to change and full visibility of the variant configuration throughout the supply chain and manufacturing processes in MTO processes.

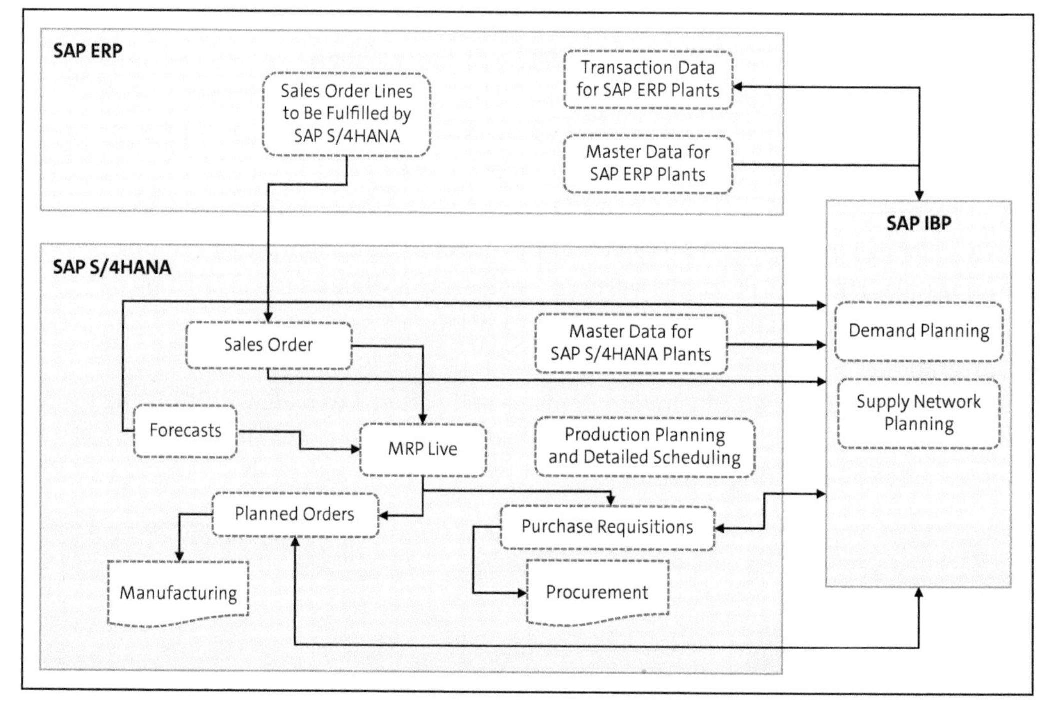

Figure 10.9 Simplified View of Split Application Architecture for MFG. Co. Ltd

10.4 Summary

In this chapter, we examined patterns for additional use cases from an enterprise architecture perspective. This completes our coverage of all use cases, in which we've provided real-world architecture decisions from our customers' transformation journeys. However, there are many other use cases—such as scale-out architecture, security architecture, and data architecture—that we can't cover in this book. Refer to the following blogs for additional use cases and patterns:

- For scale-out architecture and a customer example: *http://s-prs.co/v586307*
- For a blog on data architecture: *http://s-prs.co/v586308*

In Part III, we'll take a detailed look at SAP LeanIX and its use cases in a large transformation initiative. We'll start with the fundamental concepts of SAP LeanIX in the next chapter.

PART III

SAP LeanIX

With the acquisition of SAP LeanIX, SAP became a leader in enterprise architecture tools. The motivation for this acquisition was to better support companies in their business transformation and their move towards a cloud architecture. This part introduces SAP LeanIX. It's designed to be useful to you even if you haven't worked with the tool in depth yet. Many of the concepts we'll discuss can be reapplied to other enterprise architecture tools.

After providing an overview and explaining the core concepts, we'll give concrete guidance for working with SAP LeanIX. We'll then look into the integration and extensibility options, and close with typical SAP LeanIX use cases to examine the tool in practice.

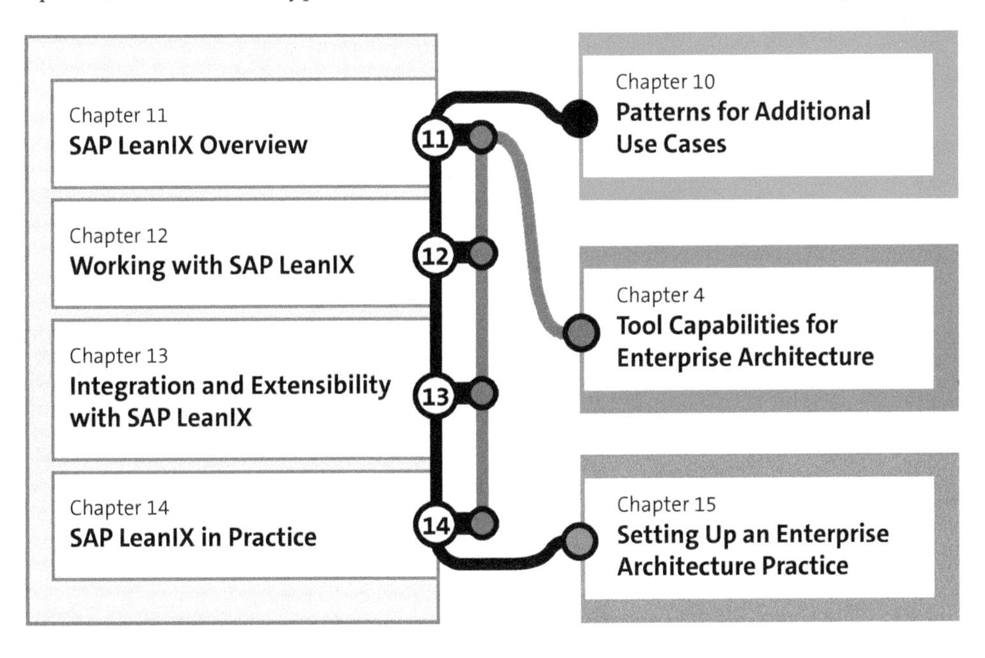

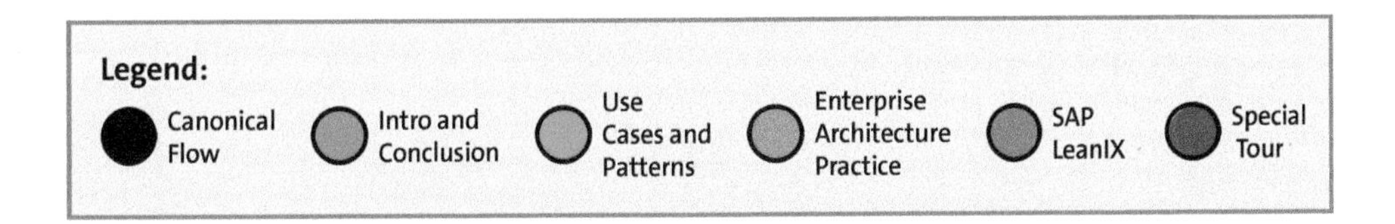

Chapter 11
SAP LeanIX Overview

SAP LeanIX is a comprehensive enterprise architecture tool that has its roots in application portfolio management. In the SAP context, its relevance to supporting architecture transformations is becoming increasingly important.

Chapter 4 provided a generic overview of how tools can support SAP-centric enterprise architecture transformations. Now, this chapter provides a more detailed look at the SAP LeanIX suite and its early integration into SAP. It'll also give you our perspective on this suite, rather than a detailed user guide or roadmap. Note that this deeper look into SAP LeanIX doesn't imply that other tools, not covered in this book, can't do the job as well.

Additional Resources

The SAP LeanIX online documentation, available at *https://docs-eam.leanix.net*, is one of the best of its kind and should always be your go-to resource for further insights and help.

Alternative approaches on how to specifically model SAP landscapes in SAP LeanIX are discussed at *https://docs-eam.leanix.net/docs/sap*.

11.1 SAP LeanIX as Part of an End-to-End Toolchain

SAP LeanIX consists of the following three products, which perform the listed functions. Of these three, SAP LeanIX Application Portfolio Management is the foundation, and neither of the other two works in isolation:

- **SAP LeanIX Application Portfolio Management**
  This product is licensed by the number of applications modeled, and it is the comprehensive base product enabling most features, including out-of-the-box integration with many other best-in-class IT management solutions such as SAP Signavio, Jira, and Collibra.
- **SAP LeanIX Architecture and Road Map Planning**
  This is a key product for modeling enterprise architecture transformations. We'll look at this in more detail in Chapter 12 and discuss a dedicated use case in Chapter 13.

- **SAP LeanIX Technology Risk and Compliance**
  This product includes special support for managing technical obsolescence risks, and it comes with a standard integration with ServiceNow. Chapter 13 provides one section on the use of this module.

Many companies use SAP LeanIX for its core enterprise architecture domain or, in narrower fashion, as a tool for application portfolio management. As discussed in Chapter 3, integration with other functions is important. In Chapter 13, we'll look at a number of integrations with best-of-breed solutions that support other functions.

This section focuses on SAP's end-to-end transformation toolchain, as depicted in Figure 11.1. The RISE with SAP offering, which supports transformations into the cloud, combines three main solutions: SAP LeanIX, SAP Signavio, and SAP Cloud ALM.

SAP LeanIX has built-in capabilities for autodiscovery of landscapes from SAP Cloud ALM and can also load reference architectures, as described in Chapter 13.

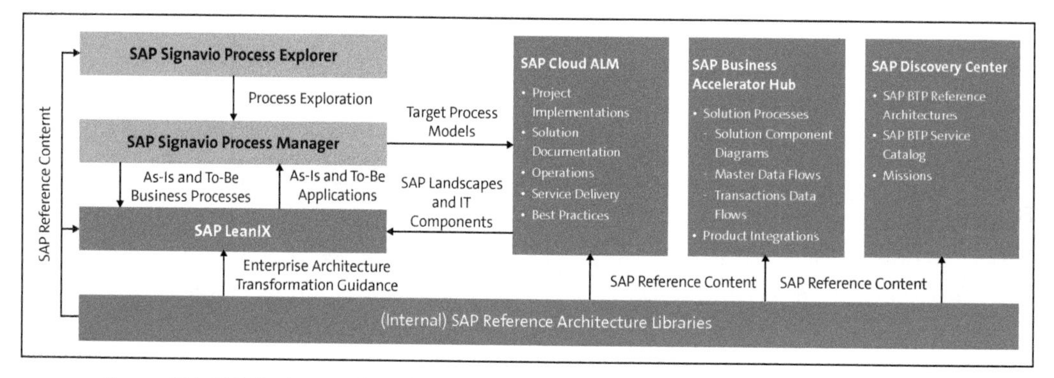

Figure 11.1 SAP End-to-End Toolchain

The integration of SAP Signavio Process Manager and SAP LeanIX enables a bidirectional exchange of business capabilities, applications, and business processes.

Exploring Reference Architecture

You can explore additional SAP reference architecture content on SAP Business Accelerator Hub and SAP Discovery Center:

- **SAP Business Accelerator Hub**
  This resource covers details on solution components and standard integrations, right down to the application programming interface (API) level.
- **SAP Discovery Center**
  This resource deals with SAP BTP reference architectures.

In the future, this content may be exposed in SAP LeanIX as well.

Figure 11.2 maps out key transformation roles in a two-dimensional schema: on the left are business and business-related roles, on the right are IT solutioning roles, and in the center are architecture-related roles. In the vertical direction, roles are spread between execution roles at the bottom and strategic C-level roles at the top.

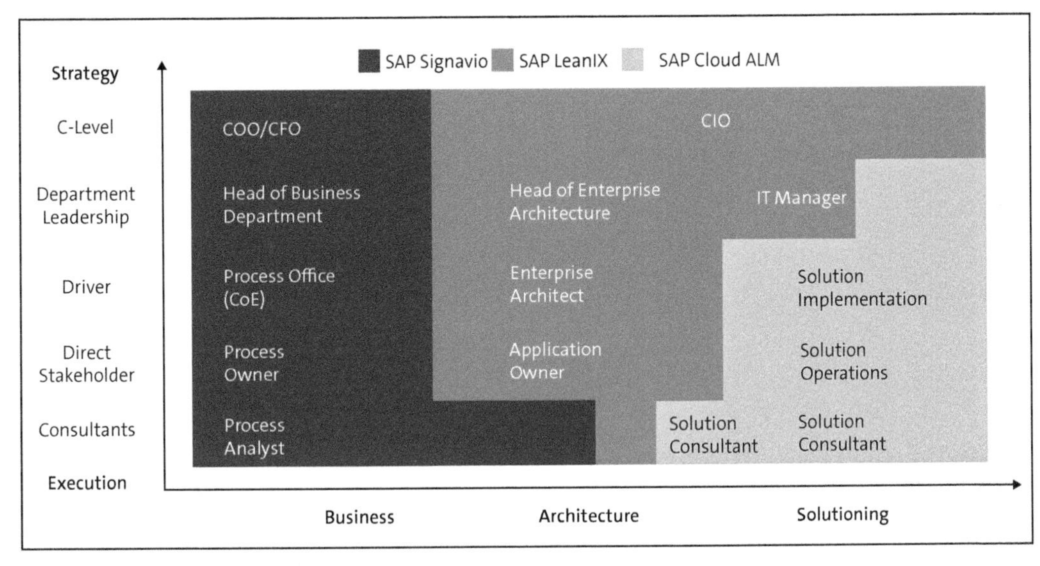

Figure 11.2 Mapping Personas to Transformation Solutions

As people often fulfill multiple roles, the individual roles won't relate to corresponding single tools. Three of these ambiguities are depicted in Figure 11.2, which is intended to show the focus of work, rather than diving deep:

- In the department leadership layer, IT managers need to both consider the solutioning perspective and engage with the architects. They'll access SAP LeanIX as well as SAP Cloud ALM.
- As direct stakeholders, the major focus of solution operations is on SAP Cloud ALM, as depicted in the orange area. In addition, there can be business process operations tasks for which SAP Signavio Process Insights is a critical enabler.
- On the execution level, some solution consultants will deal with integrations and might therefore access and/or document SAP LeanIX's integration diagrams.

11.2 Fact Sheets and Meta Model

One key principle of SAP LeanIX is to provide comprehensive enterprise architecture support via a predefined set of fact sheets. The predefined fact sheets of the SAP LeanIX meta model version V4 consists of twelve fact sheet types depicted in Figure 11.3. These types are clustered in four groups:

- **Strategy and transformation**
  This group contains the overarching concept of *objectives*, which are supported by initiatives. *Initiatives*, in turn, are the anchor points from which to plan transformations, and they can therefore be linked to most other fact sheets to document their impact. A *platform* is a strategic aggregation of applications and IT components that enables higher-level depictions of strategic architectures.
- **Business architecture**
  This group includes the typical business architecture concepts: organization, business capability, and business context. The business context fact sheet type is most often used for business processes.
- **Application and data architecture**
  On top of its core elements, application, and data object, this group holds the interface fact sheet type, which is used to document the information flow among applications.
- **Technology architecture**
  The core fact sheet type in this group is the IT component, which can be grouped into tech categories. The fact sheet provider stands for the company providing a given IT component (a hardware or software provider).

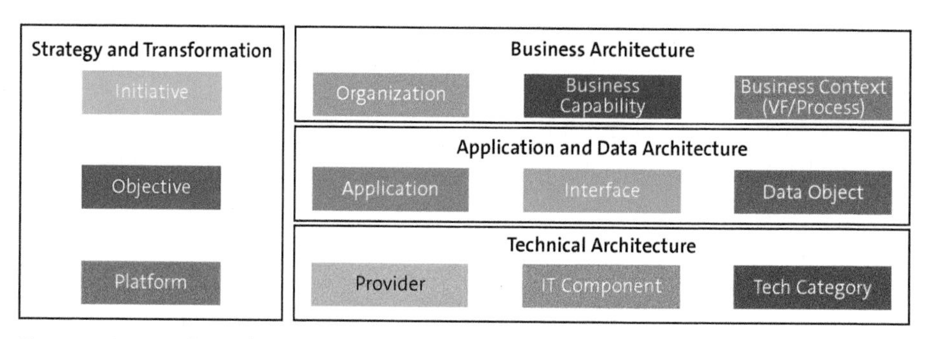

Figure 11.3 Fact Sheets by Category

Fact sheet types can have subtypes that inherit the fields, tags, and relationship definitions from the parent fact sheet but can also have their own fields. We'll list the predefined fact sheet subtypes in Section 11.2.2.

11.2.1 Relationships among Fact Sheets

As in entity relationship modeling, the different fact sheet types have different relationships among each other. Figure 11.4 shows the predefined possible relationships.

Entity Relationship Model

The concept of entity relationship modeling goes back to Peter Chen in 1976, and it provides a visual representation of relational database structures. Relationships and attributes are depicted graphically. Over the years, multiple variations have evolved, increasing the expressiveness of modeling cardinalities for the relationships and dependencies among attributes.

Relationships can be hierarchical parent-child or associative many-to-many relationships. In the SAP LeanIX applications, these relationships are used in many contexts:

- Within the details of a fact sheet, relationships can be defined and manipulated. The UI offers a section called **Relationship Explorer**, in which the user can navigate through the relationship network in multiple steps.
- You can use relationships to filter lists and fact sheets for reports.
- Relationships are also often the most important elements used in layouts for reports.
- Relationships themselves can have fields that, again, you can use for layouts and filtering.

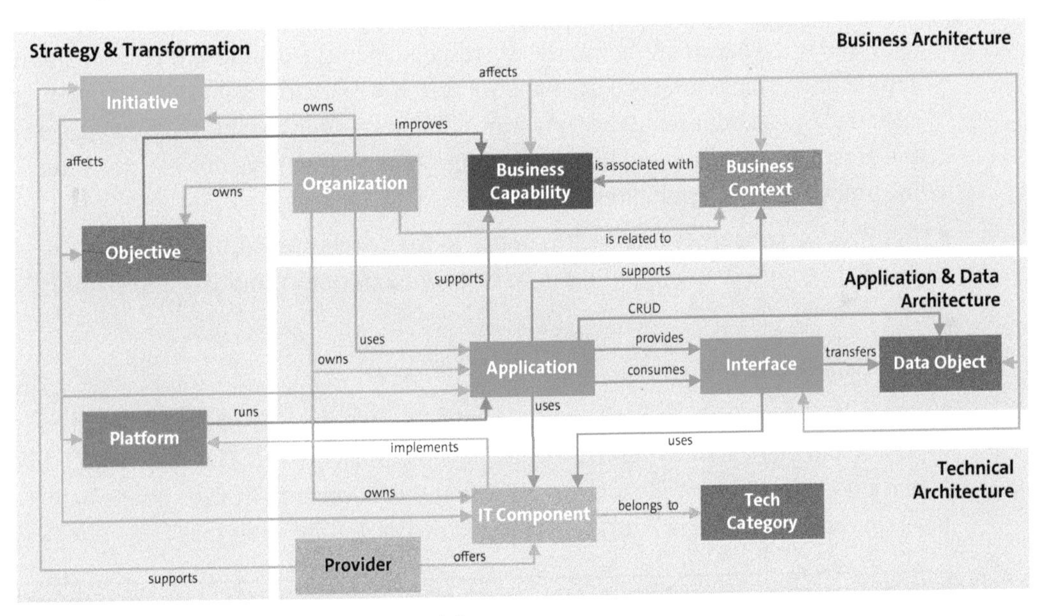

Figure 11.4 Relationships in the Meta Model

You can organize most fact sheet types in parent-child hierarchies, and many can have a successor/predecessor relationship. These are not shown in Figure 11.4, but in Figure 11.5, they're displayed in the form of an exemplary **Application** fact sheet.

In SAP LeanIX, you can evaluate relationships from both directions and name them individually. The cardinality of each relationship is defined in the meta model. Most are defined as many-to-many, but they can also take the form of many-to-one, one-to-many, and, in some cases, one-to-one.

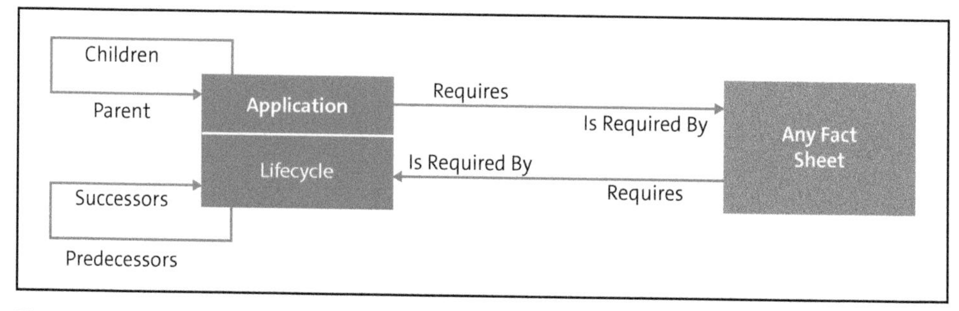

Figure 11.5 Generic Relations Shown on Application Fact Sheet

Ideally, relationships not only repeat the name of the related fact sheet type but also describe the relationship using a specific verb. This is especially necessary for fact sheet types that can have different types of relationships, such as those between application and interface:

- **Application provides interface (a one-to-many relationship)**
  The provider relationship determines the application that's accountable for the logic and governance of the interface. Irrespective of the data flow direction, there can be only exactly one provider application. Note that the data flow direction is specified independently in the **Data flow direction** field in the interface object from the perspective of the providing application. This field can have one of three values: incoming, outgoing, or bidirectional.
- **Application consumes interface (a many-to-many relationship)**
  A specific interface (e.g., an API) can be consumed by one or multiple applications.

11.2.2 Fact Sheet Subtypes

SAP LeanIX's objective is to work with a relatively small number of fundamental concepts and fan out the subtypes instead, as depicted in Figure 11.6. The idea behind this is that a subtype will typically inherit its fields and relationships from a specific higher-level concept and thus will help keep the meta model definition simple.

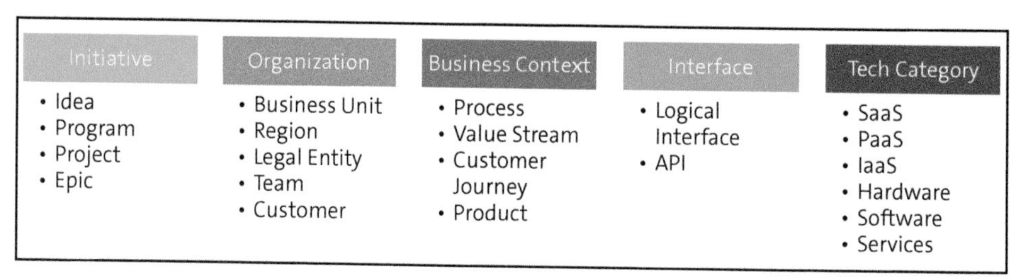

Figure 11.6 Fact Sheet Subtypes

In the instance of the organization fact sheet type, the region and team subtypes don't necessarily depict a "real" classical organization (business unit). In our enterprise architecture modeling, we often use these subtypes (e.g., to describe in which context or by which user group a given application is used).

In Chapter 13, we'll learn how to define new subtypes as well as subtype-specific relationships and fields.

11.2.3 Fields and Tagging

Fields (also named *attributes* or *properties* in other tools) are used to describe a fact sheet. A field that each fact sheet has is the **Name** string field, and another field that most fact sheets have is the **Description** text field.

Possible field data types are as follows:

- **Text**
- **Text Area**
- **Date**
- **Integer** (with the **Costs** and **Number** display options)
- **Double** (with the **Costs**, **Number,** and **Percentage** display options)
- **Single Select**
- **Multiple Select**

You can also code the **Single Select** and **Multiple Select** value options with individual colors, which you can use for data-driven color coding (heatmapping) in diagrams and reports.

Instead of adding fields, you can store information in tags. Tags, in fact, are more visual, as they appear in color at the head of each fact sheet or on inventory lists. Hence, you can make important information stand out. Be careful, however, not to tag information that you can already store in fields and/or relationships. In addition, you should limit the number of tags. An administrator can predefine tags or allow free tagging of certain fact sheet types, and we'll discuss the benefits and downsides of tagging in Chapter 12.

11.3 The SAP LeanIX Application

When logging in to SAP LeanIX, you'll always be connected to a specific workspace. Companies will typically have one single productive workspace and one or possibly multiple test workspaces. The workspace defines the different types of objects that need to be modeled, and in SAP LeanIX terminology, these are named fact sheets. As explained in the previous section, each fact sheet type has several fields and relationships, which themselves can have fields too.

Figure 11.7 shows the top navigation bar that is always visible and allows navigation to the different application sections.

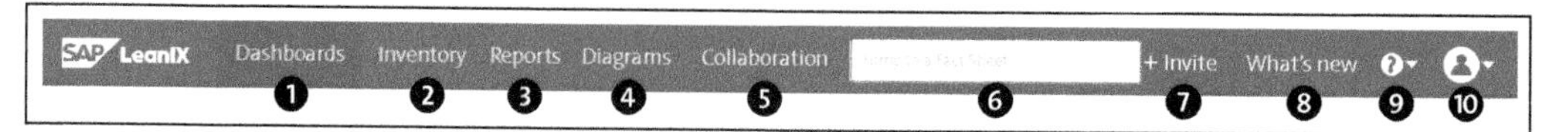

Figure 11.7 SAP LeanIX Navigation Bar

Let's walk through each option on the navigation bar:

❶ **Dashboards**
These combine selective data and reports, and you can use them to branch into deeper content. A dashboard can serve as a starting point for a specific use case and typically combines several graphical summarization reports. Even detailed object lists can be shown on dashboards.

❷ **Inventory**
This provides access to all fact sheets. Inventories can be shown in list or table format, and flexible filters can reduce the number of specific fact sheets.

❸ **Reports**
These are data-driven visualizations of fact sheets and their relationships. They come with drilldown capabilities and heatmapping, and they offer multiple layout options that fit into different use cases. Two generic report layouts are the most relevant ones because they are usable for all fact sheet types:

- **Landscape (cluster report)**: This layout allows clustering of fact sheets by fields, parent hierarchies, or associated fact sheets, which themselves can be grouped into hierarchies.
- **Matrix**: While landscape reports allow one-dimensional clustering, a matrix allows two dimensions (rows and columns).

You can apply three other report types—**Portfolio**, **Roadmap**, and **Radar**—to multiple fact sheet types in the same manner, but they are typically used for more specific use cases. Figure 11.8 visualizes a total of five default report types.

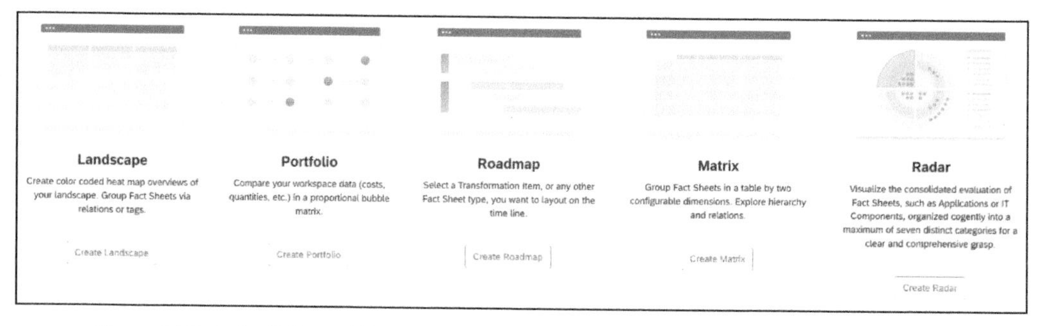

Figure 11.8 Default Report Types

In Chapter 12, we'll dive into examples and see an additional, more specialized report, while Chapter 13 will cover the development of your own reports and browse through the partner ecosystem of additional reports.

❹ **Diagrams**
These offer another way of visualizing fact sheets and their relationships, and users can usually lay them out manually. Filtering and heatmapping work in the same way in diagrams as in reports. While individually composed diagrams are widely adopted in other enterprise architecture tools, SAP LeanIX users should always ask themselves if they can't depict the same semantics via a fully generated report.

❺ **Collaboration**
This groups the following two functions:

- TO-DOS: This provides access to questions and answers or to-dos. The to-do menu entry guides you to a page listing the to-dos you created or to those assigned to you. We'll take a closer look at this option in Chapter 12.
- SURVEY: This guides you to surveys assigned to you for further data collection and to the survey you created, so that you can further edit them and see the status of the collected responses.

❻ **Search (Jump to a Fact Sheet)**
This allows you to search fact sheets of any type by name. It shows you a preview of results while you type, and it lets you jump to a specific fact sheet or to a filtered fact sheet list that matches the typed search string.

❼ **Invite**
This sends an invitation to coworkers that identifies you as a viewer, member, or administrator, depending on your user rights. For example, a member can only invite other members or viewers but can't invite someone else to become an administrator.

❽ **What's new**
This provides information on the latest product updates and bug fixes, and it highlights upcoming webinars.

❾ **Help**
This offers help from multiple sources. Help can take the form of on-screen tours, online documentation, community feedback, and more.

❿ **Settings**
These provide access to the following options:

- Personal settings like language preferences
- Switching to another workspace
- An administration area for comprehensive settings like meta model configuration, automation, notifications, and user administration

- Access to the SAP LeanIX Store, from which third-party extensions can be downloaded
- Switching from an administrator role to a member or viewer role to better test the security settings

One key selling point for SAP LeanIX is its out of-the-box, easy-to-use preconfiguration.

The areas the majority of users will navigate the most in are these ten navigation targets. They're typically displayed in an adjustable navigation area on the left side of the screen to easily direct users to the content that's relevant to them. When you're accessing a specific inventory list, report, or diagram, a configurable filter panel replaces the navigation category area and allows you to filter down lists, reports, and diagrams to relevant subsets.

The panel on the right-hand side is typically used as an action panel that refers to additional data or recently viewed items. Within a diagram or a report, you can use the panel on the right to display the most important details from the selected fact sheet.

Chapter 12 will dive deeper into common elements, such as filters, and explain how to create diagrams and configure reports.

11.4 The Time Dimension: Lifecycles and Transformations

As mentioned in Chapter 4, the representation of time is a critical element of enterprise architecture modeling. Not only do roadmaps give multiyear transformation plans, but each object in a given architecture also evolves over time. Moreover, a good tool will allow traveling back and forth in time. We'll see in Chapter 12 how SAP LeanIX uses the time dimension to define roadmaps and, in Chapter 14, when defining transformations.

11.4.1 Lifecycle Fields

In SAP LeanIX, some of the most important groups of fields are *dates*, which define the different lifecycle phases. Figure 11.9 shows the five predefined date fields that are used to document the lifecycle of a fact sheet: **Plan**, **Phase In**, **Active**, **Phase Out**, and **End of Life**. The bold vertical bar indicates the current date. For some fact sheet types, the semantics are obvious, so we clearly recommend using these fields. It is equally clear that you don't have to maintain all of the fields; some can stay empty.

Most enterprise architectures programs will define time-limited applications and initiatives by defining their start and end dates. In SAP LeanIX, this enables filtering and using lifecycle fields to "travel back in time" and show the application landscape of two years ago. In the other direction, you can show any future state of the landscape as seen from a given point in time.

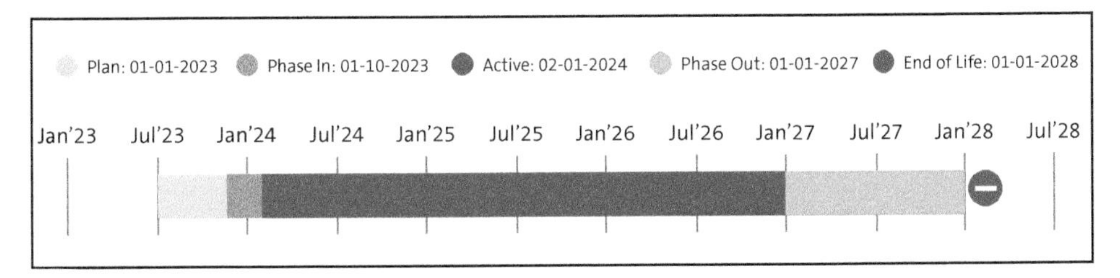

Figure 11.9 Lifecycle States as Gantt Chart

Carefully consider what additional effort you intend to put into documenting the **Plan**, **Phase In**, and **Phase Out** dates. Figure 11.10 depicts both complete and incomplete maintenance states, and the box on the left shows semantically correct definitions. Note that the last state will extend indefinitely if the **End of Life** date is not maintained. This makes the most sense for **Active** objects, for which there's no end of life as yet. It's equally OK not to provide an **Active** or **End of Life** date for objects that are in the initial planning state.

The box on the right side lists lifecycle definitions that don't make sense, such as a **Phase Out** state without a preceding **Active** state or a **Phase In** state without a subsequent **Active** state (to name just two).

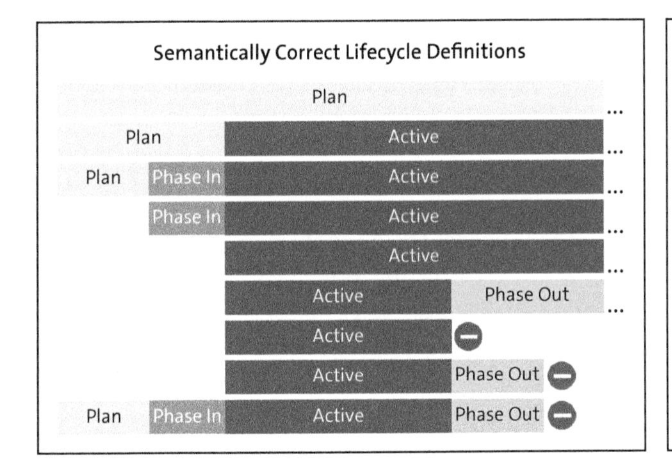

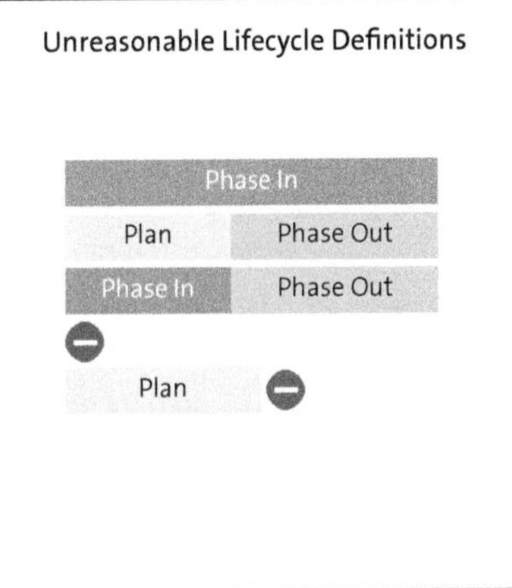

Figure 11.10 Partial Maintenance of Lifecycle Fields

11.4.2 Transformations

When you're planning transformations, and particularly at the outset of an enterprise architecture evaluation, a lot of things will be unclear: not just the timeline but also the transition path or even the target architecture. There will, in other words, be many uncertainties and the need to explore multiple alternative options.

A straightforward approach would be to lay out different architectures consisting of a variety of applications and IT components and to implement different business capabilities in different diagrams. However, this can lead to conflicts because you can plan to implement the same application in both options and possibly in different years. It would be difficult to properly define these alternative implementation timings with a single application fact sheet, and the workaround of defining two fact sheets with different names and different planned lifecycle dates would start polluting the workspace with fact sheets, so that soon, any transparency would be lost.

Instead, SAP LeanIX provides the powerful concept of planning these alternative transformation paths in combination with initiative and roadmap planning. Figure 11.11 gives a structural overview. To get a deeper understanding, we'll run through a detailed tutorial-like example at the end of Chapter 12.

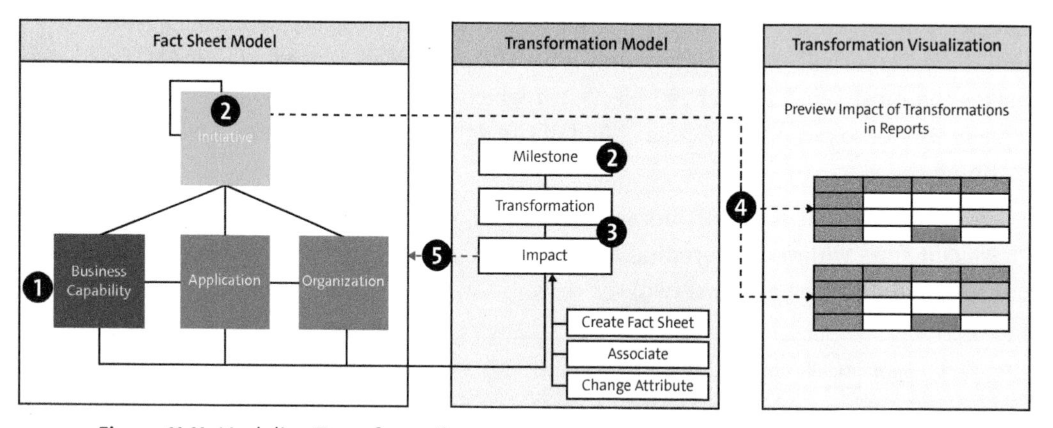

Figure 11.11 Modeling Transformations

Modeling will progress in the following phases:

❶ **Modeling the baseline architecture in the fact sheet model**

The fact sheet model depicted on the left is an aggregation of past and future facts. A future fact is to be understood as a fact sheet for which a lifecycle phase transition is defined for the future. If you are the architect and have a clear knowledge of what will and should happen in the future, then you can enter these facts directly into the fact sheet model.

❷ **Setting up the initiative's structure and milestones**

If the roadmap and target architecture are not yet defined in an early phase of a transformation program, you should *not* model the plans into a fact sheet model yet! Instead, you should start by defining the planned transformation program as an initiative or (often) as a hierarchical initiative tree. You can create different branches for alternative options, and it's best practice to define milestones for all future change events in each branch. Examples of change events are the sunset of an application and different rollout waves implementing a set of applications. The benefit of working with a milestone-based model is that planned dates are modeled at one central location and can therefore be changed easily. Make sure to attribute clear names to all of your milestones.

❸ **Planning the transformations**
Once you've created initiatives with milestones, you need to specify transformations like implementing a new application or interface. If you plan to investigate different transformation options, you need to model these options in distinct branches of the initiative tree.

❹ **Simulating the transformation's impact in reports**
When evaluating the planned transformations (e.g., in landscape, matrix, or roadmap reports), you can specify the initiatives or initiatives' branches that will need to be simulated.

❺ **Executing the transformation**
Later in the architecture planning cycle transformation, firm decisions will be made (e.g., in an enterprise architecture board meeting or a steering committee for the transformation program). Once a decision is taken, the respective transformation's state can be executed. This will finally change the fact sheet model, which means that the changes will be transferred into any reports without you needing to select the transformation program.

Architects can define customer-specific templates, which can include the automated tagging of objects and the setting of default field values. This can accelerate transformation maintenance.

11.5 Workspace Setup

In SAP LeanIX, all relevant objects of an enterprise architecture are modeled together in a single physical workspace (a.k.a., a *tenant*). For large enterprises, these workspaces can become quite large and contain thousands of objects. Figure 11.12 shows the option of introducing virtual workspaces, which have the advantage that they allow fact sheets to be grouped, which means that different user groups see only what is relevant to them.

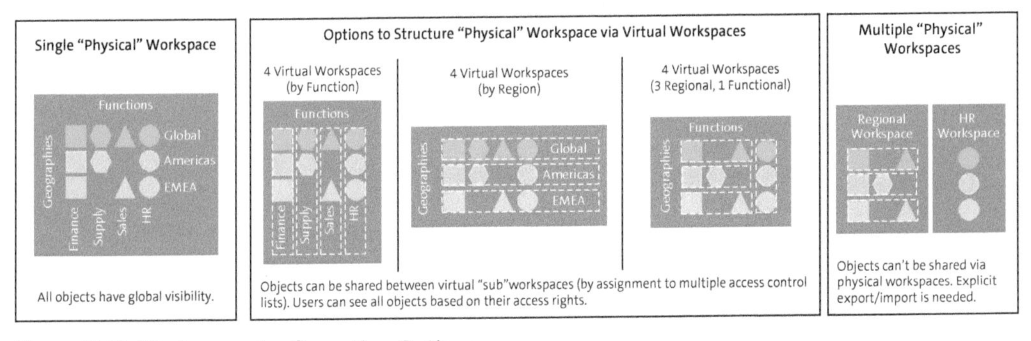

Figure 11.12 Workspace Configuration Options

Figure 11.12 uses two dimensions for clusters:

1. Symbols that represent different functions: finance, supply chain, HR, and sales
2. Colors that represent different regions (in this example)

We think that working with a single physical workspace is best practice if the workspace doesn't grow too much in size.

Defining two (or more) physical workspaces only makes sense if the enterprise architecture can be partitioned into separate areas of concern so that there are different user groups and there's therefore no need for objects to cross from one workspace to another. Such workspaces not only should have individual modeling clusters but also need an adapted metal model. They also come with the downside of the extra effort that is necessary to keep the (possibly) few common objects aligned.

The three options in the middle box all have in common that they define virtual workspaces by either functions or geographies, which are then realized via access control lists (ACLs). Individual fact sheets of each type can be linked to their respective ACLs. This means that, depending on the access control entities (ACEs), a user will only see a subset of the physical workspace.

Don't Create Too Many Physical and Virtual Workspaces

To avoid redundancies, first try to work from a single global workspace. The objective should always be to have a *single source of truth*. When working in multiple physical repositories (workspaces), you need a much more elaborate governance to avoid having conflicting information. Totally separate physical workspaces only make sense if you can ensure that there's no overarching architecture concern you need to address. If there is, it will typically raise the need to physically copy fact sheets from one workspace to another.

Virtual workspaces are flexible and allow for constellations where some user groups have access to only some of the fact sheets while other or all user groups have access to all (in all virtual workspaces). An overly granular distinction of virtual workspaces will, however, come with an additional administrational effort. To a certain degree, this contradicts the spirit of enterprise architecture, which fosters enterprise-wide evaluation and assessments.

11.6 Summary

This chapter introduced the main principles of SAP LeanIX: fact sheets and the relationships among them. The inventory, dashboards, reports, and diagrams are cornerstones of the SAP LeanIX solution and will form a starting point for exploring and visualizing the fact sheet model. We'll deal with these areas in more detail next, in Chapter 12. In

Chapter 13, we'll cover integration with other IT solutions as well as SAP LeanIX extensibility options. Chapter 14 will present a few specific use cases and explore them in a more tutorial-like style.

Chapter 12
Working with SAP LeanIX

This chapter will walk you through all major parts of the application and explain typical work patterns, such as creating fact sheets and other artifacts. It will also explain how to analyze existing workspace content.

This chapter deepens our coverage of major functional areas of SAP LeanIX, not by showing each single feature but by highlighting the most important ones. We'll lay the foundation for using SAP LeanIX, which will come in handy in Chapter 14, when we'll build an example scenario from scratch.

First, let's start our exploration of SAP LeanIX with a closer look at dashboard functionality. Then, we'll turn to other key features like tasks, reports, diagrams, presentations, collaboration, and more.

12.1 Dashboards

You can reach the **Dashboards** section of SAP LeanIX via the top navigation menu we explained in Chapter 11. A *dashboard* is a free, configurable page that groups multiple reports, lists, graphs, and diagrams, typically to address a specific use case or user group. You can then use the dashboard as a launchpad to detail content. Figure 12.1 shows a **demo** dashboard in edit mode. It contains a simple pie chart and a **Diagram Panel** that is being dragged onto it. Once you drag an element and drop it into the grid, you have to create a detailed dialog to specify details such as which diagram to display.

The following content elements are available, and we'll discuss them in more detail throughout this chapter:

- **Custom Message**
- **Diagram Panel**
- **Diagrams**
- **External Content**
- **Fact Sheet Chart**
- **Inventory**
- **Inventory Search**
- **KPI Panel**
- **Latest Fact Sheet Updates**
- **Metrics—Time Series**
- **My Survey**
- **Report Pane**
- **Reports**
- **Saved Search**
- **Subscriptions**
- **To-Do Panel**

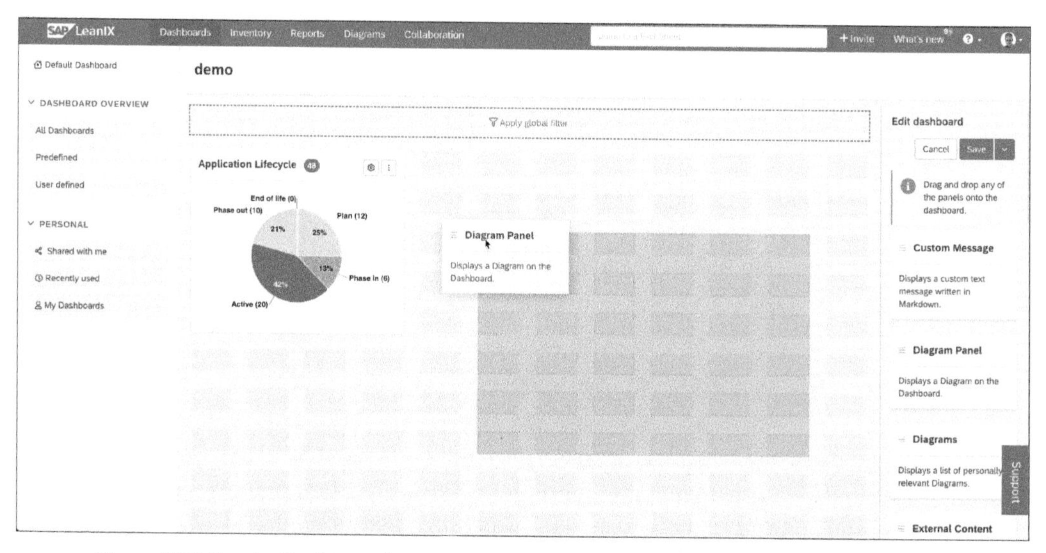

Figure 12.1 Empty Dashboard

The interactivity of a dashboard is limited in the sense that you can't change filters or adapt any report parameters on the dashboard itself. To do so, you need to navigate (drill into) a specific dashboard element. In this sense, the dashboard can be seen as a launchpad for detailed reports. It makes sense to show primarily aggregated information on the dashboard and combine a number of elements that tell a story. For example, you can use a dashboard in stakeholder meetings to give a comprehensive overview without needing to navigate back and forth to the workspace.

If, instead of just giving an overview, your intent is to give a presentation covering several reports and diagrams, then the presentation feature described in Section 12.8 will be a better choice.

12.2 Common Tasks

In this section, we'll describe tasks that apply to many parts of SAP LeanIX. These include filtering objects and performing ad hoc analysis, as well as a common approach to structuring the uses of dashboards, reports, and diagrams.

12.2.1 Filters

Filters don't have a dedicated tab in SAP LeanIX, but they play a major role in restricting inventory searches, reports, and diagrams. A dedicated filter pane is accessible on the left side within the **Inventory**, **Reports**, and **Diagrams** tabs. Under **Reports**, there's an additional filter area at the top summarizing the detailed filters and giving options for easily manipulating them.

Filters can be grouped into the following two categories:

- Filters directly linked to the fact sheet:
  - Tags
  - Fact sheet fields (single or multi select)
- Filters via related fact sheets:
  - Fields or tags of related fact sheets
  - Fields defined by their relationships to other fact sheets
  - Relationships of related fact sheets

An example of using a filter would be showing how an initiative is improving applications used by the organization. The typical approach is narrowing down the list of filtered fact sheets.

12.2.2 Exploration

Later in this chapter, we'll look in more detail at how to define customization beyond filtering in inventories, reports, and diagrams. For all three types, the SAP LeanIX way of working is *explorative*, as shown in Figure 12.2. Right after running a report, opening a diagram, or entering inventory, you can start your exploration by changing the layout, applying filters, switching the color-coding view, and more, until you get to the desired exploration state.

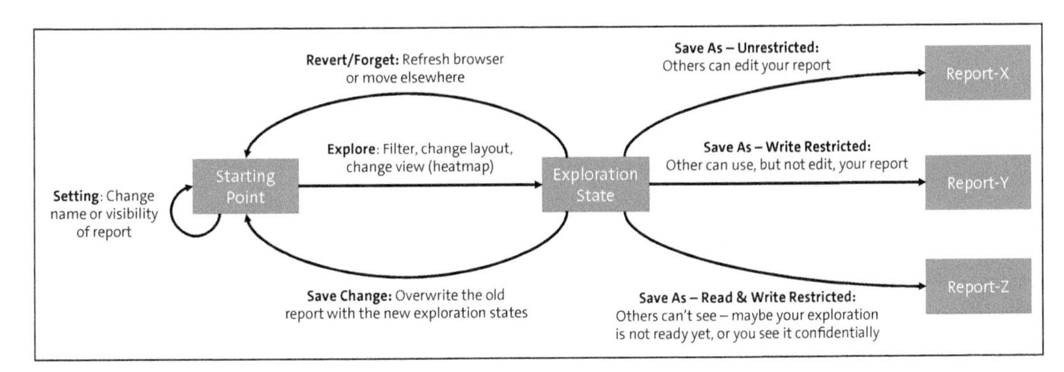

Figure 12.2 Exploration States

All these exploration steps are temporary and only apply to your local session, until you make one of the following possible decisions:

- **Save**
  If the exploration has led to a state that is "better" than the starting point, you can save it, and with that, overwrite the starting point with the new exploration state. Note that if you explore a report that is write protected, **Save as** will be the only option available. Figure 12.3 shows the **Save** and **Save as** buttons in the upper-right corner of the screen.

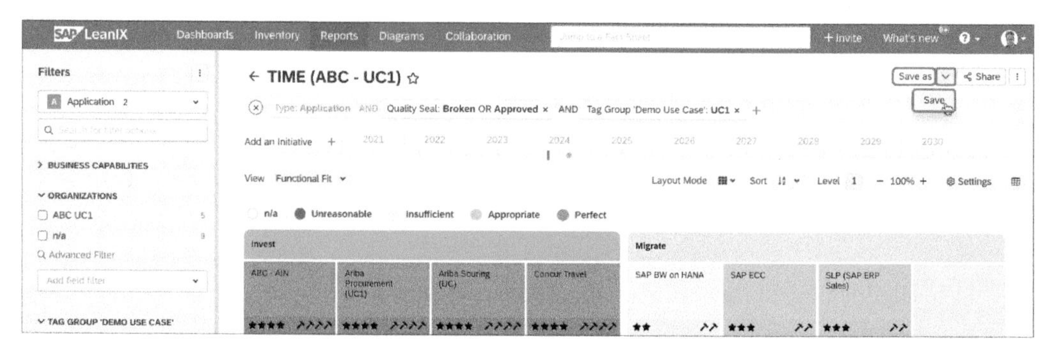

Figure 12.3 Save as and Save Options

- **Save as**
  If you'd like to keep both the starting point and the new exploration state, then you need to make **Save-as** the exploration state with one of the three alternative permission types also shown in Figure 12.4:
  - **Unrestricted**: This lets anybody use the new exploration state under the new name.
  - **Write Restricted**: In a workspace with many users with editing rights, you use this to protect important assets so that other users don't unintentionally overwrite specific reports.
  - **Read & Write Restricted**: This option typically indicates that you're not yet done with your exploration and would not like to confuse others by publishing a semi-finished exploration. Note that SAP LeanIX currently has no autosave function, which means that when you do long explorations, consider saving a temporary copy. You should also use this **Read & Write Restricted** mode if the new exploration state is only relevant to you and doesn't bring much value to others. This keeps the number of visible reports, lists, and diagrams small.

  Be careful not to save too many variations of the same report, as these can confuse other users. An additional variant should always come with a meaningful title and description.

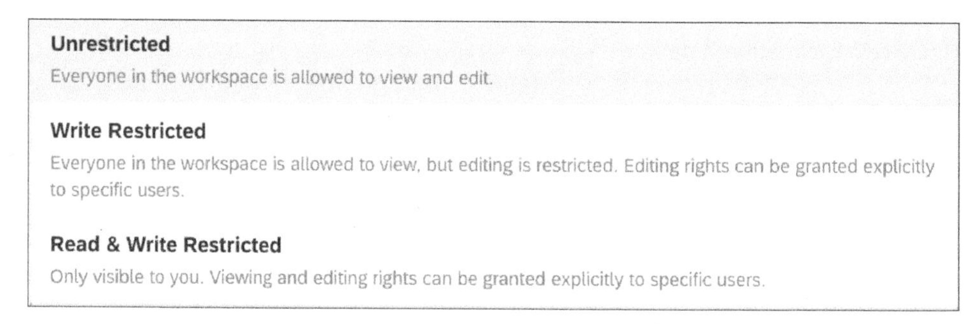

Figure 12.4 Permission Types

- **Revert/forget**
  If the exploration was a temporary one that you don't wish to keep or share, you can move on by either switching to another SAP LeanIX area without saving or by refreshing the browser page. In the latter case, the view will revert to the starting point of your exploration. (Note that this is an implicit option, as there's no UI element called "Revert/forget.")

Note that while you explore reports, lists, or diagrams, you'll often edit the underlying fact sheets themselves. These changes in the fact sheet model will always be saved and visible after each single change (e.g., changing a field, adding a relationship). Only when you're editing an inventory table view do you confirm changes via a specific **Save** button.

12.2.3 Structuring Dashboards, Reports, and Diagrams

As we've already illustrated for dashboards in Figure 12.1, you can search and organize reports and diagrams in very similar ways via *collections*. Figure 12.5 shows the configuration of collections and collection groups within the **ADVANCED SETTINGS** (in the **Administration** area).

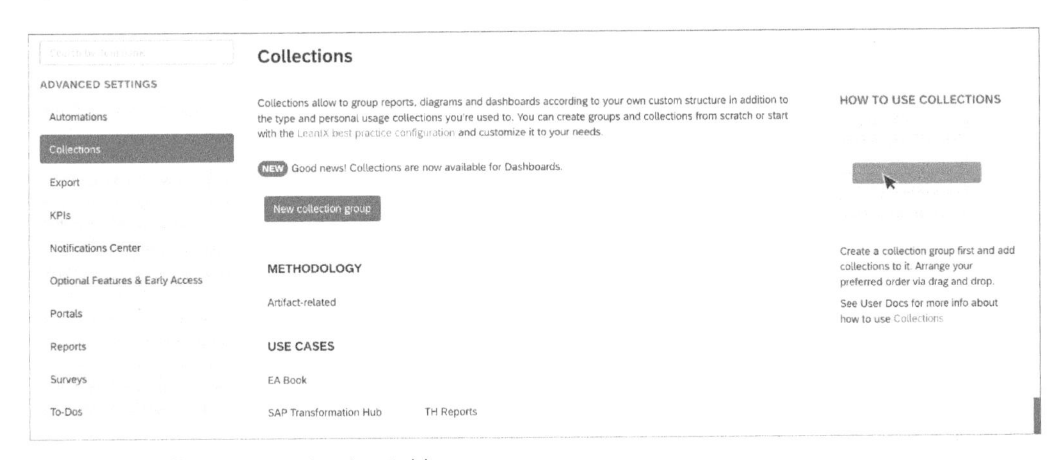

Figure 12.5 Collections: Navigation Folders

Figure 12.6 shows how you can filter down **My diagrams** to two diagrams by typing "com" in the search bar. The context menu button of the **Application Component Diagram** allows you to add this diagram to the **EA Book** collection.

Other entry points besides collections are as follows:

- Grouping by diagram and report type
- Favorites
- Share with me
- My diagrams

The options for reports and dashboards are comparable.

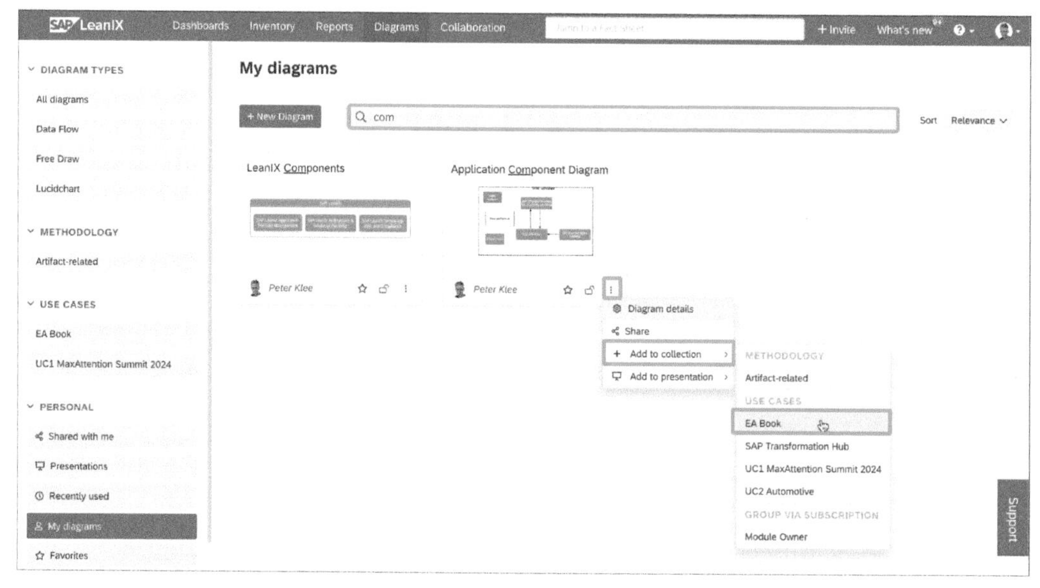

Figure 12.6 Assigning a Diagram to a Collection

12.3 Inventory

The **Inventory** option in SAP LeanIX is the primary area where you can access all fact sheets. Typically, users will first select one fact sheet type and then filter the inventory further as described in Section 12.2.

Inventories can be shown as lists as shown in Figure 12.7. These lists include the following:

- The fact sheet's full display name
- The description, if defined
- All specified tags (Section 12.4)

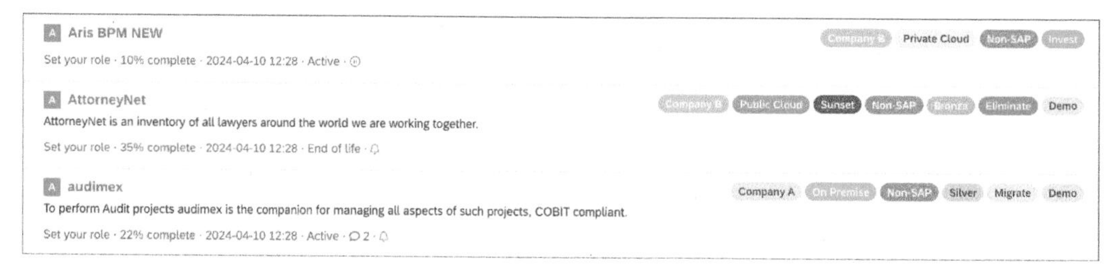

Figure 12.7 Inventory: List View

- A list of addition information:
  - Actions to set the personal role to observer or responsible
  - Information of data completeness in %

- The last time change stamp, with the possibility to branch into the change history within that fact sheet
- The current lifecycle of the object (if defined)
- An indicator of whether the quality of the fact sheet is not approved

As an alternative to the list-oriented inventory, there's a table view, in which you can configure the fields and relationships shown via the right panel as shown in Figure 12.8. In edit mode, you can prepare bulk changes without the need to visit every fact sheet. The example in Figure 12.8 allows editing the three **6R Strategy**, **TIME Classification**, and **Hosting Type** fields, for a set of twenty-one applications. Field changes are marked with a light-yellow background and will be stored simultaneously when you confirm them by clicking **Save**.

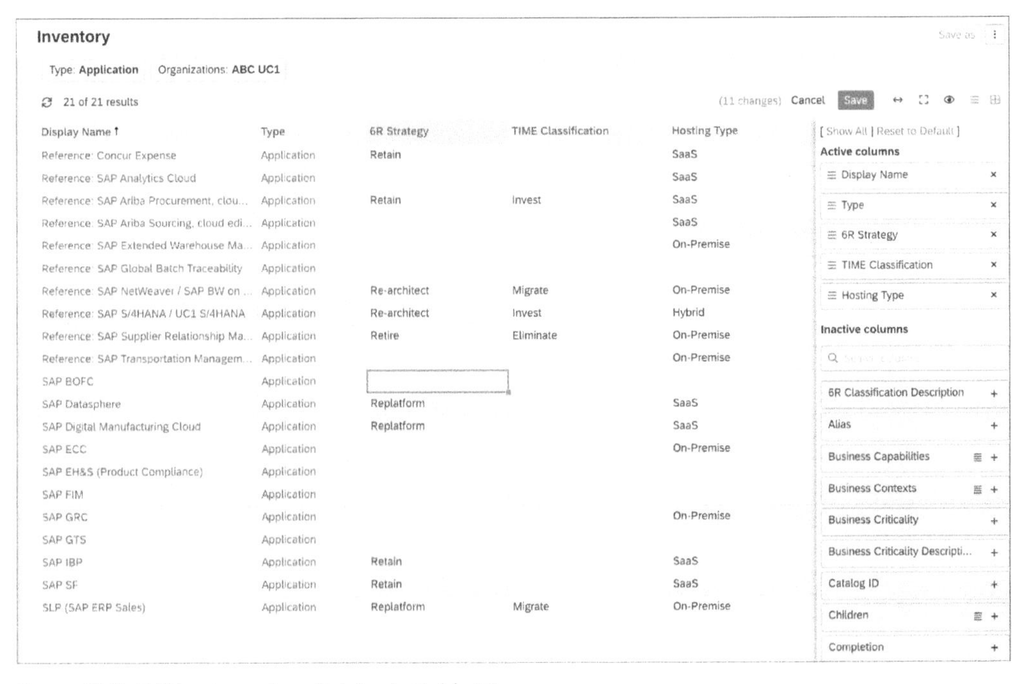

Figure 12.8 Editing Inventory Entries in Table View

12.4 Tagging

Fact sheets are primarily described via their fields. Using tags is an alternative way to collect additional information on fact sheets; you can use them more loosely in the sense that you can define them on the fly. Tags are always text strings that are color coded for easy identification. Tags are always shown at the top of each fact sheet and therefore have high visibility. You should reflect carefully on which type of information to use tags for and where it is better to use fields. Using many tags can easily become too much trouble.

You need to configure tagging in the **Administration • BASIC SETTINGS** area, as shown in Figure 12.9.

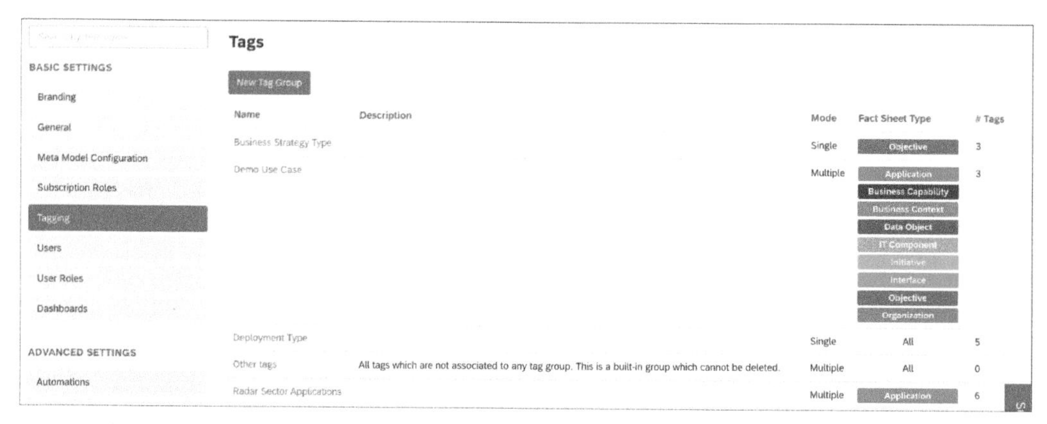

Figure 12.9 Tagging Definition

You can group tags into **Tag Groups**, and each tag group can be defined if an object can only get a **Single** or **Multiple** tag of the group assigned to it. This concept is identical to the use of the **Single Select** and **Multiple Select** fields in the meta model definition, which we'll look at in Chapter 13.

Figure 12.10 shows a custom-defined tag group that is restricted to fact sheet types and allows for fact sheets to be assigned to multiple tags of this tag group. The specific tags and their colors are defined on the **Tags** tab as shown in Figure 12.11. The tag list also provides a counter by tag in the **# Fact Sheets** column.

We'll see in Section 12.5 that you can use tags in reports and diagrams in the same way as fields.

Figure 12.10 Sample Tag Group Definition

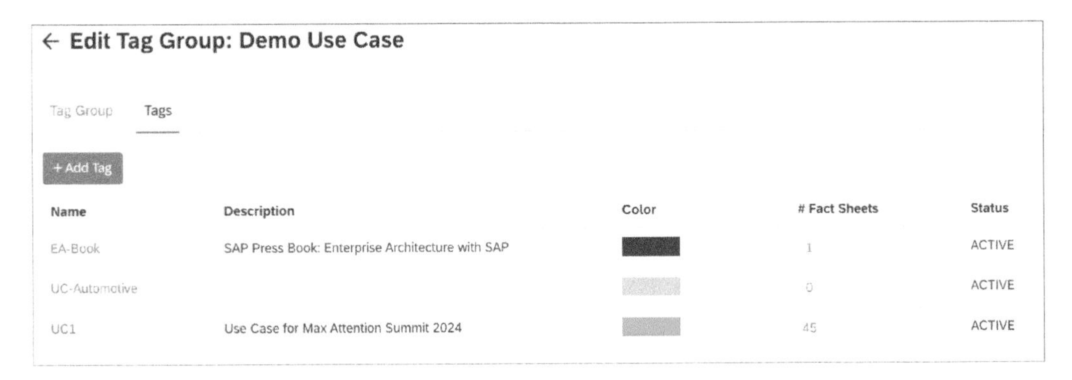

Figure 12.11 Tag List

Table 12.1 gives recommendations on when to use and not to use tags.

Use Tags	Don't Use Tags
▪ Use tags for important information relevant for most users. ▪ As tag groups can be assigned to multiple fact sheet types, you may want to use tags to classify fact sheet of different types with a common categorization schema. However, keeping equivalent fields consistent across fact sheet types requires manual attention.	▪ As tags in tag groups are always sorted alphabetically, don't use tags if the sequence of tags has a specific ascending semantic, so that this sequence should be used in reports. ▪ Don't create tags for information you could express in the default SAP LeanIX meta model. For example, to document which countries or regions an application is used in, we recommend that you model regions and countries with organization fact sheets, which are associated with applications. If you're using tags, it won't be possible for you to model geographies as hierarchies. This is easily doable by modelling a hierarchy of fact sheets, which can be nicely evaluated in reports.

Table 12.1 Dos and Don'ts for Tagging

If the number of tags in a tag group is very high (e.g., greater than twenty), check if there's a simpler alternative. There's no technical limit, but selecting the right tags will be more challenging if there are too many. A very long list of custom field values can also be difficult. Especially when the number grows to fifty or more, consider creating a new fact sheet type and associating the values. Industries or supported business models could be items for you to define as fact sheets rather than tags.

12.5 Reports

A *report* is an automated way to show associations among fact sheets and to use heatmapping (color coding). *Automated* means that a new fact sheet will be added to the report if it matches the filter criteria. In the same way, attributed value changes or changed tags can impact heatmapping and/or the layout of the report.

In Section 12.2.3, we introduced the assignment of reports to specific collections. The straightforward way to do this is to access the reports by their type, as shown in Figure 12.12.

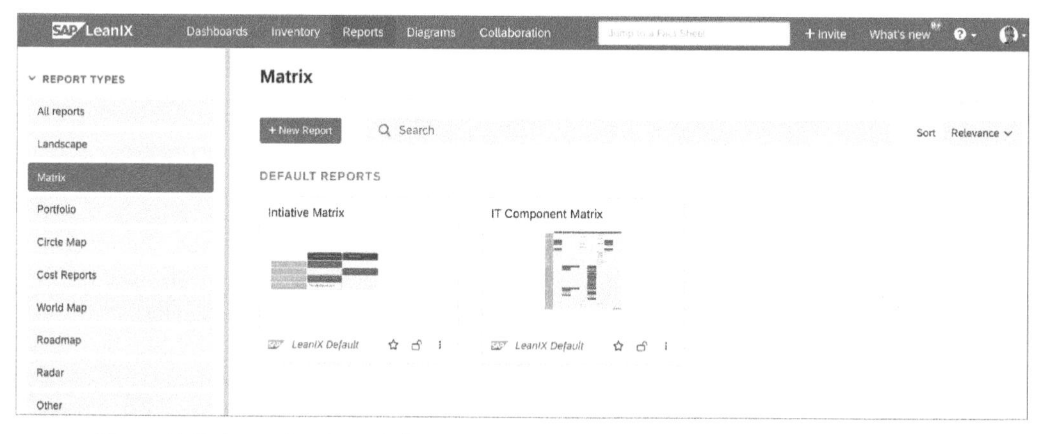

Figure 12.12 Accessing Reports by Report Type

Table 12.2 lists the most common reports and describes how users can configure and use these reports in an interactive way. We'll describe them in detail in the next sections.

Report Name	Number of Fact Sheets	For Which Fact Sheets?	Description
Landscape	1–2 (and as a group by hierarchy)	Any	The second fact sheet is used when grouping via an association. See Section 12.5.1.
Matrix	1–4	Any	Rows and columns can show related fact sheets (and their parent hierarchy). Additionally, a further relationship can be embedded in the cells as a drilldown. See Section 12.5.2.

Table 12.2 Number of Possible Fact Sheet Types per Report

Report Name	Number of Fact Sheets	For Which Fact Sheets?	Description
Roadmap	1–3	Any	This is a Gantt view with the option to cluster (as in tables) and drill down. See Section 12.5.3.
Portfolio	1	Any	This counts two criteria values and shows them as bubbles in a grid. See Section 12.5.4.
Circle map	2	Interface and application	This is an interactive circle with applications all around and interfaces in the form of connecting lines. See Section 12.5.5.

Table 12.2 Number of Possible Fact Sheet Types per Report (Cont.)

In Section 12.2.1, we explained the use of filters. For all reports, the system gives a warning if there are fact sheets for which filter criteria can't be evaluated due to missing field values. In such cases, the system can ask fact sheet owners to complete the data. The warning simply states this: **Attention: xx Fact Sheet(s) not included in this report due to missing data.** It then offers to access and maintain those fact sheets via a hyperlink.

12.5.1 Landscape Reports

The section title suggests that this report type is designed to primarily depict application landscapes. A better name for it would have been *cluster reports*, as this report allows you to choose any fact sheet then cluster it (group it) either by a field or by a relationship to another fact sheet type. If that related fact sheet type has a hierarchy (a parent-child relationship) defined, then by default, the respective columns or rows represent their hierarchy.

Figure 12.13 shows a *landscape report*, which clusters objectives by the defined hierarchy **Objective → Goal → Strategic Priority**. In this demo workspace, these three values have been defined as fact sheet subtypes, using different shades of the same color in a default hierarchical display.

To achieve that, the administrator needs to configure subtypes with different color values and set the flag so this field can be included in views.

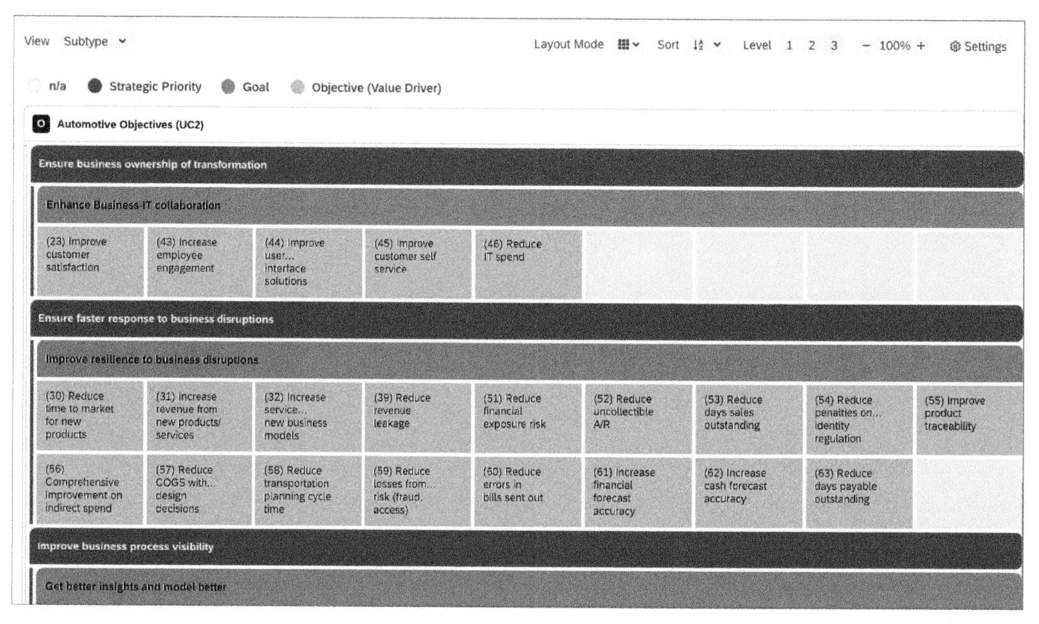

Figure 12.13 Landscape Report: Clustering Objectives

12.5.2 Matrix Reports

In a *matrix report*, objects of one fact sheet type are shown in a grid (matrix). To avoid showing huge grids with too many entries, you'll use filtering to only show a smaller subset. Instead of one dimension (as used in the landscape report), you need to define two clustering criteria: one for the horizontal *x* axis and one for the vertical *y* axis.

As in clustering for landscape reports, the clustering options can be fields, tags, or relationships to other fact sheets.

Figure 12.14 gives a schematic overview of all possible layout patterns:

- **1 type**
  The simplest layout pattern uses one type of fact sheet and uses fields or tags for the rows and columns.
- **2 types**
  If relationships are used for row or column layout (green areas in the depicted patterns), these will enable a hierarchical dynamic drilldown if a parent-child relationship is defined.
- **3 types**
  In this layout, you can either combine relationships of both columns and rows or use the drilldown feature for the cells in combination with either a row or a column relationship.

- **4 types**
  This represents the most complex layout, in which relationships of rows and columns are combined with the drilldown into a fourth type of fact sheet.

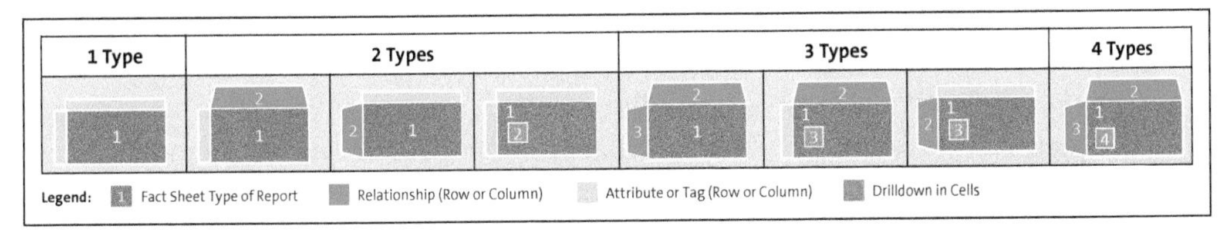

Figure 12.14 Layout Variants of Matrix Reports

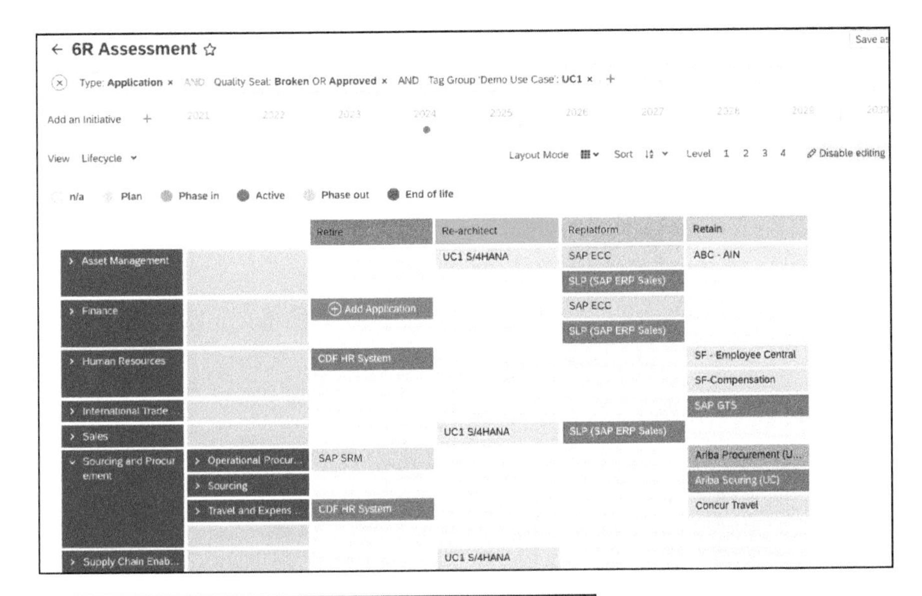

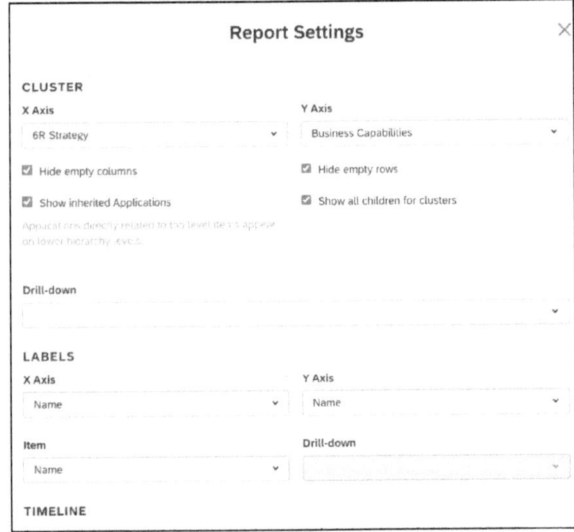

Figure 12.15 Matrix Report and Settings Dialog

The report shown in Figure 12.15 uses the left-of-middle layout variation showing two types of fact sheets, with the following structure:

- **Application**
  The application is represented as a fact sheet filling the matrix.
- **Vertical (y axis)**
  The relationship to business capabilities is used to determine the rows of the report. The business capability hierarchy, which groups business capabilities into business areas and further into business domains (see Chapter 2, Section 2.2.3), allows a hierarchical expansion of that hierarchy via the expand controls (the small white arrows) in the dark blue row headers.
- **Horizontal (x axis)**
  The 6R assessment field is used to determine the columns.
- **View**
  The lifecycle is used to color code the applications within the matrix.

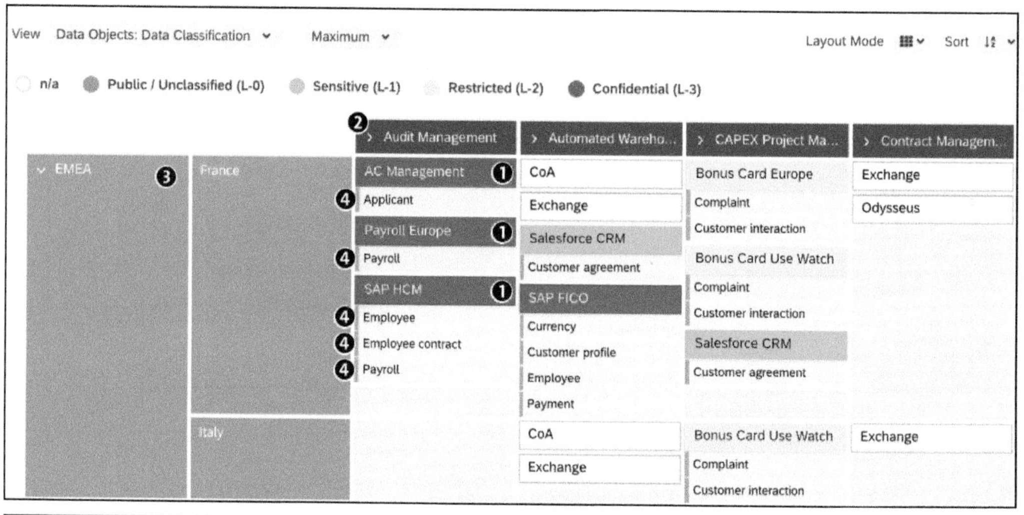

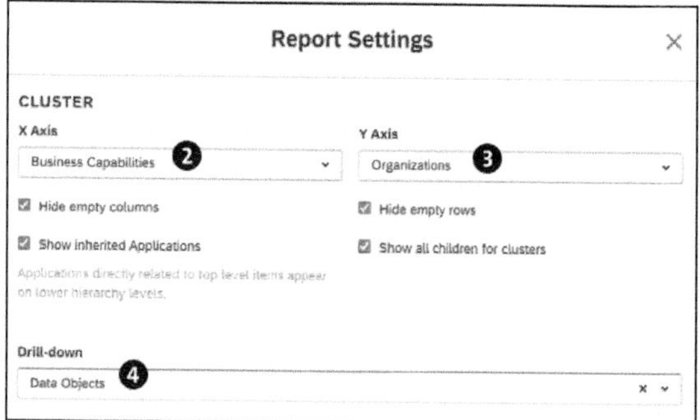

Figure 12.16 Application Matrix Report with Drilldown to Data Objects

In Figure 12.16, we see an example combining four types of fact sheets, showing two relationships (horizontal and vertical) and an additional drilldown. Let's walk through the different parts:

❶ The report is created as a matrix report showing applications.

❷ On the *x* axis (horizontally), applications are grouped by business capabilities (with the option to drill down to more detailed capabilities).

❸ On the *y* axis (vertically), applications are grouped by the regions they are used in.

❹ The drilldown dimension shows the data object modeled by the application.

12.5.3 Roadmaps

The time dimension in SAP LeanIX builds on top of the lifecycle described in Chapter 11, Section 11.4. Roadmaps show a classical Gantt view of the different phases and milestones and can be enriched to show additional fields.

Figure 12.17 shows an initiative roadmap. We typically recommend that initiatives primarily model the two **Active** and **End of life** lifecycle fields as these two dates demark the start and end dates of projects. However, in this example, the future **X RISE Transformation** program is still in the preplanning phase.

The **Phase in** option could be interpreted as the project preparation or exploration phase, while the **Phase out** phase may depict the hypercare phase of a project. It's important that when planning a large portfolio of initiatives, you fully align the semantics of these phases with the project management office (PMO). We'll show a mapping of these phases with the SAP Activate methodology, which is used in RISE with SAP projects.

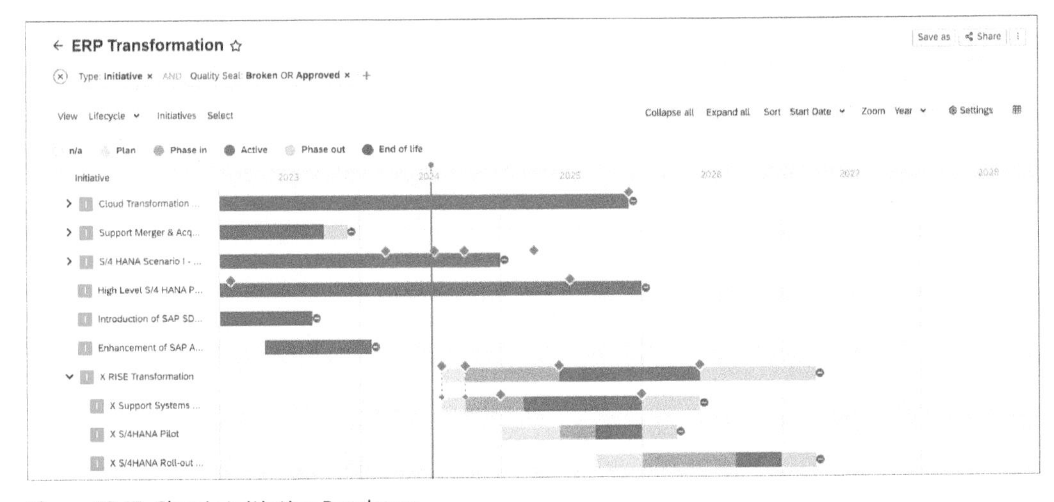

Figure 12.17 Classic Initiative Roadmap

As you've already seen in the matrix report in the previous section, roadmaps can be clustered by any field or relationship defined for the initiatives. Figure 12.18 shows a grouping by objectives. Objectives are structured in a hierarchy (e.g., the **Green Line** top-level objective is broken down to **Reduce carbon emissions** and **Reduce energy usage**, which map to different supporting initiatives).

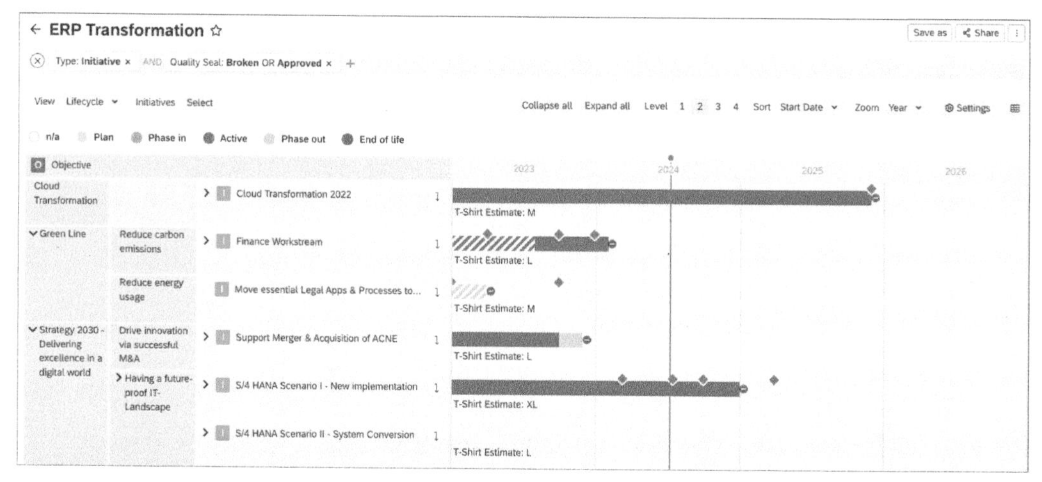

Figure 12.18 Roadmap Grouped by Objectives

In SAP transformation programs and projects, SAP Activate is often the default implementation methodology. The most recent version is SAP Activate methodology for RISE with SAP S/4HANA Cloud Private Edition or SAP Activate for SAP S/4HANA Cloud Public Edition (for two- or three-system landscapes, respectively). Several other versions support other cloud products. When planning initiatives in SAP LeanIX, an easy way to reflect SAP Activate phases is to define project milestones reflecting the SAP Activate phase names (as defined on the RISE with SAP roadmap):

1. **Discover phase**
   The purpose of the discover phase is for the customer to understand the breadth, depth and functionality of RISE with SAP S/4HANA Cloud Private Edition and learn about the benefits it can bring to the customer's business.

2. **Prepare phase**
   The purpose of the prepare phase is to provide the initial planning and preparation for the project. In this phase, the project is started, plans are finalized, the project team is assigned, and work begins to start the project optimally.

3. **Explore phase**
   The purpose of the explore phase is to develop detailed plans in all workstreams of the implementation to be executed in the realize phase. The plans should cover all aspects of the implementation project, including scoping, testing, end user enablement, and operations, to how the solution will be configured and extended.

4. **Realize phase**
   The purpose of the realize phase is to execute on all plans made and signed off on in the explore phase. The solution is configured according to the backlog that was collected and approved previously. Users are prepared for the switch, and end user training is prepared and executed. The productive environment is prepared for the cutover.
5. **Deploy phase**
   The purpose of the deploy phase is to cut over to the production system. After the confirmation of organization readiness, the business and the operations are switched to the new system.
6. **Run phase**
   The purpose of the run phase is to generate value and continuous learning.

In SAP-centric transformations, we recommend that for initiatives, you map the SAP LeanIX lifecycle phases to the SAP Activate phases as depicted in Figure 12.19.

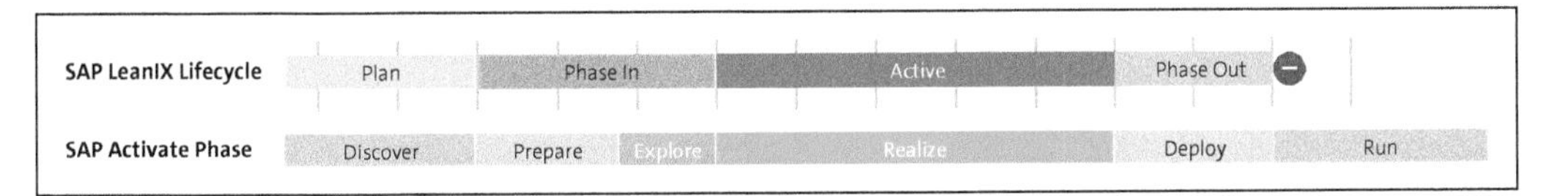

Figure 12.19 Possible Mapping to SAP Activate Phases

In this proposed mapping, we combine the prepare and explore phases from SAP Activate with the phase-in phase on the SAP LeanIX side. As we said earlier, architecture management needs to align with the PMO.

Note

Don't confuse these initiative phases with the lifecycle of the to-be-implemented applications: while the *initiative* is phasing out, the *application* is getting deployed!

In Figure 12.20, you can see the lifecycle translations for the initiative fact sheet type exposing SAP Activate phases in parentheses. Colors can't be changed, and we don't recommend making configuration changes in any other type of fact sheet (like application or IT component), because SAP Activate phases are only applicable to initiatives and don't deal with the IT component or application lifecycle management. Our recommendation is therefore to just add the SAP Activate phase in the respective translations for initiatives.

With the proper SAP Activate-defined milestones and the meta model adjustment done, you can define SAP Activate-conforming initiatives, as depicted in Figure 12.21.

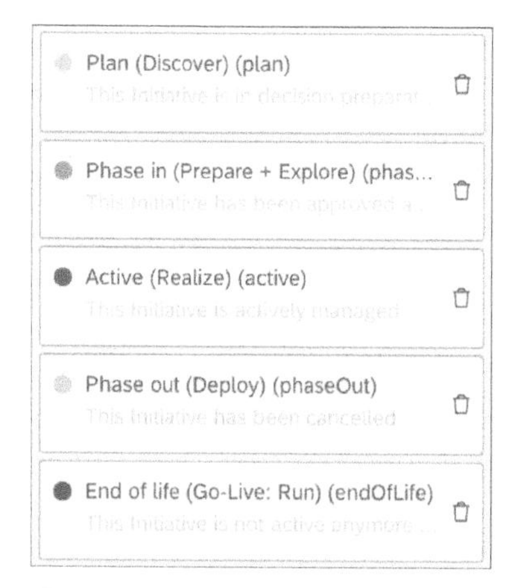

Figure 12.20 Adopting SAP Activate Terminology

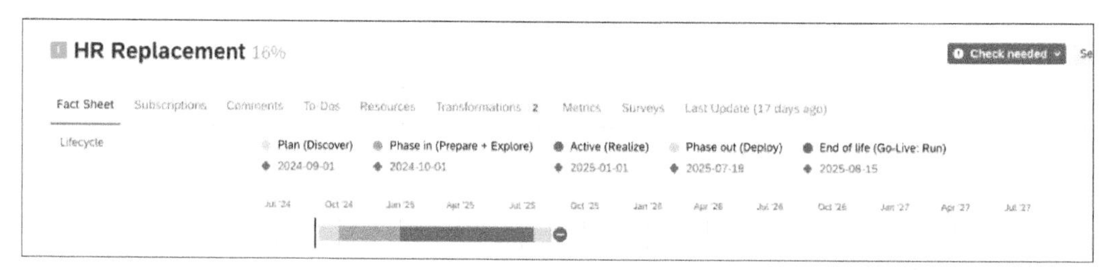

Figure 12.21 Use of SAP Activate Milestones and Terminology in Initiatives

As lifecycle fields are maintained not only for initiatives but also for applications, IT components, and (possibly) business capabilities, it's possible to show roadmaps for these fact sheet types too. Our recommendation is to only focus on a few visualizations and test them with key stakeholders.

For application owners, an application-centric roadmap, as shown in Figure 12.22, makes sense. These views are part of application portfolio management, which we'll refine in Chapter 14.

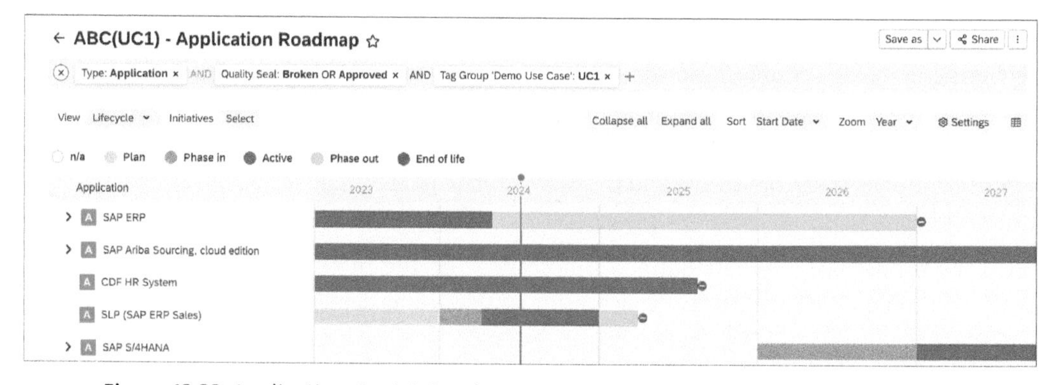

Figure 12.22 Application-Centric Roadmap

12.5.4 Portfolio Reports

A portfolio report summarizes fact sheets of a given type and classifies them in two dimensions (the *x* and *y* axes). The size of the bubbles could either reflect a pure count of fact sheets or (as shown in Figure 12.23) be an aggregation of numeric values defined in the fact sheets. This is in contrast to matrix reports, which are primarily used to show the fact sheets instead of just counts.

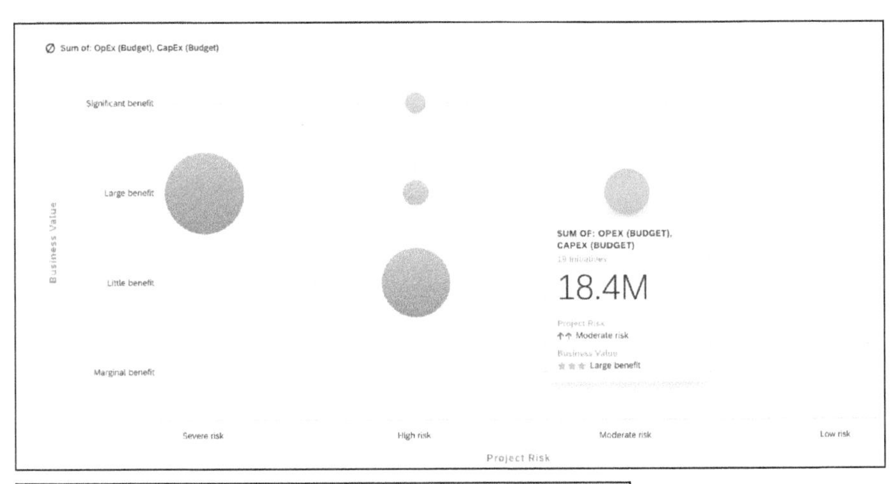

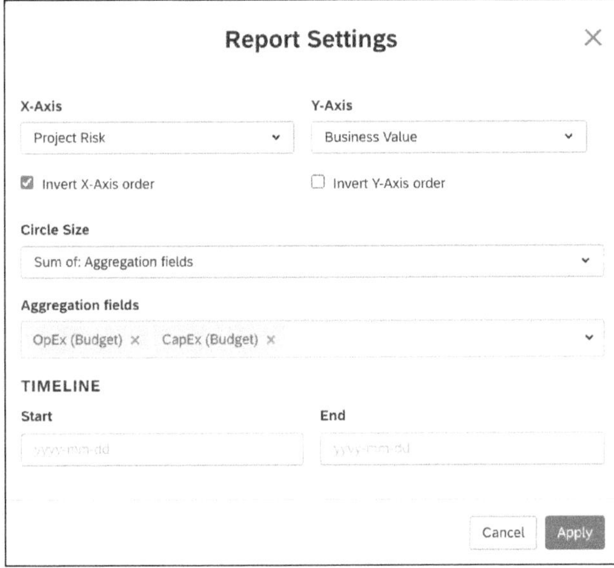

Figure 12.23 Initiative Portfolio Assessment (Risk versus Value) with Evaluation of Financial Investment

The example shown in Figure 12.23 evaluates initiatives by the **Business Value** and **Project Risk** fields. The third dimension (the **Circle Size**) is the aggregation of CapEx and OpEx budgets defined in the fact sheet.

12.5.5 Interface Circle Maps

An easy, interactive way to explore integrations is by using the interface circle map, which is one of a few reports that is designed for the specific fact sheet type interface and (in its current form) is not used to analyze other associations. Around a circle, all applications providing or using interfaces are depicted. Each line within the circle represents a provider-consumer relationship between two applications. In Figure 12.24, you can see the following three exploration states of one identical circle map:

- **Left: Unfiltered**
  Without selecting an application, the report is too busy to derive insights.
- **Center: Selecting a specific application**
  After selecting the **Case management** application (indicated by the bold red frame), the specific integrations for this application become well explorable.
- **Right: Filtered**
  Via the filter mechanism explained in Section 12.2.1, we've filtered for high-complexity data integrations and can immediately identify event management and configuration management as the most complex integrated applications.

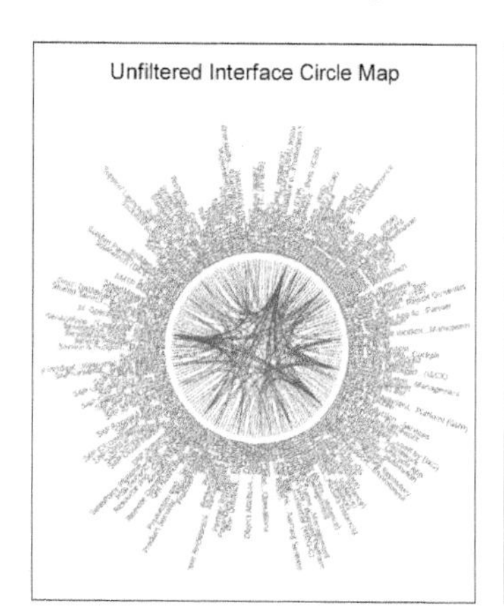

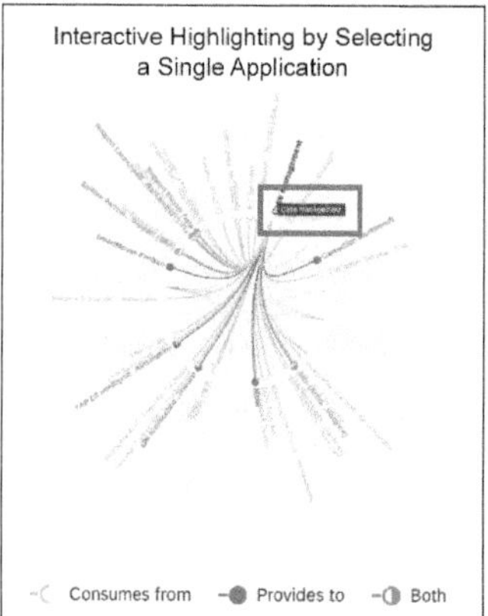

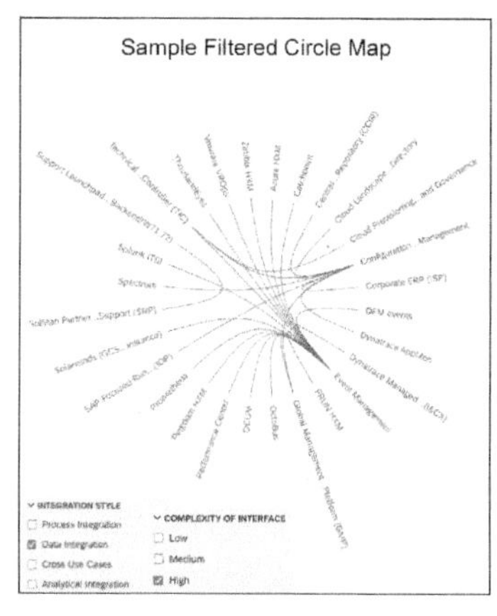

Figure 12.24 Interface Circle Map: Three Exploration States

We'll see in Section 12.6 how integrations with interfaces can be depicted in data flow diagrams.

Other Reports

There are more standard and third-party reports. We recommend you browse the online documentation as well as the SAP LeanIX store at *https://store.leanix.net/en/browse*.

In addition, we recommend reading Chapter 16, Section 16.4, which includes examples from SAP internal enterprise architecture practice.

12.6 Diagrams

While reports are 100% data-driven visualizations of the repository without the possibility to influence positions, shapes, and colors, *diagrams*, in contrast, can be enriched with custom shapes, and positions can be defined freely.

Internally, SAP LeanIX embeds the engine of *draw.io*, which enables you to draw a large variety of best-in-class diagrams. One very useful advanced feature of draw.io is the use of layers, which can be shown and hidden independently.

Avoid Drawing Shapes without a Link to Fact Sheets

As we discussed in Chapter 4 in the section on the benefits of a model-based approach, drawing shapes without linking them to any fact sheet creates individual visuals, which can't be explored via any type of reports and for which consistency across multiple diagrams can't be guaranteed either.

In the following sections, we'll explain the basics of creating diagrams and examine the two different types of diagrams: data flow and free draw. Which one to choose is the first decision you need to make and is depicted in Figure 12.25.

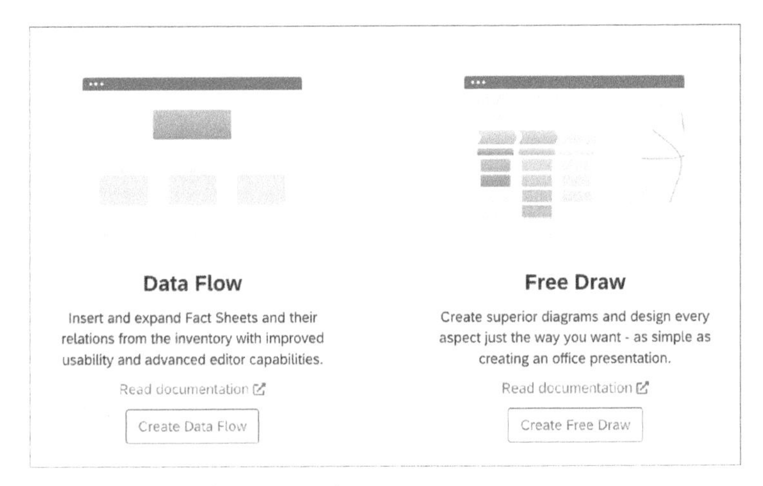

Figure 12.25 Creating a New Diagram

There's a third option to integrate Lucidchart diagrams as well, but this is a less fluent integration and hence not recommended.

12.6.1 Creating and Editing Diagrams

In all diagrams (free draw and data flow diagrams), you can draw using a rich palette of graphical objects to be dragged onto the drawing canvas. Figure 12.26 shows a number of such shapes and arrows. Connecting shapes with lines, embedding shapes in other shapes, and using colors express an implicit semantic which is not supported by the fact sheets layer of the model.

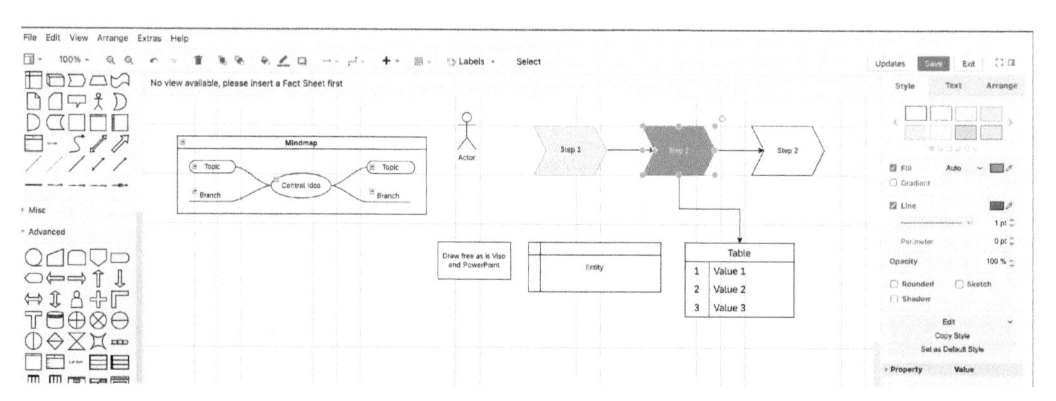

Figure 12.26 Graphical Elements from Palette

While depicting dedicated additional content that goes beyond the fact sheet meta model might be acceptable, you must avoid drawing objects existing in the inventory (like an application) as purely graphical objects *without* the link to the inventory. You'd miss the chance of using the heatmapping function, and you'd also run the risk of creating inconsistencies, as you can't propagate the renaming of such applications without having the explicit link to the shape.

Diagrams appear to be static visualizations in which, unlike in matrix and cluster (landscape) reports, associations can be visualized via inclusion *and* connecting lines. The power comes when using diagrams as a dynamic exploration space. When a diagram is created and saved, it can form a starting point for further explorations.

When you open a diagram, it opens in display mode in the exact state you saved it in during the last edit. In display mode, you can still do explorations by applying filters and changing colors (heatmapping).

The display mode is static in the sense that positions can't be changed. By navigating to edit mode, you can work with the diagram dynamically in the following ways:

- **Insert a fact sheet from inventory**
  Via the black bold + sign in the top toolbar, you can insert a fact sheet from the inventory. This is the preferred way to add any kind of fact sheet to a diagram. The alternative way is to first create a nonassociated shape and link it to a fact sheet in a second step.
- **Add related fact sheets**
  You can add fact sheets in the following ways, by selecting these functions from the right-click context menu:
  - **Drill down**: Add related fact sheets within the body of the shape.

- **Drill down by filter**: Not all related fact sheets will be embedded, but you can select dedicated ones or apply filters to downselect the ones you wish to add. This is shown in the upper part of Figure 12.27 ❶, drilling down from SAP ECC to the related IT components.
- **Roll up**: Add one or multiple "parent" shapes. The object from which this operation is initiated will be the child of the roll up shape(s).
- **Roll up by filter**: As in the drilldown by filter case, there's an intermediate state for restricting the number of objects the roll-up is executed on.
- **Show dependencies**: This adds the related fact sheets with dotted lines. It's shown in the lower part of Figure 12.27 ❷, in which the dependent IT components are added with dotted lines.

All three actions are available in the context menu of fact sheet shapes in two flavors: one "by filter" and one nonfiltered. The former is relevant if the number of associated objects is very high and only a subset of related objects that are relevant in the current context should be shown. The latter will add all fact sheets.

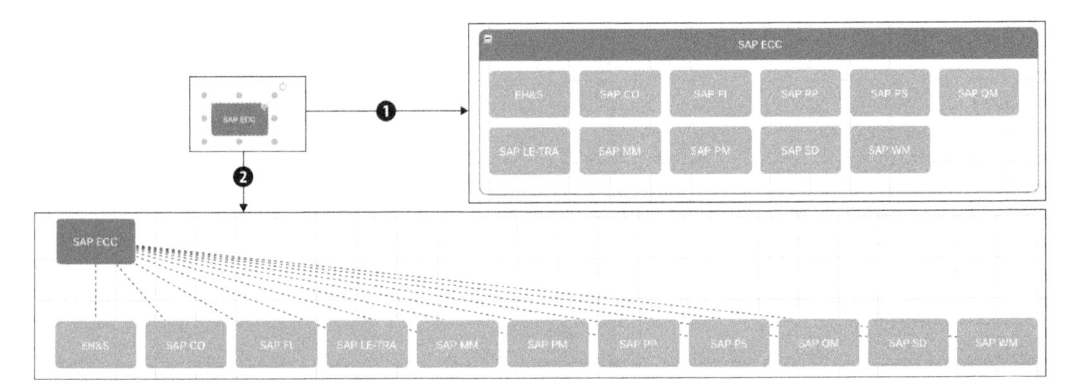

Figure 12.27 Adding Related Objects as Embedded Objects or as Linked Objects

- **Create relationships**
  This follows steps 1 through 4 in Figure 12.28:

  ❶ Draw a line between two fact sheets and create a relationship between them via the **Create new relation** context menu in the line. This draws an arrow between the two application fact sheets.

  ❷ The context menu recognizes the connected fact sheet types and offers to create one of the six relationships between the applications. In this case, a **Predecessors to Successors** relationship is the desired association.

  ❸ A pending update is indicated via a circled box with a pending updates counter. You can select this to trigger the update.

  ❹ In the final step, all pending updates from diagram to inventory are shown, and you need to confirm them before the changes will be applied to the inventory.

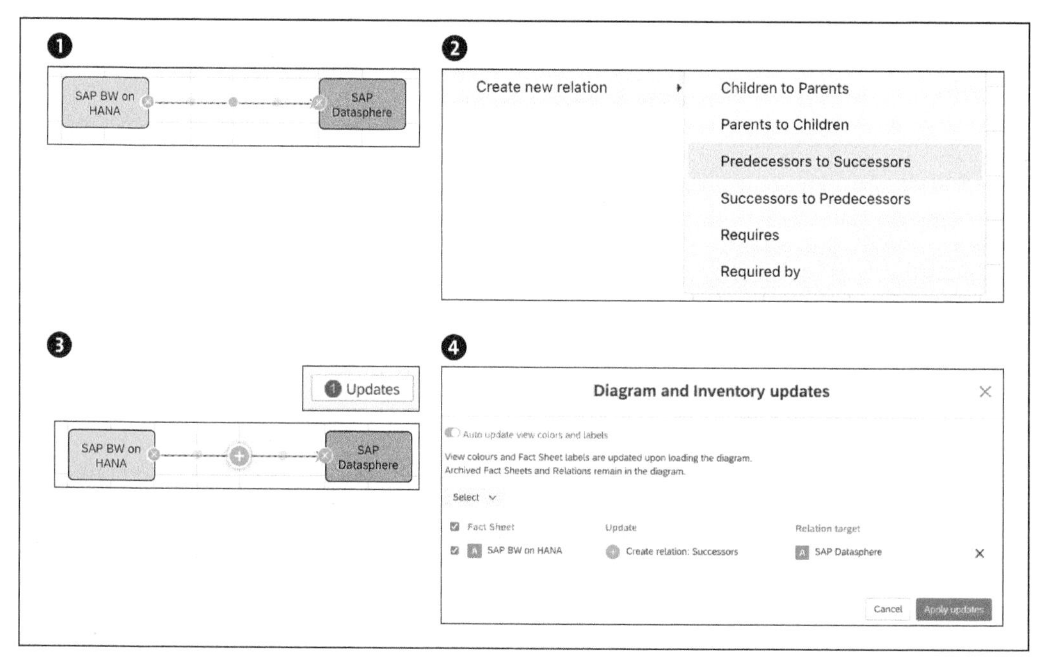

Figure 12.28 Creating Relationship by Connecting Line in Diagram

12.6.2 Data Flow Diagrams

The data flow diagram is in between a report, for which the user typically has little influence on the layout, and a free draw diagram. In data flow diagrams, you can explore the integrations between applications defined in interface fact sheets in a dynamic and interactive way.

Exploration in data flow diagrams will often show the data objects connected to applications and or interfaces. The power of exploration and the autolayout algorithms are intended to help you get a good understanding of the data flows through an entire landscape. Depending on the depth and accuracy of the documented interfaces, there's a risk that diagrams may get too complex and therefore become useless. We therefore recommend modeling integrations more selectively by including only the most important applications and integrations or focusing on specific parts. For these diagrams, it can make sense to switch from the autolayout to manual layout, as this enables creating optimized diagrams that document a specific use case.

Figure 12.29 shows a data flow diagram in edit mode. The circled arrow icon indicates that there's an additional interface from the SAP LeanIX application not depicted yet on the diagram. By clicking on it, you can add the interface and a possibly connected application.

When you load a diagram, it will be checked automatically if the depicted inventory items are still valid.

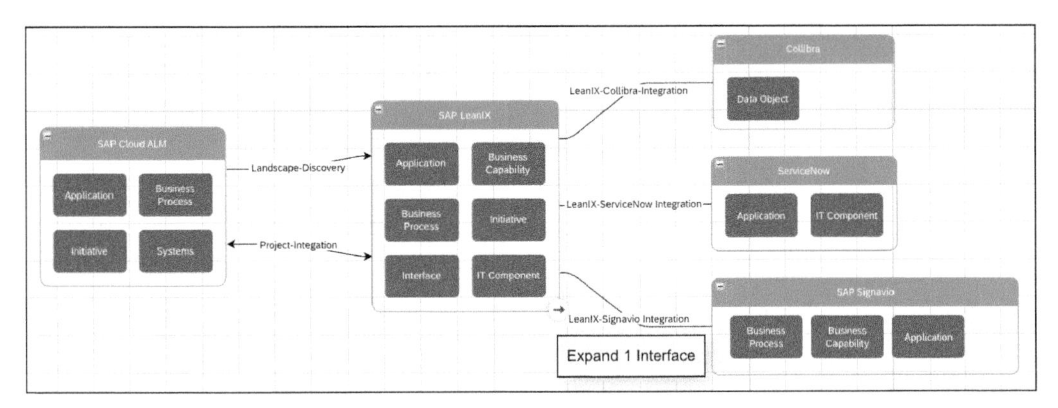

Figure 12.29 Data Flow Diagram with Drilldown to Data Objects

Note that, different from reports, in diagrams (as of the time of writing), there's no time-slider feature to shift the time perspective from baseline to target architecture. We therefore recommend modeling explicit baseline, optional interim, and target data flow diagrams.

Avoid Overly Complex Data Flow Diagrams

When you have a well-documented interface landscape, avoid depicting the maximum number or integrations or even all integrations in one diagram, as this typically leads to an unmanageable viewpoint. For an example, see Figure 12.30, which contains two such antipatterns. You won't be able to read and deduct anything, and even zooming in won't help a lot.

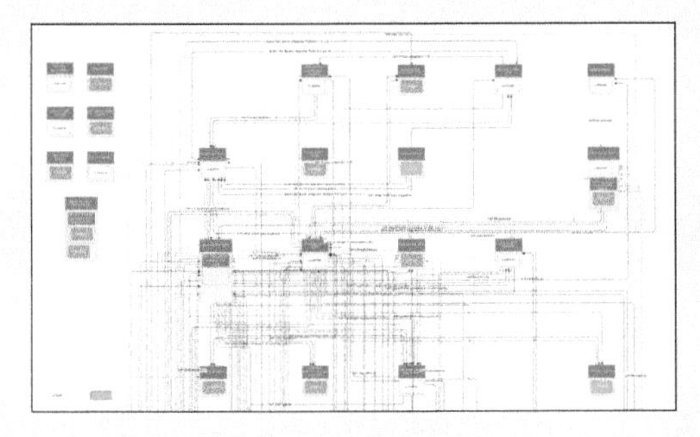

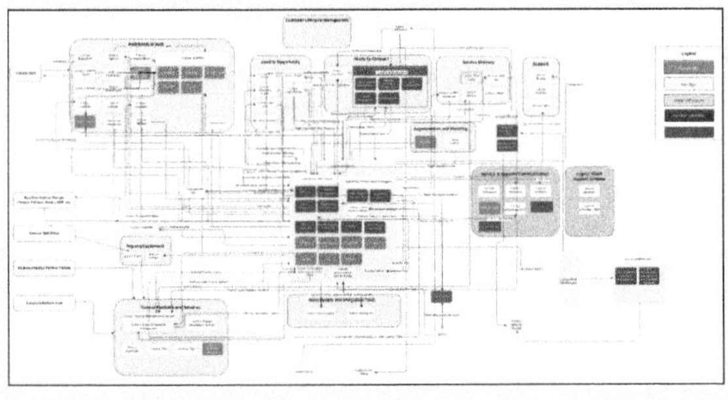

Figure 12.30 Overly Complex Data Flow Diagrams with Limited to No Value

12.6.3 Free Draw Diagrams

If the main purpose of a diagram is not showing data flows through the landscape, first consider if your documentation need can't be fulfilled by a landscape (cluster) or matrix report.

When building up a free draw diagram, don't think of painting some semantics, but rather use the **Show dependencies** and **Drill down** features explained previously. Each visual composition (e.g., a line or arrow from one object to another or any overlay of an object) should be backed by relationships of different types.

Figure 12.31 shows that you can activate a large set of SAP-specific shapes just by clicking the **+ More Shapes** button at the bottom of the left tool pane and scrolling to the **SAP** entry as shown in the top part of Figure 12.31. These icons include, among others, all the official SAP BTP icons.

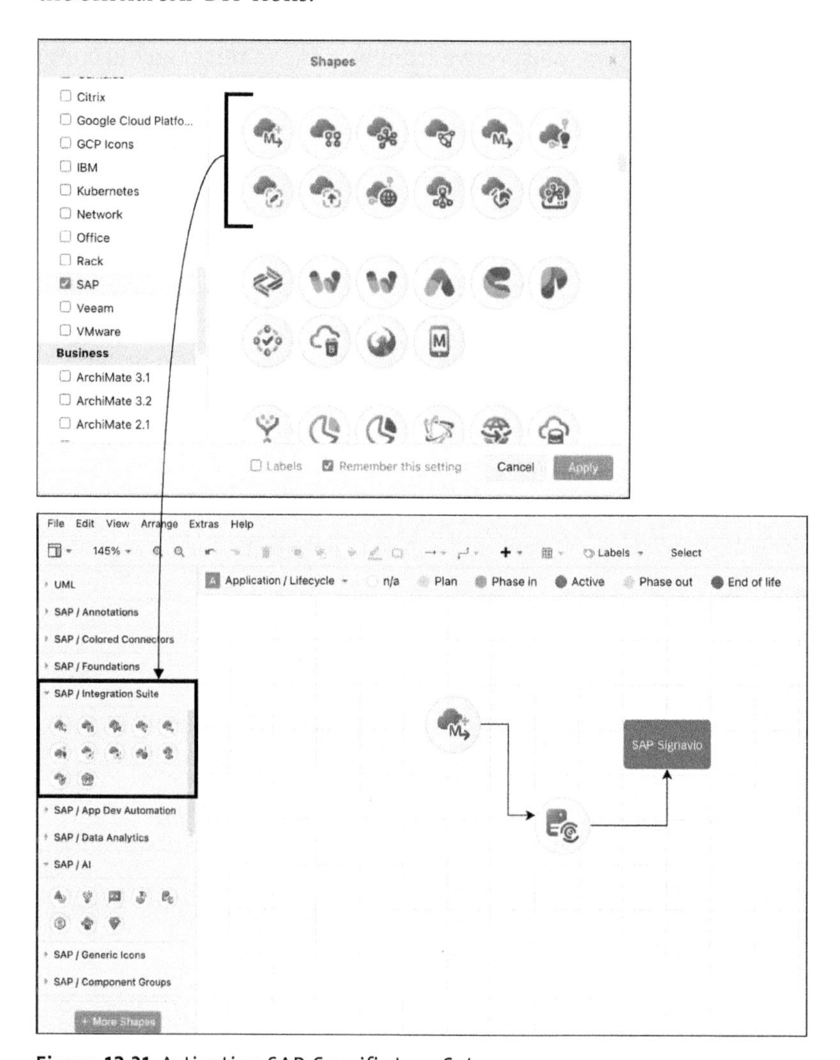

Figure 12.31 Activating SAP-Specific Icon Sets

12.7 Artificial Intelligence

We touched on the topic of artificial intelligence (AI) already in the Preface as its relevancy not only to SAP LeanIX but also in enterprise architecture will continue to grow. In this section, we'll focus on the use of AI within SAP LeanIX.

Before you can use AI features, they need to be activated by an administrator and AI terms need to be agreed to as shown in Figure 12.32. After you trigger this request, this needs to be done in the **Optional Features and Early Access** section under **Administration • ADVANCED SETTINGS**.

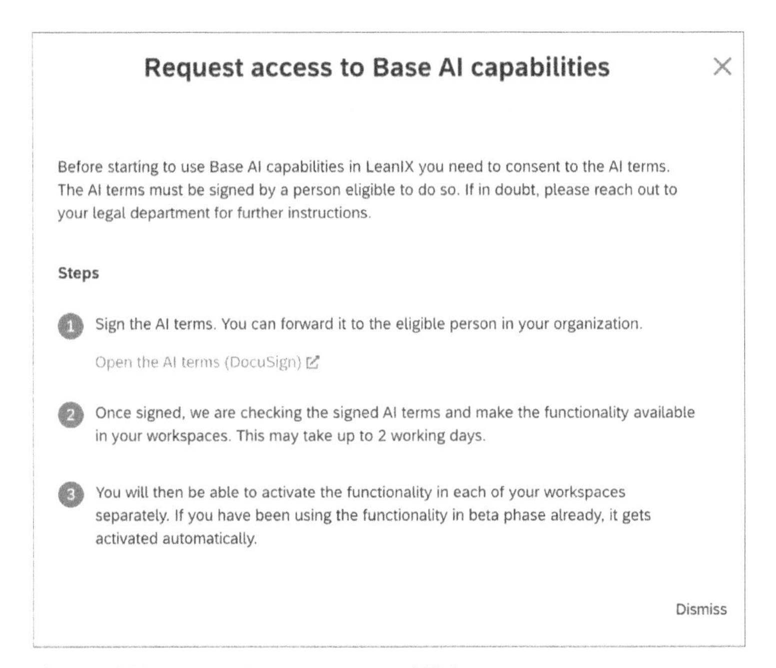

Figure 12.32 Requesting Base AI Capabilities

The AI features of SAP LeanIX are currently twofold:

- **Prompts**
  You can choose **Start AI Prompt** from the top of the inventory view for several fact sheets. On the left side, Figure 12.33 shows ten predefined prompts for different fact sheets. On the right side, the custom prompt "List value drivers, details as bullets" was entered and has led to responses in the **AI Response** column. In the current version, the blue checkmark would overwrite a possibly existing description. This is why the **Fill descriptions** default prompt will by default filter only applications that don't have a description.
- **Synthesis**
  The second use of AI allows you to synthesize information when editing a single fact sheet. As with interactive AIs, you can ask the AI to rephrase, summarize, or format

the response as paragraphs or bullet points. You can activate this function by pressing a light purple icon that appears when you're editing a long text field.

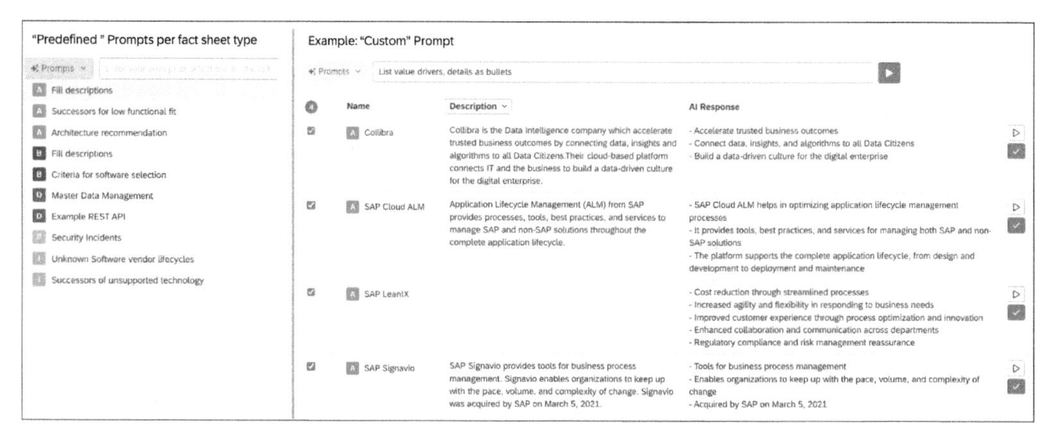

Figure 12.33 AI Prompting from Inventory View

The potential to use AI is broad, and within SAP, we're exploring multiple specific options in the area of enterprise architecture. Read through Chapter 17, in which we'll pick up the discussion of AI in the context of future trends in enterprise architecture.

12.8 Presentations

Presentations constitute a recent functional extension that allows the composition of a flow of diagrams and reports. They can be created from either the **Reports** or the **Diagrams** tabs in SAP LeanIX. As shown in Figure 12.34, you can access existing presentations or create new ones from the **PERSONAL** section in the left navigation panel.

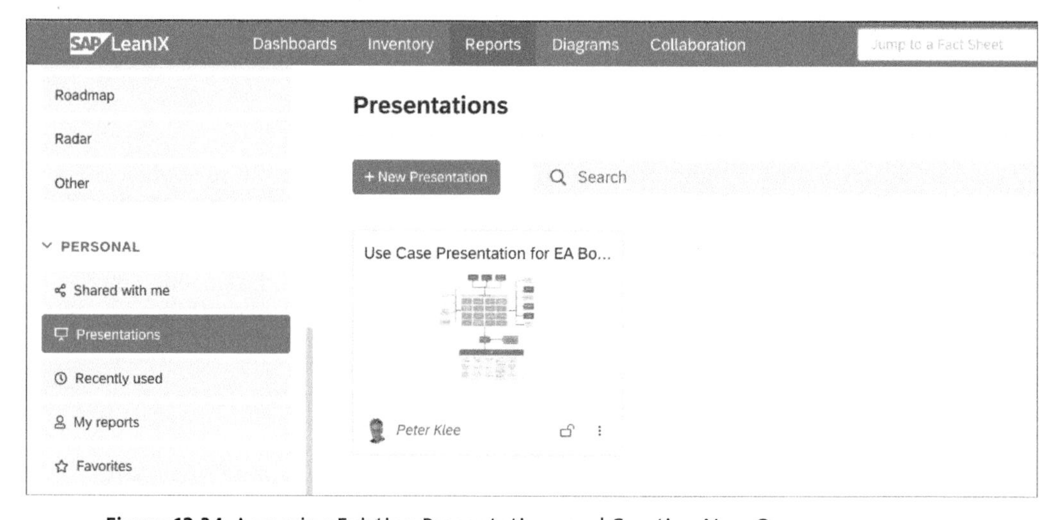

Figure 12.34 Accessing Existing Presentations and Creating New Ones

Unlike in dashboards, in presentations, only one diagram or report is visible at a time. The user is presenting a storyline, and a dashboard is a launchpad in which a user dynamically decides which part of the storyline is most relevant and can branch to the underlying report. While the user is in a dashboard, the size available for a dashboard element is fixed and typically smaller; however, each element of a presentation can use the full real estate and functionality of the screen, including scrolling.

Figure 12.35 shows how to add selected reports or diagrams to the presentation by clicking the blue **Add** button. You can easily change the order of these presentation elements via drag and drop in the left overview panel. When presenting in this editing mode, like in Microsoft PowerPoint, you can switch among the individual presentation elements easily. In **Preview** mode, the left navigation area is hidden so that reports and diagrams can use the full screen.

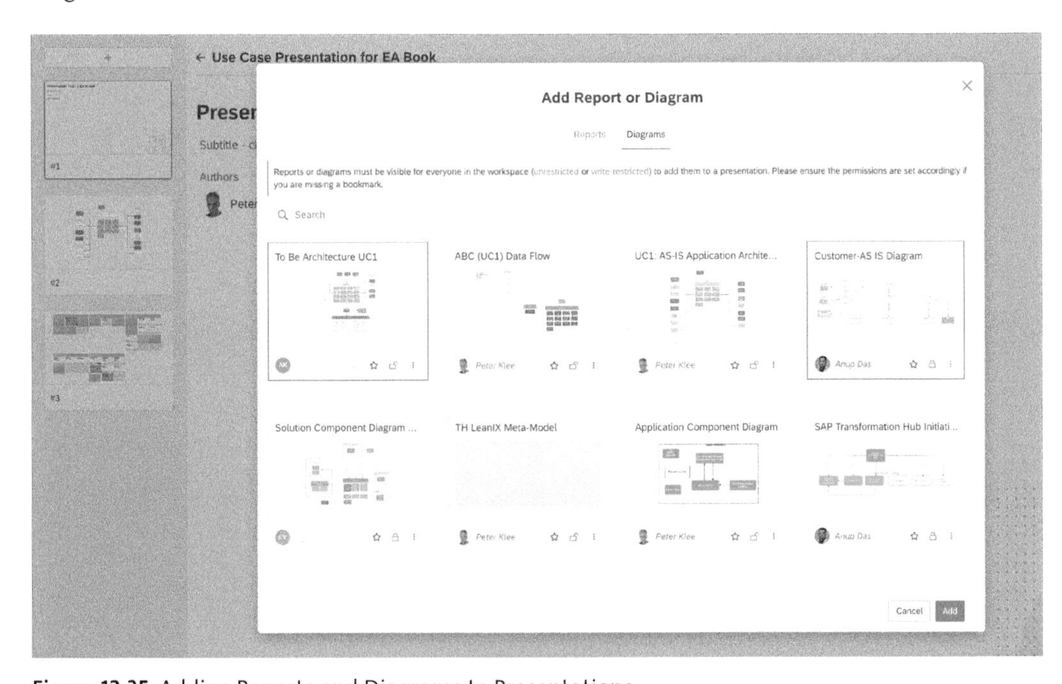

Figure 12.35 Adding Reports and Diagrams to Presentations

You can add notes to each element (slide), and the notes will be shown on the right side while you're running the presentation. As these notes are shown in a large text font, you can easily use them as major text-based explanations.

Figure 12.36 shows an active presentation with navigation controls at the bottom. You can select displayed fact sheets and allow further navigation. In the current version, the defined heatmapping can't be changed.

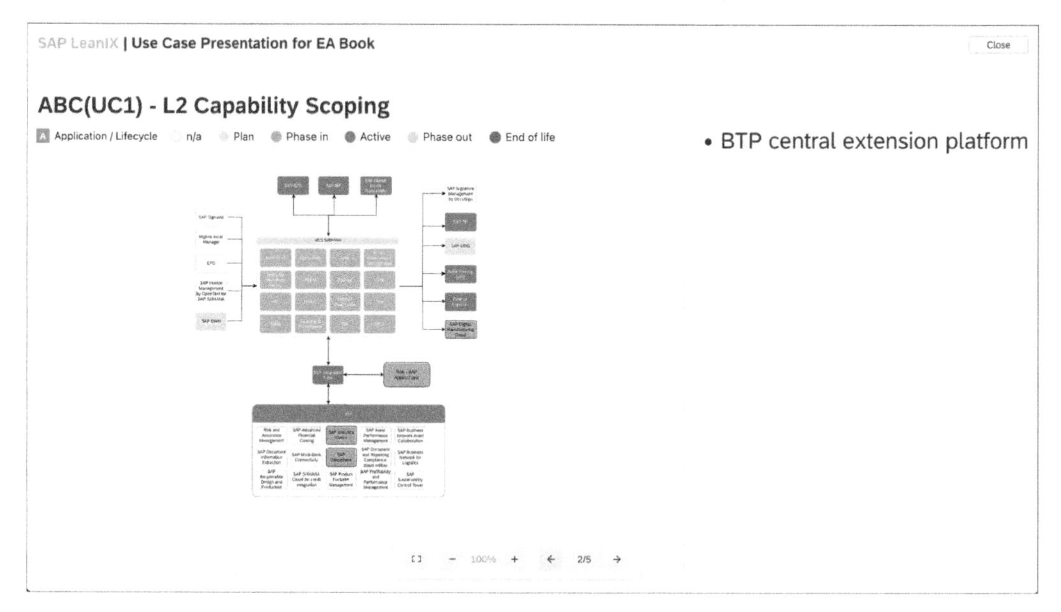

Figure 12.36 Running Presentations with Navigation Controls

12.9 Subscriptions, User Roles, and Authorizations

Users can subscribe to any change of a fact sheet in the model. Besides simply being interested in the information, users can have other reasons for subscribing, and subscriptions in SAP LeanIX can be configured in a flexible way. Figure 12.37 shows that subscriptions are first of all a link between user and fact sheet. They can be annotated with a configurable subscription role.

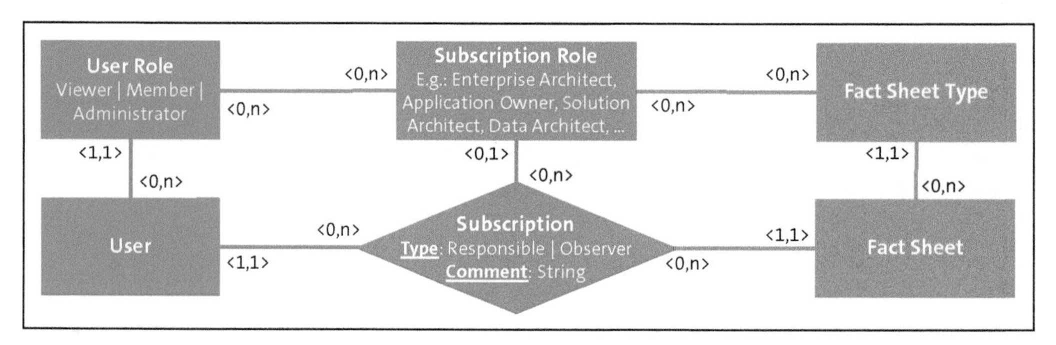

Figure 12.37 Subscriptions in SAP LeanIX

User roles determine the overall access rights a user has in the workspace: as a viewer, member, or administrator. You can maintain these user roles in the **Administration** area under **BASIC SETTINGS • Users**. When you invite a new user via the **+ Invite user element** on the top toolbar, you can assign the new access rights up to the level you own yourself.

Subscription roles describe the semantic role a team member has for a specific object. You can use them in "push" and "pull" modes and configure them in the **Administration** area under **BASIC SETTINGS • Subscriptions Roles**:

- *Pulling* means that you are adding yourself as an observer to all fact sheets you are interested in.
- *Pushing* comes into play when you are registered as responsible for a fact sheet fulfilling a dedicated role. Having all important roles maintained via subscriptions will allow you to regularly run quality checks, which can be triggered by surveys.

For example, for an application, you could have the following subscription roles: application owner, business owner, developer, and tester. We recommend not specifying too many roles. There's a risk that this information may get outdated, and the focus should be on key roles, who have accountability in the application portfolio management or enterprise architecture efforts. Figure 12.38 shows all subscriptions within an application fact sheet.

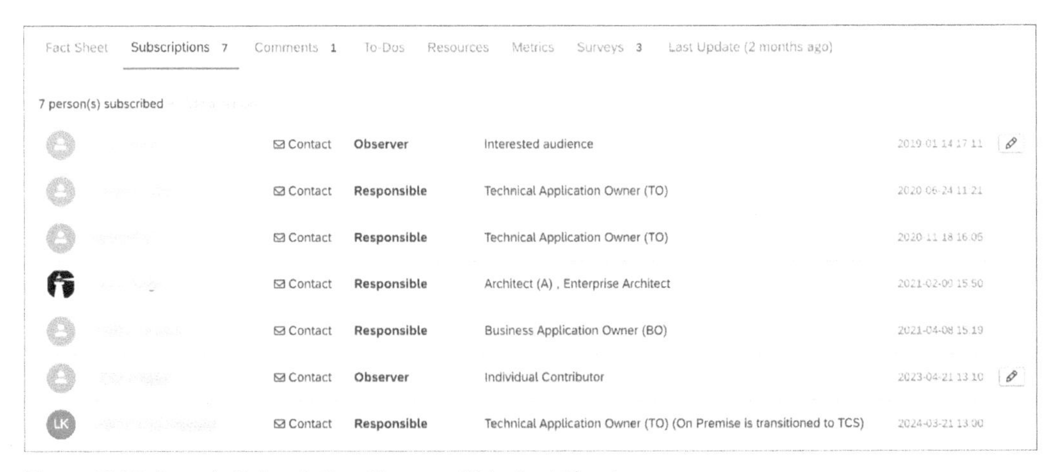

Figure 12.38 Sample Subscription Shown within Fact Sheet

If you have the administrator user role, you can temporarily switch user roles to test configuration access rights.

12.10 Collaboration

On any type of fact sheet, there can be a collaboration via comments, action items, and questions defined. Comments are simple discussion threads where neither the initial question nor the replies have any status. User-specific accountabilities by fact sheet are documented via subscriptions, while just sharing or marking relevant reports and diagrams is done via favorites and sharing. We'll look at these specific functions in the following sections.

12.10.1 Favorites and Sharing

You can mark reports and diagrams as favorites and share them with others. In Figure 12.39, the blue star right of the report title indicates that this report is in your favorites. The **Share** button at the top right will enable the coworkers you share your report with to find the shared item easily in their personalized lists.

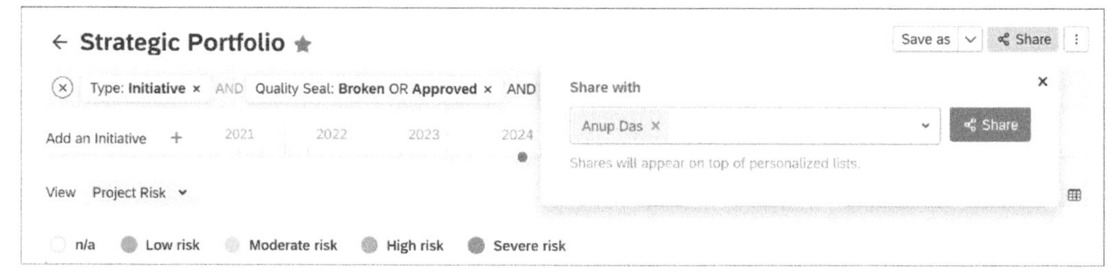

Figure 12.39 Sharing Reports

Favorites and *sharing* are the two ways to make exchanges on the report and diagram level. You can access reports and diagrams shared with you via the left pane in the **PERSONAL** grouping, as shown in Figure 12.40. This section also includes the **Presentations** feature introduced in Section 12.8, and it lets you access recently used reports and all reports you have created.

Figure 12.40 Accessing Shared Items and Favorites

On the fact sheet level, you use the concept of *subscriptions*. It's superior to favorites and sharing in the sense that it can define a granular specification to document why a fact sheet has relevancy for a particular user.

12.10.2 Comments

Commenting is a feature that's well known from many standard applications such as Microsoft Office. You can make comments on the fact sheet level, and you can access them in the **Comments** tab of each fact sheet. As usual in such a feature, you can address coworkers by typing the @ sign, followed by their email address.

12.10.3 Action Items and Questions

Questions and action items are very similar with regard to assignment and status management. The difference between them lies more in semantics that in the action items described as tasks to be done on the fact sheet they're assigned to. Questions may be directed to the responsible person to give some additional information to the requestor.

Figure 12.41 depicts the conceptual collaborations model, in which questions and action items are specializations of to-dos. Each has a status, a description, and a due date, and it is always linked to at least one fact sheet.

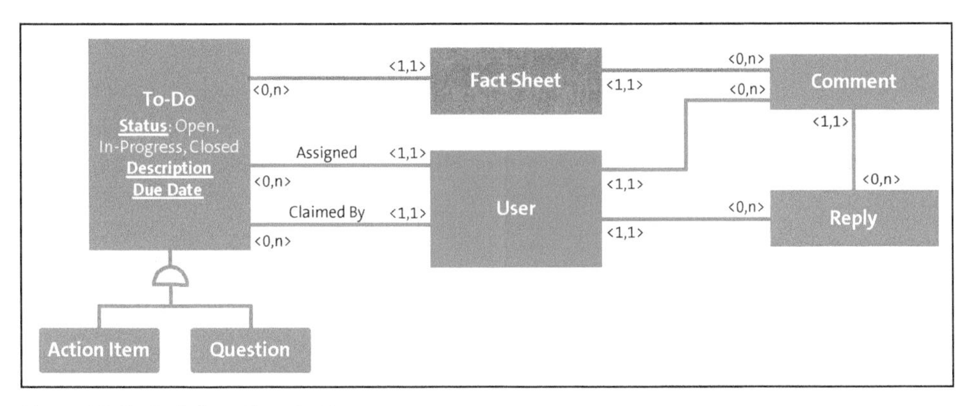

Figure 12.41 Collaboration Options

Adding action items or asking questions within SAP LeanIX are simple straightforward tasks and are not intended to replace more elaborate task planning and management solutions. Restricting these questions and answers as well as the to-dos to the context of a fact sheet enables others to gain further insights.

Figure 12.42 shows how to define an action item, assign it to a user, and specify a due date.

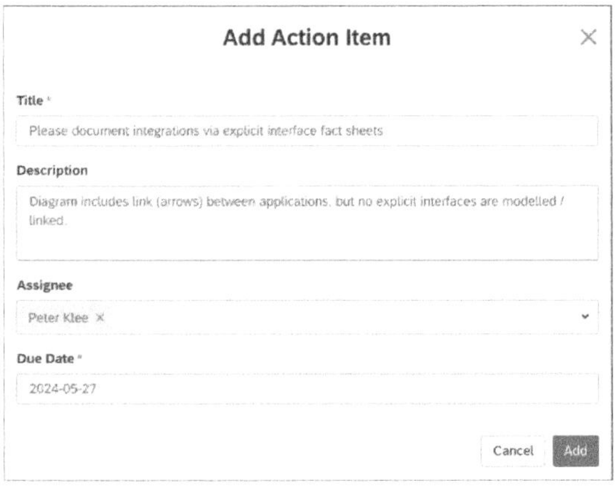

Add Action Item

Title *

Please document integrations via explicit interface fact sheets

Description

Diagram includes link (arrows) between applications, but no explicit interfaces are modelled / linked.

Assignee

Peter Klee ×

Due Date *

2024-05-27

Cancel Add

Figure 12.42 Adding Action Items

12.10.4 Surveys

Conducting surveys is a targeted mechanism for collecting information and helping to improve fact sheet quality. You perform survey configuration in the **Collaboration** tab under **SURVEYS,** following a stepwise approach:

1. **Scope**
   Determine for which fact sheets you should execute the survey. This step uses the usual filter mechanisms known from the inventory and reports:
2. **Design**
   The most important items to design are the fact sheet elements because they directly relate to fact sheet fields, so that when the survey is being run, the responses are stored directly in the fact sheet. The base elements consist of freely defined text with single or multioption elements. These as well as the answers to specific fact sheet fields are accessible in custom code you can put behind advanced calculated fields. Figure 12.43 shows the survey editor. Elements are dragged from the left to the right.

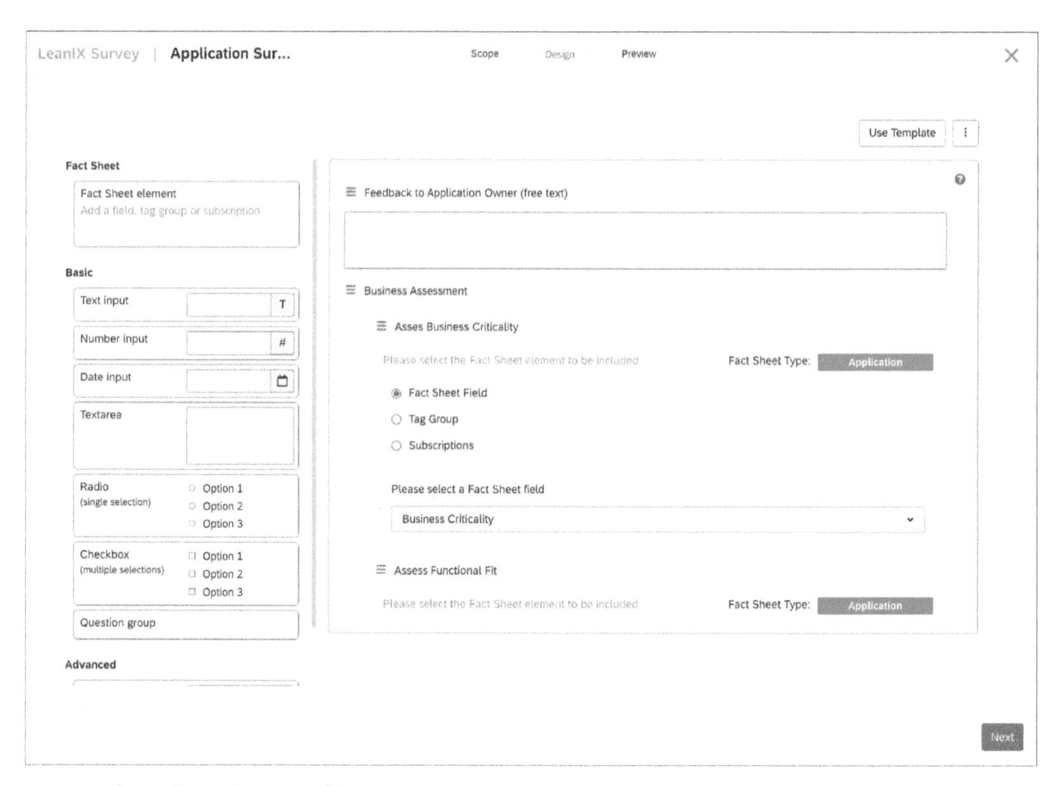

Figure 12.43 Survey Editor

3. **Preview**
   This provides the visual preview, as shown in Figure 12.44.

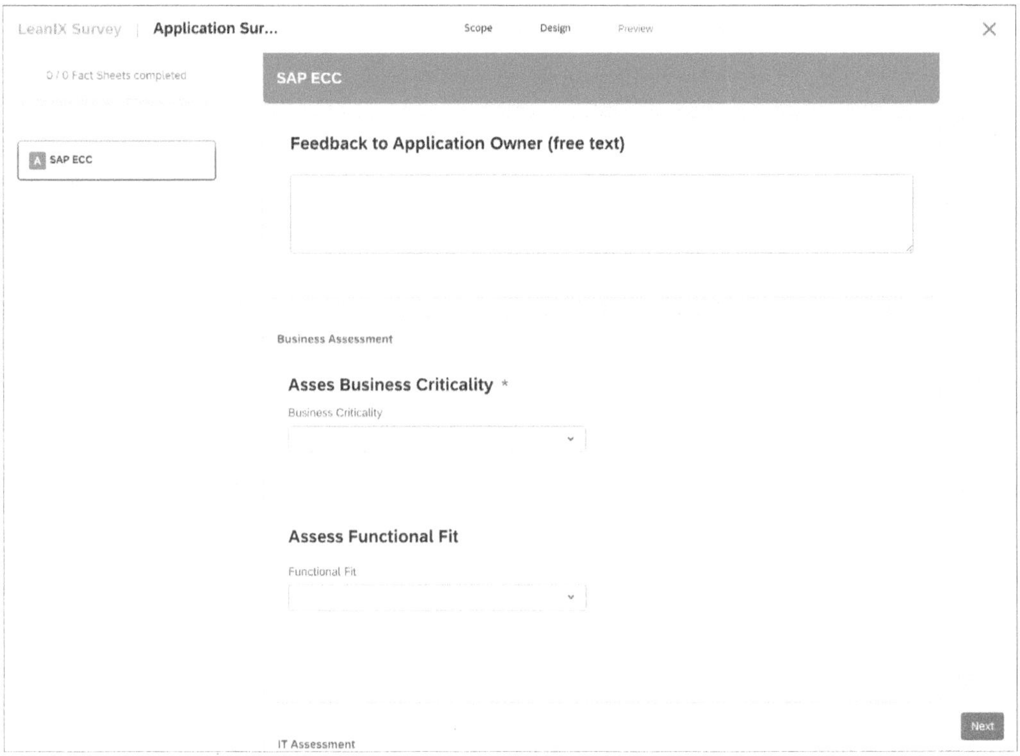

Figure 12.44 Survey Preview

4. **Share**
   You can share surveys with specific subscriber roles on the fact sheet. For example, for an application, there could be the following subscriber roles: application owner, business owner, and architect. It would not make sense for you to send the same survey to all three roles. Rather, you should bundle questions that are specific to their individual accountability. Before you send, the application will check how many of the scoped fact sheets don't have the specified subscription role defined and to how many different receivers survey requests will be sent. Figure 12.45 shows the sharing step, in which you can specify a parameterized email for the recipients.
5. **Insights**
   The insight view shows how many fact sheets have been processed by how many users, as shown in Figure 12.46.

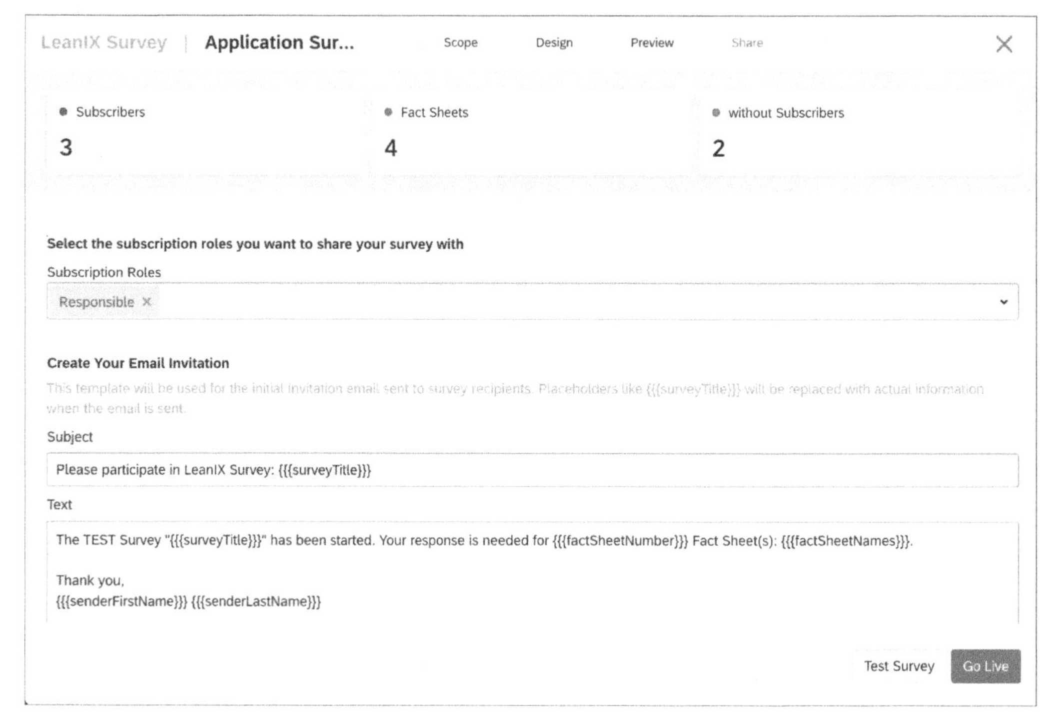

Figure 12.45 Sharing Survey

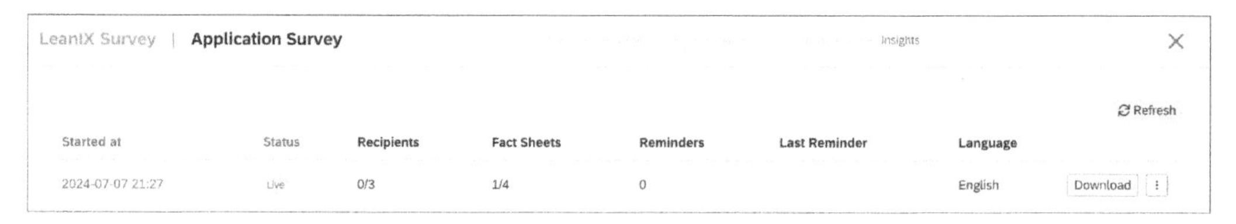

Figure 12.46 Survey Insights

You can download results, and via the context menu (...), you can stop the survey or send reminders to those users who haven't taken action yet.

12.11 Quality Management and Governance

SAP LeanIX offers a set of approaches to caring about data quality. For all fact sheets, it calculates a completeness score based on the sections that the administrator has defined as being included in the calculation. In general, you should set realistic quality targets for your team and not shoot for perfection.

In addition, the use of quality seals and the definition of mandatory fields and relationships offer fine granular quality management capabilities. If fields or relationships are defined as mandatory but also conditional, it allows for finer granular control.

Figure 12.47 shows a completeness score of **18%** right next to the fact sheet name, while the green **Approved** quality seal is displayed farther to the top right, next to the **Observer** subscription type, which indicates an existing subscription. Look back to Section 12.9 for more details on subscriptions.

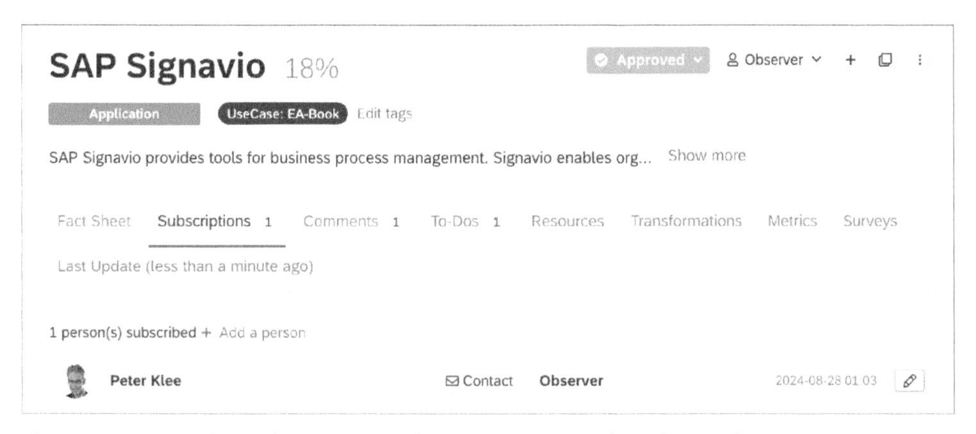

Figure 12.47 Fact Sheet Showing Complete % Measure and Quality Seal

12

The quality seal states and their transitions are shown in Figure 12.48. They start with the draft status and then can be either approved or rejected. By fact sheet type, you can configure that after a period of time (e.g., six months), the seal will switch to **Check Needed**. It's also possible that fact sheets will need to be reapproved after a predefined time interval.

To manually assign a new state, simply click on the quality seal and choose the new seal you want. Note that first, an administrator in the meta model needs to activate the use of the draft seal and the transition from draft to rejected.

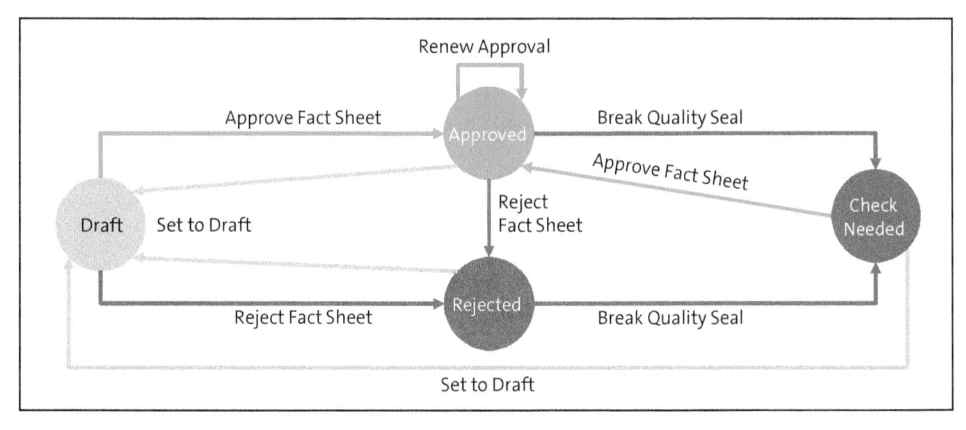

Figure 12.48 Quality Seals State Diagram

You can use the survey feature we discuss in in Section 12.10.4 for data gathering and improvement of fact sheets in the **Check Needed** status.

Regular Quality Checks

Checking the quality of SAP LeanIX should be a continuous task that should be in the DNA of each user working with the tool. If everyone feels accountable, there should be a positive culture in which using collaboration features to address quality issues is appreciated.

On top of ongoing peer quality awareness, it's best practice to have the core team proactively perform quality inspections.

Incompletely maintained data can often exclude quite a lot of fact sheets from reports. In these cases, the application gives the **Attention: 323 Fact Sheet(s) are not included in this report due to missing data** warning below the report header. When sharing the report and using it for decision-making, you should check whether the missing data can be maintained so as to make the report results more reliable. Alternatively, you can use filtering to try to narrow down the scope of the report to a well-defined subset of fact sheets for which data is more complete.

12.12 Summary

In this chapter, we explored most of SAP LeanIX's main functions, including the following:

- Dashboards and presentations that reveal insights in an easy-to-consume format
- Working with the inventory for initial orientation
- Running and designing a variety of report formats (matrix, layout, roadmap, etc.), data flow diagrams, and free diagrams
- Collaboration and quality management via action items, surveys, comments, and quality seals

Note that with agile and frequent innovations, some of the areas will quickly evolve further.

The following chapter will investigate integration and extensibility, and in Chapter 14, which will conclude Part III on SAP LeanIX, we'll look at a few relevant use cases.

Chapter 13
Integration and Extensibility with SAP LeanIX

We've introduced the concepts of SAP LeanIX and covered the typical way of working with the tool. In this chapter, we go one step further and look at how to extend the meta model, how to integrate other tools, and what other opportunities exist to customize the environment to your personal needs.

To support your integration needs, SAP LeanIX offers simple Microsoft Excel-based import and export of data. To have continuous and possibly bidirectional integrations, SAP LeanIX provides configurable integration options to multiple best-in-class IT solutions and allows you to develop custom integrations. Particularly, the integration of SAP reference architecture content, which is a key pillar in SAP EA Framework, is supported. To document the current baseline architecture, existing SAP landscapes can be imported into SAP LeanIX with an autodiscovery function.

Extensibility is another important feature we cover in this chapter, and it ranges from adapting the meta model, to writing custom reports, to implementing specific automation tasks.

In the following sections, we'll start with simple integrations from and to Microsoft Excel and the use of reference catalogs. Then, we'll show you how to enhance the meta model we introduced in Chapter 11. Complex prebuilt and custom integrations with external tools will round out the chapter supplemented with explanations of other advanced customizations such as automations and metrics.

13.1 Export and Import

In the **Inventory** tab, export and import fact sheets from and to Microsoft Excel are triggered via **Export** or **Import** commands in the **ACTIONS** section in the right-hand side panel (see Figure 13.1). As a best practice, you define or open a table view that includes all fact sheet attributes and relationships you wish to see in the export to Microsoft Excel.

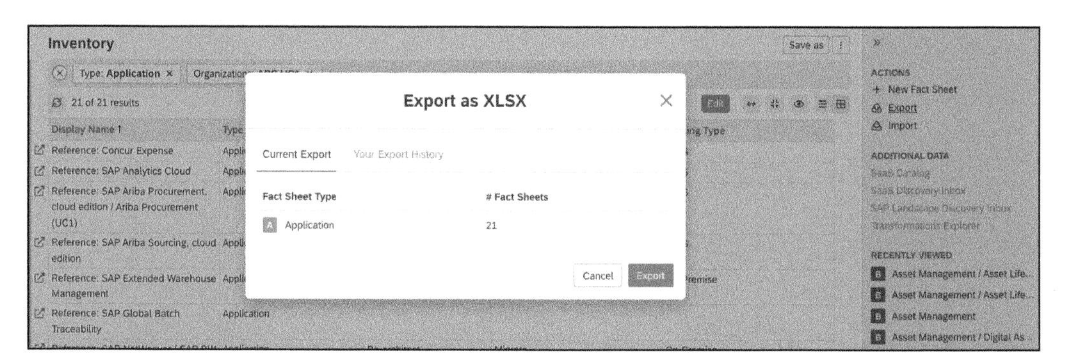

Figure 13.1 Triggering Export to Microsoft Excel

You can reuse the generated export for an import. The table view and herewith the export file can include a relationship and the attributes defined on this relationship. Figure 13.2 shows the relationship of **Application** to **IT Component** and its **Technical Fit** field. As multiple IT components are assigned to the SAP ERP application, it is repeated six times in lines 15 through 20.

Before you can import fact sheets, you need to understand how to handle IDs. The following three fields are important in this context:

- **ID**
  This is the internal ID, which is not shown in the user interface but is contained in the fact sheet's URL. In the example shown in Figure 13.2, **Concur Expense** has an internal UUID ID of **4eeee452-5cb9-4c55-bd3d-c53bd3da67d6**. The full URL for the fact sheet would look like this:

  https://eu-6.leanix.net/<xxxxx>/factsheet/Application/4eeee452-5cb9-4c55-bd3d-c53bd3da67d6.

  Note that during imports, you can't use these IDs to identify objects in relationships—only the anchor object identified by each of the lines.

- **Display Name**
  This is the concatenation of names in the hierarchy from the root to the current object. For objects without a parent, **Display Name** equals **Name**. In reports, you typically won't display the full **Display Name** but rather the "local" one. But when loading relationships to other fact sheets, you must use the full **Display Name** and can't use the internal ID nor the external ID of the related fact sheet.

- **External ID**
  This relates to external IDs coming from other sources located outside of SAP LeanIX, and it provides a stable, unambiguous ID for each fact sheet. When performing uploads into SAP LeanIX, you can use this external ID to check in a preprocessing

step whether the external ID already exists. You'll then need to decide if you wish to overwrite attributes of the existing fact sheet or skip the record.

id	type	name	displayName	externalId	lxSixRClassification	lxTimeClassification	relApplicationToITComponent:displayName	relApplicationToITComponent:technicalSuitability
ID	Type	Name	Display Name	External Id	6R Strategy	TIME Classification	IT Component: Display Name	Technical Fit
4eeee452-5cb9-4c55-b	Application	Concur Expense	Reference: Concur Expense	CONCUR-EXPENSE	retain	Invest		
02d03737-12f0-4a6c-b	Application	SAP Analytics Cloud	Reference: SAP Analytics Cloud	SAP-SAC	retain	Invest		
e3f26342-4142-41cd-a	Application	Ariba Procurement (UC1)	Reference: SAP Ariba Procurement, cloud edition / Ariba		retain	Invest		
28f2d83b-739d-4917-b	Application	SAP Ariba Sourcing, cloud edition	Reference: SAP Ariba Sourcing, cloud edition					
3f16175c-4a02-4048-8	Application	SAP Extended Warehouse Management	Reference: SAP Extended Warehouse Management					
69e52a53-c5f6-411f-88	Application	SAP Global Batch Traceability	Reference: SAP Global Batch Traceability					
a2c06780-575e-4ff9-9b	Application	SAP BW on HANA	Reference: SAP NetWeaver / SAP BW on HANA		rearchitect	Migrate		
8db74750-7fc1-4549-9	Application	SAP SRM	Reference: SAP Supplier Relationship Management / SA		retire	Eliminate	SAP SAP SRM 7.0	
07e65151-6408-4a27-t	Application	SAP Transportation Management	Reference: SAP Transportation Management					
9308f760-98c1-4ce1-b	Application	SAP BOFC	SAP BOFC					
a802a1f5-b1f1-4367-93	Application	SAP Datasphere	SAP Datasphere		replatform		SAP Datasphere SaaS Hosting	
3cda19c3-24d4-4b6d-a	Application	SAP Digital Manufacturing Cloud	SAP Digital Manufacturing Cloud		replatform		SAP Digital Manufacturing SaaS Hos	
3cbcc7ac-d765-4a4c-b	Application	SAP ECC	SAP ECC				SAP ERP / SAP CO	unreasonable
3cbcc7ac-d765-4a4c-b	Application	SAP ECC	SAP ECC				SAP ERP / SAP FI	adequate
3cbcc7ac-d765-4a4c-b	Application	SAP ECC	SAP ECC				SAP ERP / SAP LE-TRA	unreasonable
3cbcc7ac-d765-4a4c-b	Application	SAP ECC	SAP ECC				SAP ERP / SAP MM	adequate
3cbcc7ac-d765-4a4c-b	Application	SAP ECC	SAP ECC				SAP ERP / SAP PM	adequate
3cbcc7ac-d765-4a4c-b	Application	SAP ECC	SAP ECC				SAP ERP / SAP PP	adequate
4edca409-08b5-42ee-8	Application	SAP EH&S (Product Compliance)	SAP EH&S (Product Compliance)					
1610eaef-7a13-4515-9	Application	SAP FIM	SAP FIM					
5c32e07a-2bb6-421a-9	Application	SAP GRC	SAP GRC					
ffe54baf-acf9-4dfc-a8fb-	Application	SAP GTS	SAP GTS					
141fa593-ce83-4eee-b7	Application	SAP IBP	SAP IBP		retain			
6304642b-bf71-4e5a-9	Application	SAP SF	SAP SF		retain			
39dbb8ce-ea2a-4518-b	Application	SLP (SAP ERP Sales)	SLP (SAP ERP Sales)		replatform	Migrate	SAP ERP / SAP FI	
39dbb8ce-ea2a-4518-b	Application	SLP (SAP ERP Sales)	SLP (SAP ERP Sales)		replatform	Migrate	SAP ERP / SAP LE-TRA	
39dbb8ce-ea2a-4518-b	Application	SLP (SAP ERP Sales)	SLP (SAP ERP Sales)		replatform	Migrate	SAP ERP / SAP SD	
39dbb8ce-ea2a-4518-b	Application	SLP (SAP ERP Sales)	SLP (SAP ERP Sales)		replatform	Migrate	SAP ERP / SAP WM	

Figure 13.2 Export/Import Format

The **ReadMe** tab in the export, shown in Figure 13.3, provides further explanation of field types and possibly values, which is helpful if you plan to use the template for a future upload.

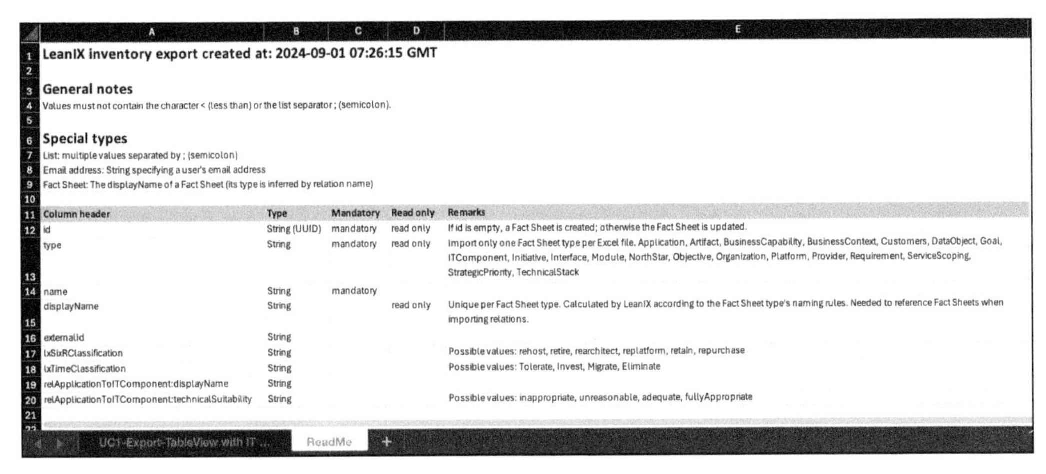

LeanIX inventory export created at: 2024-09-01 07:26:15 GMT

General notes

Values must not contain the character < (less than) or the list separator ; (semicolon).

Special types

List: multiple values separated by ; (semicolon)
Email address: String specifying a user's email address
Fact Sheet: The displayName of a Fact Sheet (its type is inferred by relation name)

Column header	Type	Mandatory	Read only	Remarks
id	String (UUID)	mandatory	read only	If id is empty, a Fact Sheet is created; otherwise the Fact Sheet is updated.
type	String	mandatory	read only	Import only one Fact Sheet type per Excel file. Application, Artifact, BusinessCapability, BusinessContext, Customers, DataObject, Goal, ITComponent, Initiative, Interface, Module, NorthStar, Objective, Organization, Platform, Provider, Requirement, ServiceScoping, StrategicPriority, TechnicalStack
name	String	mandatory		
displayName	String		read only	Unique per Fact Sheet type. Calculated by LeanIX according to the Fact Sheet type's naming rules. Needed to reference Fact Sheets when importing relations.
externalId	String			
lxSixRClassification	String			Possible values: rehost, retire, rearchitect, replatform, retain, repurchase
lxTimeClassification	String			Possible values: Tolerate, Invest, Migrate, Eliminate
relApplicationToITComponent:displayName	String			
relApplicationToITComponent:technicalSuitability	String			Possible values: inappropriate, unreasonable, adequate, fullyAppropriate

UC1-Export-TableView with IT ... | ReadMe

Figure 13.3 Explanation of Fields in Export Microsoft Excel

In the **Administration** section of SAP LeanIX, administrators can trigger an export of the entire workspace via **Export** in the **ADVANCED SETTINGS** section of the panel on the left, as shown in Figure 13.4.

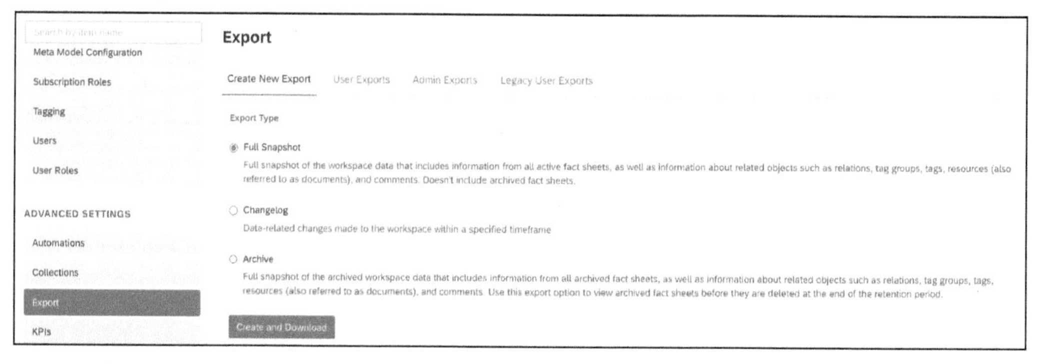

Figure 13.4 Full Snapshot in Administration

13.2 Reference Catalogs

In SAP LeanIX, you can use the **Reference Catalog** area to mass-load either all of or parts of an external model into the workspace.

Admins can bulk-load refence content via the administration area by selecting **Reference Catalog** in the left-hand panel under the **DISCOVERY & INTEGRATIONS** heading, as shown in Figure 13.5.

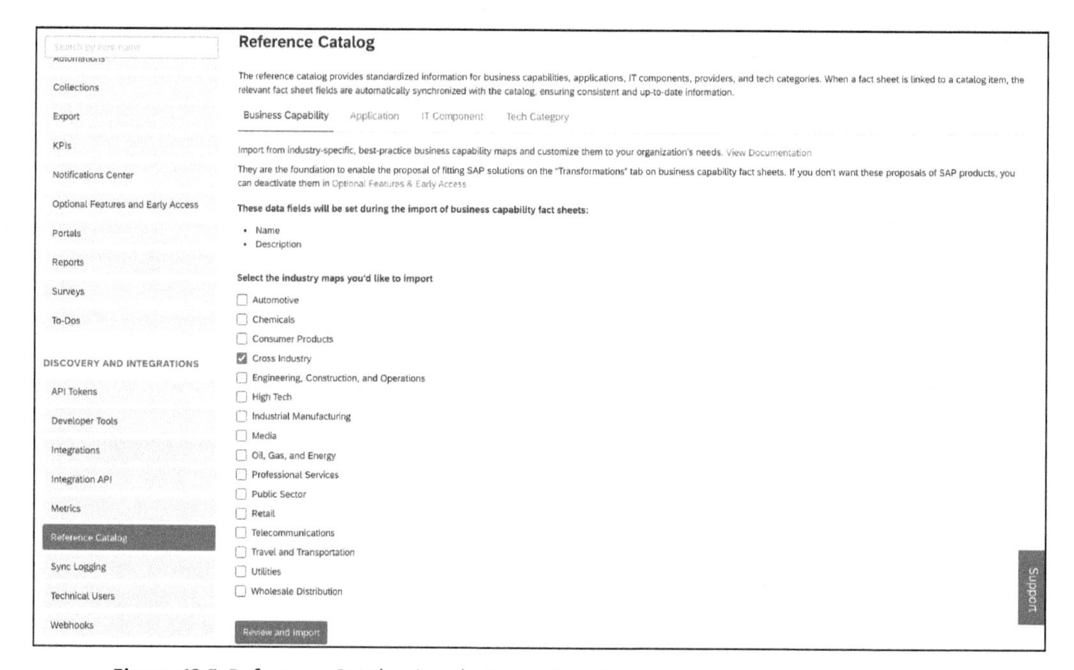

Figure 13.5 Reference Catalog in Administration Area

The following four tabs are currently available in the **Reference Catalog** area, which will continue to grow in the coming years:

- **Business Capability**
  The **Business Capability** area contains a capability model with industry-specific content. The core is a cross-industry model that is reused among industries to keep industry-specific adaptions to a minimum. This implies that industry-specific models for industries with long-established, highly specific capability and process models (e.g., telecommunications) can vary from the SAP reference models. Figure 13.6 shows the import of three level-3 capabilities below **Digital Asset Operations**.

Figure 13.6 SAP Reference Business Architecture

After an import is triggered, the imported fact sheet will show the connection in the **REFERENCE CATALOG** section in the panel on the right, via a chain symbol and a **Linked** hyperlink, which will bring you back to the catalog. Figure 13.7 shows the linked **Maintenance Execution** business capability fact sheet.

For business capabilities that are linked to the reference catalog, the **Transformations** tab branches into a suggestion of SAP applications. In most cases, you can switch between a private and a public cloud view, which can differ in terms of the recommendations they give. In the example, the **Hybrid** option (private cloud/on-premise) is selected, after which, the following two options are displayed. In both options, SAP S/4HANA is the first application to be suggested. The adjacent yellow dot indicates that this application is mandatory, and the blue **In Workspace** rectangle to the right of its name says that it's already available in the workspace.

- Option 1 suggests a second, optional **SAP Geographical Enablement Framework** application.
- Option 2 suggests a second mandatory application, **SAP Service and Asset Manager**, as well as four further optional applications. Only two of them are immediately visible; the other two require some scrolling.

The blue **Plan transformations** buttons can trigger the transformation in the course of which the missing application is created and linked to the business capability. In Chapter 14, we'll explain the transformation planning and the later transformation execution step. Initiatives (projects) can group multiple transformations together.

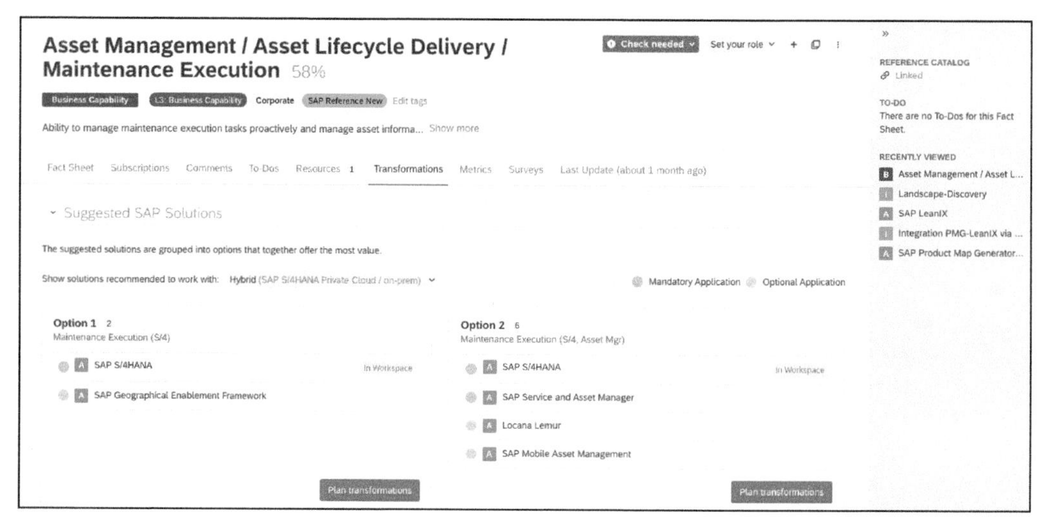

Figure 13.7 Imported Capabilities with Suggested Transformations

- **Application**
  When you're linking a fact sheet to the **Application** catalog, centrally available information like **Description, Hosting Type, Single Sign On – Available Providers**, and **Single Sign On – Availability** can be synchronized automatically. The **Provider** fact sheet type can group a company's multiple application offerings by their different providers. You can configure these options as shown in Figure 13.8.
- **IT Component**
  This is a central repository with information on IT components' support periods. When you're synchronizing IT components, their lifecycle (date) attributes will be completed, as discussed in Chapter 11, Section 11.4. Figure 13.9 shows the **SAP HANA 2.0 SP6** database linked IT component.

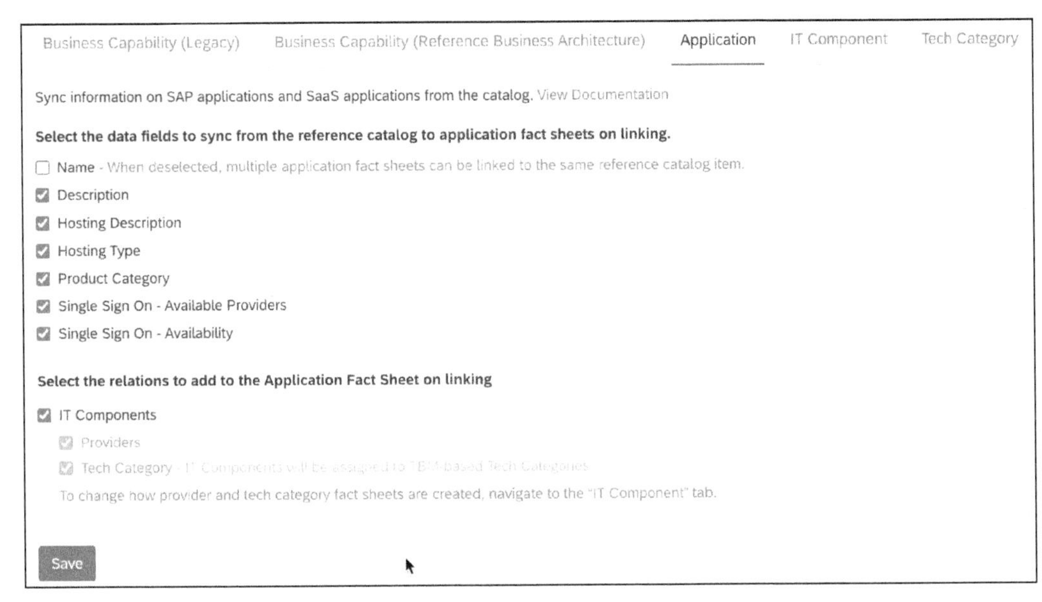

Figure 13.8 Configuration of Application Catalog Synchronization Options

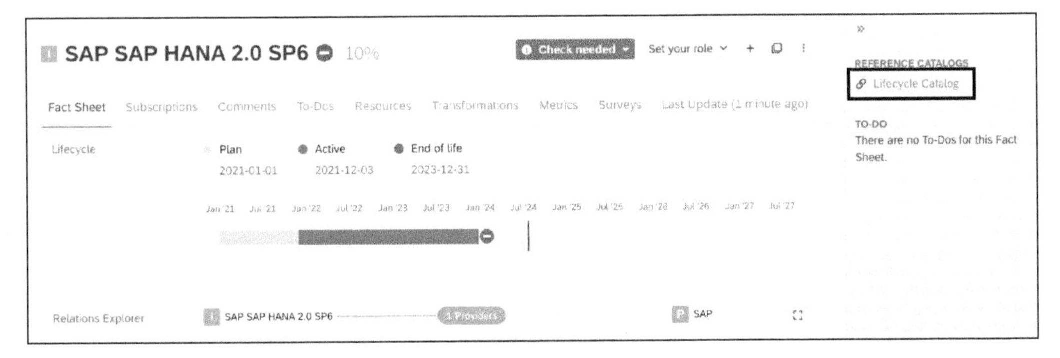

Figure 13.9 Lifecycle Catalog: Linked IT Component

- **Tech Category**
  SAP LeanIX offers to load technology categories from technology business management. This categorization provides the following six top-level categories: workplace, business, shared and corporate, delivery, platform, and infrastructure. These categories are further broken down into three levels. Before importing, we recommend that you possibly align the business, shared, and corporate technology business management categories with your business capability model. Many of the proposed subcategories of these categories may overlap but may also differ in the way your business capability hierarchy or a used reference capability hierarchy is built. SAP LeanIX will also scan existing IT components and propose mapping them to the respective subcategories.

13.3 Extending the Meta Model

Fact sheets, in a simplistic sense, are combinations of fields (attributes) and a set of relationships to other fact sheets.

While SAP LeanIX offers admins full flexibility to freely configure the meta model, we recommend that you stay as close to the predelivered meta model as possible, as this ensures the use of predefined out-of-the-box reports. We also recommend starting with only a subset of fact sheet types and not using all of the fact sheet fields if all possible. Their use should be triggered by the practice's needs. A more complex meta model often requires a higher maintenance effort.

In this section, we'll create a new fact sheet type, **Requirement**, with two attributes and relationships to initiatives, business capabilities, and applications. To start, navigate to the **Administration** area and choose **Meta Model Configuration** from the **BASIC SETTINGS** section in the navigation panel on the left. Figure 13.10 shows that SAP LeanIX's meta model configuration groups fields into sections and subsections. The panel on the right shows the configuration of the **Criticality** fields, and there, it shows the **Single Select** field type with three fixed and color-coded values. A field's width correlates with the size used when editing single fact sheets.

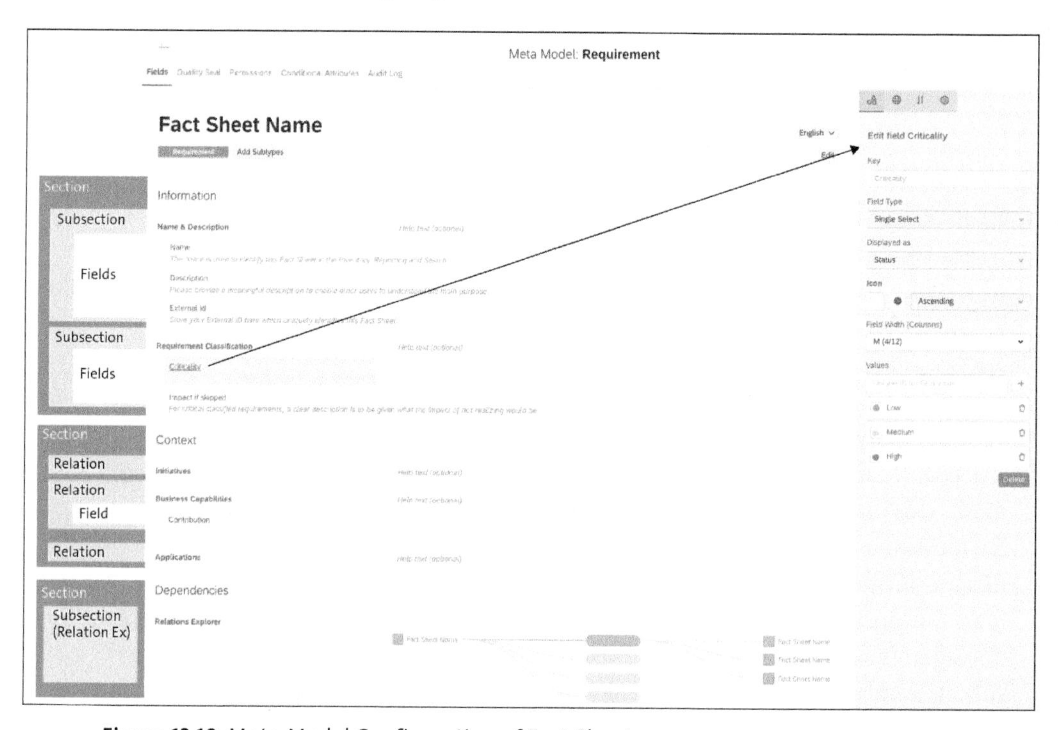

Figure 13.10 Meta Model Configuration of Fact Sheets

There are seven possible data types for fields:

- **String (Date)**
- **String (Text)**
- **String (Area)**
- **Double**
- **Integer**
- **Single Select**
- **Multiple Select**

Relationships link to other fact sheet types and can themselves have attributes, like the **Contribution** attribute of the relationship between **Requirements** and **Business Capabilities**.

Some fields only make sense if another field has a specific value. This behavior can be modeled in SAP LeanIX via conditional attributes. Figure 13.11 shows the **Conditional Attributes** tab, in which an **Impact if skipped** description field will only be shown if the **Criticality** attribute, which is defined for the same fact sheet, is set to **High**.

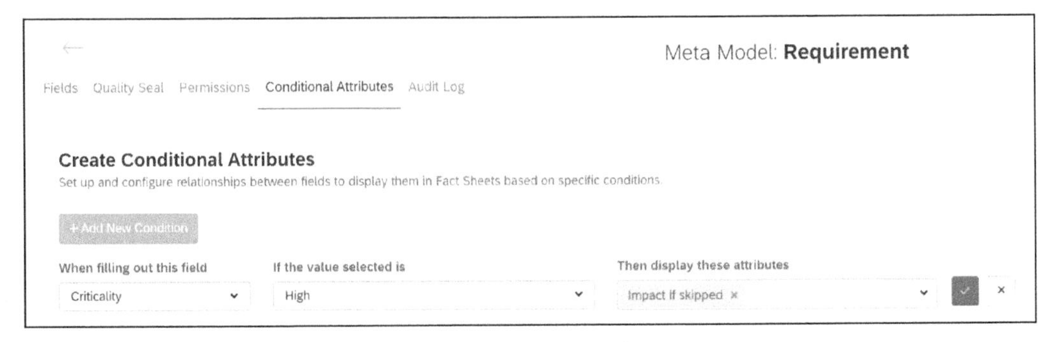

Figure 13.11 Criticality Triggers Visibility of Conditional Attribute

Adding relationships within the meta model will enable you to relate the respective fact sheets to each other and to evaluate them in reports. Figure 13.12 shows the six steps for creating a relationship between the newly created **Requirement** fact sheet type and the **Initiative** fact sheet type.

Let's walk through the steps in detail:

❶ Select the section in which you wish the new relationship to be shown.

❷ Click the **+ Add relation** link at the bottom of the section.

❸ In the right panel, select the **Target Fact Sheet type** for which you wish to create the relationship.

❹ Specify a **Descriptor** to describe the semantics of the relationship, if they're not obvious.

❺ Specify the **Multiplicity** of the relationship: many to many, one to many, many to one, or one to one.

❻ Specify in which **Section in target Fact Sheet type** the reverse relationship is to be shown.

Figure 13.12 Creating New Relationship in Meta Model

Once a relationship is defined in the meta model, it can be evaluated from both sides. This means that it doesn't matter from which end the relationship was initially defined.

13.4 Integrations

To enable the best possible collaboration with other functions, as discussed in Chapter 3, SAP LeanIX supports out-of-the box integrations with a number of best-in-class IT management solutions.

Figure 13.13 shows some of the most prominent integrations. The purple boxes in the blue applications specify the data objects handled in these applications. The more integration points are implemented, the more important it becomes to know where the data is created, read, updated, and deleted (CRUD).

There are three integration categories:

- **Enterprise architecture toolchain**
  These are integrations with tools that contain related enterprise architecture information (e.g., on applications, data, and processes). Likewise, these tools can be receivers of information modeled in SAP LeanIX. You'll need to configure the specific flows according to your needs.

- **Discovery**
  This set of integrations performs autodiscovery against operational systems to discover different parts of your application landscape.
- **Collaboration**
  Integrations in the collaboration category link with documentation and project management platforms. The OData integration can expose fact sheets to business intelligence tools that support OData as a data source.

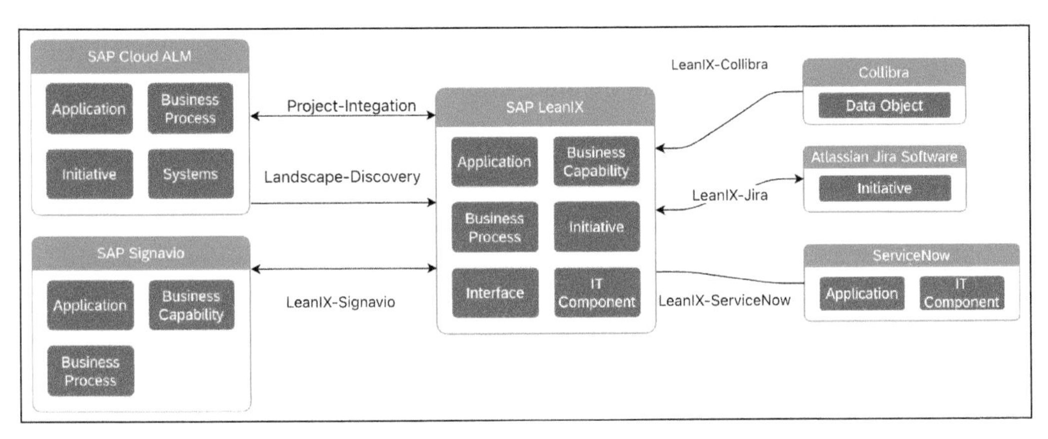

Figure 13.13 Data Flow Diagram of Integrations with SAP LeanIX

Figure 13.14 gives a view of the currently available integrations grouped by these categories.

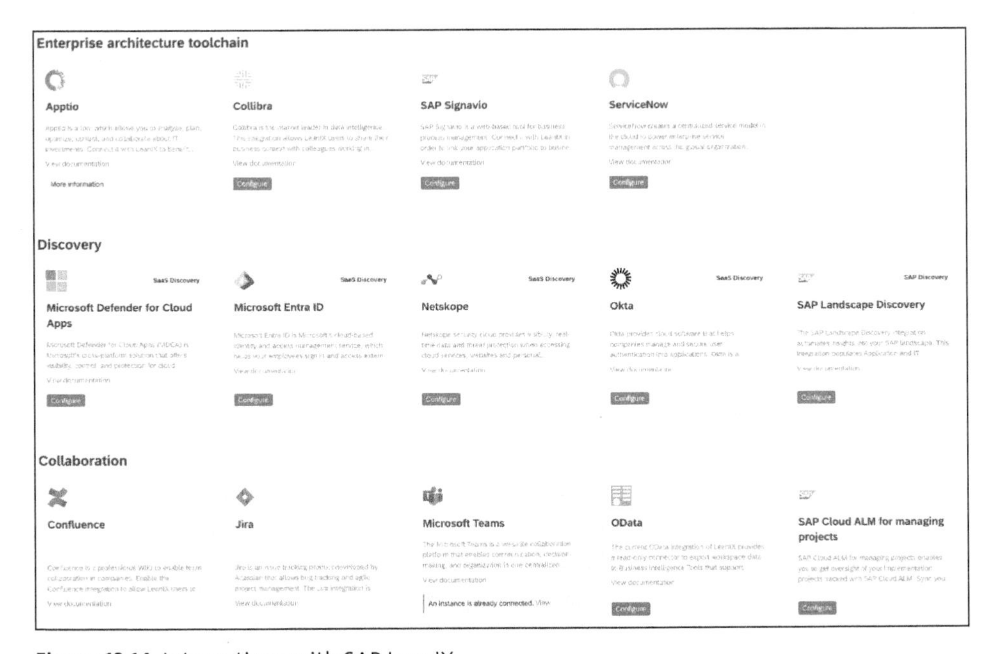

Figure 13.14 Integrations with SAP LeanIX

Now, let's dive deeper into the autodiscovery process and examine both prebuilt and custom integrations.

13.4.1 Autodiscovery of SAP Landscapes

Modern cloud-based SAP landscapes are documented in SAP Cloud ALM, which can be accessed by SAP LeanIX. To avoid manual documentation of SAP-centric landscapes, an autodiscovery capability from SAP Cloud ALM has been introduced.

The **Discovery Inbox** in Figure 13.15 shows a list of discovered items. Each list entry is classified as **SaaS**, **ERP SaaS**, or **On-prem**. While you're managing the discovered items, the status will move from **Action needed** to either **Linked** or **Rejected**. If the to-be-linked fact sheets don't exist in the workspace yet, you can create them on the fly.

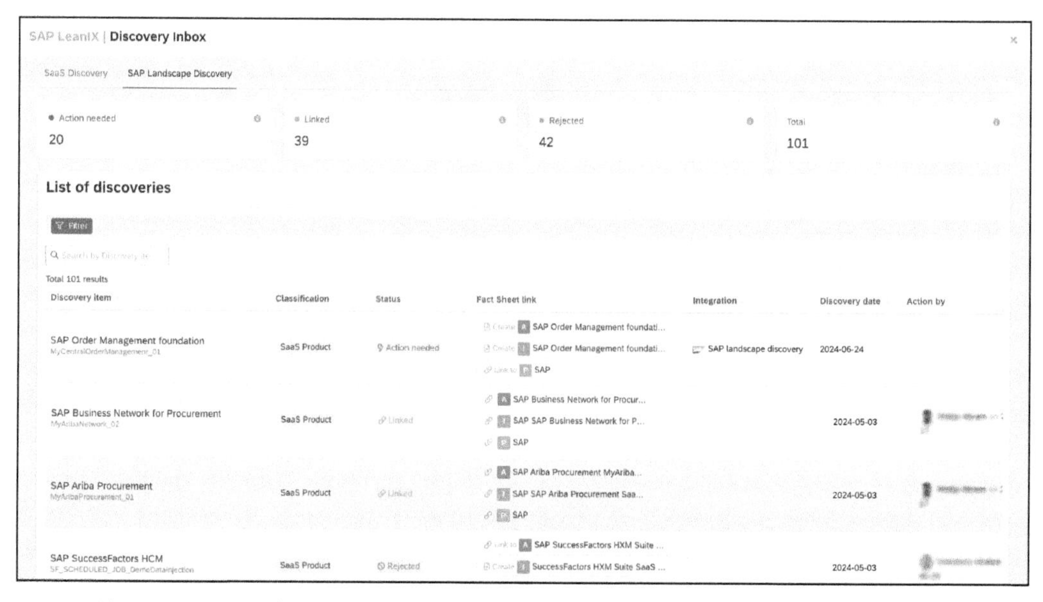

Figure 13.15 Autodiscovery of SAP Landscapes

13.4.2 Prebuilt Integrations

SAP LeanIX provides prebuilt integrations rather than aiming to be the best-in-class solution for every related IT function. These integrations can receive and send fact sheet information to ensure consistency with the integrated tools. Let's list those tools from the large set of integrations shown in Figure 13.14, which are typically used the most:

- SAP Signavio integration for business process management
- Jira, an enterprise project management tool that started out as ticketing system and is largely adopted for managing project portfolios and requirements

- Collibra, a best-in-class data governance tool that, in the SAP ecosystem, is designed to play a role in the context of SAP Datasphere
- ServiceNow, a widely adopted tool for IT service management that builds around a configuration management database (CMDB) holding all important IT assets.

Let's take a closer look at ServiceNow, as an example. ServiceNow serves several configuration management and other IT operations management functions. A synchronization model like the one shown in Figure 13.16 synchronizes applications and business capabilities from SAP LeanIX with ServiceNow. Likewise, tech categories and IT components can be sourced from ServiceNow and then interfaced into SAP LeanIX.

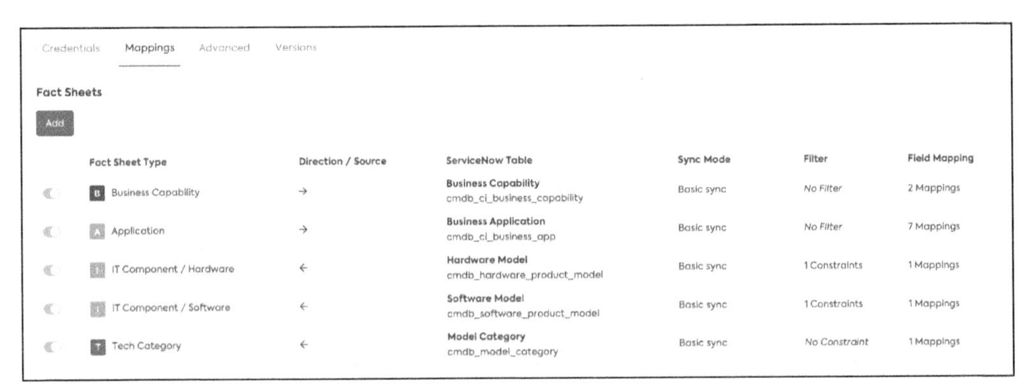

Figure 13.16 Mapping of Fact Sheets to ServiceNow Tables

13.4.3 Custom Integrations

Even though the list of prebuilt integrations is long, there can be a need for other solutions too. SAP LeanIX provides the possibility of creating and configuring custom integrations either with other packaged solutions or even with custom solutions.

Custom integrations are maintained in the **Administration** area under **Integration API** in the **DISCOVERY & INTEGRATIONS** section, as shown in Figure 13.17. Besides your individually configured integrations, you can access three exemplary ones. From these, we'll take a closer look at the **CMDB Connector Example**.

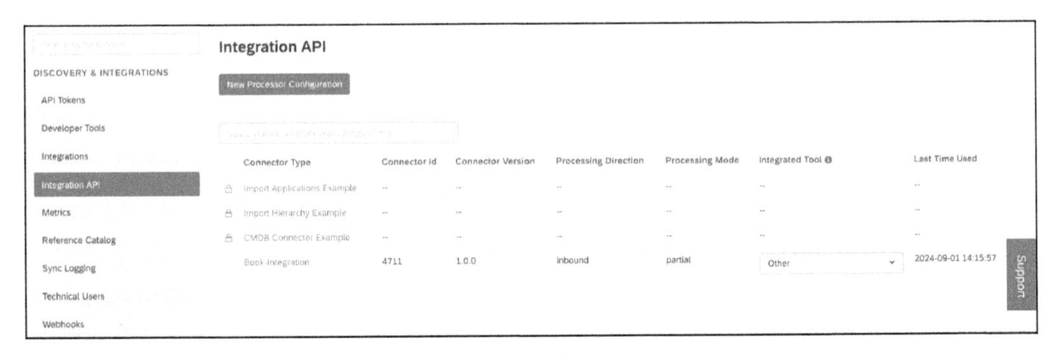

Figure 13.17 List of Custom Integration and Example Templates

This is where the generic integration framework comes into play. It allows you to create additional bidirectional integration scenarios. Figure 13.18 shows a template for a custom CMDB integration.

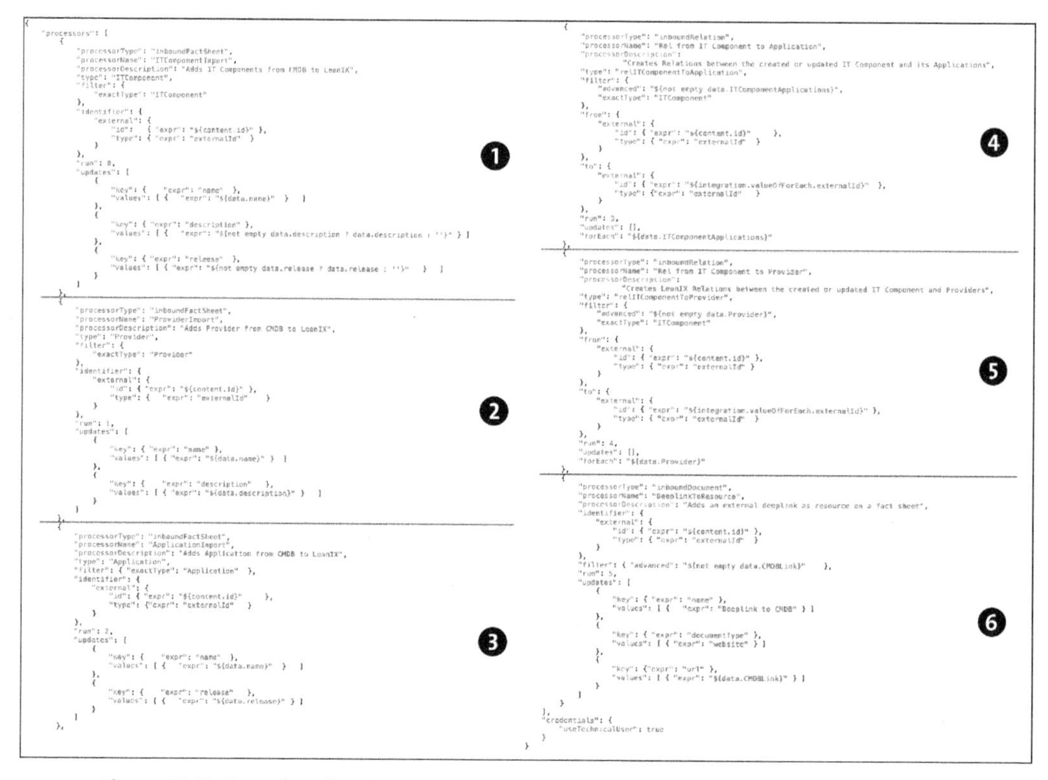

Figure 13.18 Template for Custom CMDB Integration

The example lists six processing blocks for the following:

❶ **IT components**

Retrieving IT components from a CMDB is the most important integration because the CMDB will most likely be the leading system and SAP LeanIX will hold the replica.

❷ **Providers**

You can use software or hardware providers when qualifying IT components and possibly applications. You'll need to decide which should be the leading one.

❸ **Applications**

Applications are center stage for our architecture work, and we therefore recommend keeping them maintained in SAP LeanIX because you'll need to be able to deal with planned new applications that a CMDB can't autodiscover yet.

❹ **Relationships between IT components and applications**

Where these relationships should be maintained will likely depend on your organizational responsibilities and the defined governance processes.

❺ **Relationships between IT components and provider**
See remark under providers.

❻ **Deep link to the CMDB**
Leveraging deep links in the CMDB will help to position SAP LeanIX as an integral part of a seamlessly integrated end-to-end toolchain.

For more details, see *https://docs-eam.leanix.net/reference/integration-api.*

13.5 Advanced Customization

In this section, we'll see in what ways key performance indicators (KPIs) and metrics can provide analytical insights. They're both displayed in dashboards. We'll also explain how automation can increase productivity for certain tasks.

Custom Reports

Note that it's beyond the scope of this book to instruct you on how to program custom reports. You need JavaScript web development experience to program custom reports, and for detailed instructions, including a step-by-step tutorial, see the SAP LeanIX documentation at *http://s-prs.co/v586309*.

13.5.1 Key Performance Indicators

KPIs are maintained in the administration area via **KPIs** in the **ADVANCED SETTINGS** section, as shown in Figure 13.19. There are a good number of examples to pick from, and you can adjust them or use them as inspiration for your own KPIs.

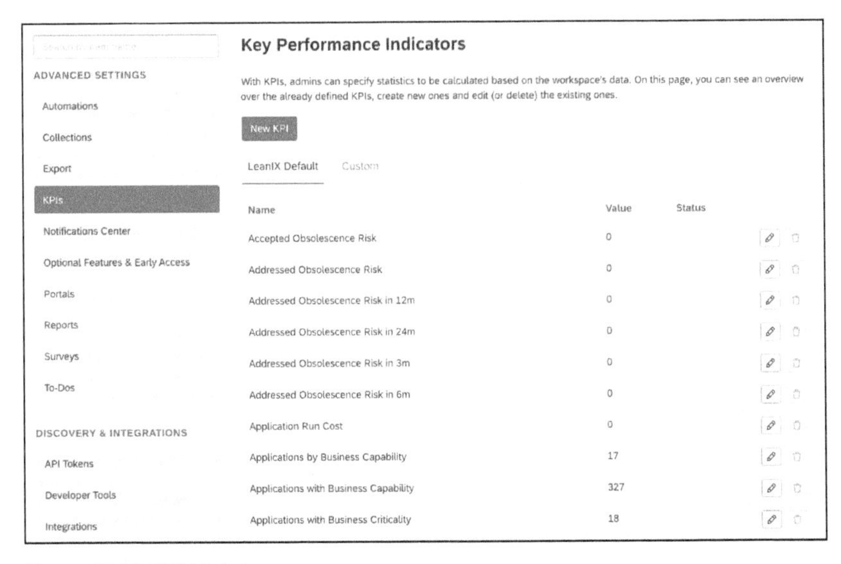

Figure 13.19 KPI Maintenance

The calculation logic of KPIs is defined in JavaScript Object Notation (JSON) format. Figure 13.20 shows the JSON example for the **Applications with Business Capability** KPI.

In this the example, the JSON code specifies five filters on applications:

- Filter `Application` type fact sheets.
- Applications that are already `endOfLife` won't be considered (type `none`).
- Application fact sheets that are `DRAFT` or `REJECTED` won't be included (type `none`).
- Applications that are in `planned` state won't be included (type `none`).
- Applications that have at least one relationship to a business capability (filter on `relApplicationToBusinessCapability` with type `any`).

The calculation logic in this case is specified in the `aggregations` section and is a simple application of the `count` function. The `totalFilter` repeats the same filter logic except the one checking the availability of the relationship to business capabilities. With this total number, it's possible to show this KPI as a percentage value, which is typically much more expressive than a plain number.

```
{
  "filters": [
    {
      "type": "factSheetType",
      "types": [
        "Application"
      ]
    },
    {
      "type": "none",
      "filters": [
        {
          "type": "equals",
          "fieldName": "lifecycle",
          "fieldValue": "endOfLife",
          "path": "$.currentPhase"
        }
      ]
    },
    {
      "type": "none",
      "filters": [
        {
          "type": "equals",
          "fieldName": "lxState",
          "fieldValue": "DRAFT"
        },
        {
          "type": "equals",
          "fieldName": "lxState",
          "fieldValue": "REJECTED"
        }
      ]
    },
    {
      "type": "none",
      "filters": [
        {
          "type": "equals",
          "fieldName": "lxTransformationStatus",
          "fieldValue": "planned"
        }
      ]
    },
    {
      "type": "any",
      "filters": [
        {
          "type": "forAnyRelation",
          "relation": "relApplicationToBusinessCapability",
          "filters": []
        },
        {
          "type": "contains",
          "fieldName": "naFields",
          "fieldValue": "relApplicationToBusinessCapability"
        }
      ]
    }
  ],
  "aggregations": [
    {
      "name": "count",
      "operation": {
        "type": "count"
      }
    }
  ],
  "totalFilter": [
    {
      "type": "factSheetType",
      "types": [
        "Application"
      ]
    },
    {
      "type": "none",
      "filters": [
        {
          "type": "equals",
          "fieldName": "lifecycle",
          "fieldValue": "endOfLife",
          "path": "$.currentPhase"
        }
      ]
    },
    {
      "type": "none",
      "filters": [
        {
          "type": "equals",
          "fieldName": "lxState",
          "fieldValue": "DRAFT"
        },
        {
          "type": "equals",
          "fieldName": "lxState",
          "fieldValue": "REJECTED"
        }
      ]
    },
    {
      "type": "none",
      "filters": [
        {
          "type": "equals",
          "fieldName": "lxTransformationStatus",
          "fieldValue": "planned"
        }
      ]
    }
  ]
}
```

Figure 13.20 Logic KPI Applications with Business Capability

KPIs can easily be embedded in dashboards (see Chapter 12, Section 12.1).

13.5.2 Automations

The term *automation* is well known from business process automation. The same idea applies in enterprise architecture modeling in SAP LeanIX: identify repetitive tasks that can be automated.

Admins can create automations via the **ADVANCED SETTINGS** section in the left-hand panel, as shown in Figure 13.21.

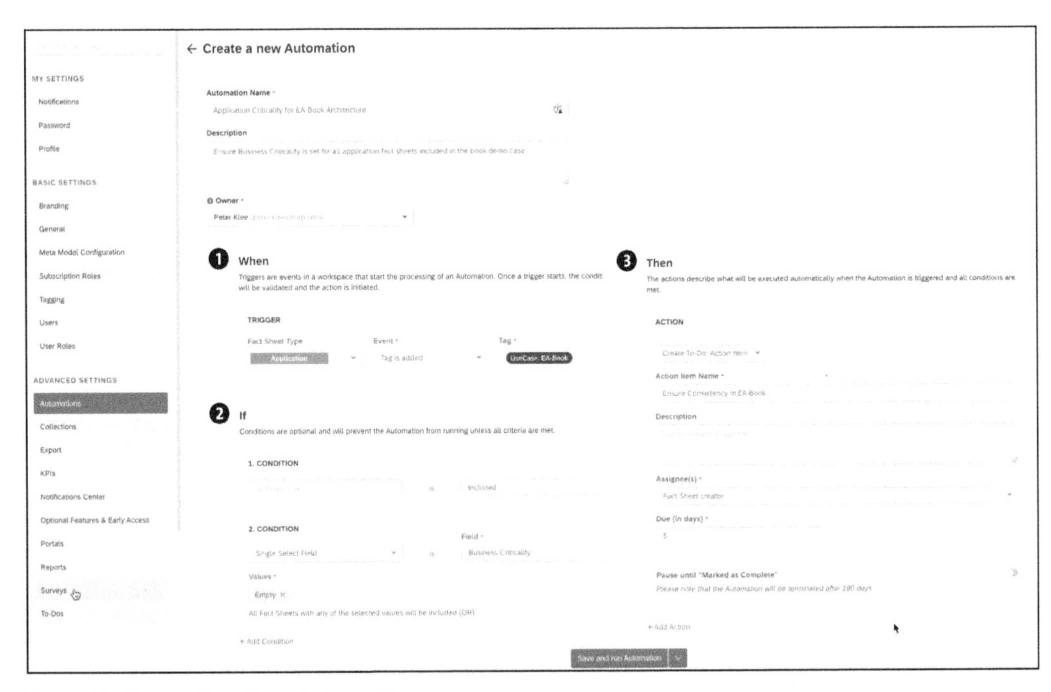

Figure 13.21 Creating New Automation

Figure 13.21 shows that an automation has a title, owner, and description and that its core definition consists of three parts:

❶ **When**

This part specifies when the automation will be triggered. In the example, if the **Use-Case: EA Book** tag is added to an **Application** fact sheet, the automation is triggered.

Possible triggering events are adding/deleting a tag, creating a fact sheet, changing a field, altering the quality seal, and changing the subscriptions of a fact sheet. For the last of these, see Chapter 12, Section 12.9.

❷ **If**

Additional conditions can restrict the selected automation to a subtype or limit the validity of the automation to a specific set of field values or tags.

❸ **Then**

This part of the automation defines one or multiple actions. The example shows the **Create To-Do Action Item**. Overall, the following action types are supported:

- Creating to-do or approval action items; With these two options, any possible further actions can be put on pause until the to-do or approval is marked as complete. They can be used, for example, to automatically set the quality seal after a "quality check" to-do is marked as complete.
- Setting a field value.
- Adding a subscriber and a specific subscription role to a specific user or creator of the fact sheet.
- Adding or removing a tag.
- Triggering a web API via **Send Webhook**. This action type will trigger further processing in a different web application, which offers such hook.

13.5.3 Metrics

Metrics can visualize KPIs over time. They can be configured in the **Administration** area within the **DISCOVERY & INTEGRATIONS** section in the left-hand panel. Figure 13.22 shows a custom metric based on the predefined SAP LeanIX functionality for object counts: **leanIXv4FactSheetCounts**. We often name these reports *time series reports* because the measurement is shown over time, and you can define different series for different kinds of measurements and define the charting type (line, bar, or area). Time series aggregation buckets can be days, weeks, months, quarters, or even years. As with the aggregation function, you can choose maximum, minimum, mean, count, or sum.

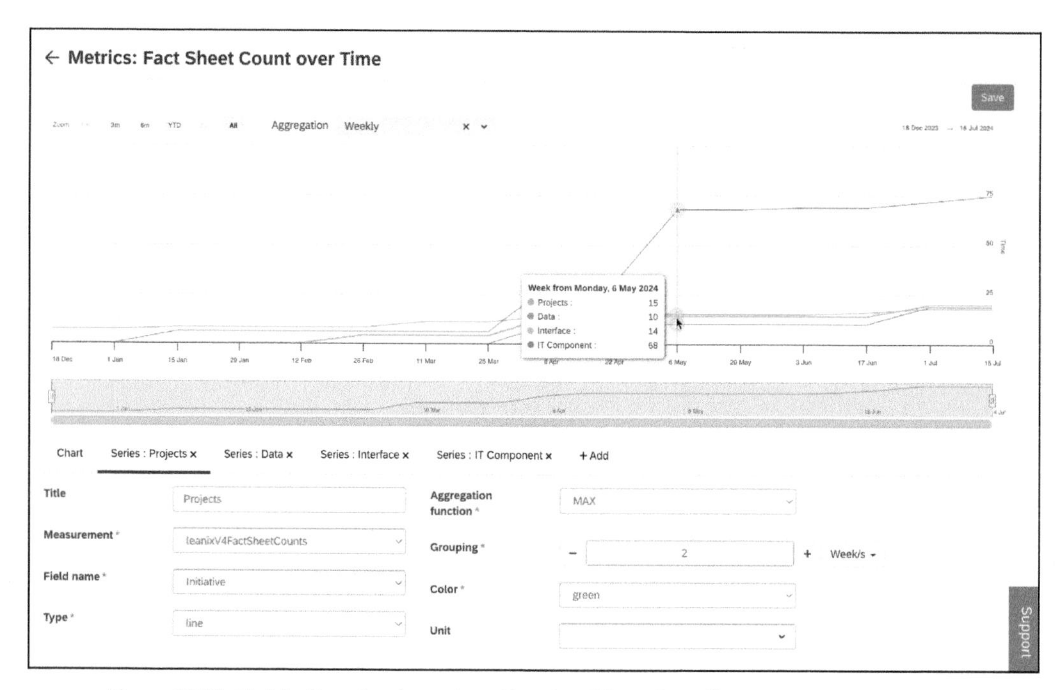

Figure 13.22 Metric Showing Inventory Counts of Four Fact Sheet Types

13.6 Summary

In this chapter, we've seen different ways to extend and integrate SAP LeanIX. A meta model can be extended with new tag groups, fields, and fact sheet types (as well with as relationships) in minutes, without any development complexities. We recommend not overconfiguring the workspace by adding too many fields and tags, since that may not have a clear business case or clear accountability in terms of maintenance.

More advanced configurations—including the definition of custom KPIs, integrations, or reports—require a deeper technical understanding.

In the next chapter, we'll finish up our deep dive into SAP LeanIX with an example practical scenario.

Chapter 14
SAP LeanIX in Practice

In the first three chapters on SAP LeanIX, we focused on concepts and specific features. Now, we'll look at specific end-to-end cases or practices that show SAP LeanIX in action.

In this chapter, we'll start to explore SAP LeanIX in the context of application portfolio management. Then, we'll give a comprehensive example of how to handle transformations in SAP LeanIX. We suggest that you run through this example as a tutorial because we present it as a step-by-step approach. We'll close the chapter by suggesting ways to model parts of SAP's clean core principles in SAP LeanIX.

14.1 Application Portfolio Management

Application portfolio management is one practice that many customers have adopted, and SAP LeanIX offers a series of assessment tools that we'll discuss in this chapter. How many assessments make sense depends on the overall health of the portfolio and its focus. If this is more on the IT/hosting side of the business, a 6R assessment may be advisable. If assessing health from a business support perspective is at least equally important, a tolerate, invest, migrate, and eliminate (TIME) assessment framework would be advisable.

If you use an existing assessment framework, you'll have the advantage of not needing to develop a new one from scratch. However, you may need to adjust it by extending its range of criteria, for example. A mature practice will do this and will typically aspire to move from a high-level qualitative scoring model to more measurable, quantitative scoring models.

As an example, we'll dive into the TIME assessment in this section. The concept of TIME is depicted in Figure 14.1, which shows the four different values that make up TIME as quadrants. The horizontal axis correlates with functional fit and the vertical axis with technical fit. This concept typically shows the best score in the top right corner and the worst at the bottom left. The level of functional fit is specified as one of four values: unreasonable, insufficient, appropriate, or perfect. Similarly, the level of technical fit is specified as inappropriate, unreasonable, adequate, or fully appropriate. Each of the TIME quadrants can be split into four subquadrants, as depicted on the right-hand side of Figure 14.1.

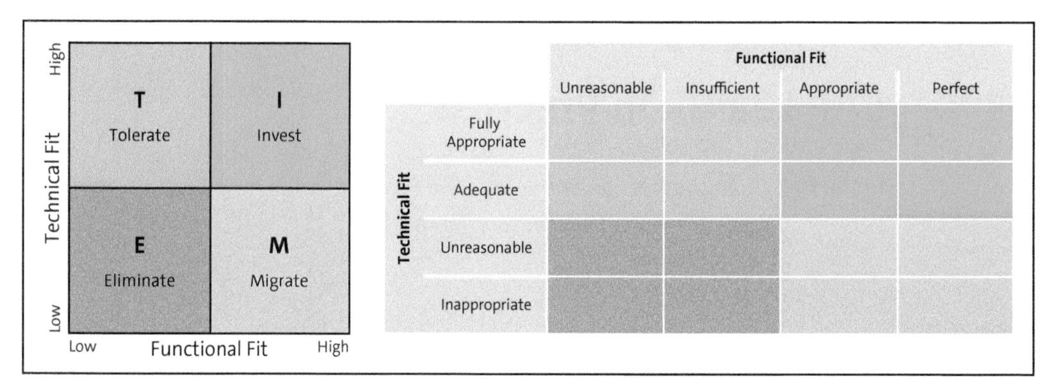

Figure 14.1 TIME Assessment

In SAP LeanIX, you can use multiple reports to implement TIME assessments. The most straightforward approach is to use a matrix report, as shown in Figure 14.2.

Figure 14.2 Detailed Matrix Report with Drilldown to Defined Successors

Here, you use the TIME attributes for a heatmapped view, while you use functional fit for the *x* axis and technical fit for the *y* axis. In the version of SAP LeanIX as of the time of writing, the attribute names (in this example, **Functional Fit** and **Technical Fit**) are unfortunately not visible in the report, which is shown on the top of Figure 14.2. They are only fully visible when you open the report settings, shown on the bottom.

You can see the more aggregated portfolio "bubble" report in Figure 14.3. The circle size in this report represents the count of the fact sheets having the respective values for **Functional Fit** and **Technical Fit**.

Figure 14.3 Portfolio Report Evaluating Functional and Technical Fit

Another way to depict the same information is to use a landscape report as configured in Figure 14.4:

- It clusters applications by **TIME Classification**.
- It shows the technical and functional fit attributes as **Left property** and **Right property** attributes.

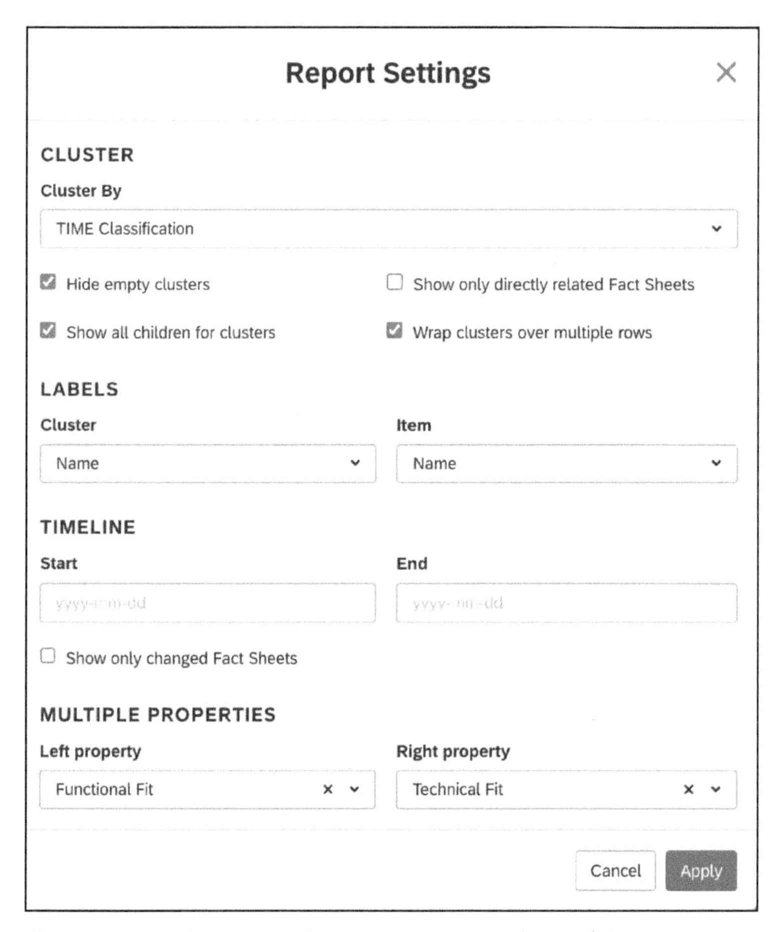

Figure 14.4 Configuration of TIME Report via Landscape/Cluster Report

This configuration leads to a report that is shown three times in Figure 14.5 and analyzes three different time horizons using the **Lifecycle** attribute for heatmapping. In Figure 14.5, the time slider at the top is dragged from Q1 2023 in the left instance to 2026 in the middle and finally to 2027 in the right-hand side report. With the time progression from left to right, you'll notice that two SAP Ariba applications grouped in the **Invest** bucket turn from **Plan** (gray) to **Active** (dark blue). Similarly, the two applications in the **Eliminate** category turn from **Active** in 2023 to **End of Life** (red) in 2026 and 2027.

Note that the **Technical Fit** as the **Right property** of each application shape is depicted as a hammer, not as a star icon, and that this is a customization in the meta model and is not report specific.

Figure 14.5 Three TIME Cluster Reports with Different Lifecycle Heatmappings Depending on Top Time Slider

14.2 Roadmap and Transformation Management

In this section, we'll further discuss the modeling of transformations, which describe one or multiple architectural changes before they get fully decided on. We'll then go through a use case step by step.

14.2.1 Use Case Description

No enterprise architecture is constant over time. This is the main reason for the lifecycle concept, which says that users can define validity periods for any given fact sheet. This section will introduce how you can model alternative transformation designs, if you don't know yet which one will be the best.

The left-hand side of Figure 14.6 shows three applications with a country association and two capabilities (demand forecasting and constraint-based planning) implemented by SAP Supply Chain Management (SAP SCM). The right-hand side depicts the target state of two applications (SAP Integrated Business Planning for Supply Chain [SAP IBP] and SAP S/4HANA Cloud).

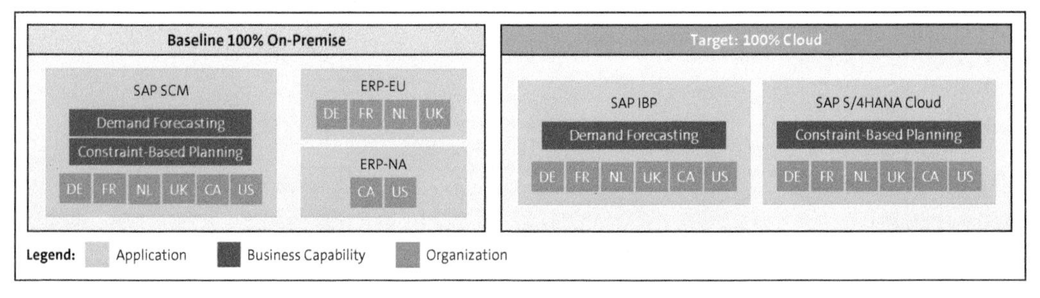

Figure 14.6 Use Case: Transformation from On-Premise to Cloud

We typically model two transformation options, as shown in Figure 14.7. In this case, we've opted for a very high level of abstraction, rather than detailed consideration. In particular, we haven't modeled detailed components that show all the activities required for a transition to SAP S/4HANA, such as business partner introduction or general ledger activations.

Scenario 1 is a conversion-based approach and consists of three phases:

- **Phase 1**
  Move forecasting from SAP SCM to SAP IBP and consolidate the two ERP systems systemically.
- **Phase 2**
  Convert the consolidated ERP system to SAP S/4HANA while running the constraint-based planning still in SAP SCM.
- **Phase 3**
  Move constraint-based planning into SAP S/4HANA.

Scenario 2 shows a greenfield approach in which SAP S/4HANA is initially implemented for just a pilot country (Canada) and subsequently rolled out to the other countries in subsequent waves:

- **Phase 1**
  Build a global template and run this in a pilot implementation just for Canada. Implement the forecasting process via an SAP S/4HANA adoption in SAP IBP.

- **Phase 2**
  Implement the United States and the Netherlands as European pilot markets.
- **Phase 3**
  Complete the rollout for Europe.

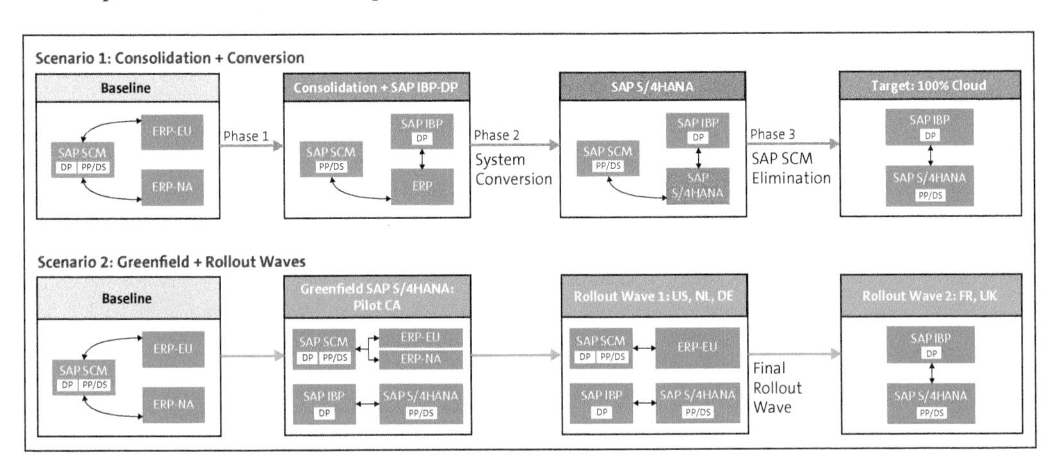

Figure 14.7 Two Alternative Transition Paths

14.2.2 Modeling the Baseline

Before describing the modeling of transformations, we'll show you how to maintain the baseline architecture. Follow these steps:

1. Create fact sheets in the inventory by doing the following:
   - Create three applications: SCM, ERP-EU, and ERP-NA.
   - Specify the active lifecycle date to be January 1, 2015, for all three applications.
   - Create the IBP-Demo Inc application, for which you specify the planned lifecycle date as January 1, 2023.
   - Create two business capabilities: demand forecasting and constraint-based planning.
   - Assign the two business capabilities to the SCM application
   - Create nine organizations: Demo Inc., Demo Inc. – Europe, Demo Inc. – NA, CA, US, DE, FR, NL, and UK.
   - Build the organizational hierarchy visible at the top of the report shown in Figure 14.8 via parent/child relations. Starting from a root fact sheet for Demo Inc., add the nodes for NA and Europe in the **Fact Sheet Dependency** section under **Children** and repeat accordingly for the country level.

2. Create a matrix report for the **Application** fact sheet type via the configuration specified on the bottom of Figure 14.8. The configuration specifies the depicted **Organizations** in columns (**X Axis**) and the **Lifecycle** field as rows (**Y Axis**). The **Drill-down** selection field specifies the relationship of the applications shown in the grid to their supported **Business Capabilities**.

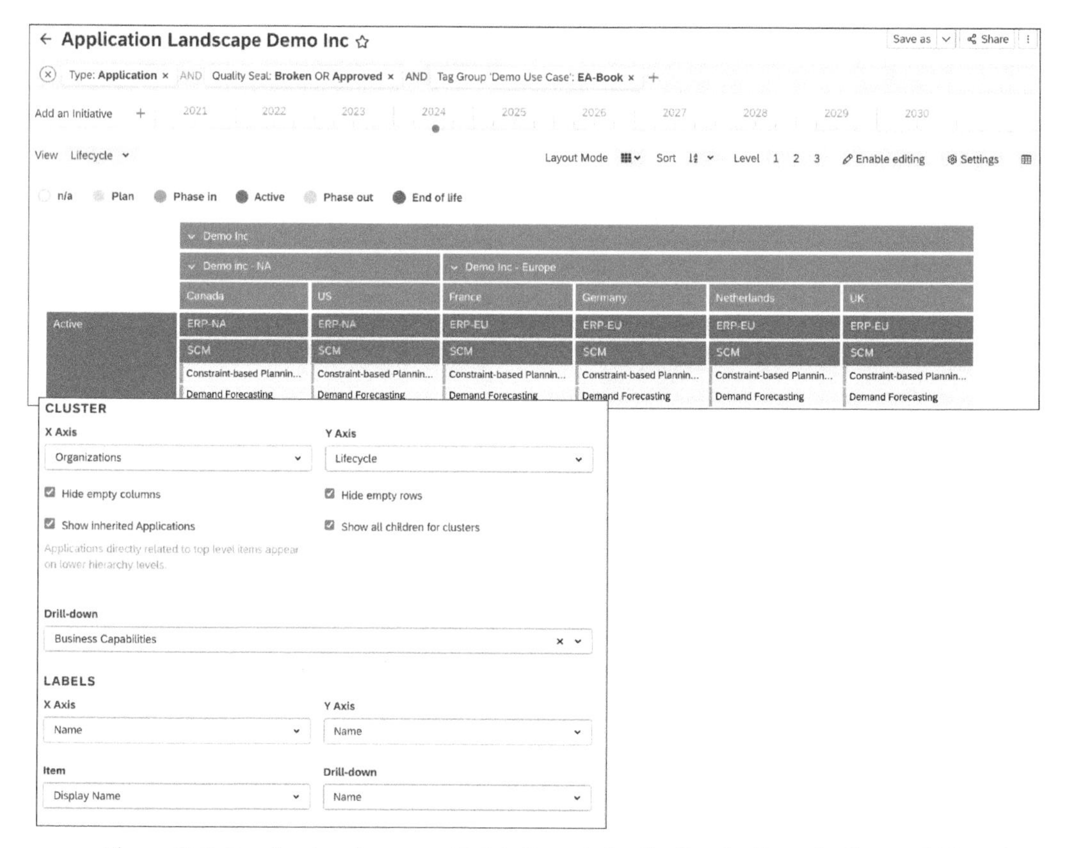

Figure 14.8 Baseline Landscape as Matrix Report: Application by Organization and Lifecycle

ERP applications typically have a long list of associated business capabilities. Therefore, you should be careful when using the relationship business capabilities as drill-down dimensions in a matrix report. This report is intended to verify the initial setup before moving on to the next step.

3. Create an application roadmap report showing the three active applications as shown in Figure 14.9. To get there, press the blue **New Report** button in the **Reports** view and select the **Roadmap** report type. In the following dialog, specify the report title and select the base **Application** fact sheet type for this report. If your workspace contains many applications, filter down to the applications created in step 1. (In our example, we used an **EA-Book** tag to do so.)

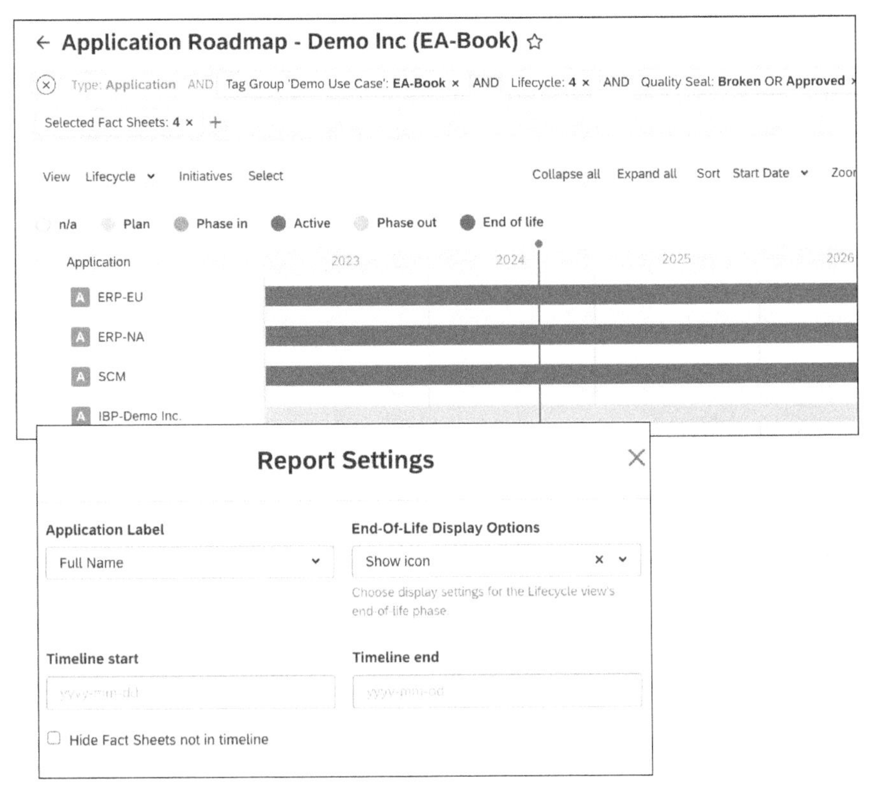

Figure 14.9 Application Roadmap

14.2.3 Creating Program Structures and Milestones

The previous section described how to model the current baseline architecture. In this section, we'll create a simple program structure under the **Demo Inc Transformation** root in the inventory, as we'll need these initiatives to model the planned transformations in the next section. We create this hierarchical initiative structure in the same way as the organizational hierarchy in the previous section.

Two alternative transformation scenarios will be in focus:

- **Scenario 1: Consolidation + conversion**
  - ERP consolidation
  - SAP S/4HANA system conversion
  - Production planning and detailed scheduling (PP/DS) migration
- **Scenario 2: Greenfield + rollout waves**
  - SAP S/4HANA greenfield template + NL pilot
  - SAP S/4HANA rollout of DE + FR
  - SAP S/4HANA rollout of UK and CA + US

In Figure 14.10, you'll see that we define all **Milestones** in the scenario-level initiatives, so you can refer to them from the lower-level initiatives. Within the scenario initiative fact sheet details, you add these milestones beneath the **Lifecycle** section. Each milestone consists of a **Name**, a **Target date**, and an optional description.

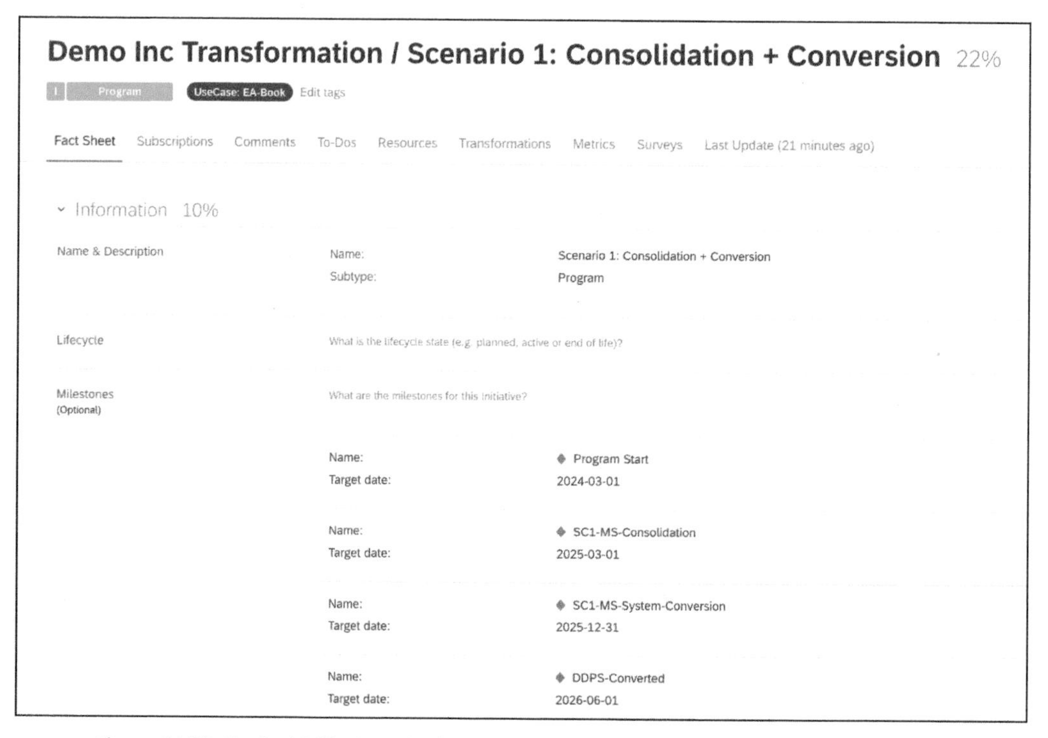

Figure 14.10 Central Milestone Maintenance

When you're defining the projects, the **Lifecycle** field can link to the milestones from the parent program, which is helpful because a multiple-child project can use (inherit) the same milestones. Figure 14.11 shows two milestone for **Active** and **End of life**, referring to two milestones from the parent program.

In parentheses after the standard SAP LeanIX names are the names of the SAP Activate phases. They are customized from the English translations in the **Administration** workspace area (see Chapter 13).

To refer to a milestone, you need to first activate the selected milestone by clicking on the blue diamond icon to the right of the **Start Date** field. The **Start Date** field the diamond is next to will switch to a milestone dropdown list.

Figure 14.12 shows a roadmap plan for the transformation process. Even though the milestones have been equipped with some semantics, this is a simple program plan without any impact on the underlying application fact sheets. Any other fact sheets like applications, business capabilities, and organizations haven't been touched yet.

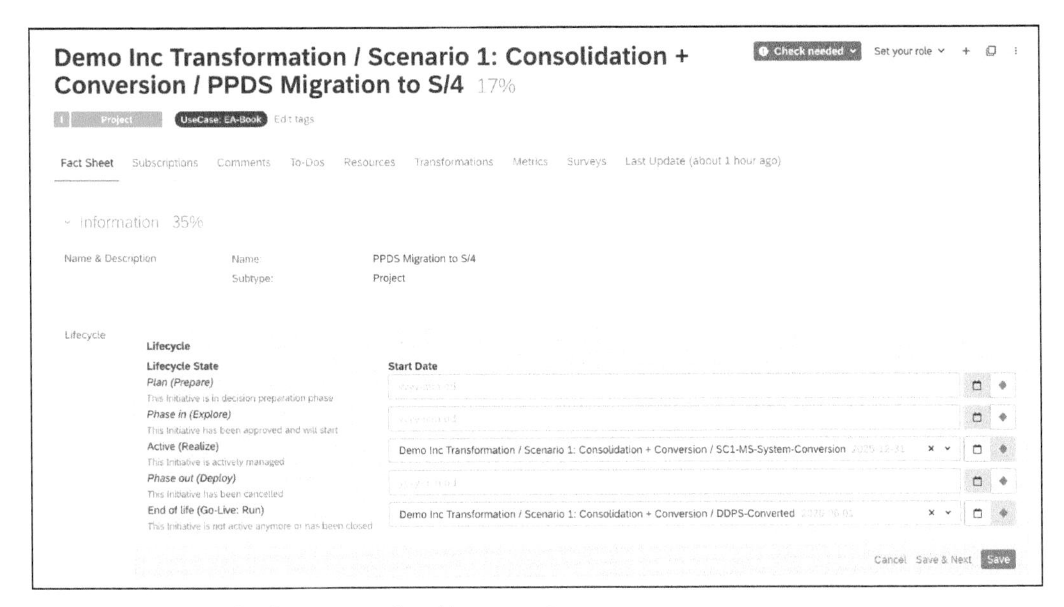

Figure 14.11 Reuse of Milestones Defined in Parent Program

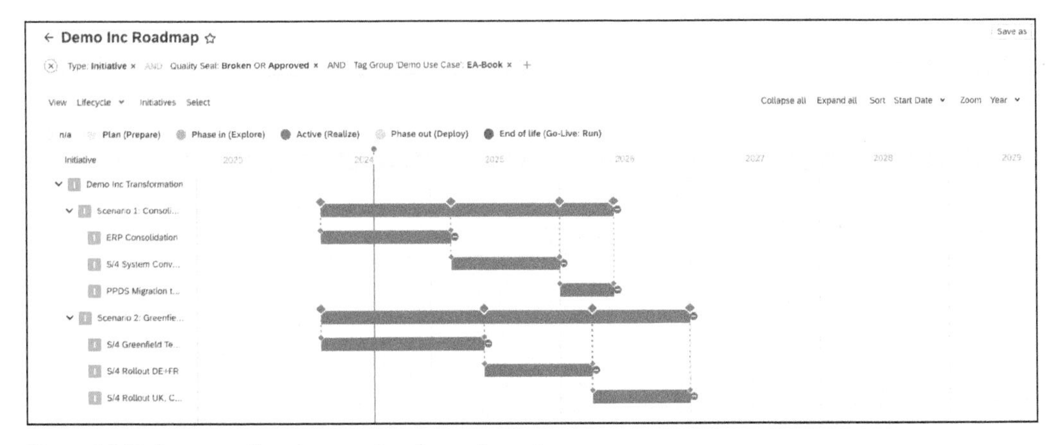

Figure 14.12 Program Structure as Roadmap Report

14.2.4 Modeling the Transformations

The next step is to model a transformation for each of the implementation projects on level 3. To do so, we'll build a transformation for each of the three subprojects from scenario 1.

Scenario 1, Phase 1

In phase 1 of this first scenario, we've modeled the consolidation instance as the decommissioning of ERP-NA and the rollout of ERP-EU to the United States and

Canada. We create the transformations from the **Transformations** tab seen in Figure 14.13, on the left side. We'll set up three parallel transformation steps.

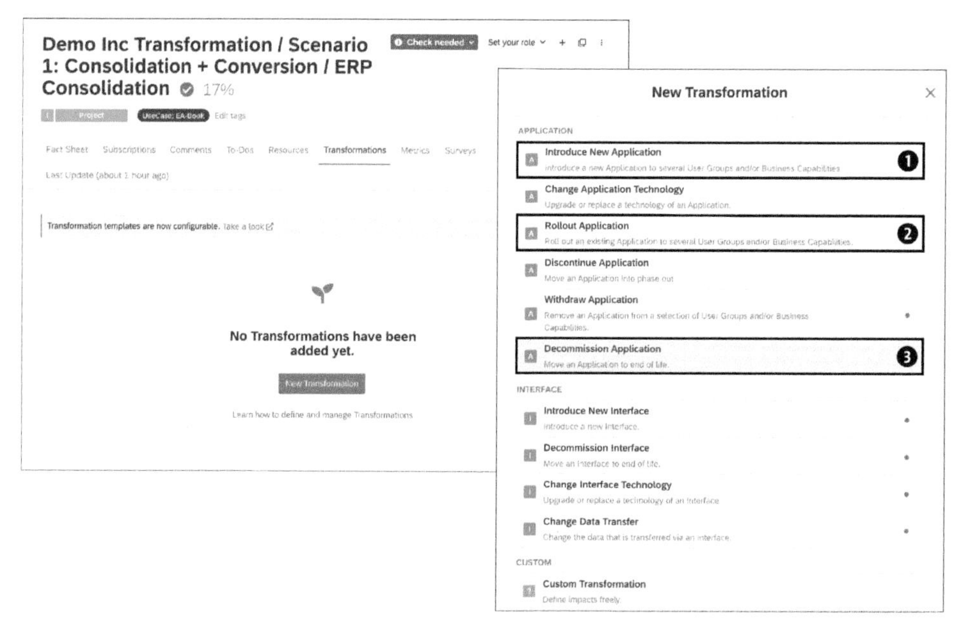

Figure 14.13 Creating Transformation from ERP Consolidation Initiative

When creating transformations from within a level 3 project, the **Completion Date** of the transformation step will automatically equal the end date of the project. As each of the three steps needs a different input, each one comes with specific UI screens dedicated to the tasks at hand. Let's walk through them (note that the callout numbers on the list refer to the numbered items in Figure 14.13):

❶ The first transformation will sunset the ERP-NA instance, as shown in Figure 14.14. Providing a meaningful description will help you and others understand the rationale of the modeling.

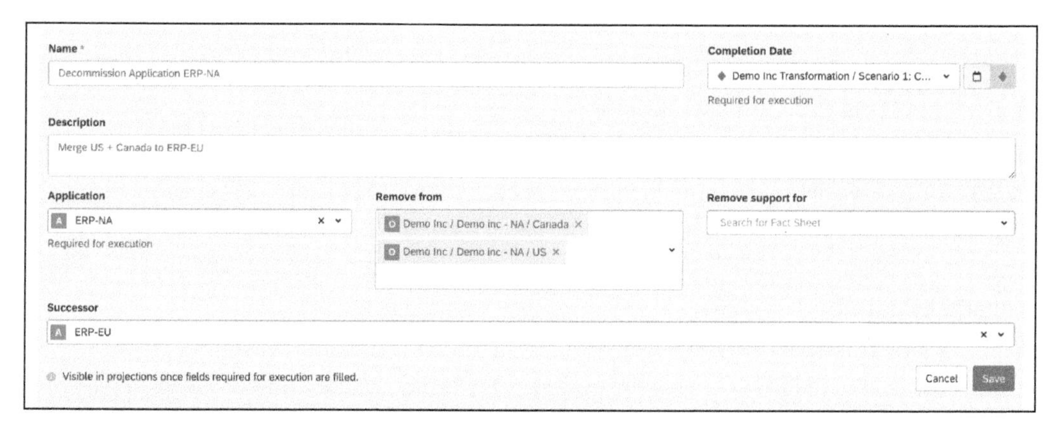

Figure 14.14 Sunsetting of NA Instance as Part of Instance Consolidation

❷ Add Canada and the United States to the European ERP-UE instance, as shown in Figure 14.15. Even though the merger might use data-based migration tools and may not be a classical rollout project, we use the **Rollout Application** template as indicated previously in Figure 14.13. We can do this because the impacts on the fact sheets for rollouts and mergers are largely the same.

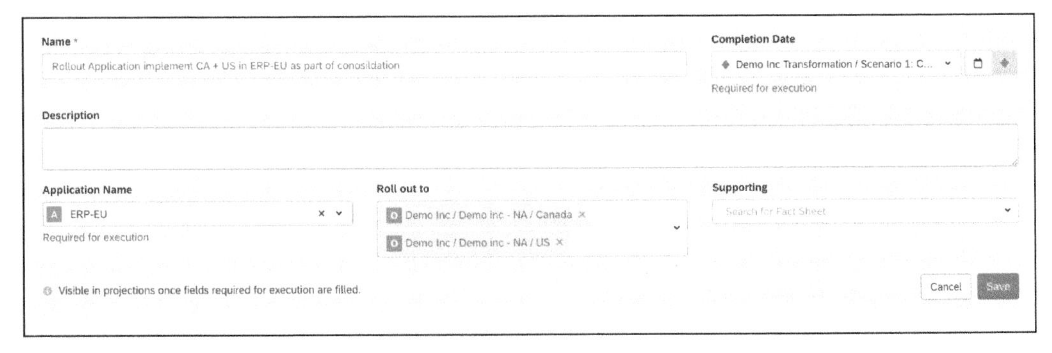

Figure 14.15 Rollout to Add Canada and United States to ERP-EU

❸ Introduce SAP IBP to replace SAP SCM, as shown in Figure 14.16:

- As in the previous steps, we name the transformation step and select the **Completion Date** from the named milestones.
- You then link from the **New application name** to the **IBP-Demo Inc** application we had created in our initial preparation step.
- Under **Introduced in**, you can specify all countries the application will be used in.
- Under **Supporting**, you can specify one or multiple business capabilities the application will support.
- The last piece to specify is the replacement of SAP SCM with SAP IBP for all regions in one step. You do this by specifying SAP SCM as the successor and by defining the **Predecessor Handling** with **Decommissioning**.

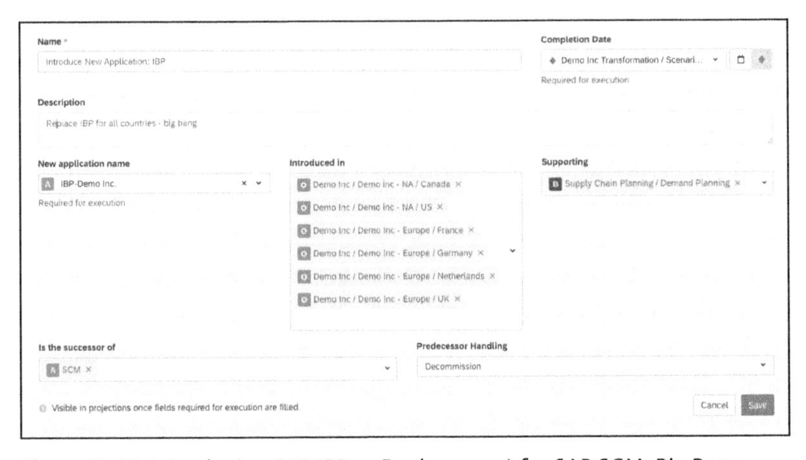

Figure 14.16 Introducing SAP IBP as Replacement for SAP SCM, Big Bang

Scenario 1, Phase 2

The second project is a big-bang system conversion. This is the least complex form of transformation modeling because we're maintaining the **ERP-EU** application and only replacing its IT component, **SAP ERP**, with **SAP S/4HANA** (see Figure 14.17).

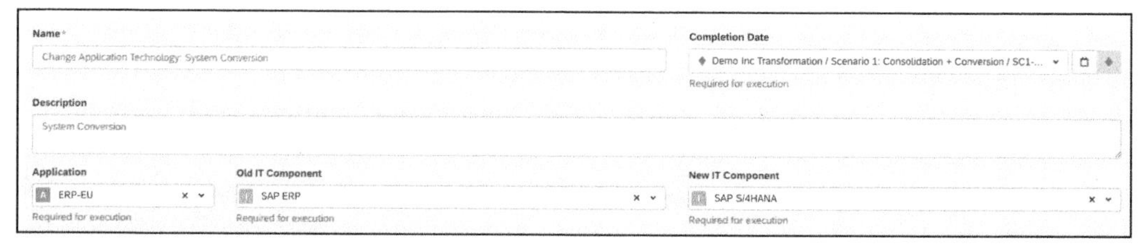

Figure 14.17 System Conversion by Way of Replacing IT Components

Scenario 1, Phase 3

Finally, PP/DS functionality moves from SAP SCM to SAP S/4HANA, as shown in Figure 14.18. Just as in the introduction step in phase 1, you need to specify a **Name** in this decommissioning step, select the relevant milestone to specify the **Completion Date**, and choose **SCM** as the **Application** to be decommissioned. Then, specify the impacted **Organizations** and the removal capability in the **Remove support for** field. Finally, name the **Successor** application as **ERP-EU** at the bottom of the dialog window.

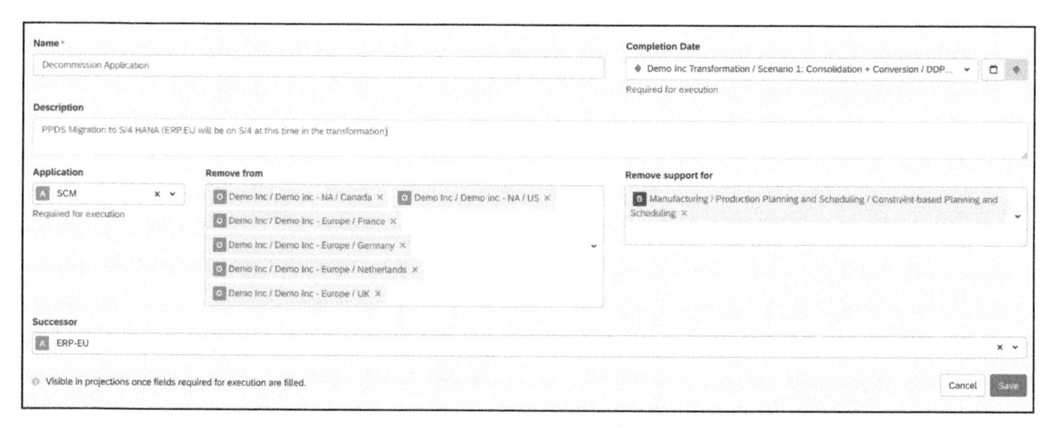

Figure 14.18 SAP SCM-PP/DS Replacement as Last Step of Decommissioning SAP SCM

Scenario 2

The second group of transformations is for the purpose of modeling scenario 2 in the form of a classical greenfield project, during which we synchronously adopt SAP IBP demand planning (DP) market by market. Scenario 2 can be configured in a comparable way, so we won't go into more detail for the purposes of this chapter.

The following sections will show you how to simulate the modeled impacts within reports.

14.2.5 Simulating Transformation in Reports

To incorporate the impacts of transformations into a report, let's go back to the application roadmap shown previously in Figure 14.9 and press the **Initiative Select** option. This brings up the selection screen shown in Figure 14.19, in which we scroll to scenario 1, select the four transformation initiatives, and confirm the dialog with the blue **Use Selected Fact Sheets** button.

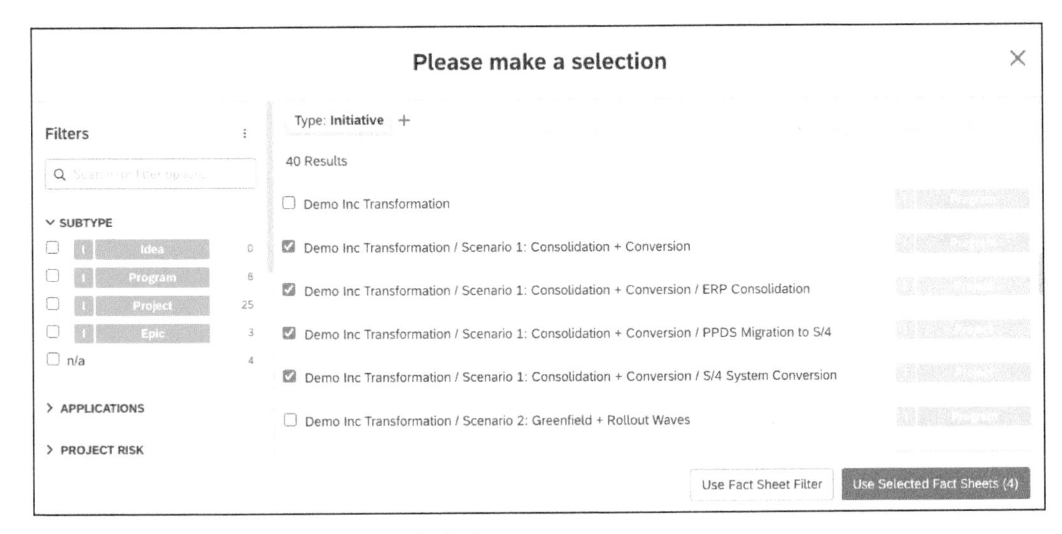

Figure 14.19 Select Initiatives to Be Included

This results in the changed roadmap shown in Figure 14.20. ERP-NA ends in 2025, the same year in which SAP IBP is introduced. SAP SCM will run until 2026, when we'll have moved PP/DS from SAP SCM in ERP-EU.

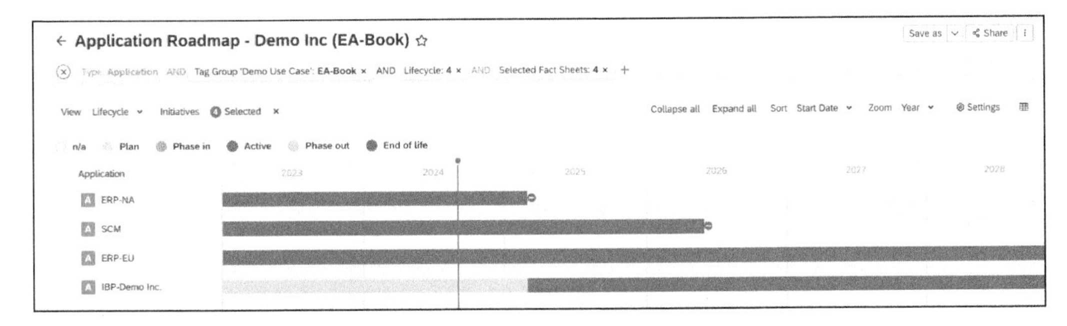

Figure 14.20 Simulation of Scenario 1

You might like to rename the **ERP-EU** system as **S/4-Global,** as shown in Figure 14.21. Here, we've used an additional custom transformation that can be employed to manipulate any attribute or relationship possible.

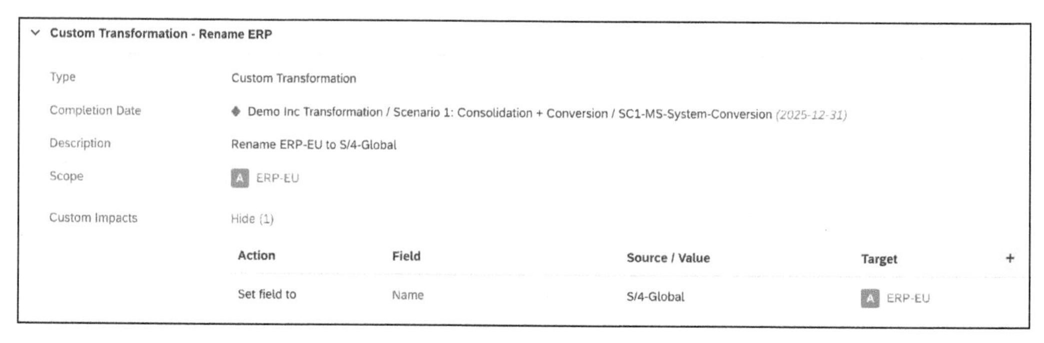

Figure 14.21 Custom Transformation

Once you've selected an initiative to include in the roadmap, you can spot the milestones of that initiative so that you can easily time travel to the interesting change events.

14.2.6 Executing Transformations

Once your decisions are final, you can execute transformations as shown in Figure 14.22.

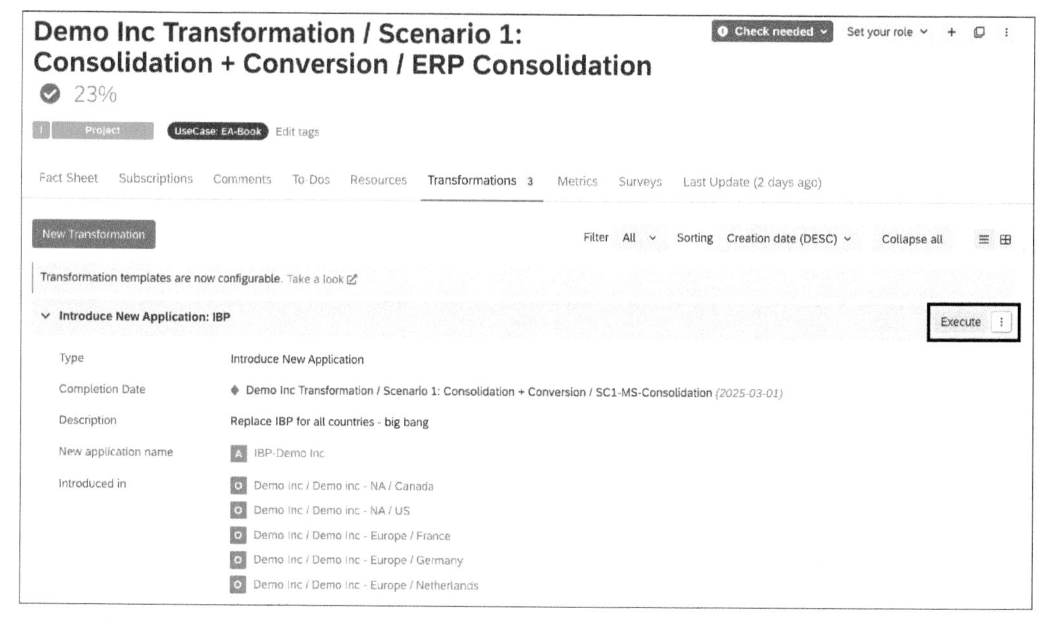

Figure 14.22 Execution of Transformation

Note that you can only execute them once. If you've modeled alternative scenarios, make sure that you only execute one of them, as an additive execution of alternative scenarios will likely lead to unwanted difficulties in predicting and correcting the changes in the fact sheet model.

You also need to take care to execute all transformations that you had modeled together as one logical group. Execution of only parts of a logical group of transformations from the same initiative will lead to unpredictability similar to that from executing conflicting transformations from different initiatives.

Transformations form an advanced but very powerful topic in SAP LeanIX. You can define custom transformation templates that can accelerate the standard use of transformations.

14.3 Clean Core Modeling

Another focal point in SAP-centric transformation initiatives is clean core, or the idea of performing ERP system upgrades of SAP S/4HANA (which is now an event) by automating test efforts and only allowing upgrade-friendly customizations.

The dimensions of the clean core are as follows:

- Process
- Data
- Integration
- Extensibility
- Operations

In this section, we'll only focus on the topic of extensibility. In old ERP systems, the number of custom code objects can be extremely high—and as in enterprise architecture, where a complex application portfolio is managed, a successful clean core program manages one or multiple custom code portfolios.

As dealing with each single custom code program is not scalable, SAP has developed an intelligent custom management solution (ICCM). The outcome of an ICCM analysis is a list of custom code clusters.

Before you choose a way forward, you should know that different assessment techniques can come into play. As SAP LeanIX has these techniques at hand, you can use it actively. Figure 14.23 uses color-coded dots to illustrate custom programs, and ICCM can cluster them to larger clusters. You can then assess each of these clusters and categorize them in terms of the action to take. For those actions that have enterprise architecture impacts, SAP LeanIX is the perfect tool to model these impacts via the transformation concept, which we discussed in the previous section.

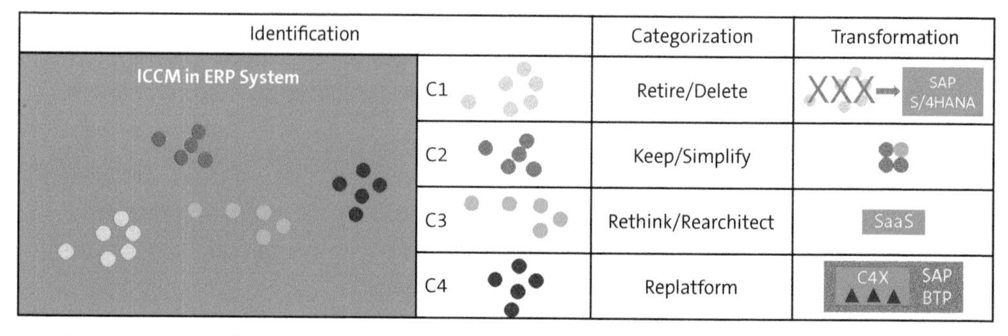

Figure 14.23 Intelligent Custom Code Management

We recommend that you model custom code clusters as IT components and link them to their consuming applications.

Some of the valid lifecycle and technical fit value combinations for IT components are shown in Table 14.1.

Action	Lifecycle	Technical Fit
Retire/delete	End of life	Inappropriate
Keep/simplify	Active	Adequate
Rethink	Phase out	Unreasonable
Replatform	Phase out	Unreasonable

Table 14.1 Lifecycle and Technical Fit

For the rethink action, the architect will need to seek the best solution. For the replatform action, the architect will need to clarify whether SAP Business Technology Platform (SAP BTP) will be the right target platform; they can use enterprise architecture principles to guide their decision.

Figure 14.24 continues the abstract example, in which you can see the following:

- C1 will be marked for deletion.
- C2 is the only cluster that will stay untouched.
- C3 can possibly be replaced by a new SaaS solution.
- C4 will be imported into SAP BTP.

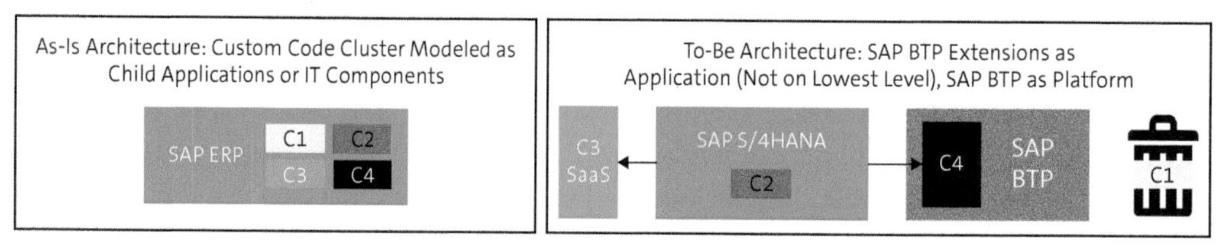

Figure 14.24 Clean Core Modeling Example

14.4 Summary

In this chapter, we looked into three main use cases to showcase SAP LeanIX in practice:

- Application portfolio management
- Transformation planning
- Clean core modeling

With that, we've closed our coverage of SAP LeanIX. Now, we move on to Part IV of the book, in which we'll focus on enterprise architecture practices. We'll start with information on how to set up your own practice.

PART IV

Enterprise Architecture Practice

An enterprise architecture practice is the organizational implementation of enterprise architecture. In this part, we'll provide guidance on how to set up and improve an enterprise architecture practice. While much of our discussion is independent of SAP, we'll include aspects that bring SAP and non-SAP subpractices together. We'll also explore the different independent, yet connected, enterprise architecture practices run by SAP.

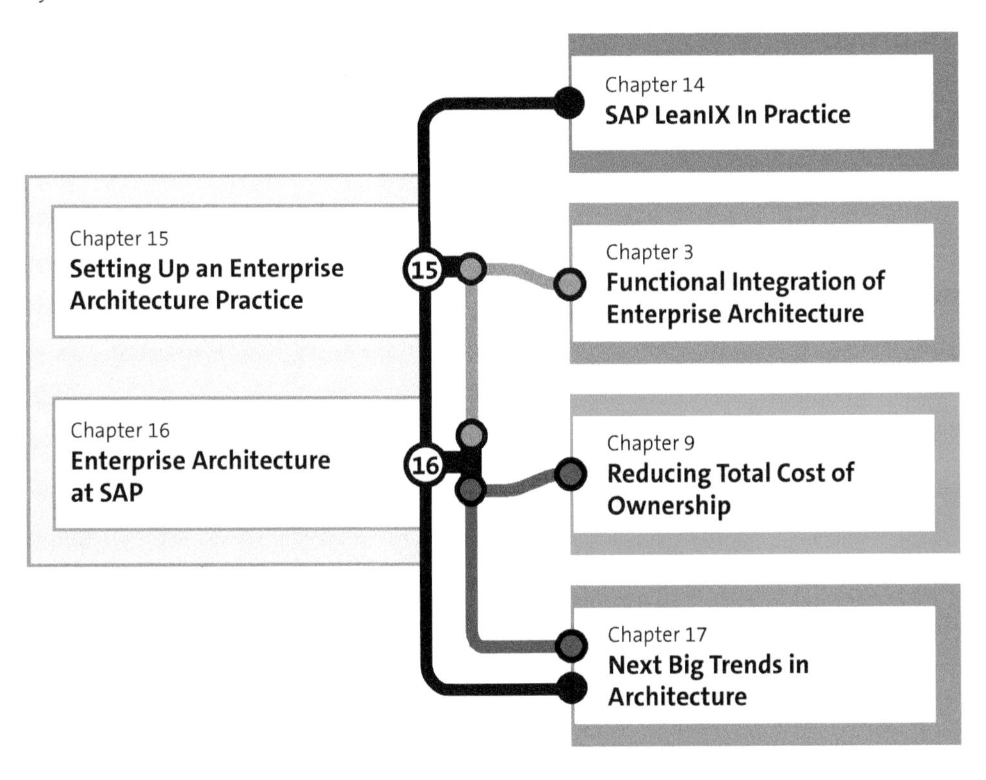

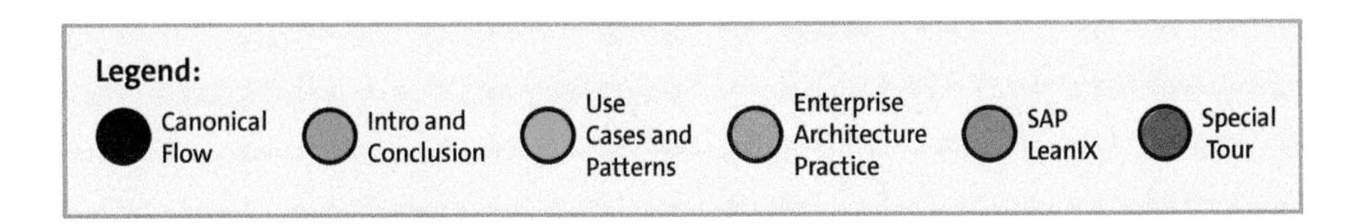

Chapter 15

Setting Up an Enterprise Architecture Practice

An enterprise architecture practice is the organizational implementation of enterprise architecture. It operates enterprise architecture within the organization by adopting methodology and establishing governance processes. How to best put enterprise architecture into practice and improve it depends on the organizational context, the management support, and the desired outcome.

Part II described the *what* of enterprise architecture by looking into its methodology and five different use cases. This chapter investigates the *how*: how to run, set up, or improve an enterprise architecture practice. We define a practice in an organizational context, it can be implicit or follow more strict organizational accountabilities and processes. You'll learn how such a practice can be anchored in the organization: as a dedicated organization, as federated groups, or implicitly, as part of initiatives. Besides the organizational aspects, you'll learn to define the outcomes/deliverables expected from this function. We'll also discuss assessing the maturity of the practice to start planning its progression.

To ensure that a practice functions well, we often use the term *governance* to describe a set of defined rules, processes, and standards on how to run the practice. A practice without an explicit organization and without a governance structure will typically have low maturity and rely on informal alignments. If the primary focus of governance is to ensure proper decision structures, it could result in a "policing" culture and might not lead to the best outcome either. This leads to the insight that the objective needs to be to find a proper balance between an open collaborative approach and a given set of governance rules.

This chapter is laid out as follows:

1. We first provide a definition of *enterprise architecture practice* and look at two extremes: a fully "implicit" enterprise architecture with an underlying proactive approach and an enterprise architecture that is run as a central organizational unit under the chief information officer (CIO)'s office.
2. We then introduce a maturity model and explain why that the highest level of maturity is not always desirable.
3. We then look at organizational aspects and processes.

The next chapters will be dedicated to two explicit examples that are similar to the use cases in Part II.

15.1 Enterprise Architecture Practice Overview

The term *enterprise architecture practice* was already introduced in Chapter 2, Section 2.4, as one of the five elements of SAP EA Framework. Defining *practice* as an explicit dimension underlines how important it is to not only have a good methodology and tools in place but also to explicitly care about how enterprise architecture work is run within an organization.

While many companies might agree on methodology and tools, the implementation of their practice may range from one that primarily focuses on technology and infrastructure to one that emphasizes applications and processes. We also see major differences in how interconnected enterprise architecture teams work and what relevance and visibility these teams have.

Many enterprise architecture techniques can be applied when defining the enterprise architecture practice. For example, The Open Group Architecture Framework (TOGAF) contains "the TOGAF Leader's Guide to Establishing and Evolving an EA Capability" (see *http://s-prs.co/v586310*). This guide follows TOGAF's own Architecture Development Method (ADM) cycle when establishing an enterprise architecture function in an organization, and we would recommend partially following this procedure. Like the Open Group IT4IT reference architecture for managing the business of IT, this would form an *enterprise architecture for enterprise architecture* (EA4EA) approach. But the nature of an enterprise architecture practice is distinct from running an enterprise architecture transformation program, so we'll limit our attention to this EA4EA thought model to the following deliverables that are key to building the enterprise architecture practice:

- **Stakeholder map**
  Who is interested in and will benefit from the enterprise architecture practice? How do we engage with the business, who are the decision makers, and what is the overarching company culture?
- **Capabilities**
  What are related functions that are required within the practice, and what are the services they offer?
- **Processes**
  How will the practice work in collaboration with other parts of the organization?
- **Data**
  Which data is collected, processed, and stored?
- **Applications and technologies**
  What enterprise architecture tools and linked repositories are needed?

- **Transition planning roadmap**
  How do we define a clearly staged adoption and improvement plan?

Small and midsize companies that operate in a single country and follow a single business model can often be organized without an explicit enterprise architecture function. If the current enterprise architecture is in good shape, having an implicit shared responsibility within the organization can be sufficient. We still recommend that key stakeholders come together and document their implicit principles and architecture approaches.

As an organization grows, it may need to specialize and/or duplicate within one or multiple dimensions. Being aware of these dimensions is important because they often are the root cause of more complexity.

Figure 15.1 illustrates three of these dimensions: geography (in columns), divisions (in rows), and functions (designated by color).

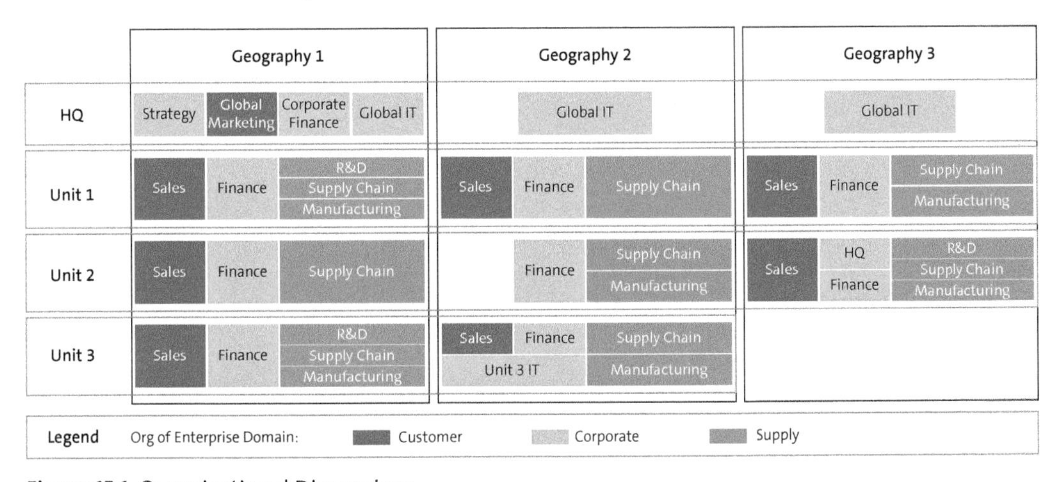

Figure 15.1 Organizational Dimensions

We'll now assess how these four and additional dimensions can impact how to best organize an enterprise architecture practice:

- **Geography**
  The spread of the business to multiple regions doesn't imply that the enterprise architecture function should be spread in the same way. What's most important is to understand the current and the desired business as well as the IT operating model: if a company acts globally with a single business model, then typically, a single IT organization will be sufficient to enable the business. Hence, such a global IT organization should run only one enterprise architecture practice. Team members of that practice can nevertheless be distributed among multiple regions, as regional initiatives may benefit from an equally regional presence.

Typical factors that relate to the geographical dimension are time zones, legal requirements, skills available, and differing cost structures. All these factors might trigger the cross-regional spread of a function.

Some companies have brought together different geographical areas through acquisitions, and their business and IT operating models may differ and therefore require different target architectures. See Chapter 8 on mergers and acquisitions (M&A), where we describe the magnitude of required changes, primarily with respect to the type of synergies expected. When harmonization needs are high, it's beneficial to unite the architecture function early, as it can play an instrumental role. Often, the strategy is to assimilate the acquired part of the business—but key stakeholders may be tempted to protect the current state and therefore postulate that no architecture work is needed.

If the enterprise architecture practices are distributed and organized according to companies' evolution, then the risk is that different parts and regions of the organization will oversee convergence needs.

- **Size of the business/company**

  Typically, the number of employees is a first good indicator of the company's complexity. However, this can vary from industry to industry.

  Small and midsize companies may tend to think they don't need an enterprise architecture practice, and growing ones might recognize the need only if serious problems occur. We recommend making enterprise architecture part of overall IT governance when complexity is still low. The effort required to set up a small enterprise architecture practice is low, and aligning enterprise architecture with other functions as described in Chapter 3 can lay a good foundation for success.

  If a company grows by acquiring another large company, then immediate, intense enterprise architecture work is needed, as discussed in the earlier paragraph on geographies and in the use case in Chapter 8.

- **Lines of business**

  If different lines of business follow different business models or operate in different industries, then it implies an explicit instance of differing functions. In this case, synergies among different units may be limited. These limited synergies will need to be reflected in the real architectures as well as when setting up a common or distributed enterprise architecture practice.

- **Business functions**

  Business functions integrate with one another through end-to-end business processes. We therefore recommend not splitting up enterprise architecture practices by business function.

  While the idea of enterprise architecture is to apply a common approach to all functions, the business and IT strategy might look different from function to function. In Section 15.9.1, we'll analyze a use case in which e-commerce is seen as a separate business function.

The enterprise architecture practice needs to be aware of specific functional needs like local and regional requirements in finance and HR, as both typically face specific local regulations. But these needs still don't justify splitting off the enterprise architecture function.

- **Application platforms**
  This can be the typical split of teams between an SAP and a non-SAP setup. We discourage using the application platform as a dimension for setting up enterprise architecture practices, as this can lead to competitive fights and can negatively impact technology selection.

15.2 Enterprise Architecture Maturity

You can easily apply the five-level maturity model depicted in Figure 15.2 to the enterprise architecture function. Later, we'll discuss how difficult it is to reach maturity level 4 (Measured) by introducing quantified enterprise architecture measures later. This is why many companies only reach level 3 (Defined).

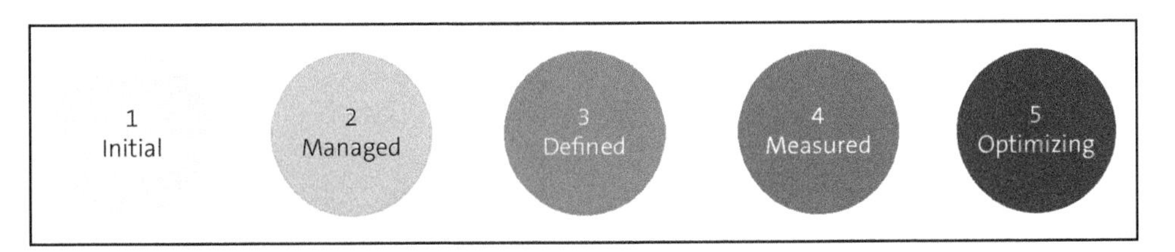

Figure 15.2 Five-Level Maturity Model

To get to a more differentiated view of enterprise architecture maturity, it makes sense to not only assess the practice from the top down but also to drill into domain-specific perspectives.

Collecting unbiased input is important if you want to have meaningful outcomes. The reasons to run a maturity assessment can be multifold, but they typically include the following basic elements:

- Understanding the current state of the practice.
- Benchmarking with other organizations. To do this successfully, you need to hire an independent organization to collect assessments and run anonymized benchmarking.
- Defining target maturity level progressions over time.
- Deriving improvement plans to get to the agreed upon target levels and reassessing them after a given period of time. The derived actions can also function as input for budget planning.

But let's start with a top-down description of the maturity levels. During an assessment, you should use different qualification criteria (as depicted in Section 15.3) to gain more actionable results:

1. **Initial**
   Enterprise architecture efforts of this maturity level may sometimes be executed by multiple teams working in isolation. But if the coordinating organization fails to come up with an agreed upon working model and defined processes, outcomes will often be incomplete and nonrepeatable. This can lead to unaligned business and IT strategies as well as nonscaling and partially redundant architectures.

2. **Managed**
   At this level of maturity, enterprise architecture work is only done when needed, which is both a strength and a weakness. The various groups and projects run their own architecture work differently, and the products therefore don't fit together when the need to collaborate arises. This will often lead to extra alignment work or the need for reworks further down the line. Without clear definitions, collaboration with other related functions of the organization can only be informal, and the synergies among these functions as described in Chapter 3 can't be realized. The result is that enterprise architecture work will have limited impact.

3. **Defined**
   At this level, the enterprise architecture practice is established, and the enterprise architects work within a defined operating model that specifies the agreed-upon roles and organizational setup, the processes that need to be executed, the tools that are required, etc. This enumeration mirrors the structure of the entire book, which makes sense because both center on the definition of an enterprise architecture practice.

   In following sections, we'll see that having a maturity level of Defined doesn't necessarily equal having one single global practice.

4. **Measured**
   At this maturity level, enterprise architecture's key performance indicators (KPIs) are defined and can be measured. Ideally, they also cover multiple domains. Measurement results say something about the quality of an architecture (e.g., the number of redundant applications) or are process measures that assess the quality of the enterprise architecture development and governance processes (e.g., the number of statements of enterprise architecture work, the number of enterprise architecture decision records).

5. **Optimizing**
   At the highest level of maturity, the company derives optimizing steps from the quantified measures' results. As mentioned before, defining relevant and easily quantifiable architecture measurements is not easy. A first step can be to focus on

process measurement, possibly aligned with related functions. During budget portfolio planning, for example, the enterprise architects running each major initiative would be required to prepare an architecture vision document.

Real quantitative measurements could be linked to reducing a landscape's complexity (e.g., by paring down integration complexity or the number of legacy applications). Other quantifiables are implementation and ongoing maintenance costs or changes in projected costs and benefits—but it can be hard to quantify those.

Our recommendation is therefore to focus less on isolated enterprise architecture measures and more on those with an overall benefit for the company.

You should not base your determination of your organization's maturity level solely on the internal judgement of your architecture team. You should consider input from multiple stakeholders.

An organization should also take care not to give rise to unrealistic expectations as to the speed with which an enterprise architecture's maturity will progress. Moving from Initial to Measured, as shown in the example in the first row of Figure 15.3, typically involves multiple improvement steps over possibly multiple years.

Practice Dimension	Initial	Managed	Defined	Measured	Optimizing
Business and IT Alignment	As-Is →	→	→	To-Be	
Stakeholder Involvement	As-Is →	→	To-Be ⋯→		
Action and Impact	As-Is →	→	To-Be		
Architecture Development		As-Is →	To-Be		
Architecture Process		As-Is →	To-Be ⋯→		
Organization and Governance			As-Is+To-Be		
Enablement	As-Is →	→	To-Be		
Community	As-Is →	To-Be ⋯→			

Figure 15.3 Progression Plan Covering Multiple Maturity Levels

Instead of aiming for perfection, which would require an unrealistic amount of time and resources, organizations should carefully assess and plan their enterprise architecture's progression.

In companies where SAP and non-SAP areas operate independently, SAP-oriented organizations may only have an implicit enterprise architecture practice. One reason for this pattern is the predefined SAP guidance and/or best practices, which entail less architecture-related decisions. A less mature, implicit enterprise architecture practice in the SAP space may, however, be as successful as one operated in a best-of-breed/non-SAP solution setting.

Our recommendation is to strive for a higher maturity level when running an SAP-centric solution landscape. The reasons for this are the following:

- One driver of establishing a common enterprise architecture practice is that it helps bridge the gap between business and IT. Since the early days of SAP R/3, SAP has been providing reference process models. The event process chains of the past have evolved into the current, "real" Business Process Model and Notation (BPMN) models for end-to-end processes, and detailed best practice contexts are available for fit to standard analysis in implementation projects. Getting business access involved through business processes is extremely helpful, but a higher level of abstraction should be sufficient in enterprise architecture. Functional capability models and nested process value flows have been included in the SAP enterprise architecture reference content. With these, SAP now propagates using the process perspective as a possible alternative starting point for enterprise architecture.
- SAP solutions never run in isolation. Integrating them with other external and internal solutions should be one of the foci of the architecture practice.
- Aside from public cloud solutions, SAP solutions follow a typical lifecycle, as all non-SAP components do. Managing a healthy application portfolio therefore is equally beneficial to SAP and non-SAP areas.
- Historically, an SAP-centric architecture often meant that mid- to large-size companies operated an SAP ERP system and only peripherally complemented that with specialized solutions. Modern SAP-centric landscapes not only complement the core ERP with specialized functional cloud solutions for supply chain, procurement, and human resources (for example), but they also make use of SAP Business Technology Platform (SAP BTP). Having a well-architected, company-wide platform strategy is another good reason for stepping up enterprise architecture maturities that align SAP and non-SAP areas.

15.3 Organizational Setup

Let's start with a provocative statement: there's no ideal positioning for the enterprise architecture function in an organization. In this section, we'll look at a range of such positionings, and we'll show that the answer to the question of which is best suited for hosting the enterprise architecture function is not a clear and simple one.

Let's walk through the major enterprise architecture positioning options:

- **Linked to business and IT versus linked to IT only**
  - *Linked to business and IT*: The enterprise architecture function is a combined (virtual) function that establishes an architecture board or practice that includes key thought leaders and stakeholders. Such a function can be very powerful, and to ensure the success of such a hybrid setup, decision power and delivery capacity (in terms of full-time equivalents or budgets) should be granted.
  - *Linked to IT only*: Despite the undisputed fact that a major role of enterprise architecture is the alignment between business and IT, enterprise architecture teams

are typically located with the IT organization. The missing linkage to business thus needs to be mitigated, but forming an explicit enterprise architecture team is probably easier in this type of setup.

- **Part of the project management office (PMO) versus separate from the PMO**
  - *Part of the PMO*: Changes will be project driven, so enterprise architecture will be primarily driven from a shorter-term transformation project perspective.
  - *Separate from the PMO*: Having enterprise architects do their work outside of a project perspective will keep a focus on midterm and long-term company objectives as well as on possible short-term project criteria.
- **Part of the C-level versus not exposed to the C-level**
  - *Part of the C-level (e.g., the office of the CTO or CIO)*: Placing enterprise architecture in line with the organization's C-level, where major investment decisions and other strategic decisions are typically made, can give enterprise architecture a stronger backing within the organization and an especially strong integration with the overall portfolio, plus smoother strategic funding processes.
  - *Not exposed to the C-level*: C-level offices can be far removed from the realities of IT project work, and architects who are positioned at the C-level can easily be perceived as disconnected from the organization and operating in an ivory tower. It's therefore essential that a team that's anchored at the C-level should demonstrate its value to many initiatives.
- **Permanent team versus rotating assignments**
  - *Permanent team*: Collecting enterprise architecture experience takes time, and skill sets will improve over longer periods. It's often stated that experienced enterprise architects should have a job history of more than fifteen years, with at least five of those years in enterprise architecture. Clearly, the best way to aggregate long-lasting enterprise architecture expertise is to establish a long-lasting fixed team.
  - *Rotating assignments*: Rotating assignments bring fresh experiences and perspectives to the architecture team, and if architecture team members rotate back into their functional or technical teams, they'll naturally inject architectural thinking into the rest of the organization over time.
- **Organizational unit versus virtual team**
  - *Organizational unit*: Leadership and budgets can be assigned and named easily. Conversely, members of a virtual team will often have a conflict of interest due to dual accountabilities and lines of reporting. Also, higher priorities from their primary organizations will impact their assigned enterprise architecture work.
  - *Virtual team*: In virtual teams, all relevant units of the organization can collaborate and contribute the required expertise. The organization responsible for building and dispatching the virtual team will feel more involved and might

therefore support a virtual team's decisions more readily than a fixed central team's decisions.

- **Co-sourcing (selective outsourcing) versus complete in-housing**
  - *Co-sourcing (selective outsourcing)*: The skills needed in an enterprise architecture team are typically very broad, and it can be a challenge for a relatively small team of architects to amass such skills while trying to stay on top of all relevant changes. While it's recommendable to keep final decisions in house, it can be very beneficial to take systems integrators or key software partners such as SAP on board when driving larger transformations, or it can be helpful to engage them in the ongoing governance. This is even more the case if the internal architecture function is small and not very mature.

    SAP services, for example, can help build target architectures and roadmaps and provide superior insights into how to best leverage SAP's solutions to achieve more value with a lower total cost of ownership. As this book is written with the SAP service perspective in mind, let's list a few ingredients of SAP enterprise architecture services (which should be familiar from our discussion throughout the book):

    - Use of the SAP Reference Business Architecture content
    - Experience from multiple customer engagements
    - Access to SAP's strategy and development units
    - Use of transparent, decision-enabling architecture methodology

    Using SAP as enabler makes particular sense if it has been selected as strategic partner.
  - *Complete in-housing*: Co-sourcing strategic tasks needs to be done with care, and final decision-making power should always remain within the company.

The assessment of each of an organization's principles will largely depend on the overall principles a company uses when forming its organization. These principles—along with the more general company culture—can vary, ranging from a centralized top-down approach to a highly federated and consensus-driven one. Trying to introduce enterprise architecture via an organizational model that is incompatible with the rest of the organization will probably lead to low acceptance or even failure.

For large organizations, the organizational decision criteria above can be assessed globally or in decentralized fashion, by organizational structures that could be regions, business units, or both.

Before moving on to the positioning of enterprise architecture within an organization, let's look at the following three organizational IT setups in Figure 15.4. Note that many very large organizations can run combined and nested versions of these:

- **Decentralized IT**
  In these setups, IT is primarily run in different entities organized either by division

or by region. Often IT is broken into clusters that manage services and solutions independently as products. Product ownership often comes with more freedom to make architecture decisions. In Section 15.9.1, we'll see an example of a commerce platform that's run independently of most of the other enterprise architecture work, which is conducted from within the decentralized sales department.

A product-oriented form of IT that we often see is one in which SAP solutions are managed within a company's specific SAP organization. There are many good reasons for such a setting, such as the required skill sets. But we also see clear downsides when an SAP architecture is managed in a manner that is totally independent from a non-SAP enterprise architecture. Often, such compartmentalizations have evolutionary causes, for example, when they go back to a set of SAP ERP systems as standard software that was managed in isolation. In a modern SAP-centric architecture, the ERP systems are part of the end-to-end value chain, and it is crucial to integrate them with external business partner systems and multiple non-SAP solutions.

Table 15.1 shows the dos and don'ts of a decentralized enterprise architecture practice.

Dos	Don'ts
▪ Collaborate on common enterprise architecture principles and standards. ▪ Assess whether a common enterprise architecture methodology should be adopted. Allow for adaptation by unit. ▪ Leverage a common enterprise architecture repository (with possibly independent tenants per unit). ▪ Identify critical initiatives that require cross-unit collaboration.	▪ Don't force a single enterprise architecture practice. ▪ Don't force a cross-unit decision board, if leaders of the units run their businesses fully independently.

Table 15.1 Dos and Don'ts of Alignment among Enterprise Architecture Practices in Decentralized IT Organizations

- **Hybrid/orchestrated IT**
  Which functions are centralized and how much autonomy is given to divisions, regions, and functions can vary a lot. The degree of centralization should typically match the need of the business operating model: if, for example, the supply chain runs globally and overarches the regions, it won't make sense to organize its IT by regions. It would rather make sense to run the project portfolio management and operations of global IT solutions centrally.
- **Central IT**
  Smaller companies and those running a standardized global business model will often opt for a centralized IT organization. Such organizations typically need less

governance and alignment layers, but in very large but still centrally managed organizations, it can be difficult to act in an agile way.

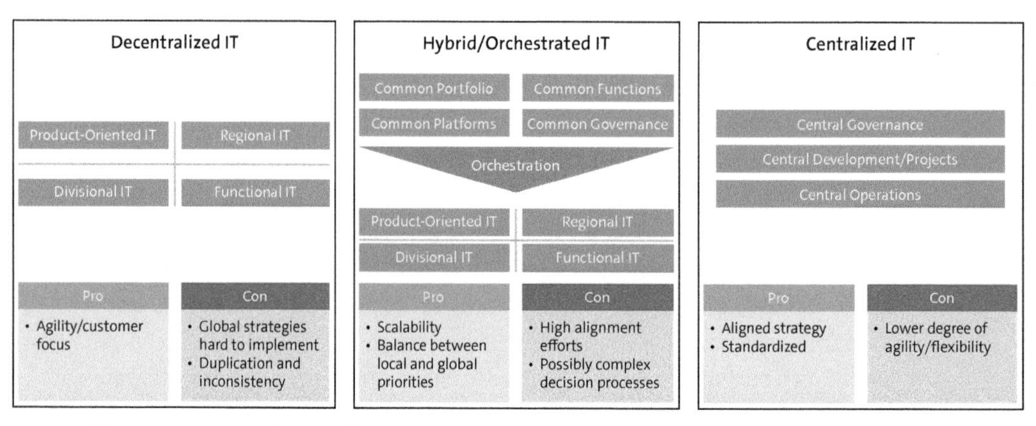

Figure 15.4 Organizational IT Setup

In any of the three organizational models that we've described, enterprise architecture might be run implicitly, without an explicit team or named global owner of the enterprise architecture function. In such a setup, IT leaders will need to have a good implicit architectural mindset and incorporate that into their strategic planning as well as their operational governance.

When coming from an implicit function, our recommendation is clearly to make the enterprise architecture function explicit by doing the following:

- Building a case for the value of the enterprise architecture function, with clear objectives and outcomes
- Aligning the function with the strategic transformations
- Starting small but requesting business support and collaboration from the beginning

We'll discuss two use cases in Section 15.9, which shows the possible setup and evolution of an enterprise architecture function.

If your IT runs in a decentralized fashion, we don't recommend trying to change the organization via an "isolated" global enterprise architecture program. Having separate architecture practices and establishing collaboration between them will have a much greater impact. Such a collaboration should focus on synergy areas and should not result in a battle for control.

Figure 15.5 shows the different options for placing an enterprise architecture team in the organization, depending on the overarching organizational layout:

- **Informal**
  There are no dedicated enterprise architecture teams, but the enterprise architecture profession and skills are integrated into other roles within the organization.
- **Isolated/siloed**
  If organizational units operate in a more local mode, then this separated model would be appropriate.
- **Federated/distributed**
  Each organizational unit establishes an enterprise architect role. All local enterprise architects build a virtual enterprise architecture team.
- **Centralized**
  Establish a central enterprise architecture team responsible for the whole organization.

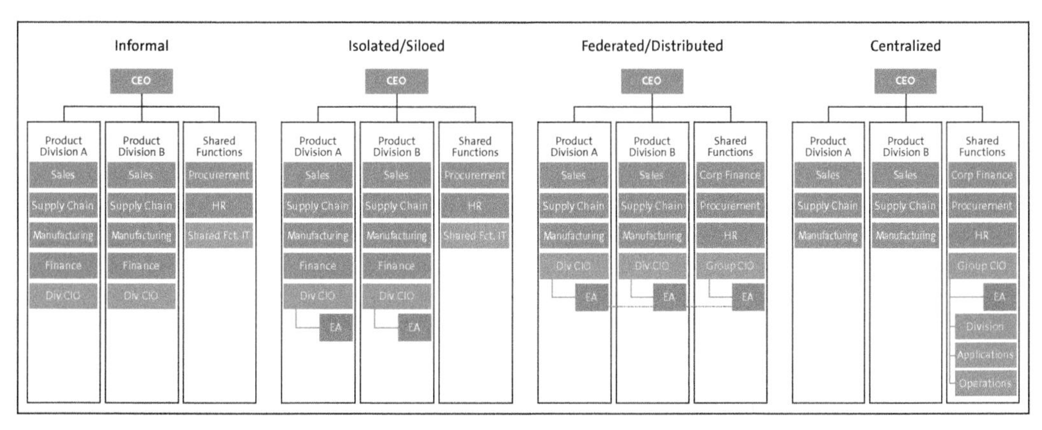

Figure 15.5 Anchoring Enterprise Architecture Practices in Organization

Table 15.2 discusses the pros and cons of each of the options.

	Pros	Cons
Informal	In the early phases of enterprise architecture adoption, it may make sense to act as an informally collaborating virtual team that helps build a case for a more formal establishment of enterprise architecture within the organization at a later point in time.	In the long run, the nonexistence of an official enterprise architecture function entails missing out on many of the benefits such a function can have.

Table 15.2 Pros and Cons of Different Ways of Anchoring Enterprise Architecture in Organization

	Pros	Cons
Isolated/Siloed	If divisions are loosely coupled and follow different architectures and different architecture principles (as the division might support quite different business models and/or industries), it can be more effective to run separate architecture practices and teams.	Often, separate enterprise architecture functions can be leftovers from former acquisitions or just the result of strong leaders who like to run IT separately. In both cases, not even establishing a federated collaboration is a miss, as synergies won't be recognized.
Federated/Distributed	Compared to centralized models, distributed enterprise architecture practices/groups have the benefit that the enterprise architecture function is closer to the business function they are collaborating with. The federated structure can help the company avoid redundant methodologies and architectural conflicts.	Federating and distributing an enterprise architecture function works fine as long there are no conflicts of interest. Problems arise when divisions have no clear conflict resolution paths in place, and this can lead to competition or even fighting. In such cases, the federation will no longer work properly, and the company might relapse into separate working modes.
Centralized	If centralized/corporate functions are core elements of a company's culture, running a centralized enterprise architecture function in a matrix approach across functions can work very well.	In a very large organization running only one centralized architecture function, there can be risks. For example, a team may have a hard time keeping close enough ties to other functions and divisions. Typically, such a team isn't adequately staffed to support many large transformations, and if that team consequently focuses only on governance, it could be seen as policing and therefore risk the losing the acceptance of the centralized enterprise architecture team.

Table 15.2 Pros and Cons of Different Ways of Anchoring Enterprise Architecture in Organization (Cont.)

So far, we've discussed multiple organizational aspects. The next step is to compile the findings in a hybrid approach, as we see that large organizations generally try to balance a centralized approach with utmost agility and regional IT functions with business unit-specific ones. Figure 15.6 depicts a central enterprise architecture team typically reporting to a central C-level function (CIO or CTO). An architecture board is included to connect regional and business unit stakeholders. One objective is to increase bandwidth by using rotated assignments on top of a central team. This team can then act as a group providing architectural services to initiatives and other organizational units. We would classify this as a hybrid setup.

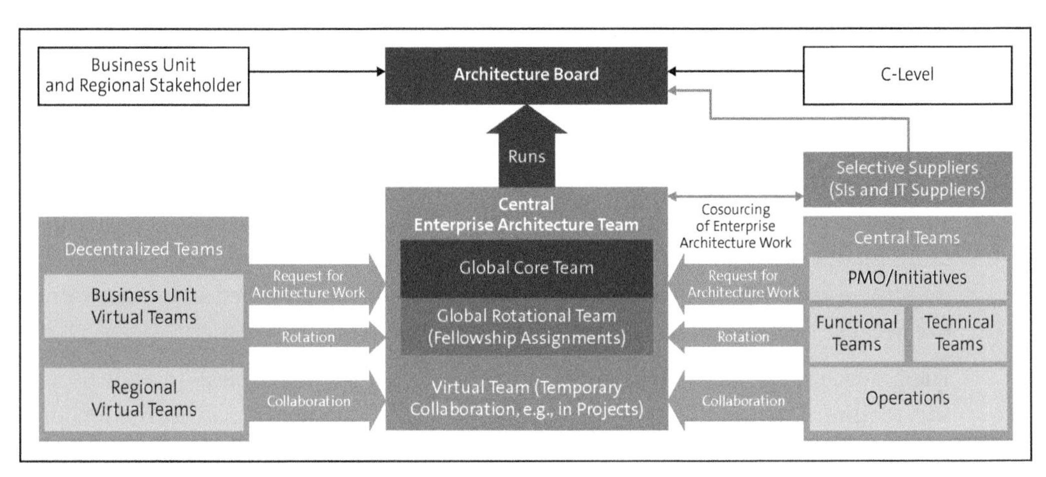

Figure 15.6 Hybrid Enterprise Architecture Team Setup

Establishing the depicted architecture board needs backing from the C-level. We'll return to the aspects of enablement and communication in Section 15.6 and Section 15.8, respectively.

15.4 Enterprise Architecture Processes

When setting up the practice, it should be clear which business processes the practice owns and which processes are run by other functions, where the enterprise architecture practice merely plays an active role. Note the high relevance of governance processes in this context; this is why we've devoted an entire section to them in Section 15.5.

The primary purpose of the practice should be to create enterprise architecture artifacts. We call this process *architecture development*, and it is typically aligned with larger initiatives and collaborates with other functions. If the maturity level at any given time is Defined or higher, the variations on the role of enterprise architecture in initiatives will be equally well defined. A lower maturity status of Managed or Initial implies that there's a case-by-case assessment of if and how enterprise architecture should be used during the setup of an initiative. We'll discuss architecture development in detail in Section 15.4.1.

In Chapter 3, we explored a wide variety of business functions, and in this section, we'll take a look at cross-functional collaboration. For this purpose, we've chosen three specific examples:

- IT investment and program portfolio management (Section 15.4.2)
- Business and IT alignment (Section 15.4.3)
- Solution and process documentation (Section 15.4.4)

15.4.1 Architecture Development

Explicit processes are typically those defined in the architecture framework. In Chapter 2, we described services as part of SAP EA Framework and introduced a modular approach. Figure 15.7 shows a subselection of these modules and typical patterns of these modules mapped to types of initiatives. These patterns can be the basis for a repeatable architecture development process. This implies a Defined architecture practice because it guides the degree of enterprise architecture involvement.

The practice can predefine the architecture work needed. A technical upgrade project will require much less architecture work than more complex transformation programs, but not all possible project types can be predicted and prepared with a ready-to-use methodology. A modular architecture development approach will therefore allow for adapting the architecture development process to the actual needs.

A reasonable objective is to be prepared for 90% of the architecture work ahead. This includes how you scope the work; what steps you need to perform to execute it; and what artifacts you must create to complete the development process and, possibly, seek approval from the architecture board.

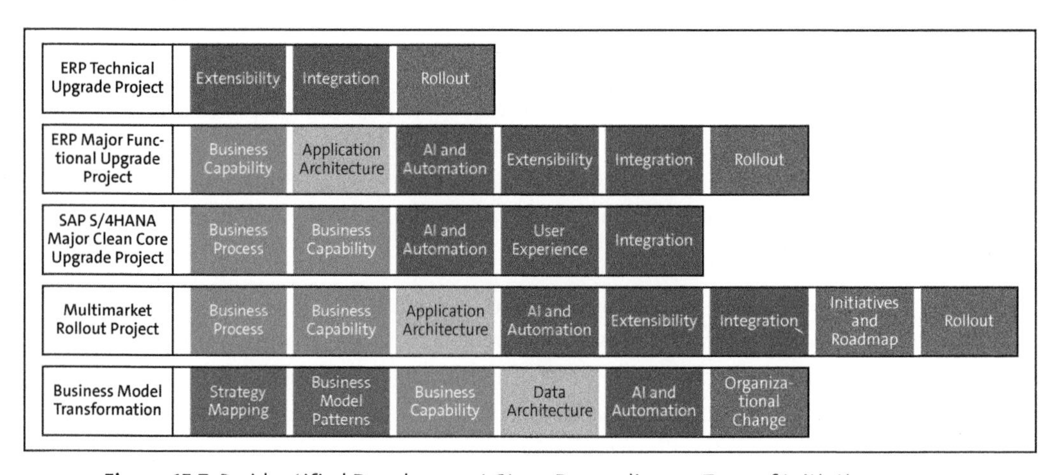

Figure 15.7 Preidentified Development Steps Depending on Type of Initiative

It isn't possible to be 100% prepared for all the work ahead, either methodologically or from a cost perspective. Therefore, we recommend starting the development process for each initiative by reflecting briefly on the availability of a well-prepared, reusable pattern. In some cases, it will be necessary to create a number of specialized visualizations to illustrate differences among architectural options.

Using patterns developed up front comes with two benefits: it supports less experienced architects in the tasks ahead, and it can help convince project stakeholders of the usefulness of practicing enterprise architecture in the transformation. It does the latter by letting you show stakeholders sample results and a structured approach to getting the result they are looking for.

As your enterprise architecture practice becomes more mature, you will execute the architecture development process steps in a repeatable way, and any investment you make in a common enterprise architecture tool can strongly contribute to even greater maturity. Figure 15.8 is a schematic illustration of the tool layer in each of the maturity phases. Initial and Managed practices typically won't have an explicit enterprise architecture tool. In the Managed state, business capabilities (Cap), applications (Apps), and technologies (Tech) are managed independently and might (or might not) have a decentralized store other than Microsoft Excel lists.

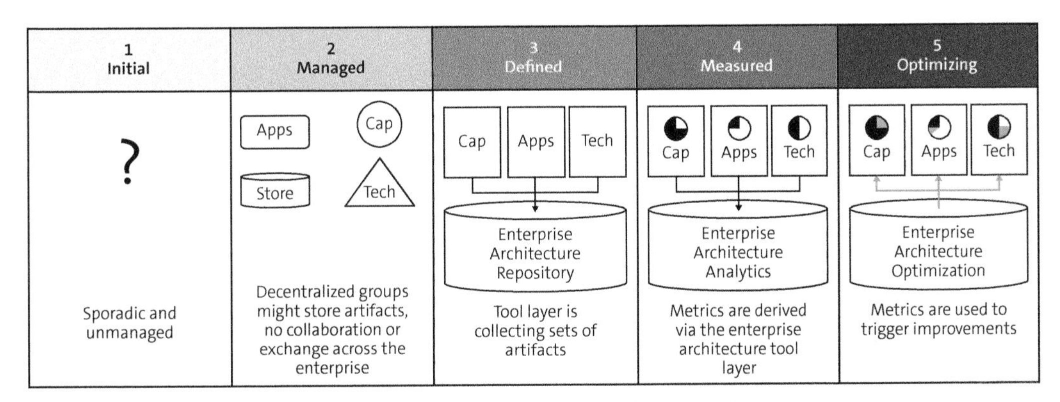

Figure 15.8 Increasing Relevance of Common Enterprise Architecture Tooling

In the Defined state, a common repository is established and enables the linkages among capabilities, applications, and technologies. On top of that, metrics get defined in the Measured state and are then used to trigger improvements in the Optimizing state.

15.4.2 IT Investments and Program Portfolio

The head of enterprise architecture or the chief enterprise architect won't be the chief financial officer (CFO) or the CIO of the company. The enterprise architecture practice doesn't form a governing body that decides on strategic investments, but the CIO, the CFO, and/or the PMO all need to manage risks and understand current and future cost structures. Therefore, an enterprise architecture practice that has achieved the Measured maturity level will provide decision support to the C-level.

Single investments, at least those above a certain threshold, are typically made by the central procurement department. For goods and services, this department will typically run a sourcing process (e.g., what laptops to buy where).

When your company is acquiring a large IT solution (e.g., for HR), it won't be procurement that decides which one to buy. The decision-making process will rather take the form of a collaboration between IT and the business (in this case, the HR department). The enterprise architecture practice will typically be involved in the evaluation process.

In some cases, the overall IT footprint might grow without anybody noticing it. Here are some things your company may do that can cause this to occur:

- Subscribing to a smaller cloud service for collaboration purposes
- Developing a small Python program for data analysis of customer segments
- Using Microsoft Excel for budget planning

In large organizations, the sum of such additions will lead to a growing, initially unnoticed "shadow IT." Best practice is to have clear rules on how enterprise architecture is engaged to better govern investment decisions. The following list shows some types of investments and the considerations they require:

- **Small one-off software expenses**
  IT end user and server software should always be procured through the IT department, as the packages should be installed via central software distribution mechanisms for security reasons. Whether an end-user software package needs to be looked at from an enterprise architecture perspective will depend on its function and the business process context it's used in. We therefore recommend extending the capability model not only to cover enterprise software–related business capabilities but also those covered by end user packages.
- **Packages with a recurring software subscription or maintenance fee**
  All of the previous considerations can be transferred to this type of investment, but the principles need to be applied more rigorously due to the recurring nature of the costs.
- **Open-source software**
  Even though open-source software can easily be downloaded and installed, it can have major implications: from a legal perspective, open-source software comes with many different licenses, which may significantly restrict the way the software can be used in a professional context. Large open-source packages often include many different subcomponents, and it's critical, from the security perspective, to anticipate any vulnerabilities that may arise from them.

We recommend that the technical IT team, the enterprise architecture team, and possibly the legal department maintain joint lists of both endorsed and prohibited open-source packages. This way, IT configuration management can check the status of packages and ensure that users uninstall prohibited ones.

- **Software bundled with an internal introduction project exceeding a threshold**
  The identified need to expend extra effort to get the software implemented is an indicator of complexity. These complexities can be due to configuration, integration, enablement, and/or operations specifics. If business departments drive the introduction of a complex package, the related costs might be hidden in the overall department costs. If an architect is involved, they can check for such possible hidden costs when assessing the package and therefore ensure cost transparency.
- **Single transformation programs touching existing IT solutions**
  Single transformations might not need new software investments per se, but they can change the total cost of ownership (TCO) of an entire landscape by splitting or combining existing solutions. Therefore, their impact needs to be assessed before an initiative is started. The related business case should assess investments and benefits from both the IT and the business perspective.
- **Annual initiative investment planning**
  Annual investment planning is clearly done at the C-level, typically under the guidance of the CFO. Enterprise architecture can have a huge, positive impact on this process if operational cost projections are derivable and capital investments are linkable to projected benefits. For this to be so, enterprise architects should be involved in setting up a cost model (ideally with a multiyear horizon) that analyzes the long-term impacts. Models can be analyzed for different dimensions such as the following:
  - Technology layer
  - Organization
  - Function or process
  - IT and business operating costs
  - Change costs

  To perform a cost/benefit analysis, you need to understand existing cost structures and project effort drivers. In terms of the benefits an investment will contribute, a financial controller will only rarely accept top-line growth (that is, selling more). The reason for this is that so many internal and external factors (e.g., competition, economic situation) have an impact on sales success that linking higher sales numbers back to a single IT investment will be difficult.

  To get finance's approval to put a number on the headcount reductions a better IT solution can enable, you need to provide solid proof points.

As we discussed in Chapter 9, enterprise architecture can be used to help assess and optimize a given initiative's impact on TCO. It can also help identify opportunities for simplifying the application portfolio (e.g., by way of rationalization) and, by doing so, optimize the TCO of the overall landscape.

15.4.3 Alignment of Business with IT

The alignment of business with IT is always quoted as one of the main drivers of enterprise architecture initiatives. In many cases, this is done in the context of a business transformation with the goal of business and IT documenting the transformation jointly.

The business capability model is a central building block of such a transformation because it identifies the gaps and opportunities that need to be addressed. The risk with this type of initiative is that good capability models are only prepared for areas that either are lined up for an active transformation program or already have one actively running.

A broader and continuous alignment of business and IT is desirable so that gaps, requirements, investment needs, and more can be managed jointly. Therefore, refining the business capability model and identifying possible gaps should not be put off until an initiative is started but should be part of the ongoing operating model. Large companies have dedicated IT-business partner managers who link IT groups with business, gather their needs, and define a change portfolio. These managers should leverage the help of the enterprise architecture practice to identify and visualize the business change drivers that are part of or linked to the business capability model.

This alignment then forms the baseline for the annual IT strategy cycle. Best-run companies not only allocate funding to initiatives once a year but also run quarterly portfolio prioritizations for adjustments.

15.4.4 Solution and Process Documentation

It's legally required in some industries to have well-documented IT solutions and business processes. This legal documentation should not be approached in isolation but in coordination with the enterprise architecture program. Leveraging an enterprise architecture tool as the overarching single source of truth will increase the acceptance and relevancy of enterprise architecture.

Furthermore, only a good enterprise architecture will enable accelerated solution adoption. Without it, every change request will need to start with an evaluation of the baseline. Making enterprise architecture documentation a mandatory output for larger initiatives is an investment in the ongoing ability to change.

We see good architectural documentation as the output of regular enterprise architecture development according to an agreed-upon enterprise architecture methodology (see Chapter 2), and we discussed its practice in all chapters of Part II. Therefore, we won't elaborate on it further in this section.

15.5 Enterprise Architecture Governance

In our understanding, *governance* is the generic framework for how to run a practice. In this sense, all sections in this chapter could be seen as covering governance. More narrowly, enterprise architecture governance defines guardrails and/or architectural principles and manages the decision process. We recommend that only a limited amount of the practice's capacity be spent on enterprise architecture governance, with the majority devoted to developing enterprise architecture deliverables as part of transformation projects. We advise limiting the time spent on governance to 10% to 20% of working hours. The actual governance needs will depend on the size of the practice and its maturity.

In Section 15.3, we looked at organizational aspects of an enterprise architecture practice. This section will focus on the role of an architecture board, architectural principles, and guardrails, and it will also provide guidance on how to establish change management for the practice. We'll conclude with a discussion of how to link enterprise architecture governance with other governance layers.

15.5.1 Architecture Board

Establishing an architecture governance board is a good idea because it moves responsibility for critical decisions to a broader audience of stakeholders beyond the architecture function. An established architecture governance board will be informed of and approve important architecture decisions, and establishing such an architecture board should bring key stakeholders from IT and business together. Major architecture decisions should be presented to the board regardless of whether the decisions were prepared by the central architecture function, a third-party group, or an individual.

Successful architecture board decisions should fulfill the following prerequisites:

- **Scope of the decision**
  It needs to be clear what decision proposals need to be presented to the board. There also needs to be an agreement among leadership that this step of the process is mandatory.
- **Members of the board**
  To foster the broadest possible acceptance of the board's decisions within the organization, the board should have good representation from technical and functional team members.
- **Decision criteria**
  The board should only approve a proposal if it's clear what criteria were used in the

decision-making process. These criteria can be architectural principles, policies, and standards, as well as the target architecture and defined roadmaps. In some situations, a proposal will explicitly differ from a previous decision, and the board may have to recommend a specific impact analysis before granting such an exception.

If it's clear that the board plans to make a specific decision during the meeting, any proposal related to that decision needs to be submitted prior to the meeting to allow for review or questions.

- **Planning**
  Finding slots in packed calendars can be a challenge, especially when people from multiple departments or units are involved. Therefore, a regular, predefined meeting schedule should be established. Ideally, all board members should receive any proposal a month before the meeting when that proposal will be discussed. This leaves time for review and gathering information for the board meeting, should there be questions.
- **Required documentation**
  To help architecture board members understand architectural proposals, it should be mandatory for people submitting those proposals to also submit definitions of typical architecture artifacts.
- **Documenting decisions**
  To keep the board from forgetting or not documenting decisions, it's a best practice to keep a register of architecture decision records (ADRs). We covered ADRs in Chapter 2, Section 2.2.7, on enterprise architecture methodology. The decision-making procedure can vary from company to company and can depend more on the company's culture than on its organizational hierarchy, and this implies different requirements for documenting a decision. The most formal way is to collect signatures on a form.
- **Rejecting a proposal**
  It needs to be understood beforehand that rejecting a specific proposal is a possible option. Therefore, the impact of such a rejection is ideally investigated up-front, so it doesn't come as a surprise and the board is fully aware of it.

  Instead of opting for a hard rejection, the architecture board will more likely ask for additional details and, possibly, a quantitative analysis. Another alternative could also pop up during a discussion, and that could then trigger another thorough analysis.
- **Communication and community management**
  It's imperative that good architecture work be complemented by well-established communication channels that distribute the results. The reasons for this are as follows:
  - New decisions have to be communicated to downstream projects so that all parties can incorporate them.

  - Good results should be made available for others to learn from them in both small and large organizations.

- **How decisions are made**

  A company is not a state with a clear constitution, elections, an executive branch, and a legislature. Companies may well be organized in hierarchies, but real decision-making is not always a top-down process. A company's decision-making culture needs to be reflected in the structure and operations of the architecture board. Consider the following points on the spectrum and decide which one fits your company culture best:

  - *Decisions by vote*: A fixed group of board members votes on proposals, and proposals that receive a majority vote are approved.
  - *Decision by prealignment*: All critical decisions are talked through, prealigned, adjusted, and realigned prior to the board meeting. This means all critical stakeholders will be on board with a decision once the architecture board comes together.
  - *Top-down and veto*: If a proposal is in dispute, the highest-ranking manager can overrule the other board members or a small core group of stakeholders may have the right to veto the proposal. As a rule of thumb, controversial proposals should be avoided, as fighting in public is detrimental to working relationships.

  It needs to be clear that any decision against a proposal is a decision in favor of the only alternative approach, which by default is the continuation of the status quo. The impact of the default or "do nothing" option should be investigated prior to every decision.

15.5.2 Architecture Principles and Guardrails

Architecture principles, policies, and standards form clear guardrails and should be taken into consideration in any architecture development process. They can also play a role when you're deciding on alternative architecture options, either during the development process or when presenting a decision proposal to the architecture governance board.

Figure 15.9 shows the dependencies and the definitions of principles, policies, and standards:

- *Principles* express the key long-term objectives of enterprise architecture and are very high level and long lived. Principles should be based on the strategic business and IT objectives driving the organizational transformation, and they should not change based on day-to-day business or IT demands.

  For example, an architecture principle could be "subscribe to software—before buying software—before building software."

- *Policies* are actionable rules that must be applied to the enterprise architecture. They may be based on current business and IT trends, but they should *not* name specific products, best practices, protocols, etc. Policies may change more often than principles but should be designed to be as long lasting as possible.

  For example, new IT capabilities will adopt cloud solutions in the following order of preference: (1) SaaS, (2) PaaS, and (3) IaaS.
- *Standards* are the specific products, best practices, protocols, etc. that support the policies. They should ideally be long lasting but may change more frequently than policies.

  For example, our standard cloud IaaS provider is Amazon Web Services (AWS).

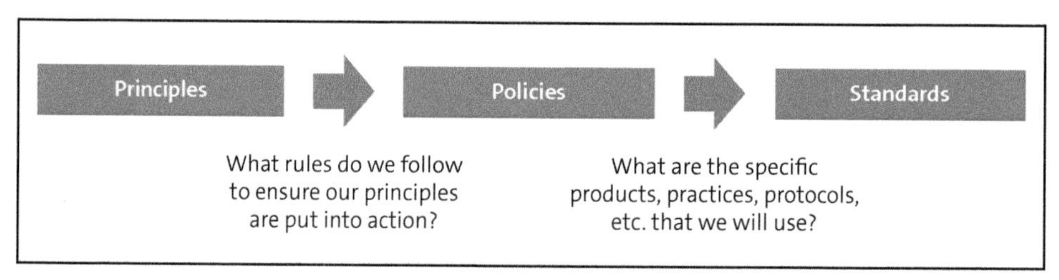

Figure 15.9 Deriving Standards from Principles and Policies

There will regularly be cases in which some of the principles, policies, or standards will appear to prevent a team from being able to go with a better option, such as a nonstandard solution that can be implemented faster, uses a less complex architecture, or has some other advantage. In this case, the board should consider following three courses of action:

1. Assess whether there's a need to change the principles, policies, or standards, as the current case doesn't seem to be a one-off exception.
2. Put the arguments together, including an assessment of the impact that diverging from the principle or standard would have. An exception process should be part of the governance process.
3. Subject certain principles and standards to review on a regular basis. Alternatively, the architecture board might trigger an explicit review of an existing principle or standard. Reviews are necessary because technology progresses quickly, and principles and standards in areas like security, development, web platforms, and AI should be reviewed biannually.

The list of principles should not be too long, and they should be easy to understand. The principles should not put forward an opinion but should point in the agreed-upon direction toward adequate architecture. Furthermore, architecture principles should not contradict each other. If they do, this should be brought to the relevant parties' attention so that they can discuss tradeoffs in the given situation. Figure 15.10 shows a set of possible architectural principles grouped by different domains.

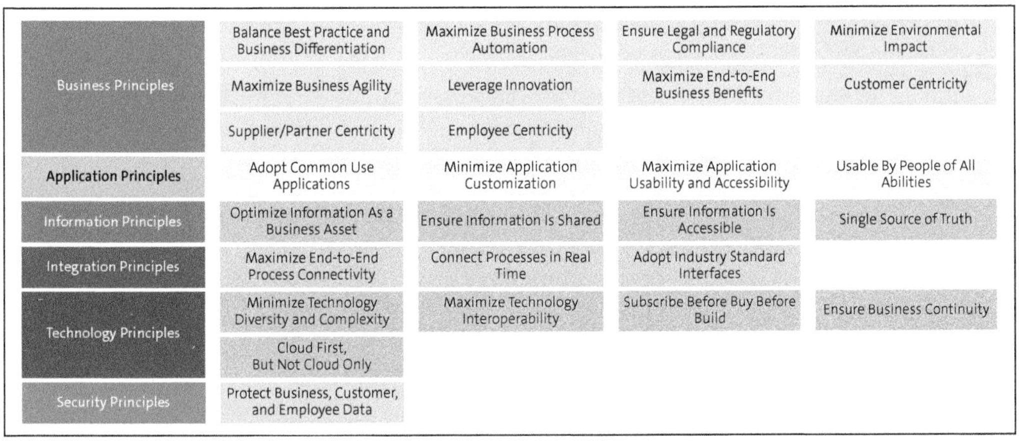

Business Principles	Balance Best Practice and Business Differentiation	Maximize Business Process Automation	Ensure Legal and Regulatory Compliance	Minimize Environmental Impact
	Maximize Business Agility	Leverage Innovation	Maximize End-to-End Business Benefits	Customer Centricity
	Supplier/Partner Centricity	Employee Centricity		
Application Principles	Adopt Common Use Applications	Minimize Application Customization	Maximize Application Usability and Accessibility	Usable By People of All Abilities
Information Principles	Optimize Information As a Business Asset	Ensure Information Is Shared	Ensure Information Is Accessible	Single Source of Truth
Integration Principles	Maximize End-to-End Process Connectivity	Connect Processes in Real Time	Adopt Industry Standard Interfaces	
Technology Principles	Minimize Technology Diversity and Complexity	Maximize Technology Interoperability	Subscribe Before Buy Before Build	Ensure Business Continuity
	Cloud First, But Not Cloud Only			
Security Principles	Protect Business, Customer, and Employee Data			

Figure 15.10 Sample Collection of Enterprise Architecture Principles

Table 15.3 shows a more specific example where clear standards are derived from SAP's clean core principles, as discussed in the use case in Chapter 5. Note that some clean core policies go beyond core enterprise architecture concerns. In Section 15.5.4, we'll discuss the governance of clean core implementation.

Principle	Related Policy	Product Standard
Balance best practice with business differentiation: Customize business processes and IT only if it helps the business stand out from the competition.	*Process standardization*: Processes for new IT solutions will be standardized, unless there are legal and/or regulatory reasons to not do this or unless we have the specific need to stand out from our competitors. There's no reason for legacy business processes, applications, or architecture to deviate from the standard.	The standard for ERP-based business processes (see SAP Best Practices at *https://me.sap.com/processnavigator*)
Clean core architecture: Achieve lower maintenance cost while allowing differentiating extensions.	*Code extensibility rules*: Prove the business need. Check alternative standard solutions. Use upgrade-friendly technologies.	No-code standard: SAP Build Apps Cloud software development kit: SAP Java software development kit Other: SAP Build Code

Table 15.3 Clean Core: Principles, Policies, and Standards

Principle	Related Policy	Product Standard
Clean core architecture: Achieve lower maintenance cost while allowing differentiating extensions. (Cont.)	*Clean core data-related rules:*	
	Use SAP standard solutions for master data.	SAP Master Data Governance, cloud edition for business partner
	Use a uniform cloud-based data warehouse platform.	SAP Datasphere
	Use a uniform embedded web-based analytics framework.	SAP Analytics Cloud
	Clean core integration:	
	Use a single cloud integration suite for SAP and non-SAP integrations.	SAP Integration Suite

Table 15.3 Clean Core: Principles, Policies, and Standards (Cont.)

We recommend storing principles, policies, and standards within the architecture repository so that principles and policies can be linked to capabilities, IT components, and other objects.

Deep Dive: Extending SAP LeanIX to Capture Principles, Policies, and Standards

In Part III, we did a deep dive into SAP LeanIX and discussed its meta model. Figure 15.11 shows a possible new type of fact sheet called *guardrail*, which could be used for the architecture principle and policy subtypes. The link to business capabilities and tech categories defines the context of principles and policies, and the policies can then connect to specific platforms or IT components to flag them as standards.

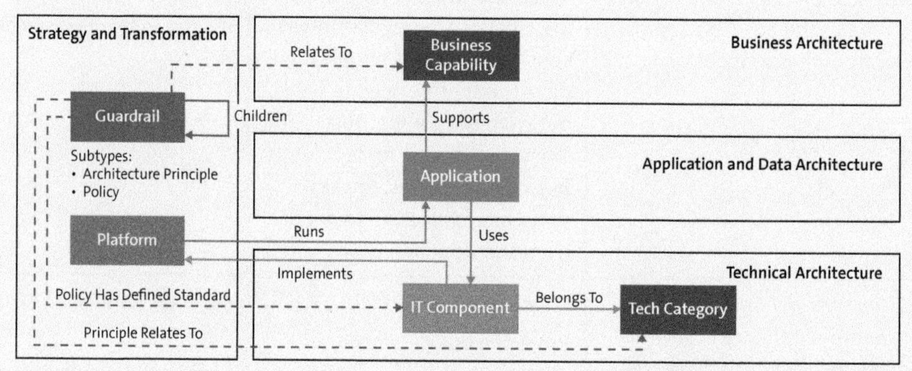

Figure 15.11 Possible Extension of SAP LeanIX Meta Model to Cover Principles and Policies

15.5.3 Planning and Change Management

As we've discussed, a mature enterprise architecture practice will document and ideally measure its own processes (not just business processes run elsewhere in the company).

The setup of and proper change management for these processes are part of overall governance and should be documented as enterprise architecture practice rules. A few such rules could be as follows:

- Document a clear change log when changing any principles, policies, and standards.
- Clear schedules for community sessions, with rotating accountability for contributions.
- Have an easily accessible register of templates.

As in Part II, where we looked at several concrete use cases and described the evolution of concrete architectures in them over time, Section 15.9 will look at two uses cases to show you how to develop a stronger enterprise architecture practice over time. The examples will illustrate that enterprise architecture subpractices are often built in compliance with the currently implemented architecture. Each of these subpractices is optimized to meet its own needs and requirements, and it therefore shows a natural resistance to change. Possible change drivers might come from a change in management or a broader need for the company to adopt the overall business model.

15.5.4 Linking to Other Governance Layers

To a certain extent, setting the context of enterprise architecture governance can be compared to the discussion we had on related functions in Chapter 3. Enterprise architecture governance can be seen as the highest level of enterprise architecture, but there's a need for orchestration among different functions. We therefore propose that you find a pragmatic governance structure that fits your organization. This is why the governance layers depicted in Figure 15.12 don't suggest a strict hierarchy.

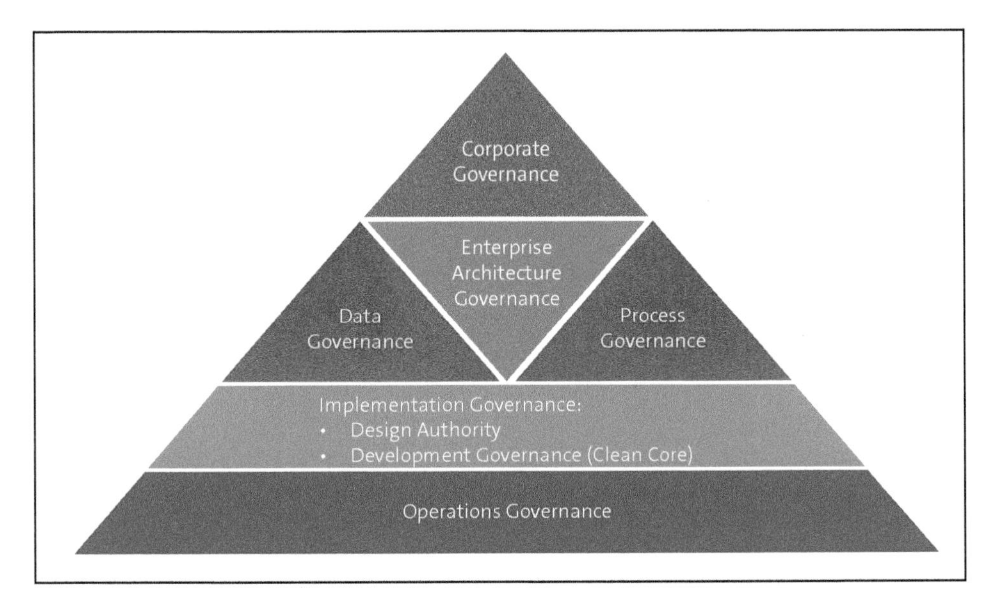

Figure 15.12 Governance Layers

Table 15.4 uses the governance layers introduced in Figure 15.12 and lists their accountabilities. As discussed earlier, the form and operation of the governance decision body will depend on the specific organization's design and culture, whereas the items on the table are typical roles or functions related to the corresponding governance layers.

Governance Layer	Accountability	Decision Body
Corporate governance	■ Corporate strategy ■ Legal compliance ■ Financial transparency ■ Management of resources and budget ■ Global risk and compliance ■ Partner governance	■ Corporate board ■ CFO ■ COO ■ CIO
Enterprise architecture governance	■ Enterprise architecture principles ■ Architecture decision making ■ Target architecture ■ Cloud adoption ■ Application rationalization ■ Transformation roadmap	■ Architecture board ■ CIO ■ CTO
Data governance	■ Data privacy ■ Data standards ■ Data harmonization ■ Data retention policies ■ Master data governance ■ Test data management	■ CDO
Process governance	■ Process performance and improvement ■ Process documentation and modeling ■ Process automation	■ Global process owner ■ Chief process owner ■ Process Center of Excellence (COE)
Implementation governance	■ Project standards ■ Risk assessment ■ Clean core implementation governance ■ Usability ■ Test execution	■ PMO ■ Design authority

Table 15.4 Different Governance Accountabilities

Governance Layer	Accountability	Decision Body
Operational governance	▪ Operational change management ▪ Cybersecurity and security engineering ▪ Environmental governance ▪ Service-level management ▪ Performance management ▪ Availability ▪ Value delivery	▪ CTO ▪ Head of operations, change management

Table 15.4 Different Governance Accountabilities (Cont.)

Any subset of governance aspects your company seeks to implement should be supported by proper tooling, as manually collecting KPIs in Microsoft Excel will be neither long term nor sustainable. Enterprise architecture tooling is covered in Chapter 4 and Part III. For other layers, consider best-of-breed solutions like SAP Signavio for process governance or ServiceNow for service and operations governance.

15.6 Enterprise Architecture Skills and Enablement

Building a practice from the ground up can be difficult, as the desired leadership is often not available and recruiting key people can be difficult. To identify good candidates for building the architecture practice, we recommend that you do the following:

- Identify a highly respected person who already has work experience as an (enterprise) architect or, at least, has work experience in multiple other architectural domains.
- If no good internal candidates are available, try hiring externally, even if this requires careful candidate selection (Section 15.7).
- Leverage external consultants and/or services.

In Chapter 16, we describe how the enterprise architecture practice at SAP is set up to help customers with their transformation projects.

Once the practice is established, it should inject architectural thinking into initiatives and other parts of the organization. This will both increase the acceptance of enterprise architecture work in general and attract more people from the organization to the practice. At the least, it will increase the number of people actively collaborating with it.

A successful enterprise architect might typically have five to ten years or experience in related functions, while a senior or chief enterprise architect might even have ten to fifteen years of experience.

We'll explain the key skills an enterprise architect needs, as well as some tips for the learning and enablement process (including relevant conferences), in the following sections.

15.6.1 Business-Related Skills

As linking business with IT is one core element of enterprise architecture, a base understanding of more than one business function is almost a necessity. Learning offerings could range from traditional training-based programs to shadowing employees for a certain time, as in, for example, "a day at your workplace." Having an MBA in economics may be a good foundation, and to gain a better understanding, a temporary rotational assignment into a business function can be an excellent option too.

Are business skills needed for *all* functions performed by a company—from R&D, sales, and supply chain manufacturing to finance and HR? We don't think so. Clearly, an experienced architect will build foundational knowledge over time.

In a mature company, which maintains good business documentation, accessing models of business functions (capabilities) and processes is another good way of complementing partial knowledge. But even if these are not documented, there are external best-practice capabilities and process models available, and they can serve as on-demand sources of explorative learning.

The extract from SAP's reference architecture in Figure 15.13 shows the level-3 business capability description for the **Procurement Collaboration** business area. The right two columns specify the reference applications implementing these capabilities.

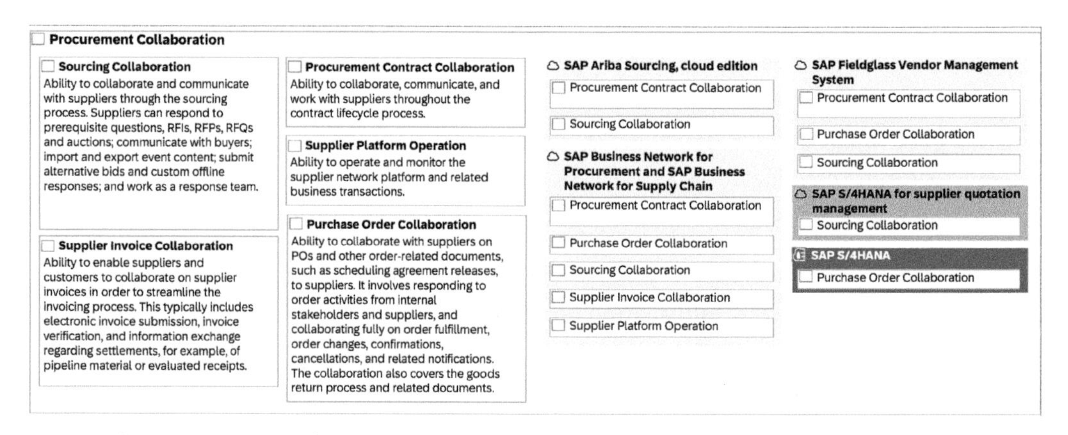

Figure 15.13 Extract from SAP Reference Architecture as Source for Learning

The level beyond understanding the business capabilities is *how* an enterprise is run. This dimension is described by the business process model, which might often imply a high number of process variants. Existing process models or external reference models like SAP Reference Business Architecture can be a good source of learning.

15.6.2 IT-Related Skills

No enterprise architect can be a specialist in all types of IT solutions, from business applications to web platforms to detailed security-related encryption technologies. Note that there's an IT-related reference content for every business architecture related reference content.

It is, however, not sufficient to merely search for the appropriate knowledge on demand, as this approach won't enable the architect to reason about content. This is why we rate the following IT-related knowledge clusters as minimum requirements. A senior architect should have deep knowledge of at least two or three of these clusters:

- SAP S/4HANA, including its main modules, integration, and extensibility options (since this book explores an SAP-centric approach to enterprise architecture)
- The difference between the public and private cloud or on-premise architectures
- Data warehousing and other analytical approaches
- Integration complexities and integration approaches
- The following nonfunctional requirements:
  - *Scalability*: Scalability and its multiple facets from the pure infrastructure layer right up to the way processes are orchestrated.
  - *Maintainability*: In the SAP context, maintainability is linked to our clean core principles, which not only try to reduce and isolate custom code extensions but also take care of clear data and clean processes.
  - *Resilience*: Fault tolerance in architecture is key. Approaches range from classical high availability and disaster recovery to large, parallelized architectures with limited consistency guarantees.
  - *Agility*: Even though agility is a buzzword, it's hard to prove which architectures best support agility. For companies with M&A as a typical growth strategy (see Chapter 8), an agile architecture allowing for faster integration of new companies is essential. Agility can also be a key need in rapidly changing market conditions and can possibly be addressed by enabling or implementing company-specific innovations on SAP BTP (see Chapter 5).
  - *Security*: Security is a long-existing concern that plays a stronger role now than ever before. It's important to understand that security concerns need to be thought through on every level of architecture, from low-level network infrastructure through data security right up to application and process matters.

15.6.3 Enterprise Architecture-Related Skills

If the practice you are working in has already chosen an enterprise architecture framework, you should get familiar with it. SAP EA Framework, described in Chapter 2, can serve as a benchmark. The practice should define learning journeys for enterprise architecture.

15.6.4 General Skills

Under *general skills*, we subsume everything an experienced knowledge worker needs to know. The following list highlights five such skills:

- **Communication skills**
  Even if an architect might have worked out the best possible target architecture, it might still fail if it doesn't reach the right stakeholder or isn't explained well enough. Communication is more than just sending the right message—it's also about listening to business and other stakeholders. See Section 15.8.3, where we talk about structuring overall communication about enterprise architecture.
- **Conceptual and analytical skills**
  Architects often need to analyze a vast amount of structured and unstructured information, draw the right conclusions, and translate those conclusions into the correct level of abstraction in architectural models. The resulting insights and models then need to be used for decision making, often by way of evaluating several alternative architectures.
- **Creativity and innovation**
  Even though proven frameworks and methodologies try to make enterprise architecture less of an art, creativity and innovative out-of-the-box thinking will help with problem solving.
- **Relationship building**
  Being widely connected and respected in the organization is a key success factor for enterprise architects. Investing in this skill and complementing it with empathy will clearly pay off.
- **Visualization skills**
  A picture is worth more than a thousand words. Using good tools will ideally ensure a better visualization result than using a manual Microsoft PowerPoint.

  Figure 15.14 is an example of a simple graph with nine nodes, which are grouped into three colored clusters: A, B, and C. Semantically, they express the same situation, but the different structuring makes one much easier to understand than the other. See the next box for tips for making clear and concise graphs.

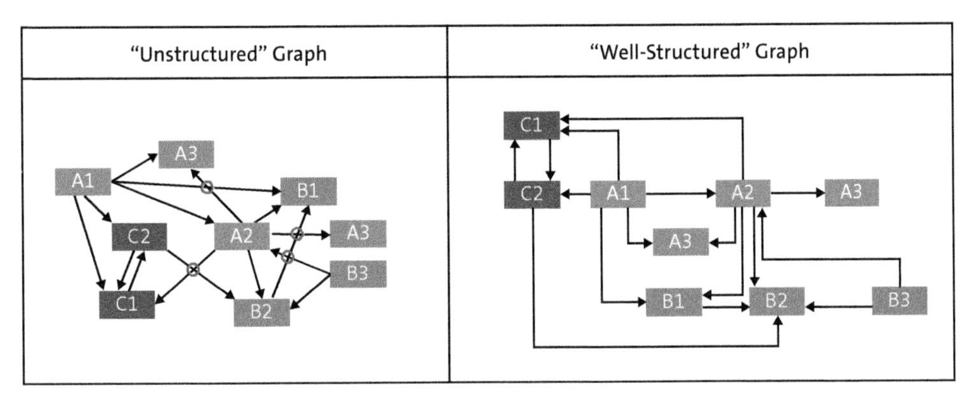

Figure 15.14 Example of Applying Graphical Skills

Creating a Graphic Layout

The following relates to the graphs shown in Figure 15.14, but our tips can be reapplied elsewhere:

1. If nodes are clustered, try to keep the nodes of each cluster close together.
2. The most complex and interconnected clusters typically fit best in the center of the graph (as with centering blue cluster A1, A2, A3, and A4). Try to push smaller clusters to one of the four sides (as with pushing pink cluster C1 and C2 to the left and orange cluster B1, B2, and B3 to the bottom).
3. Arrows should ideally never cross. Rearrange the nodes such that you end up with no or only a minimal number of intersecting arrows. In the left unstructured graph, there are four crossing arrows, indicated by a red circle. On the right side, all of the crossings have been eliminated.
4. Elbowed connectors often lead to a better-structured graph, especially if they are attached with text. In some rare cases, you can mix connector types, but you should do so with care and only if it helps.
5. Use dashes, boldface print, and colors to distinguish arrows with different semantics.
6. Out-of-the-box PowerPoint is not a good layout tool for complex graphs because most shapes have only four connector points and edges lose their adjusted routing when one of the connected nodes is moved. Your enterprise architecture tool of choice—one example being Microsoft Visio—should do better. If you can't avoid PowerPoint, you may want to "add" connector points to shapes by grouping "invisible" small shapes at the edge. By adding invisible connector shapes (which have zero width or height and invisible borders), you can split edges in two with individual routings.

15.6.5 Learning and Enablement

There are many ways in which people learn and get training. Depending on the learning needs, a dedicated learning journey can be compiled. Typical learning formats are as follows:

- Classroom training.
- Self-paced learning.
- Formal external or internal certification programs. For example, SAP offers the SAP Certified Professional—SAP Enterprise Architect certification (see *http://s-prs.co/v586311*).
- On-the-job training. Making this type of learning available should not be an excuse for not providing other dedicated learning formats.

Figure 15.15 shows a sample internal SAP learning plan combining different formats. In our experience, learning that mixes small self-paced modules is well accepted.

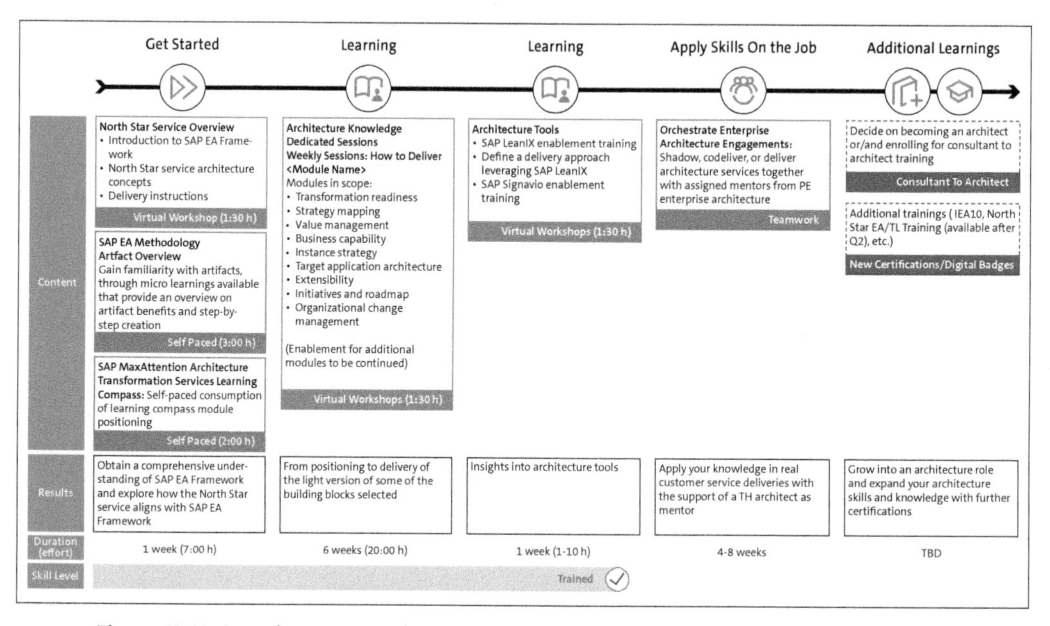

Figure 15.15 Sample Learning Plan

15.6.6 Conferences

There are not that many enterprise architecture conferences, and it's not very likely that you'll find one that is located close to your team. Nevertheless, you should consider sending the architecture lead or a senior team member to such an event, as it gives excellent external input and offers networking opportunities.

Based on the transformation challenges ahead, it can make sense to seek out conferences that go beyond the scope of enterprise architecture. These can address specific IT- and business-related topics. You should just make sure that the level of depth covered in the conference matches the needs and the current skills of the person attending and/or the overall maturity of the practice they deal with.

Make sure that whoever goes to a conference relates their insights back to the team. This could be in the form of a blog entry, a webinar, or, at least, verbal feedback in a team meeting. If attendance was planned with a view to addressing a specific challenge or initiative, the gained insights will have the best leverage.

Many formats are in person (again), some are hybrid, and some are fully virtual/online. The following list shows a number of recurring enterprise architecture conferences. Note that the list is not complete, as options do change over the years.

- Gartner Application Innovation & Business Solutions Summit (over two to three days in multiple regions at different times)
- Gartner IT Symposium/XPO (over four to five days in the fall, in Spain and the United States)
- Forrester, Technology & Innovation Summit (over days in hybrid format, in North America, Europe, and the Asia-Pacific region)
- OMG, Business Architecture Innovation Summit (over three days in hybrid format, in summer in the United States)
- The Open Group Summit (in the spring in the United Kingdom and in the fall in the United States)
- IDC Summit (over two days in the spring)
- Enterprise Architecture & BPM (IRM), (over four days in the summer in the United Kingdom)
- Camp IT: Enterprise Architecture (for one day in the fall in the United States)
- Business Change and Transformation Europe (over two days in the spring in the United Kingdom)
- Building Business Capability (over five days in the spring in the United States)
- ICEIS, International Conference on Enterprise Information Systems (over three days in the spring in Europe)
- SwitchON: The Enterprise Architecture Show (regional but fully virtual/online)
- LeanIX Connect (for one day in several locations)
- SAP Sapphire (over three days in June, in Orlando and Barcelona and virtual)
- ASUG Annual Conference (linked with SAP Sapphire Orlando)
- DSAG-Jahreskongress (over three days in the fall in Germany)

15.7 Hiring an Enterprise Architecture Team

Only in some cases will there be a C-level mandate to build an architecture team from scratch with (for example) five team positions. It's more likely that an existing architecture will need to be extended or that replacements will need to be found for people who have left the team.

Internal policies might stipulate that internal candidates need to be considered first, but in other cases, there will be a choice of internal and external as well as junior-through senior-level candidates. Our recommendation is to hire candidates from different categories. Table 15.5 shows considerations for choosing from the candidates at hand.

Seniority	Internal Hiring	External Hiring
Junior member	Hiring an internal junior-level candidate is not that different from hiring an external one, except that their problem-solving skills can be judged more accurately. If candidates quickly switch jobs early on, their motivation should be questioned. What was wrong with the initial job?	Potential candidates should have an open mind, a deeper understanding of the issues at hand, and an appreciation of business and IT. All candidates should be able to explain what a business process and an ERP system is.
Experienced member	Getting an experienced internal candidate on board can be an excellent option: ideally, they will contribute expertise on one or two enterprise architecture-relevant subjects and seek to broaden their own knowledge.	If specific areas need to be addressed specifically in coming transformations and the relevant skills aren't available internally, then hiring experienced external candidates might be unavoidable. Getting some outside-in perspective might be beneficial in any case.
Senior member	As senior architects, we see candidates who have the reputation of being thought leaders in at least one broader field (e.g., finance, integration technologies). Even if they haven't worked as enterprise architects, they should be able to pick up the relevant methodologies and expand their area of expertise. The company network they are exiting will help the architecture practice gain broader acceptance.	Hiring a senior architect externally will be difficult, as the demand typically exceeds the number of experts seeking new opportunities in the job market. Getting an external senior architect in, however, will add new perspectives and practices successfully applied elsewhere. Ideally, the candidate will fill a skill gap in the existing architecture group.

Table 15.5 Hiring an Architecture Team

Seniority	Internal Hiring	External Hiring
Head of enterprise architecture team	Hiring or appointing an internal candidate as manager of the architecture group makes sense if the candidate is well respected and has already worked as an enterprise architect. If the architect team is newly established, an equally new manager will probably be confronted with the challenge of having to hire additional team members. Their internal background and network will help the whole team to be successful.	Hiring a manager externally is the preferred approach if the practice in the process of being established and in-house experience in the field is very low. If both the manager's and most of the team's positions are to be filled, it will be critical that the incoming manager gets strong internal support in identifying strong, equally internal candidates.

Table 15.5 Hiring an Architecture Team (Cont.)

One question that arises in this context is whether hiring junior candidates in a team of typically very senior architects can be successful. We've found that new hires can be entrusted with complex architecture work after one and a half to two years if they come with strong potential, join the team with the right mindset, and are trained thoroughly. This clearly is a decision to be made thoughtfully, and it requires effort and commitment during the learning phase.

15.8 Enterprise Architecture Communication and Communities

Communities are typically groups of people who come together because they have a common interest. These common-interest groups can exist within a company or be orchestrated across companies, as are the SAP user groups.

Designing any type of community can be a challenge because the scope needs to be so specific that people can identify the learning opportunity therein and because communities that center on a single product feature will be so small that they won't meet with enough response. As membership in a community is voluntary by definition, it's up to the organizing body to make it as attractive as possible.

We'll provide guidance for finding both an internal and an external community, as well as for effective communication, in the following sections.

15.8.1 Internal Community

Forming an internal community makes sense if it brings people together from multiple functions. If the core enterprise architecture team consists of twenty members, then the community it addresses should be a much larger group, as the team meeting would be seen as equal in size to the community otherwise. Whether it makes more sense to form one global community or several regional and divisional ones depends on the overall organization. Considerations similar to those delineated in Section 15.3 can come into play when starting a community.

Considerations When Building a Community

- Start smaller than what you plan to be the community's ultimate extent.
- Brainstorm up front on what possible topics the community can address and who could advance them.
- Continuously maintain a backlog of topics and possible moderators and speakers. Try to set agenda items and speakers ahead of time, as doing so in the short term will likely fail.
- As you might be running hybrid formats (on-site and remote), make sure virtual participants are actively involved. Use the chat function to collect questions.
- If you run sessions with less than fifty people, try to foster creativity by integrating collaboration tools like Mural or others.
- Mix theoretical and practical project topics (with a higher practical share, if possible).
- Make sure that it's not always the same 10% active people who run 90% of the sessions.
- Motivate younger architects to report on their first experiences.
- Have businesspeople run sessions so that they can share their strategies and challenges.
- Try a monthly schedule; twelve sessions per year should be doable.
- If you know architects from other companies, ask them to give informal guest talks. If these external talks get noticed and exposed to a broader audience, not many guest speakers will reject the opportunity to give them.
- Record the sessions and publish slides and recordings to a central community site.
- Use the community to get feedback on early draft concepts.
- Possibly discuss difficult architectural challenges from a current transformation program in a community session to get alternative input.
- Let the community have a say in owning the agenda.
- If you struggle with the execution, ask yourself critical questions:
  - Does it have the right priority?
  - Is our scope too narrow or too broad?
  - How can we make sessions more practical, so that there's more incentive for participants to join?
  - What can we do to give the sessions more visibility?

A community should not replace a proper learning program, as topics will tend to come up during the community sessions and follow each other in an unstructured, random way. But if there are excellent community sessions, then make them part of your training curriculum. Do this by either providing recordings of sessions or going the extra mile and distilling a learning unit from the community sessions and ideally adding an exercise module on top. Training content can be created from community sessions in either a planned way (e.g., if the existing learning modules are outdated and new material needs to be prepared anyhow) or by chance (e.g., if a report back from a concrete transformation program illustrates the architecture transformation journey in a very transparent way).

If the content management system you use to publish a community's content provides for participant feedback, then you might want to use its ranking mechanisms to promote good contributions. Other, more explicit ways of promoting of good content are either via central websites or by way of newsletters (Section 15.8.3).

15.8.2 External Communities and User Groups

SAP has many local or regional user groups. The two largest are the Americas' SAP Users' Group (ASUG) in the United States (ASUG) and the German-Speaking SAP User Group (DSAG) in Germany. They regularly either have continuous working groups on enterprise architecture or address enterprise architecture in their annual events.

The Open Group (which sponsors TOGAF) not only organizes conferences, but it also has working and standards groups. It might be questionable how much return on investment you'll get, but if you like the idea of helping shape TOGAF's direction, the Open Group will welcome you to do so. Note, however, that membership with the Open Group is not free.

We listed relevant conferences in Section 15.6.6, primarily as a source of enablement, but conferences clearly have a dual role as community events as well.

15.8.3 Communication

If you have invested in enterprise architecture tooling, then ideally, the tool is the best way to communicate the latest on your enterprise architecture strategy. Good tools offer the possibility of embedding specific artifacts (such as iFrame) in an arbitrary website (e.g., SharePoint).

Given the fact that there are multiple communication needs and multiple user groups, a single (static) page or a single monthly newsletter won't be enough. We therefore recommend that you ask the following questions during communication planning:

- *Who* should receive the information? This is the most central question, as the what, when, and how will all depend on the audience. The C-level won't be interested in getting daily updates on single architecture artifacts.
- *What* is important enough to be communicated?

- *When* and how often should you communicate? Different users will have different preferences for how often they wish to be informed. Core architects might prefer daily updates, while others will prefer weekly or even monthly communications.
- *How* should you communicate? Offer alternative communication channels, if possible.

Figure 15.16 gives concrete examples of what to think about in terms of governing the communication process and the who-what-when-how dimension.

Governance	Communications Planning	Define Accountability	Collect Feedback	Measure Success	Run Communication	Adjust and Optimize	
Who	C-Level	Business Stakeholders	Project Teams	Functional IT Teams	Technical IT Teams	Enterprise Architects	External Partners
What	Principles and Standards	Learning Offerings	Project Results	ADRs	Business Architecture	Solution Architecture	Inventories	KPIs, Dashboards
When	Quarterly	Monthly	Daily	Real-Time, On Change	On-Demand (Pull)		
How	Website	Newsletter	Email	Community Meetings	Micro Video	Dedicated Meetings	

Figure 15.16 Communication Planning on Enterprise Architecture

15.9 Use Cases for Practice Evolution

We investigated multiple uses cases in Part II, and the focus there was on how to find the best target architecture and transition approach. The two use cases we'll present in this section focus on the evolution of their enterprise architecture practices. We'll see that the root cause of poor as-is architecture can be a nonexistent or poorly developed architecture practice.

The first use case, in Section 15.9.1, will highlight a company with multiple distributed enterprise architecture practices. Section 15.9.2 will discuss a second use case with one divisional enterprise architecture team, the relevancy of which has been questioned by the company.

15.9.1 Establishing an Enterprise Architecture Practice

At the starting point, BIKES-COMP EUROPE Ltd. focuses its business on bike components and runs its headquarters in the United Kingdom, from where the two other market regions, Germany and Benelux, are served, as depicted in Figure 15.17. In addition, there's a commerce-specialized branch run from Spain. The headquarters IT function covers all but the commerce division and some central HR and procurement solutions. The objective of the planned business transformation is to harmonize the business model in all four countries. As part of this transformation, the company envisions having a strengthened IT organization with a global enterprise architecture.

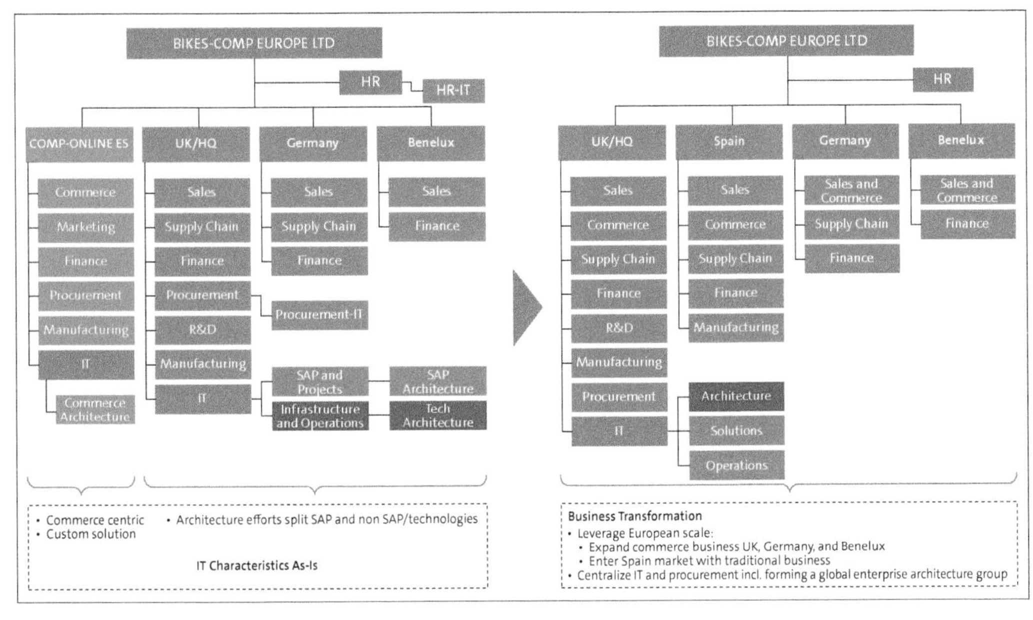

Figure 15.17 BIKES-COMP EUROPE Business Transformation

Figure 15.18 shows the maturity assessment, clustered first by the commerce-oriented unit run from Spain, second by the application centric-units (SAP, HR, and procurement), and third by the Infrastructure & Operations group.

Unit Architecture Practice Focus	Current Maturity State Assessment	As-Is Maturity Assessment
Commerce-Spain • Java • Micro Services • Web Frameworks	**2-Managed** Commerce platform development is part of business unit, therefore high business alignment. Team has grown without clear cost management. Tool support primarily from development platform (web services, ...).	Business Alignment Completeness/Quality Cost Management Enterprise Architecture Development Enterprise Architecture Review/Approval Tool Support
HQ SAP Practice • Systems • Processes • SAP BW HR Procurement	**3-Defined** Fragmented approaches in procurement and HR. SAP-centric management with strong cost governance. Clear separation of projects and support.	Business Alignment Completeness/Quality Cost Management Enterprise Architecture Development Enterprise Architecture Review/Approval Tool Support
Infrastructure and Operations • CMDB • PaaS • IaaS	**2-Managed** High-quality and comprehensive documentation of technical components in configuration management database. No business-oriented enterprise architecture efforts.	Business Alignment Completeness/Quality Cost Management Enterprise Architecture Development Enterprise Architecture Review/Approval Tool Support

Figure 15.18 Analysis of Existing Enterprise Architecture Practice

The result shows a mixed bag of experience, with the potential to leverage synergies but also clear gaps when it comes to the core enterprise architecture focus.

The vision of a future holistic practice was built after the analysis of the current state was completed. Figure 15.19 shows the future focus, which is to leverage the best approaches from the three existing practices company wide.

Value Drivers for Global Enterprise Architecture Practice	Target Maturity Levels

• Improve company-wide alignment between business and IT
• Identify strategic area for IT investments
• Improve data management capabilities and enterprise-wide integration approach
• Integrate silo solution-based IT organizations
• Introduce enterprise architecture tooling to provide transparencies on:
  • Investment priorities
  • Integration
  • Data architecture
  • Costs
• Establishment of a dedicated enterprise architecture team sourced from all relevant IT teams. Reporting line to CIO (HC initial: 10, target: 20)
	Business Alignment
	Completeness/Quality
	Cost Management
	Enterprise Architecture Development
	Enterprise Architecture Review/Approval
	Tool Support

Figure 15.19 Business Case for Integrated Enterprise Architecture Practice

In summary, it can be said that centralizing the IT function follows the business expansion and harmonization strategy. Integrating the various architecture approaches and combining the existing teams provides a good foundation for quickly establishing a strong enterprise architecture function, which can further help shape scalability and growth for the BIKES-COMP EUROPE.

15.9.2 Combining Two Architecture Practices of Different Maturities

This use case investigates a company called DUAL SPEED Inc., which previously operated in two largely independent divisions:

- The services division had a well-established enterprise architecture. Core systems were largely non-SAP, and only finance ran an SAP ERP system. Enterprise architecture was a defined practice with an established enterprise architecture tool.
- The devices division was of roughly the same size. It ran a central SAP ERP system supplemented by SAP supply chain and manufacturing solutions.

Figure 15.20 depicts the organizational perspective of DUAL SPEED Inc.'s business transformation. The change transforms all functions from a divisional organization into global functions.

With the deconstruction of the separate divisions, the two IT functions will be integrated into one single unit. This section won't discuss the specific architecture challenges but will focus on organizational aspects that should be considered when setting up the enterprise architecture function for the combined business.

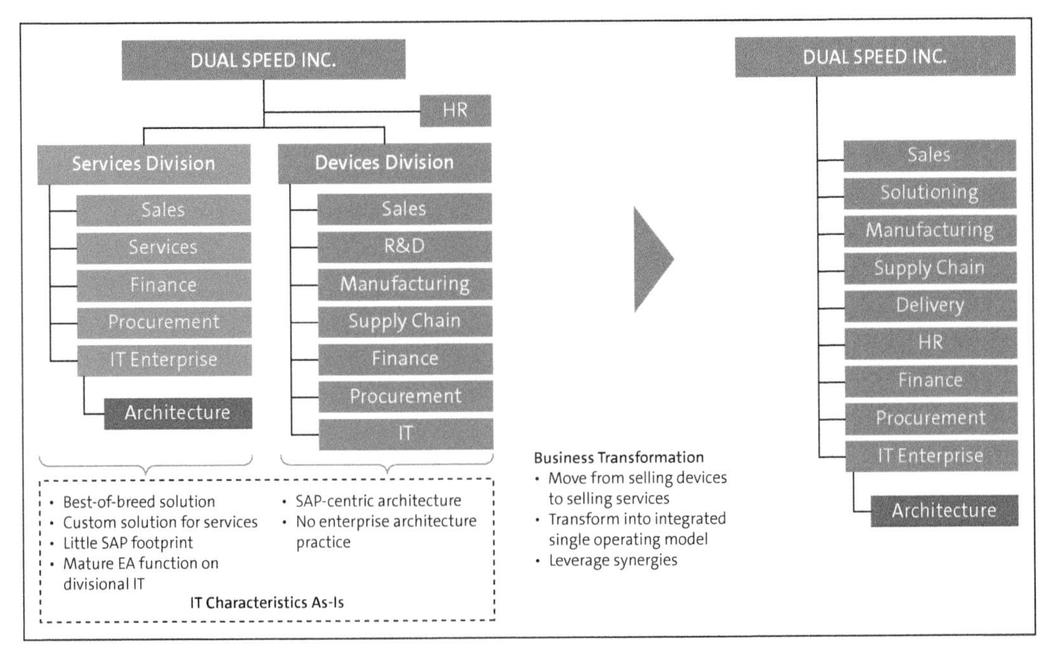

Figure 15.20 DUAL SPEED Inc. Business Transformation

A first transformation step was to assess enterprise architecture maturity, as depicted in Figure 15.21. This showed a clear maturity deficit in the SAP-oriented devices division. In the services division and despite a more mature overall enterprise architecture practice, DUAL SPEED Inc. faced an outdated core services system.

Unit Architecture Practice Focus	Current Maturity State Assessment	As-Is Maturity Assessment
Service Division • Non-SAP solution • Highly customized • Mature enterprise architecture practice	**3-Defined** For service business, a non-SAP packaged solution was selected, but multiple custom extensions had been added over the years. Good business alignment, but doubts around whether the platform still can support expected business growth. Landscape and integration are well document via mature enterprise architecture processes and good tool support.	Business Alignment Completeness/Quality Cost Management Enterprise Architecture Development Enterprise Architecture Review/Approval Tool Support
Devices Business Unit • SAP-centric architecture	**1-Initial** No enterprise architecture focus; SAP-centric management of integration and changes for finance, supply chain, and order to cash. Partially outsourced application management. Little transparency on costs.	Business Alignment Completeness/Quality Cost Management Enterprise Architecture Development Enterprise Architecture Review/Approval Tool Support

Figure 15.21 Baseline Maturity Assessment of DUAL SPEED Inc.

Initially, the idea for Dual Speed Inc. to build its future enterprise architecture function was to take the team that had previously done the job for the service division and have it cover the enterprise scope. This vision is outlined in Figure 15.22, but transformations are not always easy. The following considerations were therefore taken into account, and that led to the formation of a more differentiated enterprise architecture practice adoption plan:

- The existing team had only a few SAP skills, so it made sense to identify team members from the former Devices Division who would be able to play an equally strong role in the new enterprise architecture practice
- It did not make sense to adopt all practices that had been successfully adopted in the Service Division globally, as the Devices business followed different product standards and enterprise architecture principles. Therefore, it was recommended to rethink and refine the existing Services best practice for the enterprise architecture development process.
- Enterprise architecture still was seen as a strong enabler, especially during the transition phase of the transformation. To achieve broad acceptance, key stakeholders from all parts of the former divisions were to be engaged in and convinced of the opportunity. Identifying the right thought leaders from both sides of the organization was largely done to increase acceptance of the results.
- Adopting the enterprise architecture tooling made sense, but it needed to be validated, especially with respect to how it would be able to support the transformation. It was equally necessary to assess whether the Services tooling was sufficient and—especially for the SAP-oriented solutions—whether additional tooling and modeling guidance was needed to incorporate SAP reference architecture content.

Value Drivers for Global Enterprise Architecture Practice	Target Maturity Levels
• Business alignment wasn't a pain point in the past, but through the integration effort, a clear definition of a common strategy is needed. • Identify strategic areas for IT investments. • Considering external services to get SAP enterprise architecture best practices introduced and adopted. • Leverage existing enterprise architecture tooling from "SERVICES" and supplement if needed by additional artifacts. • Embed the needed enterprise architecture governance in the starting governance structures for the overall transformation program.	3-Defined Business Alignment Completeness/Quality Cost Management Enterprise Architecture Development Enterprise Architecture Review/Approval Tool Support

Figure 15.22 Business Case for Strengthening Enterprise Architecture Practice

15.10 Summary

For a newly established enterprise architecture practice to be impactful and successful, it needs to be well integrated with different parts of the organization, especially through collaboration with other IT functions like the project management office as well as innovation and strategy groups. Aside from organizational accountabilities, a good balance between architecture development or governance and clear communication with all relevant teams and stakeholders is a key success factor for a well-accepted architecture practice that has impact.

You've learned the building blocks of setting up your own enterprise architecture practice; in the next chapter, we'll walk through real-life example practices within SAP.

Chapter 16
Enterprise Architecture at SAP

In Chapter 2, we introduced "practice" as one dimension of SAP EA Framework. In this chapter, we'll look at multiple enterprise architecture practices within SAP that use the framework for their different purposes while further evolving it in different board areas.

SAP is an organization like any other, so its IT requires and runs an enterprise architecture function within the internal IT organization. But other business units—like development, sales, and service—join in the use of enterprise architecture to fulfill their business goals.

In this chapter, we'll start with a look at cross-unit efforts and collaboration to define and refine SAP EA Framework. The subsequent sections will look at the individual sub-practices: advisory services, product development, and internal IT.

16.1 Overview of Enterprise Architecture Practices at SAP

In this section, we'll describe the linkages and commonalities among the different enterprise architecture practices in SAP as depicted in Figure 16.1.

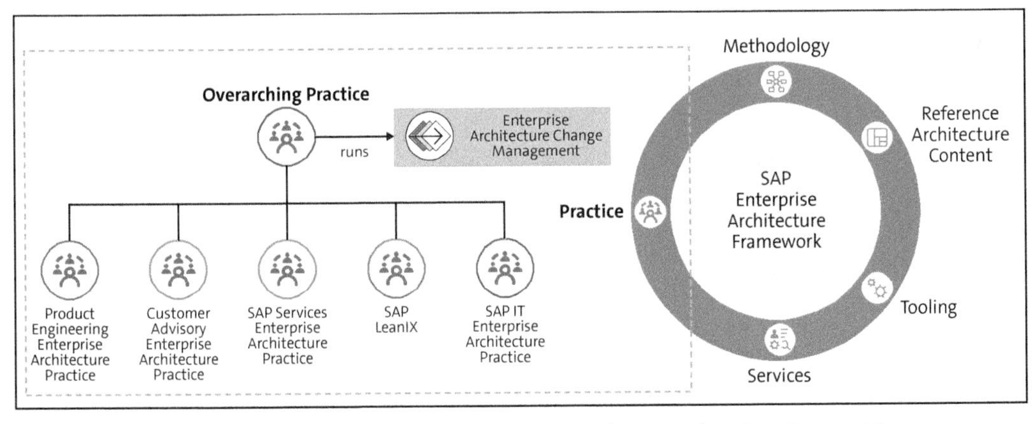

Figure 16.1 Individual Enterprise Architecture Practices Orchestrated as One Overarching Practice

The individual practices reside in different board areas, and the aggregated overarching practice primarily ensures the evolution of the common methodology and the overall framework. We'll describe these different subpractices in the course of this chapter.

Figure 16.2 specifies which areas the various enterprise architecture practices at SAP focus on, as follows:

- At the top are customer-facing teams, and the arrows reflect the phases of SAP's Customer Value Journey. Customer advisory focuses on the phases that precede a customer's decision to select a specific SAP solution, while customer service typically concentrates on the implementation phases (adopt and derive).
- The primarily blue-colored middle layer includes product engineering and technology and innovation, the two largest development-focused board areas. SAP LeanIX is depicted as a separate entity to illustrate that it offers professional services even though it's primarily a development unit.
- The purple layer at the bottom depicts the internal SAP enterprise architecture practice.

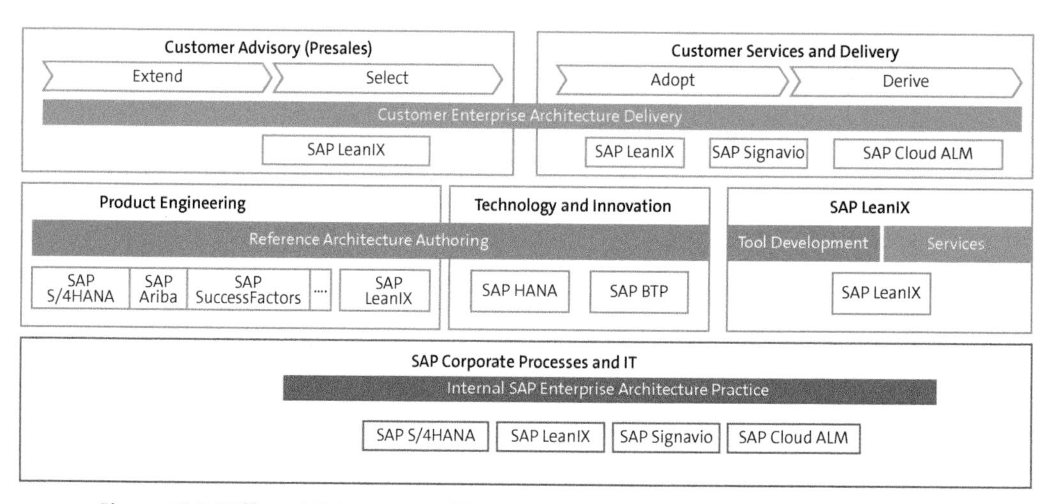

Figure 16.2 Different Enterprise Architecture Practices across SAP's Board Areas

At SAP, the different practices come together to form an overarching enterprise architecture practice provided by an enterprise architecture working group depicted in Figure 16.3. This working group is gathered under the umbrella of *cross-product architecture* and runs weekly work sessions on enterprise architecture methodology. Besides supporting enterprise architecture, the cross-product architecture initiative's objective is to develop a common target architecture spanning different product lines. This includes aspects of the user interface (UI), security, integration, and more.

The various subpractices have different focuses, offer different contributions, and benefit from each other. Figure 16.4 shows how methodology is the overarching factor pushed and co-owned by all parties.

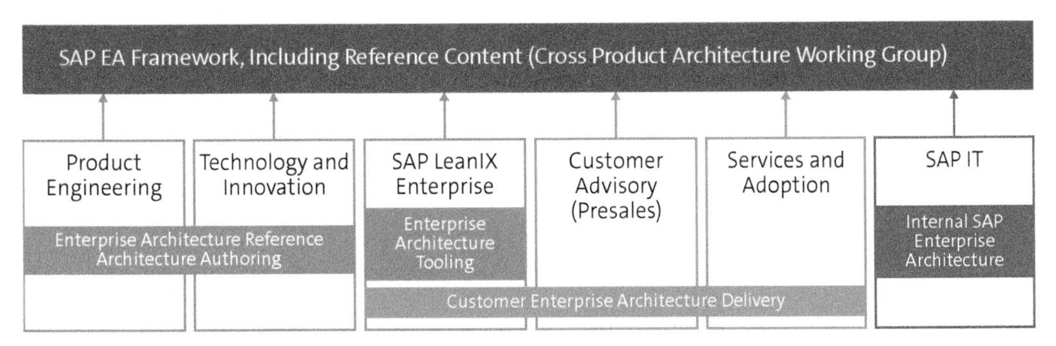

Figure 16.3 Collaboration as Overarching Enterprise Architecture Practice

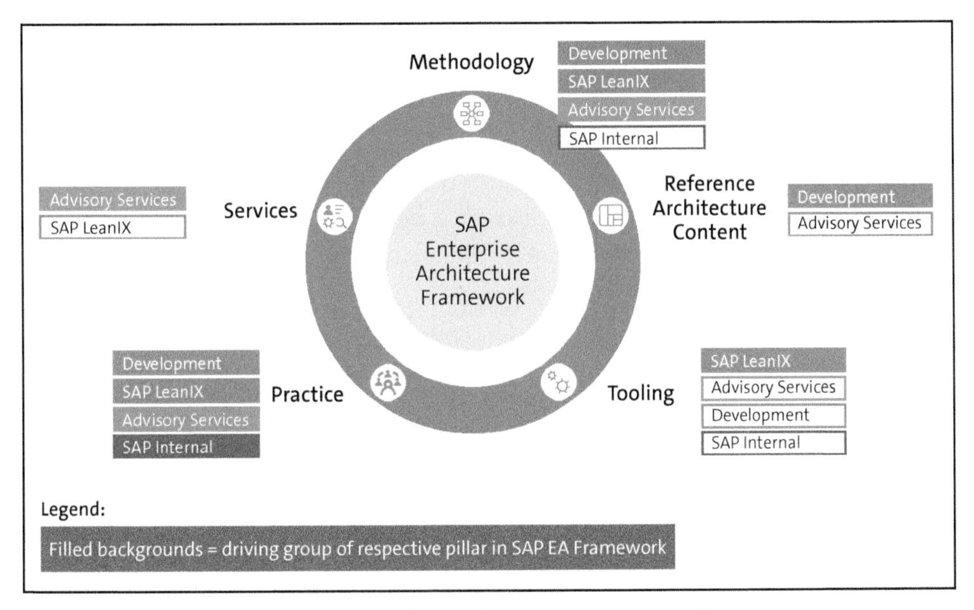

Figure 16.4 Primary Focus and Ownership within SAP EA Framework

Services, reference content, and tooling are led by one party apiece. For the practice pillar, there's no single unit taking the lead, and this entire chapter will give insight into the different perspectives within SAP. What we'd like to highlight here, by stating that all parties take part in driving this topic, is that the practices (despite all their differences) often have operational interlinkages. Here are some examples of these:

- Consulting services depend on the release cycle of the reference content and its availability within the toolchain.
- All practices need to consider adopting the enterprise architecture methodology.
- Product innovations and strategies need to be reflected in the reference content, so that advisory and services can leverage them. Advisory and services can even be the internal customers of the reference content and may therefore need to have a say in determining content development priorities.

The different perspectives become clear when you look at the relationships of practices to other framework dimensions. In subsequent chapters, we'll go through the four main subpractices in more detail:

- Section 16.2 covers the customer-facing areas of advisory and services.
- Section 16.3 combines the product engineering and technology and innovation development units.
- Section 16.4 looks into SAP's internal IT enterprise architecture practice.

We'll close with the perspective from the newly acquired SAP LeanIX unit.

16.2 Enterprise Architecture in Advisory and Services

SAP uses the idea of the *Customer Value Journey* to describe interactions with the customer through five phases:

1. **Discover**
   This is an exploration phase in the customer's needs and the future architecture are not yet clear. This is why architecture requirements and target architecture development are typically triggered in this phase.
2. **Select**
   Once requirements are clear, the concrete target architecture and initial roadmap are developed and solutions are selected.
3. **Adopt**
   In this phase of the journey, enterprise architecture is refined and implementations to adopt the solutions start.
4. **Derive**
   The solutions are put to use to maximize their value.
5. **Extend**
   This can be an innovation and growth phase, initiating a new adoption cycle

We'll discuss the enterprise architecture practice for *architecture advisory*, which typically matches presales customer interactions covering the discover and adopt phases of the Customer Value Journey. Architecture advisors often seamlessly work hand in hand with paid enterprise architecture *services*, which go deeper and reach into the adopt and derive phases.

16.2.1 Architecture Advisory (Discover and Select Phases)

The architecture advisors' objective is to explain to customers how SAP solutions will work in their contexts before they buy or subscribe to a software package. We think that leveraging enterprise architecture in a structured approach is superior to trying to win customers over with demos and standard sales pitches.

The advisory architecture leverages the SAP enterprise architecture methodology and the SAP reference content as described in Chapter 2. In comparison with most paid enterprise architecture services, architecture support via "presales" enterprise architects will focus more on a specific opportunity and less on long-term architecture services with workshops.

Such software selection processes are triggered by a software request for proposal (RFP). Even if such RFPs have a tendency to be feature oriented, the architecture advisors will use enterprise architecture approaches and deliverables because it benefits both sides: the customer gets a better end-to-end perspective on the proposed solution, and SAP gets a structured approach with which it can create the outcome that best fits the customer's needs.

16.2.2 Architecture Service Delivery (Select, Adopt, and Derive Phases)

Within the broad SAP service portfolio, enterprise architecture services intend to help the customer with their transformation initiatives. Customers can outtask architecture work or at least seek help with critical SAP-centric transformations. In addition, they can request support in setting up or improving their SAP-centric architecture practice (e.g., by establishing an enterprise architecture board, by defining best practices for capturing architecture decision records).

These customer engagements can be one-off or part of a strategic engagement in which SAP architects come back regularly and help confirm or reassess strategic decisions and principles. Within the framework of such an engagement, SAP might have a consulting voice in customers' enterprise architecture boards.

There are three delivery options:

- **North Star service**
  This option is for setups with little or no insight from former architecture work. Based on an analysis of the baseline, the reference architecture is leveraged, and a target state and transition roadmap are created, largely from scratch. The choice of modules depends on the customer's specific situation. Predefined architecture patterns can help and will be described as one element of the SAP Transformation Hub's scaling approach. For a recap of the North Star service modules, see Chapter 2, Section 2.5 (specifically, Figure 2.1).
- **Target architecture assessment**
  The modules and delivery patterns of this delivery option are very similar to those of the North Star service, with the difference being that a target state architecture is already largely available. Former work will therefore be assessed and incorporated into the selected modules. The only elements that will be filled in will be elements that were not taken into consideration in former iterations.

While such an assessment will ideally confirm the target state architecture, this is not always the case. If major risks or newly available software components are overlooked in the initial build, the assessment can come to the conclusion that the former target state and roadmap need to be revised or even rejected and replaced.

- **Architecture point of view (APoV)**
The smallest parceling units of architecture services are targeted architecture assessments that focus on a catalog of dedicated concerns addressing a specific customer need. These can be function-oriented topics like master data governance strategy or the concept of central finance. Another set of APoVs are SAP Business Technology Platform (SAP BTP) or technology-related elements like extensibility or user experience (UX) strategy.
The design of this offering is to get guidance from the outside in a short period of time (typically two to three days). The guidance will be based on best practices and include a quick customer-specific assessment.

16.2.3 Scaling via SAP Transformation Hub (Internal Scaling)

SAP Transformation Hub, depicted in Figure 16.5, is a team of global enterprise architects primarily grouped in three hub locations: Newtown Square (USA), Walldorf/Rot (Germany), and Bangalore (India).

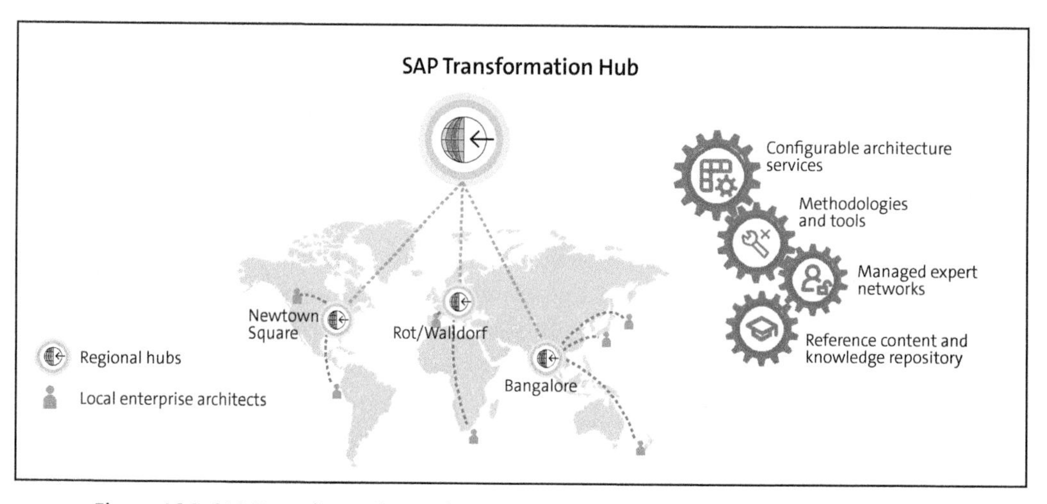

Figure 16.5 SAP Transformation Hub

SAP Transformation Hub's primary objective is to scale best-in-class enterprise architecture to premium engagement customers. These are typically all large enterprises that have a special SAP ActiveAttention or SAP MaxAttention service contract. The services delivered are described in Section 16.2.2. One responsibility of the team is to deliver these architecture engagements to customers, and another is to scale the approach in all SAP EA Framework dimensions. We distinguish three dimensions of scaling:

1. **Scaling in**
   A large global service delivery organization usually will have many slightly different approaches. Focusing on the integration and harmonization of different service offerings, methodologies, and approaches will streamline overarching development efforts and allow for the reduction of integration points and alignment with other board areas. It's important that scaling-in activities be run by well-respected teams that focus on the practicable integration of different perspectives.

   The mindset of scale-in in this context is never a "policing" one, in the sense that only one global approach is valid and no variations are allowed. Instead, it will seek to combine multiple differing approaches in a flexible, adoptable, and adaptable approach. This will increase buy-in and willingness to contribute local best practices.

2. **Scaling up**
   The scaling-up approach provides a common ground for assessing further development needs. These needs can fall into many areas such as extending the methodology or enhancing tools and automation capabilities, but they could also include a stronger quality management focus or improved engagement and delivery models.

   We subsume the enhancement of an organization's capabilities under the term *scaling up*. Efforts to implement scaling-up ideas will typically run as small projects, which can be staffed from multiple parts of the organization. Having sufficient funds for a scaling up is critical, as a collaboration based only on goodwill can slow things down.

3. **Scaling out**
   The SAP service unit operates globally and includes many delivery organizations and teams grouped by function, industry, and technology. Often, enterprise architecture is only one of the service approaches used. The scale-out focus of the SAP Transformation Hub promotes the decentralized execution of a common service delivery methodology. Enablement, communities, and central delivery support functions like quality management are a few of the elements that help with effectively using the common approach.

Scaling out at SAP also implies bringing the partner ecosystem on board. Some partners already use enterprise architecture in their customer engagements but would still like to adopt SAP reference content and tooling. Helping other specialized implementation partners better understand the SAP enterprise architecture philosophy can help pave the way for more holistic customer engagement.

On top of the three dimensions we've explained, we see *scaling down* as a fourth dimension of scaling, but we currently don't apply it to enterprise architecture. Scaling down primarily means doing less. In the area of services for enterprise architecture, it could imply covering fewer industries, not all enterprise architecture domains, and/or not all technological categories. It could also mean discontinuing services in some regions. Scaling down is not a negative strategy, per se. Even successful businesses may scale down by focusing on their core business or core markets while possibly growing their core segments.

Figure 16.6 visualizes the three scaling dimensions paired with the five dimensions of SAP EA Framework.

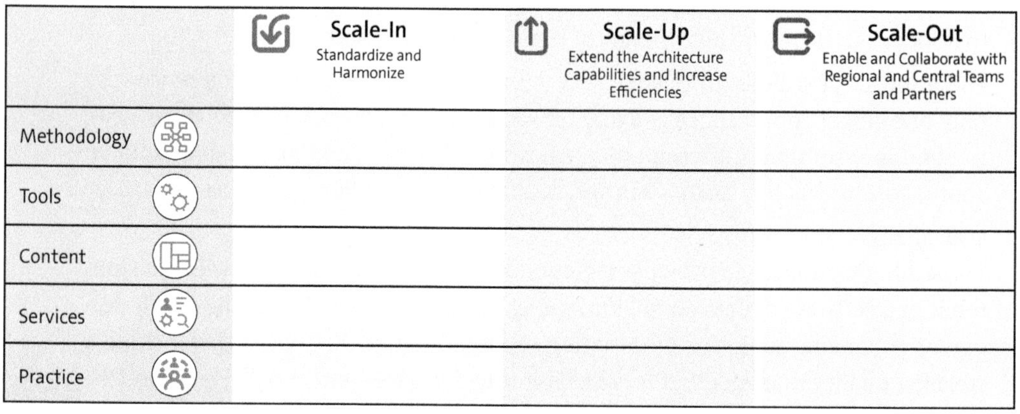

	Scale-In Standardize and Harmonize	Scale-Up Extend the Architecture Capabilities and Increase Efficiencies	Scale-Out Enable and Collaborate with Regional and Central Teams and Partners
Methodology			
Tools			
Content			
Services			
Practice			

Figure 16.6 Scaling Concepts for SAP Services Enterprise Architecture Practice

Table 16.1 gives a few examples of scaling priorities.

	Scaling In	Scaling Up	Scaling Out
Methodology	Ensure consistency from authoring to customer consumption.	Align SAP EA Methodology with SAP LeanIX meta model.	Provide methodology support via prebuilt templates in SAP LeanIX.
Tools	Distill best practices from existing tools and office-based low-key approaches.	Leverage and increase autodiscovery options and automate architecture recommendations. Collaborate with the SAP LeanIX unit on tool requirements.	Integrate transformation platforms, including SAP Cloud ALM and SAP LeanIX.
Content	Align various reference content.	Elevate learnings from customer deliveries to reference content.	Publish applicable reference content for customers and partners.

Table 16.1 Sample Scaling Actions

	Scaling In	Scaling Up	Scaling Out
Services	Eliminate redundant services and align in keeping with the North Star service approach.	Link and strengthen the data domain-related service offering. Refine the service modules.	Enable low-touch and semiautomated service consumption. Integrate enterprise architecture-related services with other transformation offerings.
Practice	Align activities among different subpractices.	Use community for early feedback on new concepts.	Establish regional internal and customer-facing communities.

Table 16.1 Sample Scaling Actions (Cont.)

The primary focus of the SAP Transformation Hub team is to run customer-facing enterprise architecture engagements. In these service engagements, the architects don't act as single heroes, but they typically build a service team that centers on the transformation hub as leading the approach but also involves functional and technical experts.

16.3 Enterprise Architecture in SAP Product Development

Pockets of product development in SAP are located in most level 1 board areas (Product Engineering, Technology and Innovation, and Customer Services & Delivery). Of these, the largest are Product Engineering (which is responsible for the core SAP S/4HANA ERP solutions) and Technology and Innovation (which is responsible for SAP HANA and SAP BTP).

Enterprise architecture techniques are relevant to designing and developing a software product. In a strict sense, SAP development doesn't create an enterprise architecture but provides building blocks that customers can use when building their own enterprise architectures. These building blocks largely form the reference content for the SAP EA Framework.

In the following sections, we'll discuss the enterprise architecture for both technical components and business functions in product development.

16.3.1 Technical Component Architecture

Within SAP, the reuse of IT components in all solutions drives simplicity and efficiency. Internally, SAP follows an explicit set of product guidelines and standards and sets up common architecture principles to ensure consistency and quality and thus reduce development efforts.

The list of topics that should, ideally, be handled in the same manner or even by employing the same software subcomponents is long and by no means exhaustive:

- User interface
- Workflow
- Extensibility
- Search
- Integration/API design
- Analytics
- Master data management
- AI and rules management
- Authentication and authorization handling

Beyond application-related considerations like these, a cloud company like SAP will focus on standardizing its cloud operations. This perspective entails a second list of concerns that is not exhaustive:

- Logging
- Monitoring
- Containerization
- Security
- Data center strategy

The majority of both application-related and operations-related components are located on SAP BTP. This enables customers to use them when composing or developing their own solutions—in the same way and based on the same SAP BTP components as SAP's SaaS solution. Chapter 5 took a closer look at an SAP BTP-enabled cloud transformation use case.

The broad exposure of SAP BTP components is why this architecture plays so strongly into a more generalized enterprise architecture for customers and has potential that reaches way beyond SAP's internal system architecture.

SAP regularly acquires new solutions to complement its product portfolio. This means that change is in progress until the new products all follow the same architecture principles and use common components. In larger acquisitions (for example, SAP Ariba and SAP SuccessFactors), the acquired solutions won't align with the criteria listed previously. Instead, you need to lay out a specific adoption roadmap that balances the upkeep of the products' core characteristics with benefits from adopting SAP architecture principles. See Chapter 8 for mergers and acquisitions (M&A) uses cases.

16.3.2 Business Function Architecture

SAP offers a comprehensive set of solutions to customers that ensure the following:

- The solutions are integrated well.
- The solutions don't overlap with each other.
- The solutions can be configured to customers' individual needs.

Developing these solutions and documenting them using a subset of enterprise architecture artifacts is not equal to developing a specific enterprise architecture. As a software vendor, SAP develops solutions that fit most customers or that are targeted to specific industry needs. Requirements to newly develop or further improve existing solutions typically come from SAP's customer base. In that sense, the development objective is to fulfill the needs of many combined customers. As SAP commits to providing documentation of reference architectures, the generation of enterprise artifacts becomes a standard of the software development process. The most important artifacts are as follows:

- Business capability models (cross-industry and industry-specific)
- Solution value flow diagrams linking business activities to solution capabilities
- Best-practice business process BPMN diagrams
- Solution component diagrams
- Data flow diagrams (including integrations)
- Data architectures

Architecture modeling for development is more than documenting "just" a company-specific architecture. What makes it so difficult is the fact that multiple industry models need to be developed in parallel and will only scale nicely if the common parts can be shareable.

The enterprise architecture term *roadmap* has a different semantic here: from SAP product perspective, it specifies the time when software components are planned to be ready. This is totally different from the customer-specific adoption roadmap, which depicts a customer-specific adoption plan.

For SaaS components, a product roadmap immediately has relevancy for customers, as in most cases, the new features and functions will be available without any further project work. Hence, it's important that reference architectures, process best practices, and product roadmaps all use the same taxonomy, which can also be used within the customer-specific models.

16.4 Enterprise Architecture in Internal IT at SAP

SAP's IT function is part of Corporate Processes and Information Technology (CP & IT). Its purpose and vision are to power SAP's next phase of business transformation to prepare the company for the AI age. For an easier understanding, we'll use the informal term *SAP IT* in the following. We'll also discuss the benefits of having IT reach beyond the internal organization.

The scope of SAP's internal enterprise architecture team is naturally broader and can only be fully covered by including non-SAP solutions. It thus covers ServiceNow and Microsoft platforms like Office, Teams, and SharePoint. From a process perspective, the architecture teams need to support SAP's business transformation into a cloud company, which includes transformations in most sales, services, and development units. The classical IT services centered on collaboration, document management, and end user devices are critical to a knowledge (worker)-based company.

In Section 16.4.1, we'll show in which ways the architecture team structures its architecture services offerings; in Section 16.4.2, we'll describe collaboration with development departments; and in Section 16.4.3, we'll dig deeper into the role enterprise architecture plays in SAP's strategic transformations.

16.4.1 Architecture Services

The internal IT architecture group has designed services for SAP's strategic transformations in the same way that the service organization offers income-generating architecture services to our customers.

Figure 16.7 shows a set of high-level transformation areas supported by enterprise architecture services. The long-term overall transformation for SAP can best be described as one from a single on-premise product company to a process-driven cloud company. The lower part of the diagram depicts the enabling enterprise architecture group services. These are structured into the following three categories:

- **Core services**
  This is where an estimated 70% of the architecture group's capacity is spent. In earlier chapters, this was described as enterprise architecture development. Major artifacts are the target and interim architectures and the roadmap. Work will typically consist of specific evaluations (architecture investigations) or of architecture work planned as part of larger transformations.
- **Foundation services**
  These services set the overall vision and strategy distilled into the business capability framework. Note the explicit feedback loop to SAP's development teams (development IT engagement). The foundation service accounts for 20% of the group's resources.

- **Architecture governance**
  These services account for a mere 10% of the overall capacity. They provide the architecture guardrails (principles) and operate the design and architecture review boards, an approach that is fully in line with our overall philosophy that governance is needed but should not be the architecture team's main purpose.

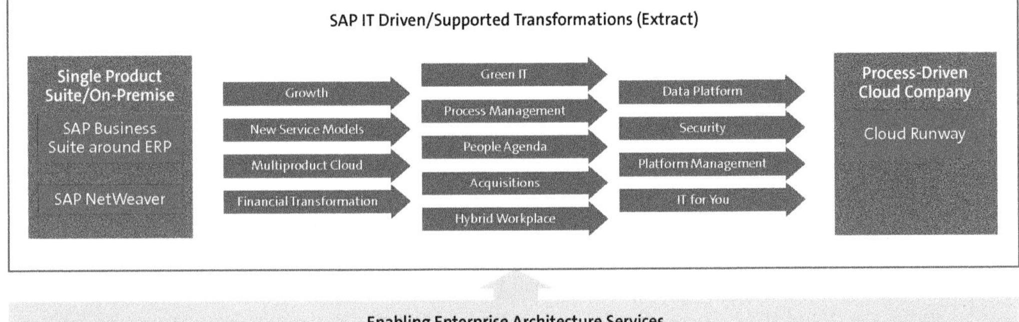

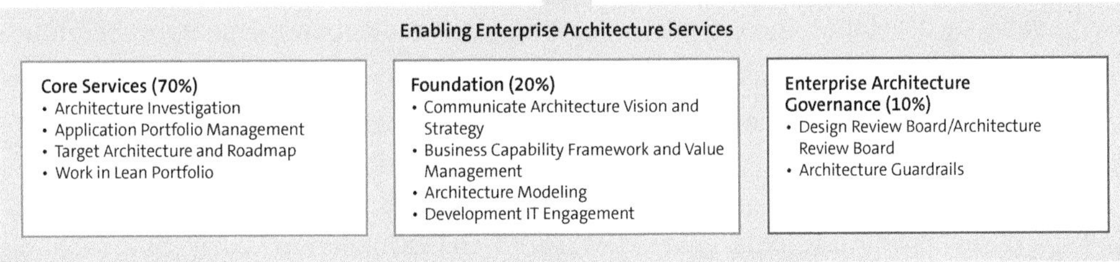

Figure 16.7 Enterprise Architecture Services Supporting Business Transformations

Figure 16.8 shows the logical sequence of and the flow among the services. The architecture board brings together stakeholders from more areas than just the office of the CEO board in which the strategic processes and IT organizations are positioned.

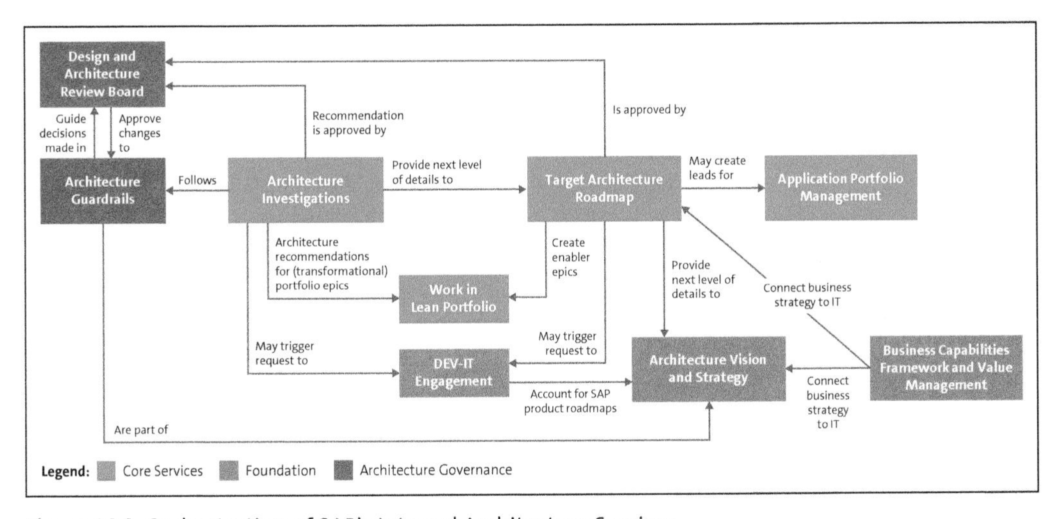

Figure 16.8 Orchestration of SAP's Internal Architecture Services

16.4.2 Collaboration with Development

In Figure 16.8, DEV-IT Engagement indicates the collaboration with SAP's development department. The fact that SAP is an early adopter of its own new solutions obviously acts as a catalyst. One motivation for this adoption is to test the value of a solution within SAP early on, as SAP customers like to see how SAP runs SAP.

SAP is large and operates globally, which comes with special requirements. These are not always matched by the capabilities provided by the SAP product portfolio, but a gap like this turns into opportunities when external products that cover it enhance SAP's products.

When prioritizing a development backlog, it's important to note that these requirements can conflict with other items supporting other industries.

Adopting solutions as a frontrunner implies listening to the initial users and reacting to their feedback and needs. This dialog within SAP is wanted, even if some long-term software developers tend to not like getting input and requirements from their internal IT department.

In some cases, the gap between what is needed and what SAP's current product portfolio can offer is too big. Let's take two well-known examples: process management and enterprise architecture management. Both functions had been implemented in SAP IT long before SAP acquired Signavio and LeanIX, and SAP Solution Manager's capabilities for process management and Sybase Power Designer's enterprise architecture features could not be easily enhanced in a way to support SAP's own transformation needs.

Table 16.2 describes the evolution from the business need (via the use of a best-in-class external solution) to both Signavio's and LeanIX's solutions being fully integrated into SAP's product portfolio.

Area	Business Process Management	Enterprise Architecture
Need within SAP's IT	Analyzing, measuring, redesigning and/or optimizing many critical business processes was required for SAP's own business transformation.	As a supporting function, documenting the needed business capabilities—including measuring their maturity—was essential. Simplifying the IT landscape by adopting a cloud-based, future-proof target architecture was equally important.

Table 16.2 Examples of Formerly External Solutions Integrated into SAP's Product Portfolio

Area	Business Process Management	Enterprise Architecture
Starting point	SAP solution manager was available in house, but it primarily addressed large IT implementation projects as opposed to business transformations.	The original SAP Enterprise Architecture Designer and SAP PowerDesigner platforms found their way into SAP with the acquisition of Sybase, whose core technologies had centered on databases and mobile data. These platforms' capabilities as modeling tools covering data and processes were legend, but the lack of data-driven visualizations and inventory assessments disqualified the tooling of these in the context of internal use cases.
Best-of-breed engagement	Signavio was selected as the best-in-class process modeling and analysis tool.	LeanIX was selected for use in the enterprise architecture domain.
Integration into SAP	After the acquisition, business process analysis and improvement capabilities, which were formerly anchored in SAP Solution Manager, got repositioned to become part of the SAP Signavio Suite. The "plug and gain" visualization metric is a good example of the synergies gained.	LeanIX was acquired by SAP in 2023.

Table 16.2 Examples of Formerly External Solutions Integrated into SAP's Product Portfolio (Cont.)

The two cases illustrate the duality of enterprise architecture activities needed in the wake of the acquisition:

- **Integrating and consolidating internal IT systems**
  These activities are in line with the use case discussed in Chapter 8. Later in this chapter, Figure 16.12 shows an SAP LeanIX matrix report for the tracking of a consolidation plan within an acquisition initiative.
- **Integrating the customer-facing product portfolio**
  These activities don't fall into the duties of the SAP IT department but will be handled by the product engineering and development teams.

16.4.3 SAP's Strategic Transformation Portfolio

At SAP, CP & IT forms a central organizational unit that is part of the Strategy and Operations department in the Office of the CEO board area. This enables SAP's enterprise architecture group, which is part of CP & IT, to play a strong role in SAP's strategic portfolio process.

Figure 16.9 describes SAP's strategic transformation portfolio process and the embedded enterprise architecture efforts. The portfolio process is split into two parts:

- **Strategic initiatives: Qualification to approval**
  This first part of the process starts with an initial review to bring ideas into the identified state. Next is an architectural readiness approval (or rejection), and only after this architectural qualification will the final decision be taken in consideration of other criteria (e.g., costs versus benefits).
- **Strategic themes: Execution mode**
  The second part of the end-to-end process brings initiatives from backlog to execution. During delivery, the enterprise architecture team gets involved via architectural assessments.

  Once the initiatives are finalized, the important follow-up needs to make sure that the enhanced capabilities are updated with their increased capability maturity realized.

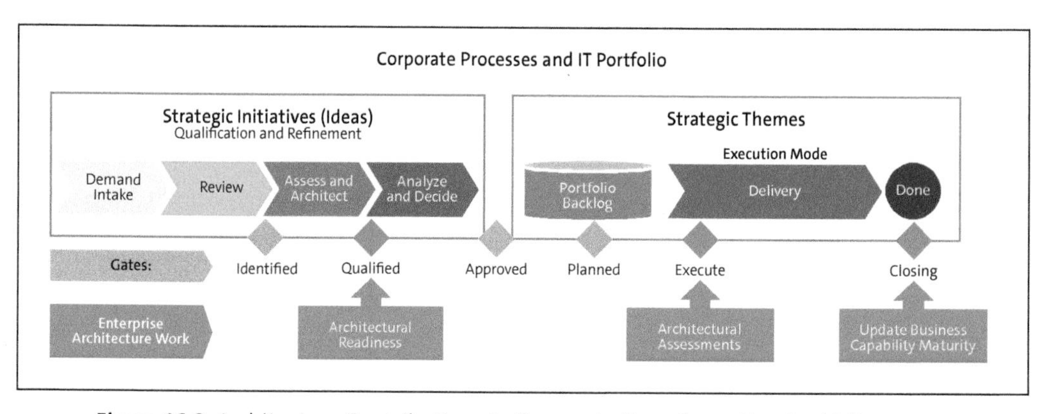

Figure 16.9 Architecture Contributions in Corporate Transformation Portfolio

In complex transformations, the SAP IT architecture team may contract with architecture services that will support it in its architecture development activities.

The following section will provide insights into how SAP LeanIX is used as an application portfolio tool, both for transformation support and for M&A.

16.4.4 SAP's In-House Use of SAP LeanIX

As described in Section 16.4.2, SAP's internal IT introduced Signavio's and LeanIX's products long before those companies became a part of the SAP family. The good experience with those products the internal IT organization had may not have been the decisive argument in favor of them, but it was clearly a strong one.

The center of Figure 16.10 depicts the fact sheets used within SAP. Interface fact sheets are mostly imported by way of an integration governance tool via SAP BTP. The bidirectional integration with SAP Signavio depicted on the right-hand side is equally important because it enables SAP Signavio to reference incoming business capabilities from SAP LeanIX and sends back business processes in return.

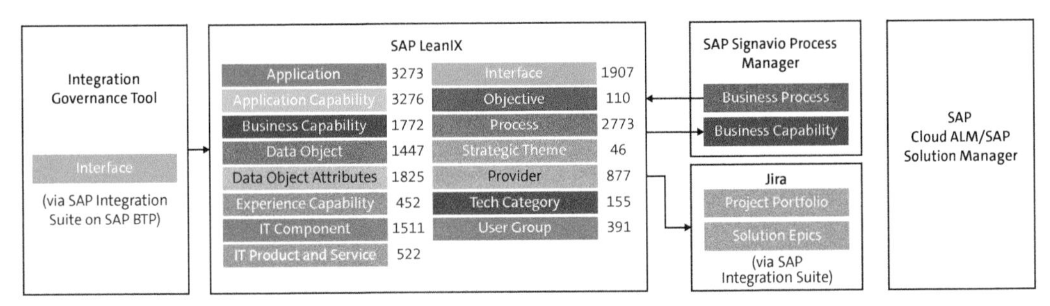

Figure 16.10 SAP LeanIX and Its Integration within SAP IT

A few specific artifacts used by SAP LeanIX are as follows:

- **Business capability model**
  Figure 16.11 shows a typical business capability model that uses a landscape report as introduced in Chapter 12. Strategic relevancy from low to critical is depicted by way of heatmapping with colors, and maturity is depicted with star icons in the bottom line of each box. The clustering in this report is realized via the respective parent so that it's easily possible to filter down to a specific function like sales or marketing.

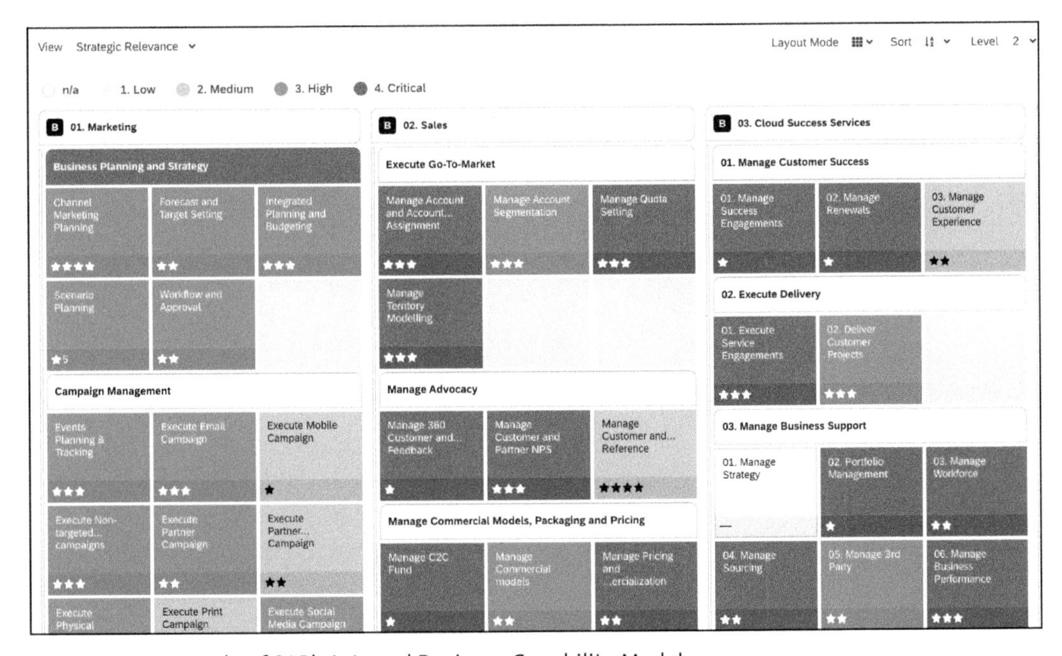

Figure 16.11 Example of SAP's Internal Business Capability Model

- **Application consolidation during M&A**
  As SAP regularly acquires small, medium, and large companies, in-depth analyses and consolidation strategies are mandatory steps on the way to possible synergies and creating a seamless operation. Vertically, Figure 16.12 shows different possible product strategies from repurchase to retire and retain. Horizontally, Figure 16.12 tracks consolidation status from in clarification (light blue) to project completion (dark blue).

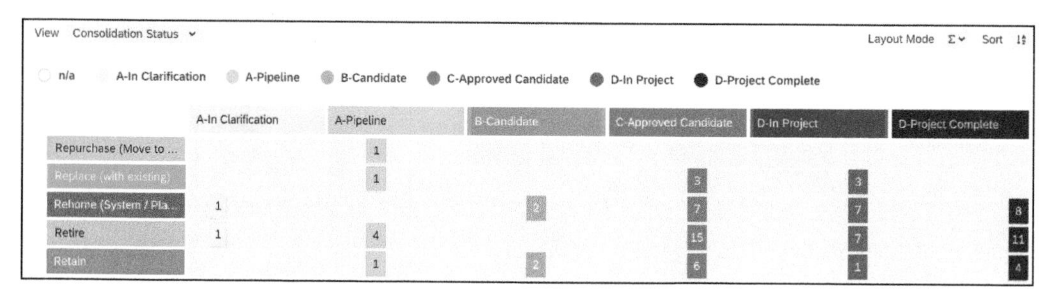

Figure 16.12 Tracking of Consolidation Strategy during M&A

- **Application radar**
  Figure 16.13 shows an application radar. The sectors depict business criticality and the ring categories match the functional fit, which is also heatmapped with colors.

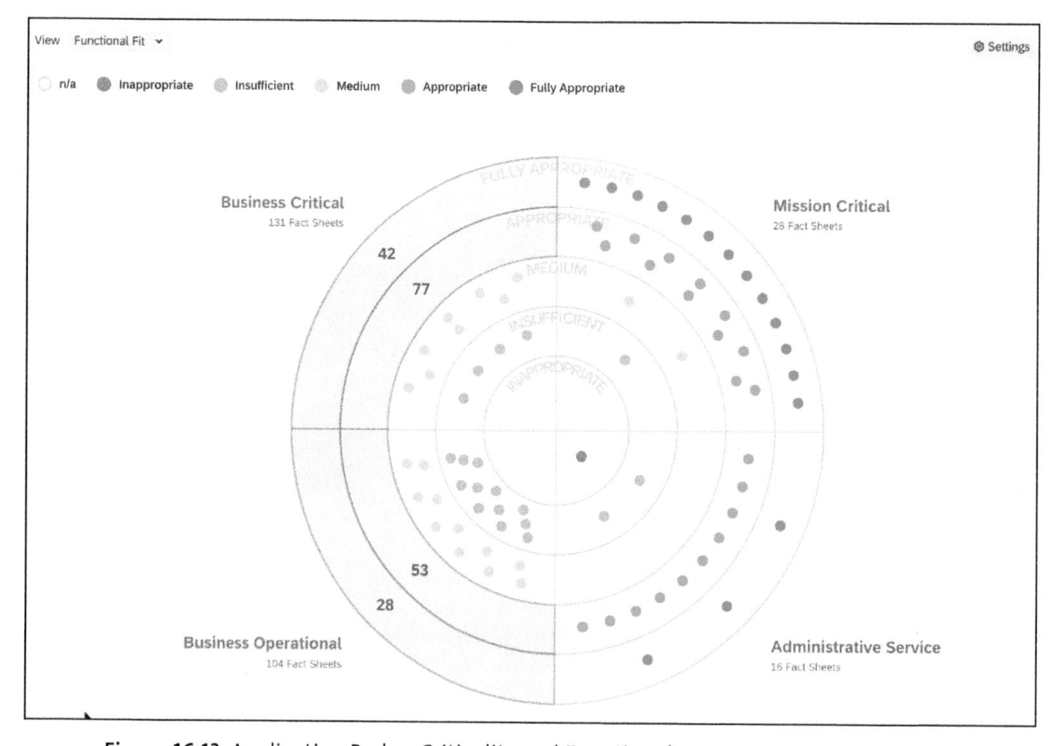

Figure 16.13 Application Radar: Criticality and Functional Fit

- **Application retirement roadmap**
  The application burndown chart depicted in Figure 16.14 shows the past and future progression of the retirement of an application. As is customary, the figure uses quarters as the horizontal time dimension. The left *y* axis depicts the number of retired applications shown as vertical green bars, while the right *y* axis and the orange line describe the total number of applications.

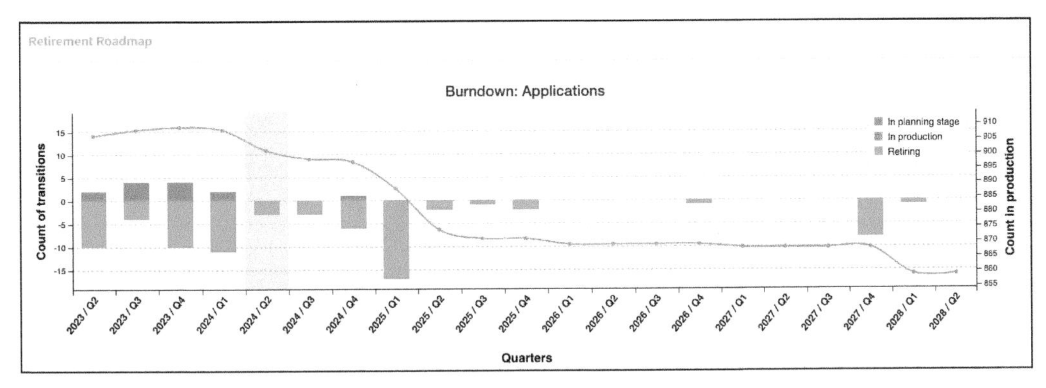

Figure 16.14 Application Retirement Roadmap

To gain qualified analyses from the SAP LeanIX repository, you need to ensure good data quality. For this purpose, the SAP enterprise architecture team has set up standards and modeling guidelines for all fact sheet types. These include the following:

- A "How to Model" guide for all major fact sheet types: application, IT component, business capability, application capability, IT product, interface, and IT product fact sheets
- A governance process defining accountabilities by fact sheet type
- Quality reporting
- A roadmap of supported use cases and integrations planned

There are clear modeling roles defined for each fact sheet type, some of them mandatory and others optional. Table 16.3 shows an example of the roles for modeling interfaces.

Role	Description	SAP LeanIX Modeling	Data Quality Role
Architect	Responsible solution architect or delivery unit architect.	Subscriber	Mandatory

Table 16.3 Roles for Modeling Interfaces

Role	Description	SAP LeanIX Modeling	Data Quality Role
Interface owner	Main technical owner of the application the interface belongs to. In most cases, this will be the application receiving the data. This role corresponds to the IT product owner.	Subscriber	Mandatory
Data architect	Responsible for data architecture.	Subscriber	Optional
IGT requestor	Contact person requesting the interface via the interface governance tool (IGT).	Subscriber	Optional
Integration developer	Implementer from the integration team or delivery unit.	Subscriber	Optional
Enterprise architect	Enterprise architect.	Subscriber	Optional

Table 16.3 Roles for Modeling Interfaces (Cont.)

Enterprise Architecture in SAP LeanIX

Today, SAP LeanIX primarily is a product division of SAP. But since it was an independent business until recently, many of the following practices described in this chapter prevail:

- Advisory (presales) and professional services leverages enterprises in a way similar to that of SAP in general, as described in Section 16.2. The primary focus of professional services in SAP LeanIX is to help customers adopt the tool. This includes data importation and setting up integrations with other tools. The focus lies less on giving opinionated recommendations in enterprise architecture assessments and more on helping customers run these assessments.
- Documentation of SAP LeanIX's own enterprise architecture is equivalent to SAP IT's enterprise architecture practice described in Section 16.4. Because SAP LeanIX has been acquired, its corporate systems will be harmonized with the standard SAP platforms.

- Finally, being a cloud software provider, SAP LeanIX has an engineering and software development department. The engineers and developers there won't act as enterprise architects, but they'll clearly have a good understanding of the domain they're developing for.

16.5 Summary

We've seen many different facets and touchpoints of enterprise architecture and its strong role in SAP. Even though these might differ from what you've seen at the companies you've been and are working for, they open up new perspectives.

As we discussed in Chapter 15 on setting up an enterprise architecture practice, the federated approach taken within SAP and described in this chapter is critical to enabling fluency among the different organizations at SAP. However, this federated approach can sometimes slow down the decision-making process, and it can't keep parts of the organization from diverging from the common federated approach.

Now, let's move on to Chapter 17, which will explore upcoming enterprise architecture trends. One such trend is the adoption of global networks, which is based on the idea that enterprise architecture is a discipline that doesn't end at a company's doorway.

PART V

Outlook and Conclusion

In this final part, we'll describe key architecture trends and conclude the book with a summary and some key takeaways.

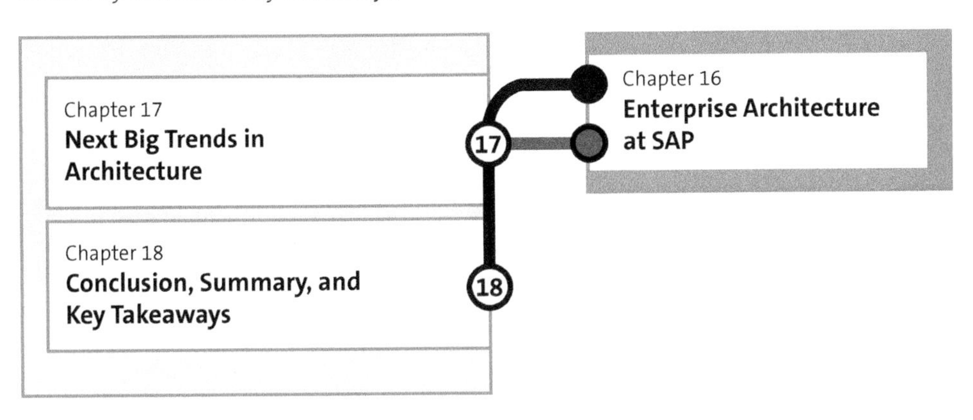

Legend:

Canonical Flow | Intro and Conclusion | Use Cases and Patterns | Enterprise Architecture Practice | SAP LeanIX | Special Tour

Chapter 17
Next Big Trends in Architecture

With core ERP applications getting more and more automated, with standard processes embedded with artificial intelligence and the differentiated capabilities moving to cloud solutions and platforms, enterprise architecture will need to be more modular, event driven, and network based.

This chapter outlines our vision for the future of enterprise architecture, shaped by our understanding of current business and IT trends. As core ERP systems become increasingly automated and standard processes shift to best-of-breed cloud solutions and platforms, enterprise architecture must evolve to be more modular, event driven, and network based, with a strong emphasis on well-defined application programming interfaces (APIs). We'll explore how the shift toward microservices, composable ERP, and open architectures is gaining momentum, potentially leading to complex ecosystems of small cloud applications built on hyperscalers and similar technologies.

Let's look at the following big trends in architecture through the lens of the enterprise architect as the practitioner:

- Artificial intelligence (AI)–supported enterprise architecture transformation
- The enterprise architecture network
- Enterprise architecture–driven innovation
- Clean architecture
- Global networks (the metaverse) for enterprise architecture
- Agile and iterative architecture
- Quantum computing

Strategic Guidance

The outlook offered in this chapter represents the author's point of view, opinions, and insights at the time of writing. The trends discussed are based on current observations and the author's professional experience in the field of enterprise architecture.

Note that enterprise architecture is a rapidly evolving discipline, and trends can change quickly due to technological advancements, regulatory changes, and market dynamics. The content herein doesn't constitute professional advice or a guarantee of future outcomes, and we encourage you to conduct your own research, consult with qualified professionals, and consider your organization's specific needs and context before making decisions based on the information provided.

This is subject to change, and we do not make any guarantee of any future functionality from SAP.

17.1 AI-Supported Enterprise Architecture Transformation

AI adoption has accelerated over the past few decades, transitioning from theoretical concepts to practical applications that are transforming all industries. The potential of AI in the field of enterprise architecture is profound, and the purpose of integrating AI into enterprise architecture is to leverage its capabilities to enhance the planning, design, and management of enterprise systems. AI can assist you in identifying inefficiencies, predicting future needs, and providing intelligent recommendations for improvement. It enables more dynamic and adaptive architecture that's capable of evolving with technological advancements and business requirements. Generative AI will be an even more disruptive force in the field of enterprise architecture, and this trend will revolutionize the way architecture decisions are made during the transformation planning phase.

AI becomes intelligent according to the training it gets, and for training, the machine is fed with datasets. In the enterprise architecture context, these datasets will be generated by solving architecture problems for our customers among the enterprise architecture domains. The data that's fed into an AI model changes according to the applications for which the AI algorithm is being developed. AI models can be broadly categorized into three domains, based on the type of data that's fed into the AI model:

- **Data sciences**
  Data sciences is a domain of AI related to data systems and processes, in which the system collects large amounts of data, maintains datasets, and derives meaning and sense from them. For example, SAP has access to business data from different industries covering finance, supply chain, manufacturing, sales, procurement, human resources, etc.
- **Computer vision**
  Computer vision is a domain of AI that depicts the ability of a machine to obtain and analyze visual information and then make some predictions with it it. The entire process involves acquiring images and then screening, analyzing, identifying, and extracting information. For example, imagine an enterprise architecture world in which computer vision algorithms analyze as-is architecture diagrams, read all the applications, enrich them with lifecycle attributes, derive their business capabilities, and then predict the target architecture and roadmaps.

- **Natural language processing**
  Natural language processing is a branch of AI that deals with interactions between computers and humans using *natural language*, meaning language that is spoken and written by people. Natural language processing attempts to extract information from the spoken and written word by using algorithms. For example, Joule is SAP's generative AI-powered assistant, which you can interact with through a conversational interface and in-app experiences.

Now, let's take a closer look at the role AI can play in enterprise architecture and the outlook for future developments.

17.1.1 Relevance of AI

AI is poised to significantly empower various domains of enterprise architecture, including business architecture, application architecture, data architecture, and technical architecture.

Figure 17.1 shows a detailed look at how AI will impact each of these enterprise architecture domains.

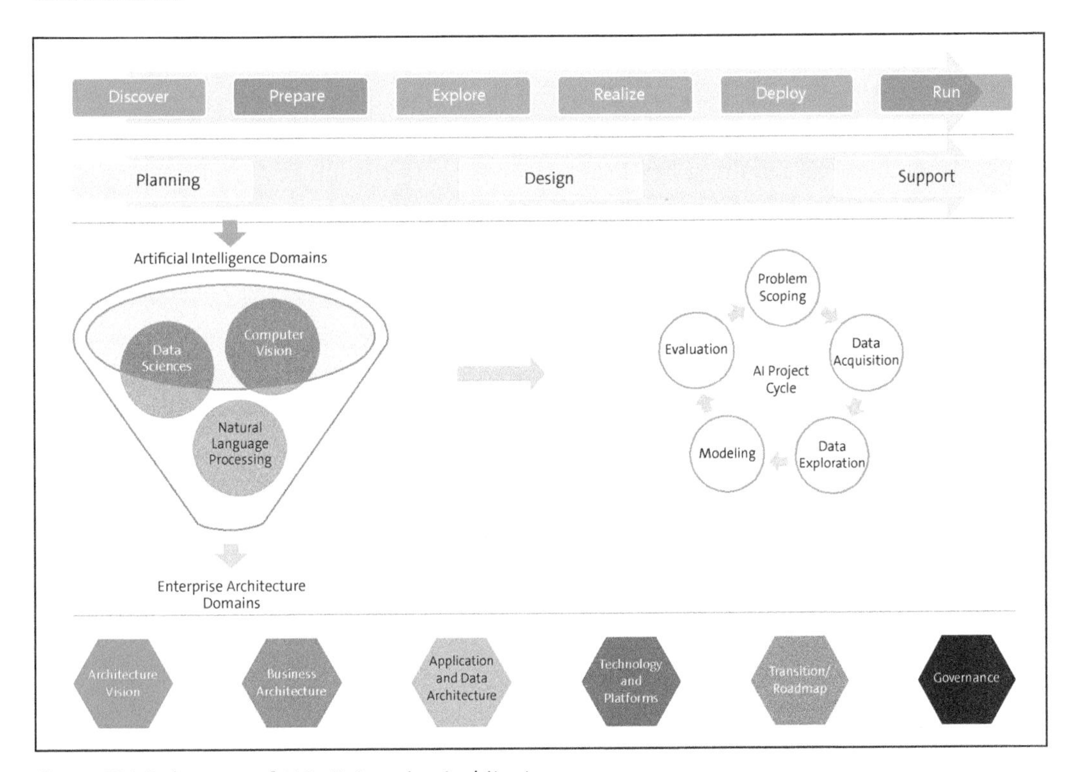

Figure 17.1 Relevance of AI in Enterprise Architecture

The different enterprise architecture tools will learn from the different types of architecture data (such as business capability, business processes, applications, and IT components) captured in different SAP Activate phases (as depicted in Figure 17.1) of a transformation initiative. This will help the enterprise architecture community identify the trends and predict the future of AI use cases in the field of enterprise architecture. For example, you'll be able to use AI to recognize and interpret various types of architectural diagrams, such as application architecture diagrams, network diagrams, process flowcharts, and system architectures.

Let's explore the different enterprise architecture domains to help us understand the future use cases for AI:

- **Business architecture**
  One of the key modules under the business architecture domain is business capability, and AI can be used to generate the business capability heatmap. AI can analyze historical and real-time business capability data to forecast business trends, identify differentiating and core capabilities, and help decision-makers plan strategically and make informed decisions. Machine learning models can simulate various business scenarios, allowing architects to evaluate the potential impact of different strategies and choose the optimal path.
- **Application architecture**
  One of the AI use cases would be to fully automate the transformation from as-is application architecture to the target architecture and roadmap. This would be a complex task requiring significant advancements in multiple AI and machine learning domains. The following list includes some of the components and approaches that can make computer vision algorithms capable of reading as-is application architecture diagrams (from any tools such SAP LeanIX, Lucidchart, and Microsoft Visio) and provide AI-driven recommendations on target architecture and roadmaps for customers:
  - *Optical character recognition*: This algorithm can extract text from the architecture diagrams and identify components, labels, and connections.
  - *Image recognition and object detection*: These algorithms can identify shapes, symbols, and connections within the diagrams and thus recognize different types of architectural components and their relationships.
  - *Natural language processing*: This algorithm can help you understand and process extracted text, especially for the purpose of interpreting labels, descriptions, and annotations within diagrams.
  - *Graph theory and network analysis*: Graph-based algorithms can help you analyze extracted components and their relationships and thus understand the current structure and dependencies within the architecture.

  - *Machine learning and predictive models*: You can train these models on historical data to predict possible target architectures and roadmaps. These models can learn from patterns and best practices in existing architectural transformations.
  - *Enterprise architecture frameworks*: These algorithms can integrate with SAP EA Framework to input the reference architecture content to guide the transformation process and provide templates for target architectures and roadmaps.
- **Data architecture**
  AI can significantly enhance the data architecture domain within the North Star service architecture by automating processes, improving data quality, and enabling more insightful data-driven decision-making. In the data architecture domain, the following are some of the future trends and use cases:
  - *Data quality*: AI can automatically identify and correct data quality issues, ensuring that the data used for decision-making is accurate and reliable.
  - *Data integration*: AI-driven data integration tools can harmonize data from multiple sources, making it easier to create a unified view of enterprise data.
  - *Automated data governance*: AI can automate various aspects of data governance, such as data lineage tracking, metadata management, and access control, ensuring compliance and security.
  - *Data privacy*: AI can help in identifying and protecting sensitive data, ensuring that privacy regulations are adhered to and reducing the risk of data breaches.
- **Technical architecture**
  In the technical architecture domain with the North Star service, AI can play a crucial role in enhancing infrastructure management, optimizing system performance, and automating various technical tasks. In the technical architecture domain, the following are some of the future trends and use cases:
  - *Resource allocation*: AI can optimize the allocation of IT resources based on usage patterns and demand forecasts, improving efficiency and reducing costs.
  - *Predictive maintenance*: AI can predict potential infrastructure failures and suggest preventive maintenance actions, ensuring higher uptime and reliability.
  - *Intelligent network monitoring*: AI can continuously monitor network traffic, detect anomalies, and respond to potential threats in real time, enhancing network security and performance.
  - *Cloud resource optimization*: AI can analyze cloud usage patterns to optimize resource allocation, ensuring cost-efficiency and performance.

As technological change accelerates, enterprise architects must adapt and evolve their skills, both technical and soft, and the tools they use. Here's are some of the future roles and tasks that AI and architects will be involved in:

- **AI as a coworker**
  AI will serve as a tutor and coach for architects, helping them navigate the increasing volume of content and information.
- **Architecting for AI**
  Architects will play a crucial role in designing systems and infrastructure that support AI capabilities, in addition to utilizing AI themselves.
- **Community engagement**
  Having a strong community of practice will be essential for architects who wish to share knowledge, collaborate, and learn together to keep up with rapid technological advancements.

The enterprise architect SAP Community is a hub where architects learn continuously through a steady stream of curriculum, mentorship, and peer-to-peer coaching. By fostering connections among organizations, this community enables rapid and meaningful discussions about current and future technology trends.

The role of the enterprise architect is becoming more collaborative, community driven, and AI augmented. Embracing these changes will position architects and their organizations for future success. Having a vibrant enterprise architect SAP Community of practice and engagement is crucial for accelerating learning and actively engaging with a peer network of architects. Members of the enterprise architect SAP community need to effectively engage, collaborate, and support each other, ensuring they stay current and drive value for their organizations. Another key dimension of focus is helping enterprise architects develop and evolve their enterprise architecture community of practice inside their organizations.

Enterprise architects, augmented by AI and supported by a robust community, are key to companies thriving in today's ever-changing world. This is a trend that will become a force in the next two years.

SAP AI Business Strategy

SAP puts AI into the context of business. SAP AI's strategy is to embed generative AI natively in SAP's business applications (i.e., SAP cloud ERP, SAP Human Capital Management, SAP Business Network and Spend Management, and SAP Customer Relationship Management [SAP CRM]) as well as into SAP Business Technology Platform (SAP BTP), expose business relevant generative AI services via SAP BTP for customer and partner extensions, and deliver trustworthy AI. SAP Business AI is relevant, reliable, and responsible.

17.1.2 Outlook

These sections outline the critical aspects of AI's role in enterprise architecture, detailing the current impact and future possibilities. AI's integration into enterprise architecture domains promises significant advancements in efficiency, security, and strategic alignment.

By leveraging AI, enterprise architects can create more adaptive, resilient, and efficient architectures that drive digital transformation and deliver tangible business value. Enterprise architects are actively seeking ways to utilize AI since their organizations strive to operate more efficiently, optimize their technology landscape, and maintain a competitive edge. Incorporating AI into your organization's architecture practice can minimize errors, handle ever-expanding datasets, and enable you to concentrate on strategic tasks.

Using SAP Business AI empowers users to interact with SAP software in the most natural way possible through Joule. Imagine the day when Joule, a copilot that will truly understand your business and the IT landscape as an enterprise architect, will predict the baseline target application architecture overview diagram based on the prompts provided by the business and the IT user. While generative AI will never be able to replace the strategic thinking and creativity of enterprise architects, the enterprise architect role will become more important than ever because it will have the power and the use cases to bring AI to both business and the IT.

In summary, the outlook for this trend is as follows:

- Impact: high
- Likelihood of happening within one year: very high

17.2 The Enterprise Architecture Network

In today's rapidly evolving business landscape, the role of enterprise architect is becoming increasingly collaborative, extending beyond the boundaries of individual organizations. Companies in any industry as well as public institutions never run in isolation. Integration therefore is not a matter to take care of within the four walls of the enterprise but through smooth collaboration with business partners.

Collaborative businesses will operate more on common network platforms. A broad adoption of public cloud standard network-enabled software will elevate process execution and control beyond the boundaries of traditional enterprises.

These trends are not new, but in the past, they happened on the periphery of enterprises. We predict that this pattern will play a strong role in core processes too.

For example, Catena-X and Gaia-X are collaboration platforms specifically designed for the automotive industry. Catena-X and Gaia-X are initiatives in the automotive and manufacturing industries aimed at improving data exchange, collaboration, and the integration of advanced technologies within this sector. Platform providers (like SAP) offer more and more "networked" solutions like these that run business processes crossing company boundaries. The idea is that the enterprise architecture networks push this boundary so that enterprise architecture can work in the same way in different companies.

In the following sections, we'll envision the role of networks in enterprise architecture and the outlook for this trend.

17.2.1 Relevance of Enterprise Architecture Networks

As enterprises adopt more interconnected and interdependent ecosystems, the traditional approach to documenting isolated architectures within a single organization is giving way to a more collaborative, cross- enterprise architecture network. This shift is driven by the need for greater agility, consistency, and alignment across the industry. In the world of enterprise architecture, we can define the following levels of collaboration:

- **Sharing reference architecture**
  In an interconnected enterprise architecture network, organizations can share reference architectures that serve as common blueprints or best practices for specific industries or domains.
- **Sharing baseline documentation**
  Organizations can contribute to and access shared baseline documentation within the enterprise architecture network. This includes foundational architectural artifacts—such as business capability maps, business process maps, data models, and technology standards—which are critical for ensuring alignment across multiple enterprises.
- **Developing target architecture and roadmaps collaboratively**
  Through the enterprise architecture network, organizations can collaboratively develop target state architectures and roadmaps that align with shared business goals.

We foresee that in the future, companies won't document their isolated architectures by themselves any longer but may instead do so in a collaborative cross-enterprise architecture network. Once most nondifferentiating software is being run by SaaS solutions, autodiscovery functions could detect most parts of the enterprise architecture and create automated enterprise architecture documentation.

We foresee two use cases:

- **Documentation**
  By using autodiscovery from all related cloud platforms, each connected company will be able to retrieve the network view relevant to them.
- **Transformation planning**
  By using multicompany collaborative transformation planning, companies will be able to produce consistent cross-company future state architecture. The services

they'll be able to model will depend on the connected network platforms. They could range from warehousing to collaborative manufacturing, selling, and talent sharing.

Let's look back at the topic of mergers, acquisitions, and divestitures (MAD) that we investigated in Chapter 8. Carving out part of a business and selling it to another company today would trigger a large number of projects in both companies, and with the enterprise architecture network modeling approach, both companies would have the same data to work on.

A key prerequisite for and enabler of all of this is a much stronger standardization of reference content. To enable this granularity, meta models need to be further refined.

The idea evolves toward a two-level architecture network, as shown in Figure 17.2. Here, the green layer on top represents the architecture network layer that dedicated company tenants are connected to. All cloud components are auto detected and documented. Companies can limit the visibility of parts of their architectures to selected business partners (partner companies). As transaction processing is shifting toward collaborative platforms, enterprise architecture discovery goes beyond company boundaries. With mutual consent, a network platform like SAP Business Network can not only reveal technical integration details to business partners but also provide an aggregated semantic analysis of the business streams (transactions) in and out of their platform.

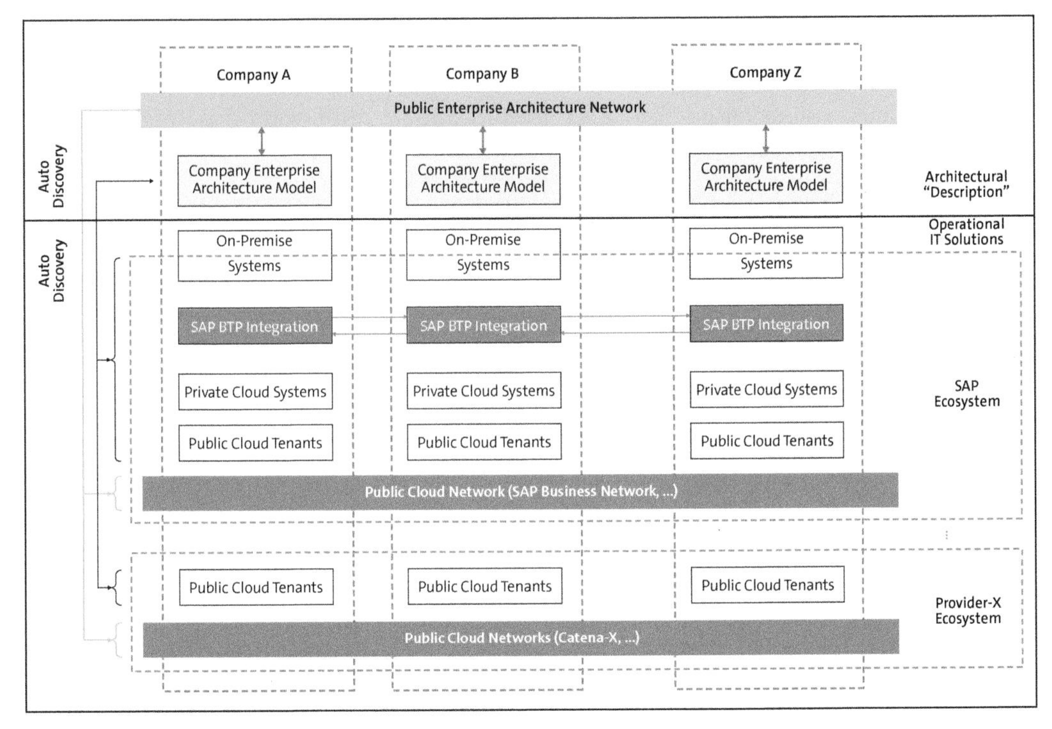

Figure 17.2 Enterprise Architecture Network

17.2.2 Outlook

The shift toward enterprise architect networks represents a significant evolution of how organizations approach enterprise architecture. By embracing collaborative models, sharing resources, and leveraging automation, enterprises can create more robust, agile, and future-proof architectures. However, this is still a long way off as we move into this new era of the enterprise architect network, where the role of the enterprise architect will be increasingly focused on facilitating collaboration, integrating diverse perspectives, and ensuring that the collective architecture supports the strategic goals of the entire network. In summary, the outlook for this trend is as follows:

- Impact: very high
- Likelihood of happening within five years: medium

17.3 Enterprise Architecture-Driven Innovation

SAP Central Business Configuration is SAP's approach to having a central configuration, in which configuration processes run in multiple public cloud solutions. Today, you start with central business configuration once the overall selection of solutions and processes is known.

With enterprise architecture-driven innovation, we like to spin the idea further by viewing the enterprise architecture models as a sort of digital twin of the system architecture.

Now, let's explore live enterprise architecture and the outlook for this trend in the following sections.

17.3.1 Relevance of Enterprise Architecture-Driven Innovation

While the idea of an enterprise architecture network, as discussed in Section 17.2, already captures auto-discovery and collaborative modeling, the live enterprise architecture approach sketched in Figure 17.3 is intended to have traceability from early enterprise architecture work like strategy, capability maps, and solution sketches down to process deployments. The upper part depicts the enterprise architecture model, which receives input from the real system layer via an autodiscovery function. These functions exist already for specific use cases like process discovery and IT component inventories in configuration management databases (CMDBs). A landscape and process configuration feature on the abstraction level of the enterprise will trigger configuration and deployment instructions to the lower layers.

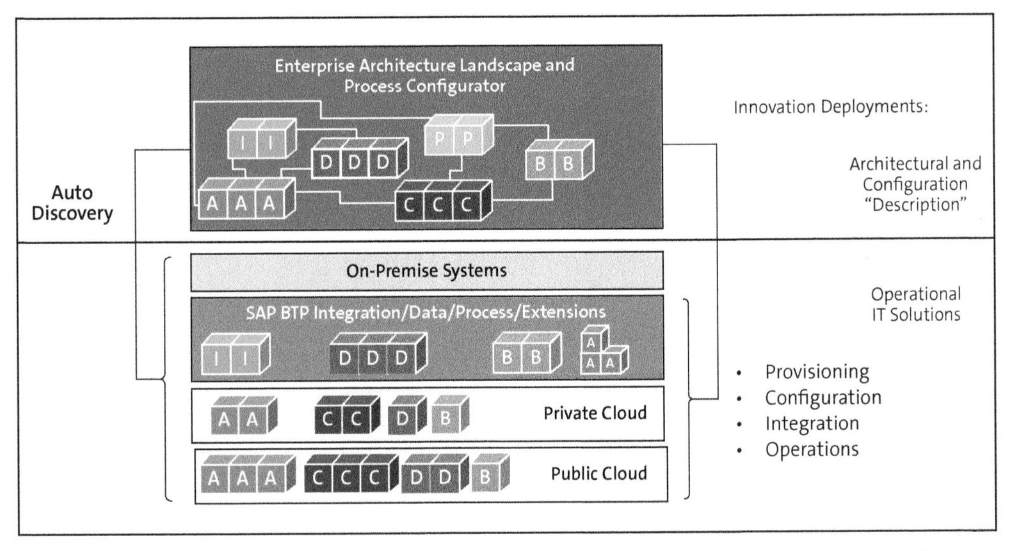

Figure 17.3 Conceptual Sketch of Live Enterprise Architecture: Architecture-Driven Innovation

17.3.2 Outlook

Enterprise architecture-driven innovation envisions that the selection of applications and processes can start while you're planning the transformation phase and can be leveraged while you're executing the projects for the transformation.

As additional dimensions are incorporated into the model—such as industries, supported solutions, data, test cases, and integrations—the complexity of managing dependencies increases. Therefore, ensuring the highest level of reference data quality requires heightened attention and specialized governance. In summary, the outlook for this trend is as follows:

- Impact: very high
- Likelihood of happening within three years: medium

17.4 Continuous Clean Architecture

We know the importance of clean core for accelerating business innovation. However, the term *clean core* triggers the following misunderstandings:

- The objective is to evolve a private cloud-deployed system into a public cloud experience that would allow frequent innovation deployments into the core. This is the opposite of a small locked-down core not receiving any updates!
- "Clean core" should *not* imply building a large SAP BTP development team and starting to develop hundreds of custom extensions of SAP BTP. This might keep the core clean but could quickly lead to a *polluted periphery*, meaning that we could develop

several line of business (LoB) apps, industry apps, and custom apps in SAP BTP and thus create islands of apps on the edges of the core that we may need to clean up later.

The continuous clean architecture approach is looking broader and is intended to equip customers with ongoing measurements and improvement proposals that on the one hand look at internal governance and change processes and on the other hand will assess "real" enterprise architecture items. For example, these can be overly complex integrations or redundant application portfolio entries that are discoverable by having multiple applications mapped to the same business capabilities.

When you look back to the architecture maturity level introduced in Chapter 15, it's obvious that when you're following the clean architecture approach, you can elevate your enterprise architecture practice to level 4 (Measured) and eventually level 5 (Optimizing).

17.4.1 Relevance of Continuous Clean Architecture

In the clean core approach, we have two phases: moving to clean core and keeping the core clean. Similarly, continuous clean architecture needs to first be established in a setup and baseline phase, as depicted in Figure 17.4. Once established, it can be bundled with possibly already existing enterprise architecture governance processes. If these don't exist, you could either seek external help through the SAP services team or invest simultaneously in the setup of your own enterprise architecture practice.

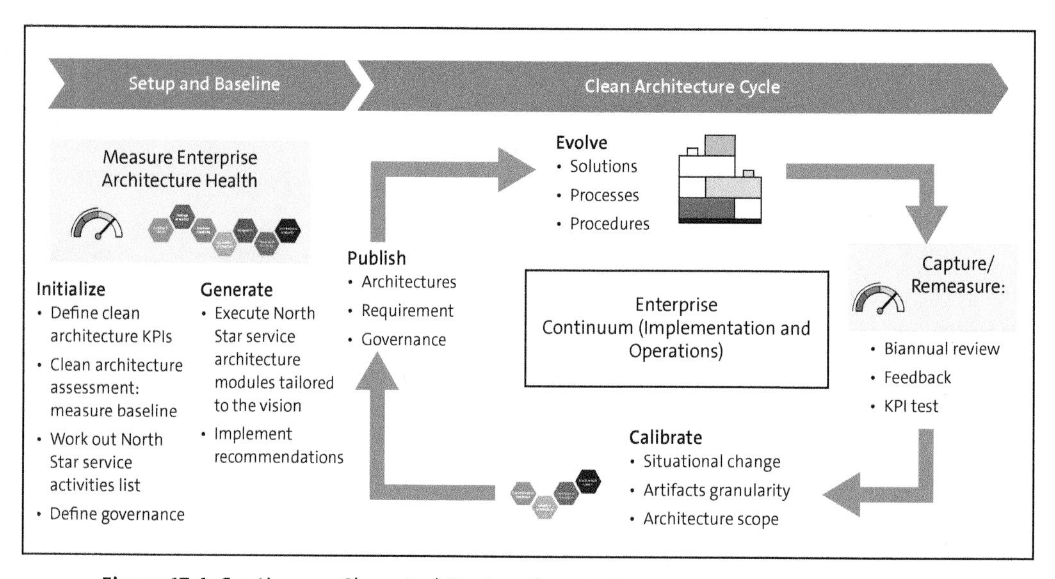

Figure 17.4 Continuous Clean Architecture Approach

During the initial setup and baseline phase, you'll select and assess company-specific key performance indicators (KPIs). During an enterprise architecture assessment, you'll assess the baseline architecture and create an initial roadmap that includes target KPI values.

We then foresee a repetitive clean architecture cycle that runs at least twice a year. After an evolution, you'll remeasure and calibrate the architecture plans. Transparent publication of measures and enterprise architecture artifacts are at the center of this approach.

As with all measuring systems, you'll want to start rather small to give you the time to calibrate on the specific field where you foresee the biggest impact.

The following KPI categories help you measure the level of clean architecture and operational maturity.

They are also sketched in Figure 17.5:

- Strategic KPIs
- Tactical (quantifiable) KPIs
- Qualitative KPIs
- Financial KPIs

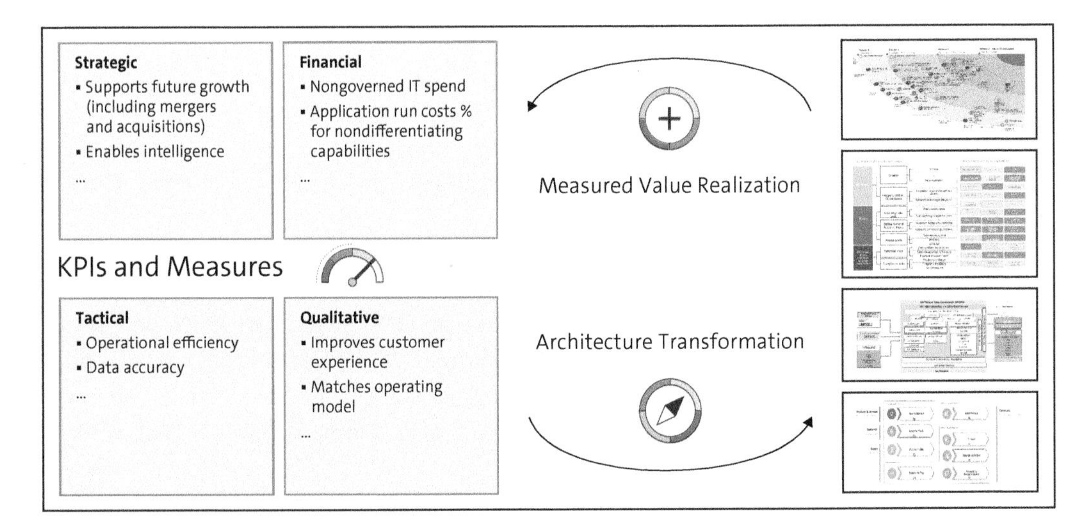

Figure 17.5 Continuous Clean Architecture Assessment

Another option is to distinguish KPIs by data domain:

- Governance related
  - Enterprise architecture board established
  - Architecture decisions documented
  - Sufficient architecture repository/support available
  - Methodology generated

- Architecture development
  - Data domain measures
  - Process domain measures (possibly via SAP Signavio)
  - Application domain measures
  - Technologies

17.4.2 Outlook

Clean architecture is a focus topic in SAP, but it will likely acquire a different name as its similarity to *clean core* could lead to misunderstandings such as thinking that architecture is part of a clean core. A working title is *continuous architecture assessment*, but it could end up being something else.

What will grow is the use of SAP LeanIX and its integration into live landscapes. The ability to make proper assessments with AI will also improve, and we believe the ingredients are there to bring this concept into reality in the next two years.

In summary, the outlook for this trend is as follows:

- Impact: high
- Likelihood of happening in within two years: high

17.5 Global Networks for Enterprise Architecture

A *metaverse* consists of a set of shared and persistent virtual spaces where people can create and explore with others who aren't in the same physical space. It aims to enable an additional dimension of internet use as an embodied internet in which people not only view content but are also part of it, for example, in the form of avatars. In its original meaning, the metaverse is one universe that is not owned by a single company. There are other technologies like generative AI that will support the wider adoption of the metaverse vision.

Over the last several years, this vision has evolved into an industrial context. The industrial metaverse is a sector that mirrors and simulates real-world systems, offering immersive, real-time, interactive, and persistent experiences. It utilizes technologies such as digital twins, AI, extended reality, blockchain, and cloud and edge computing to create a powerful interface between the real and digital worlds.

In the following sections, we'll dive into how the metaverse can impact enterprise architecture and the outlook for this trend.

17.5.1 Relevance of the Metaverse

Now, imagine the world of the enterprise architecture metaverse, where physical reality, augmented reality, and virtual reality will merge into one immersive world. For example, as soon as you put on your Apple Vision Pro, you'll get connected to all the enterprise architects in your network and discuss architecture transformation problems. Figure 17.6 shows the interdependencies of evolving technologies that can make an enterprise architecture metaverse and a global network of enterprise architects a reality in the future.

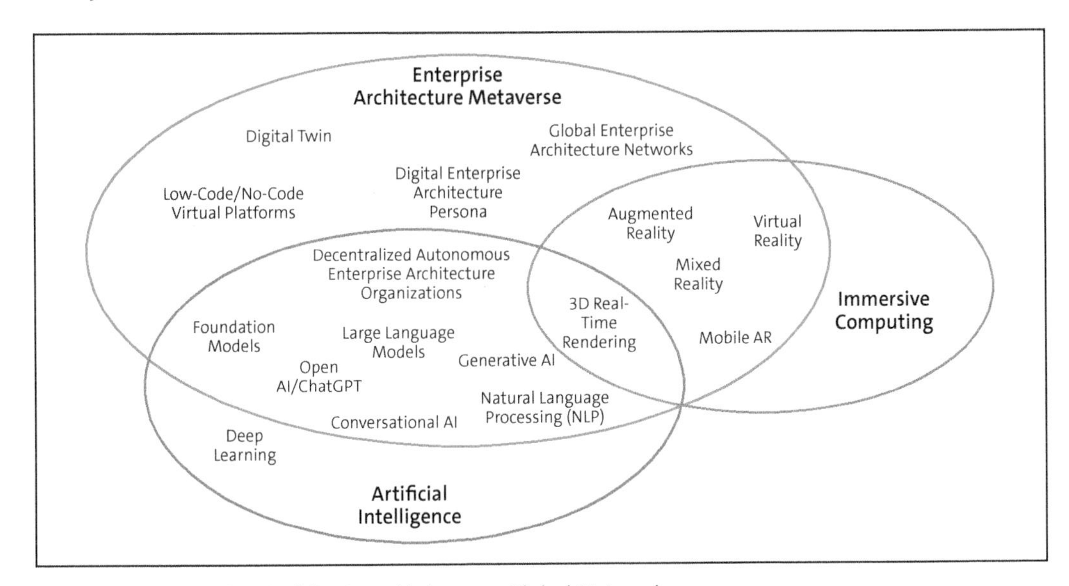

Figure 17.6 Enterprise Architecture Metaverse Global Network

The following list includes some of the use cases for an enterprise architecture metaverse:

- **Global enterprise architecture community networks**
  We need enterprise architecture community networks to share with, collaborate with, and learn from fellow enterprise architects. An enterprise architecture metaverse will make sure that the network experience is taken to the next level and make it simulate a real conference environment.
- **Enterprise architecture talent networks**
  There is increased customer demand for dynamic, subscription-based access to enterprise architecture expertise. For example, there was an announcement at SAP Sapphire 2024 that every RISE with SAP customer will be guided by an SAP enterprise architect. An enterprise architecture metaverse will ensure the digitalization of talent and will create a new business model that will have the ability to source enterprise architects from outside the boundaries of a firm and maintain high engagement levels.

SAP BTP can be leveraged to build these analytics and the insights needed to drive the enterprise architecture metaverse and global networks. This will augment the future of business decisions during phase 0 transformation planning with enhanced data visualization of digital twins using enterprise architecture domains, simulation and prototyping, and improved collaboration with networks. The future of collaboration will improve using 3D architecture diagrams, knowledge sharing, streamlined workflows, document management for the enterprise architecture artifacts, and accountability among different versions.

Examples in the Field of Metaverse Technologies

Let's highlight a few examples that apply metaverse technologies:

- *NVIDIA Omniverse* is a modular development platform of APIs and microservices for building 3D applications and services powered by Universal Scene Description (OpenUSD) and NVIDIA RTX.
- *Web3* (also known as *Web 3.0*) is an idea for a new iteration of the World Wide Web that incorporates concepts such as decentralization, blockchain technologies, and token-based economics.

17.5.2 Outlook

The enterprise architecture metaverse is a vision of a shared virtual space that acts as a persistent digital world that users coexist in and that is operated by different players in a decentralized way. While there's immense potential for the metaverse to become a reality in the future, all involved parties must strictly adhere to data privacy and protection laws for it to become a success in the long run.

Assessing the potential business value of the metaverse is very challenging, as the full-fledged vision is still very much in its early days. In conclusion, the idea of the metaverse for enterprise architecture could be just hype and may not survive or be relevant in the long term. But the sheer concept sounds promising due to its multiple potential benefits for enterprise architecture collaboration, and it will appeal to Generation Z and junior talents.

In summary, the outlook for this trend is as follows:

- Impact: medium
- Likelihood of happening within ten years: low

17.6 Agile and Iterative Architecture

Enterprise architecture has indeed undergone significant changes over the years, transitioning from traditional waterfall approaches to more agile methodologies. This evolution has been driven by the need for organizations to adapt quickly to changing business requirements and technological advancements. Here are the key stages in the evolution of enterprise architecture:

- **Waterfall approach**
  The waterfall approach to enterprise architecture follows a sequential and linear process, where each phase must be completed before proceeding with the next phase. The waterfall approach typically involves a lengthy planning phase, followed by design, development, testing, and deployment. It's characterized by a top-down, rigid structure and a focus on extensive documentation.
- **Iterative and agile approach**
  Recognizing the limitations of the waterfall model, organizations begin adopting iterative and incremental approaches. In this stage, enterprise architecture starts incorporating feedback loops, allowing for periodic adjustments and improvements. The architecture evolves over multiple iterations, with each iteration building upon the previous one.

Agile enterprise architecture is a collaborative and iterative methodology that helps businesses adapt to digital transformation. Agile enterprise architecture emphasizes responding to change, customer feedback, and rapid decision-making over following a plan. It breaks down architecture into smaller, manageable chunks called minimum viable architectures that deliver value incrementally. This approach allows the architectural design to evolve gradually as problems and constraints become clearer.

The Open Group Architecture Framework (TOGAF) uses an iterative approach to its Architecture Development Method (ADM) to help organizations develop and manage enterprise architectures. This iterative process allows organizations to adapt to change and navigate complexity.

Enterprise architecture as a living practice was described in detail in Chapter 15, and the latest stage in the evolution of enterprise architecture emphasizes the concept of continuous architecture and treating it as a living practice. This approach acknowledges that architecture is not a one-time activity but an ongoing process. It involves continuous monitoring, adaptation, and optimization of the architecture to align with evolving business needs and technology trends.

We'll explain the trend of agile and iterative architecture in more detail and provide the outlook for it in the following sections.

17.6.1 Relevance of Agile and Iterative Architecture

SAP EA Framework encourages agile and iterative architecture. In fact, this iterative approach leads to more realistic architecture design as it's very difficult to design the North Star service architecture in one go. One of the examples of how we design the target application architecture in an iterative way is illustrated in Figure 17.7.

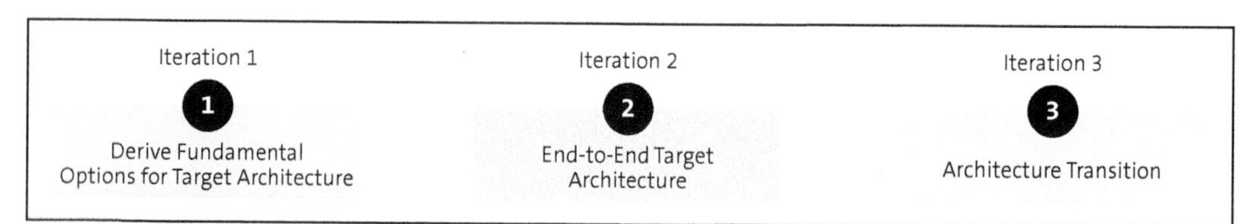

Figure 17.7 Iterative Approach to Designing Target Application Architecture

Let's walk through these iterations:

❶ **Iteration 1**

- Develop baseline application architecture based on the reference content from SAP EA Methodology, focusing on level 2 and 3 capabilities.
- Enrich (and lifecycle) the reference application architecture with the customer's as-is application architecture and include the requirements.
- Identify the target application options for the "phase out" and "retire" as-is applications.
- Choose the target application based on the criteria and a pro/con analysis and provide a high-level recommendation.

❷ **Iteration 2**

- Add the dependencies and components in and around the customer-specific target application architecture.
- Develop a target application architecture overview diagram, including a level 0 (application-to-application link) data flow.
- Identify the critical integration list (to be used as an input to create the integration architecture, probably in iteration 4).
- Identify the big-rocks architecture gaps and roadmap components.

❸ **Iteration 3**

- Phase transition steps toward the target architecture.
- Develop the interim (transition) application architecture diagram.

There are different ways we can look at this iterative approach. Let's take an example of one customer, Auto Inc., which has selected the first set of modules and created the first version of target architecture (see Figure 17.8). It has selected strategy mapping, business capability, application architecture, initiatives and road mapping, and transition scenario evaluation. Then, it performed a second set iteration using business process, rollout strategy, technical architecture deployment, and integration architecture. Auto Inc. has a plan to review the architecture every six months to one year, especially because they have a steering committee review of IT spending and the budget for future years.

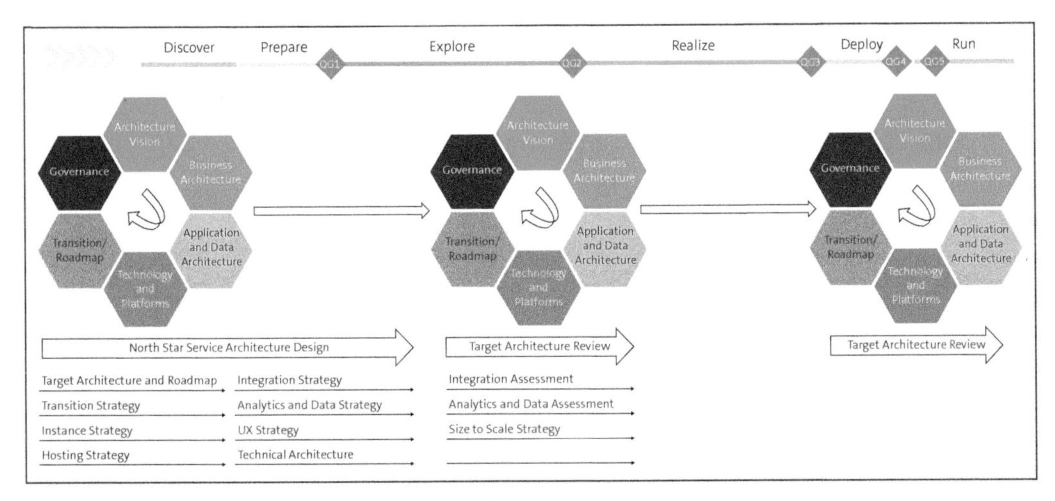

Figure 17.8 Iterative Architecture Review Plan for Auto Inc.

17.6.2 Outlook

Agile and iterative architecture is here to stay. Enterprise architecture has transformed from a rigid, waterfall-driven process into a more agile and adaptive approach, and this trend will continue in future. This evolution reflects the need for organizations to be responsive, flexible, and able to quickly leverage emerging technologies to gain a competitive advantage and empower innovation in a rapidly changing business landscape. Besides, the iterative approach in SAP Activate ensures that enterprise architecture evolves continuously to meet business needs while maintaining technical integrity. Each phase involves specific iterations to refine and validate the architecture, ultimately leading to a successful SAP implementation that aligns with organizational goals and technical standards.

In summary, the outlook for this trend is as follows:

- Impact: high
- Likelihood of happening within one year: high

17.7 Quantum Computing

Imagine a computer so powerful that it can solve problems that would take traditional computers longer than the age of the universe to crack. This is the promise of quantum computing. *Quantum computing* is an emerging technology that leverages the principles of quantum mechanics to perform computations far beyond the capabilities of classical computers. Unlike classical bits, which are binary, quantum bits (quibits) can exist in multiple states simultaneously, thanks to superposition and entanglement.

This allows quantum computers to solve complex problems in cryptography, chemistry, and optimization much more efficiently. Although quantum computing is still in its early stages, advancements by major tech companies and research institutions are rapidly pushing the boundaries, announcing a new era of computational power.

Quantum computing emerged from theoretical foundations laid in the early 1980s by pioneers like Richard Feynman, who proposed using quantum mechanics for computational purposes, and David Deutsch, who introduced the concept of a universal quantum computer in 1985. Significant advancements followed, including Peter Shor's 1994 algorithm for efficient factorization, which highlighted quantum computers' potential to outperform classical systems. Lov Grover's 1996 algorithm further demonstrated quantum computers' capabilities in database searching. Practical milestones include IBM and Stanford University's implementation of Shor's algorithm in 2001 and the release of the first commercial quantum computer by D-Wave Systems in 2010. Google's 2019 announcement of achieving quantum supremacy with a 54-quantum bit processor marked a controversial yet pivotal moment, as the field continues to evolve rapidly, addressing challenges in error correction and scalability.

Skepticism about quantum computing stems from its current limitations and the significant challenges it faces. Critics argue that while quantum computers have shown potential in theoretical and small-scale experiments, they are not yet practical for real-world applications outside research labs due to issues like quantum bit instability, error rates, and the need for extreme operating conditions. Therefore, the application of quantum computing to solve real-world business problems is still in its infancy, and quantum computing has many technical hurdles to overcome before it can deliver on its promises. This leads some to view the current excitement as premature hype.

We'll explore how quantum computing impacts enterprise architecture and provide the outlook for this trend in the following sections.

17.7.1 Relevance of Quantum Computing

Quantum computing has significant market potential, estimated between $1 trillion and $2 trillion, with rapid growth expected in the next five to ten years. Key industries such as financial services, global energy, transport and logistics, pharmaceuticals, automotive, aerospace, advanced electronics, and semiconductors could collectively see an impact of $900 billion to $2 trillion by 2035. Financial services alone may capture $400 billion to $600 billion, while other sectors like energy and logistics could each gain $200 billion to $500 billion. This highlights the transformative economic potential of quantum computing in diverse industries.

Therefore, quantum computing holds significant importance for SAP and its customers by promising to revolutionize how complex business problems are solved. It can significantly enhance optimization processes, accelerate data analysis, and improve decision-making capabilities. For SAP customers, this means more efficient supply chain management, advanced predictive analytics, and superior resource planning. The ability of quantum computing to process vast amounts of data at unprecedented speeds can drive innovations in financial modeling, logistics, and personalized customer experiences, providing companies with a competitive edge. By leading in quantum computing research, SAP ensures it can deliver cutting-edge solutions, thus maximizing the potential benefits of this emerging technology for various industries as outlined in Figure 17.9.

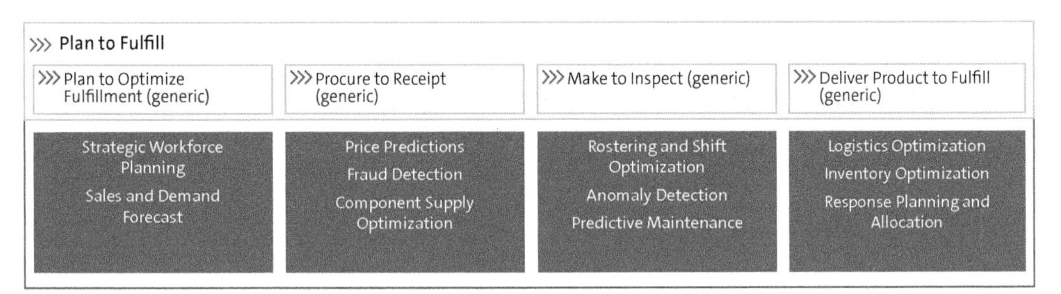

Figure 17.9 How Quantum Computing Can Impact SAP-Based Processes

SAP can leverage quantum computing to provide substantial benefits in the following business functions for its customers:

- In SAP cloud ERP, quantum computing can enhance financial portfolio optimization, simulate finance markets, and improve production planning and predictive maintenance.
- Within SAP BTP, quantum computing can generate better AI and machine learning models, develop advanced quantum applications, and bolster threat intelligence.
- In supply chain management, quantum computing aids in logistics optimization, inventory and response planning, and allocation.
- In SAP Intelligent Spend and Business Networks, quantum computing improves price predictions, fraud detection, risk management, and component supply optimization.
- In human capital management, quantum computing can optimize rostering and shift schedules, strategic workforce planning, and operational headcount planning.
- SAP CRM can benefit from quantum computing through improved sales and demand forecasts, order sourcing, product availability, and anomaly detection.
- Quantum computing can enhance sustainability efforts by optimizing a company's CO_2 footprint, business process operations, and environmental, social, and governance (ESG) performance improvement and steering.

These applications demonstrate how quantum computing can drive efficiency, accuracy, and strategic advantage for SAP customers in multiple domains.

In Chapter 2, we introduced the North Star service, which is based on several critical modules within enterprise architecture. Despite the promising advancements in technology, the comprehensive impact of quantum computing on overall enterprise architecture has not yet been thoroughly investigated in the literature. Most current research and practical applications focus on specific use cases rather than on quantum computing's overarching influence on enterprise architecture frameworks. Let's explore the potential implications of quantum computing in selected modules of the North Star service architecture, as shown in Figure 17.10:

- **Business model patterns**
  Quantum computing introduces potential for developing new business models, especially in data-intensive industries. Businesses can leverage quantum capabilities to create products and services that were previously unfeasible under classical computational limits.

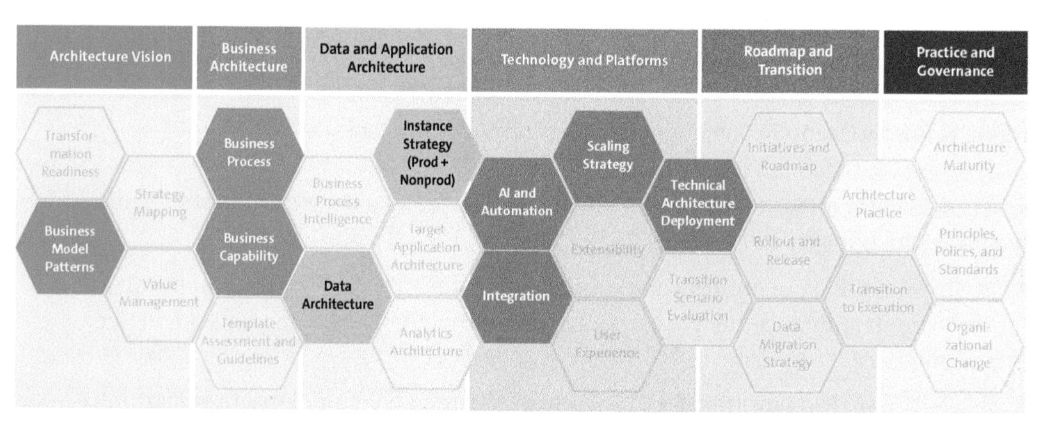

Figure 17.10 Main Modules Impacted by Quantum Computing

- **Business process and business capability**
  Quantum computing can significantly enhance optimization and logistics, offering solutions to complex routing and scheduling problems, thus enhancing both business processes and capabilities.
- **Data architecture**
  The vast computational superiority of quantum computing might necessitate reenvisioned data architecture to accommodate new types of data processing and analysis, enabling the efficient handling of larger datasets and complex simulations.
- **Instance strategy**
  In the instance strategy module, determining the optimal number of productive quantum computer instances is a new area of focus. This involves exploring how many quantum systems are necessary to maximize efficiency and effectiveness within an enterprise's infrastructure.

- **AI and automation**
  By improving the efficiency of machine learning algorithms and expanding their capability to process larger datasets, quantum computing can substantially boost AI and automation technologies.
- **Integration**
  Integrating quantum computing into existing IT systems presents a challenge that requires new frameworks and APIs capable of bridging classical and quantum computational processes.
- **Scaling strategy**
  The scaling strategy module addresses how best to scale up or scale out quantum computers, aligning their deployment with expected workloads and performance requirements. This involves strategic decisions on increasing quantum computing resources vertically (scaling up) or horizontally (scaling out) to efficiently meet the evolving needs of the enterprise.
- **Technical architecture deployment**
  Deploying quantum computing within existing technical architectures demands significant adaptations, including quantum-safe cryptography and the coexistence of quantum and classical hardware within the same IT infrastructure.

Other modules of the North Star service might also be impacted by quantum computing. For instance, the initiatives and roadmap module could see significant changes, as strategies and planning would need to incorporate quantum technologies to stay ahead of industry trends. This module, along with others like business strategy and technology and platforms, will require updates to include quantum computing considerations, ensuring that initiatives align with emerging capabilities and the shifting technological landscape. Keeping these modules in sync with advancements in quantum computing will be crucial for maintaining an adaptive and forward-thinking enterprise architecture strategy.

17.7.2 Outlook

Quantum computing holds immense potential to revolutionize various industries, promising unprecedented computational power and efficiency. However, significant challenges remain in making it practical and scalable for real-world applications. Despite these hurdles, the transformative possibilities make quantum computing highly relevant to enterprise architects. They'll need to address entirely new questions regarding architecture design, data processing, and security to integrate quantum solutions effectively. As the technology matures, its integration will undoubtedly pose complex yet exciting opportunities for future enterprise architectures, driving innovation and competitive advantage.

Enterprise architects should view quantum computing not as an immediate challenge but as a critical area to monitor for future opportunities and innovations. While the widespread adoption of quantum technologies in enterprise architectures may still be on the horizon, it's essential to stay informed about developments in this field. Being proactive in understanding quantum computing's potential impacts can position your organization to adapt swiftly and effectively when these technologies become more accessible. Therefore, it's advisable to keep a watchful eye on the evolution of quantum computing and consider preliminary strategic planning to integrate these capabilities. Embracing such innovations could provide significant competitive advantages and drive the next wave of technological advancement in enterprise architecture.

In summary, the outlook for this trend is as follows:

- Impact: medium
- Likelihood of happening within next five years: medium

17.8 Summary

Enterprise architecture is expected to see significant transformations over the next five to seven years, driven by advancements in technology, generative AI, and the ever-changing and dynamic business world. The trends we see in enterprise architecture are summarized in Table 17.1.

	Next Big Trends	Impact	Likelihood	Probable Timeframe
1	AI-supported enterprise architecture transformation	High	Very high	Within the next two years
2	Enterprise architecture networks	Very high	Medium	Within the next five years
3	Enterprise architecture-driven innovation	Very high	Medium	Within the next three years
4	Continuous clean architecture	High	High	Within the next two years
5	Global networks	Medium	Low	Within the next seven years
6	Agile and iterative architecture	High	High	Within the next year
7	Quantum computing	Medium	Medium	Within the next five years

Table 17.1 Trend Chart for Enterprise Architecture for Next Five to Seven years

In the next chapter, we'll conclude our enterprise architecture journey with a recap of what we've learned, and we'll summarize the key takeaways from this book.

Chapter 18
Conclusion, Summary, and Key Takeaways

We'll conclude the book with a look back on our enterprise architecture journey in this chapter, including chapter summaries and important takeaways.

We have based this book on our many years of practical experience in the field of enterprise architecture. We guide large enterprises through digital transformation by identifying drivers of change, mapping those drivers to enablers in terms of application components or nonfunctional qualities, and deriving industry and customer-specific target architectures. For the past six years, within the Customer Services & Delivery organization at SAP, we have been building an enterprise architecture live channel and team called SAP Transformation Hub. This team provides architecture guidance to SAP's top global and strategic customers, with a strong focus on methodology, content, and tools development tailored to the specific needs of SAP-oriented transformations. Our collaboration with SAP engineering and development teams gives us a unique opportunity to address the needs of customers and influence SAP's roadmap.

As you've learned from this book, enterprise architecture is a well-established practice in most large corporations. However, IT departments often investigate process models but don't focus much on enterprise architecture. This is often due to the perception that SAP architecture is "given" and centered on one or many ERP systems. Consequently, SAP architecture is often divided into a technology (Basis) layer and a business process layer. The application layer in an SAP-centric landscape is formed by configurable standard software, which "classical" enterprise architects don't typically delve into because they're under the impression that not much of it needs to be architected.

We've also explained that you should approach SAP architecture and enterprise architecture with a common methodology, for the following reasons:

- Changing business models and a volatile economy require companies to transform.
- The enabling IT solution portfolio is no longer a single ERP system. Networked cloud solutions and new technology trends disrupt classical IT landscapes.
- Processes cross the solution boundaries of many solutions (not only SAP solutions).
- SAP is no longer primarily a business application vendor, and its technology offerings need to be assessed and integrated alongside other non-SAP technologies.

Recognizing the need for enterprise architecture, this book has delved into typical use cases such as business transformation, cloud transformation, incorporating sustainability, mergers and acquisitions (M&A), and classical landscape rationalization and total cost of ownership (TCO) challenges, along with some additional use cases. We've utilized SAP EA Framework as the foundation of all of these use cases. Prebuilt industry reference architectures, as part of the framework, act as radical accelerators for deriving target architectures, and they eliminate the need to start from scratch. The framework is strongly oriented towards The Open Group Architecture Framework (TOGAF) and is organized by the typical domains.

The book has also emphasized that you need to carefully design any practice you wish to build within an organization and that during the design process, you must address several key questions:

- How do you best integrate SAP enterprise architecture into an overall enterprise architecture approach?
- How do you build a case for enterprise architecture in the SAP context?
- How do you avoid a negative "policing" perception while making enterprise architecture an essential element in key architecture decisions?
- How do you link enterprise architecture to portfolio management?

You've also learned that having good tools for architecture doesn't guarantee that you'll create good architecture, but good enterprise architecture tools can be strong accelerators of enterprise architecture work. This book has focused on SAP LeanIX and compared it with different tools, and it has also provided an inside view into how SAP sets up its architecture practices. A key focus of this book is on creating a practitioner's view of enterprise architecture learned while working with colleagues on SAP Transformation Hub, which serves as the "home" within the Customer Services & Delivery organization. The book explained how we do the following:

- Interact with customers during service delivery in a scalable way.
- Enable other consulting and services teams.
- Work with SAP's internal IT to build the enterprise architecture for SAP as an intelligent and sustainable company.
- Collaborate with SAP development and product engineering, the groups where SAP's software products are designed and built.

Let's take some time to recap what we've learned. This book has eighteen chapters divided into five parts, covering all the key topic areas and enterprise architecture domains based on our diverse experiences in the field. In Figure 18.1, we illustrate the schematic flow of the chapters to help you better understand and connect the dots between the topics.

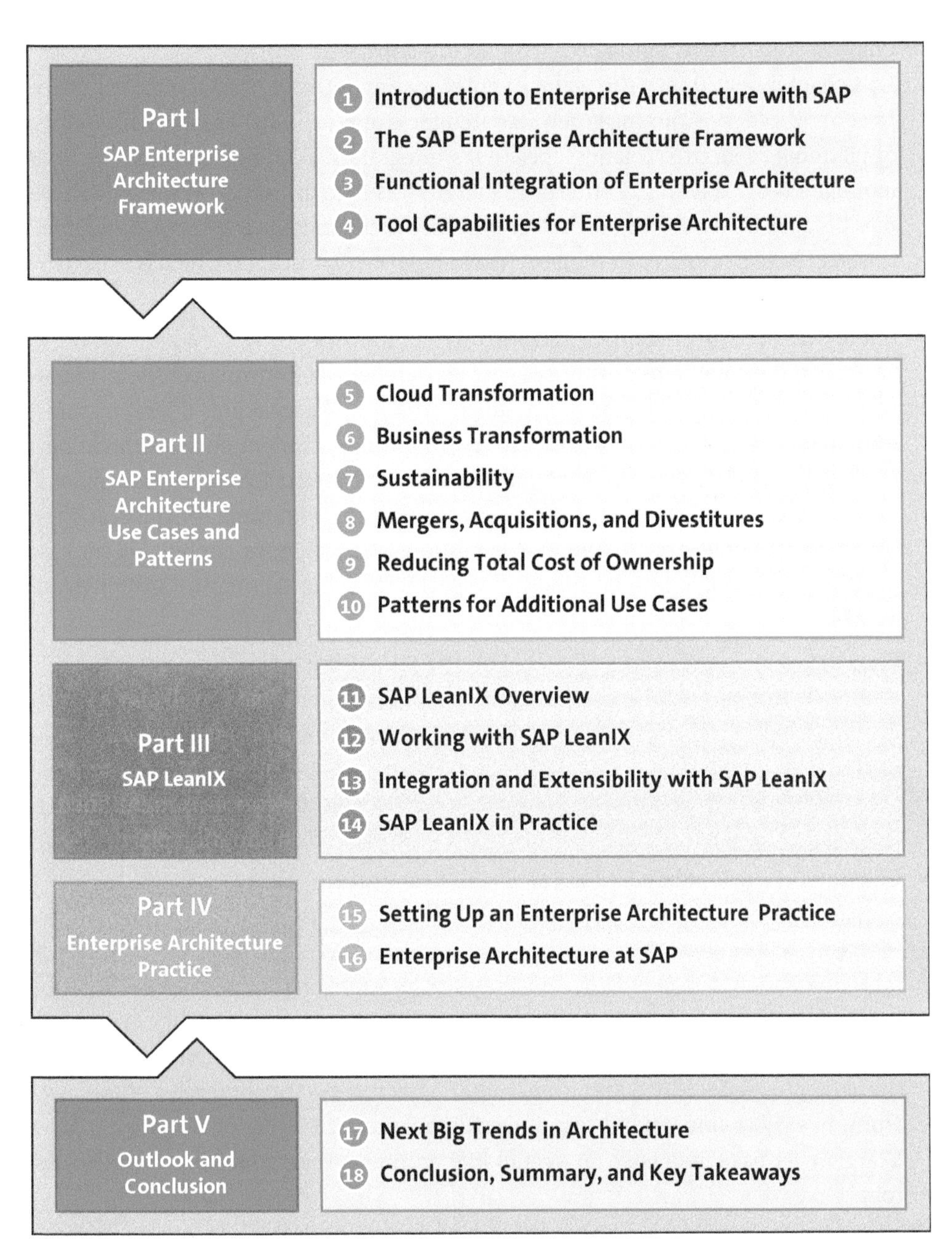

Figure 18.1 Chapter Summary and Connecting the Dots

In Part I, you learned about enterprise architecture as it relates to SAP-centric landscapes, along with the importance of SAP EA Framework. You also experienced a deep dive into various functional integrations of enterprise architecture transformation with SAP. Here are chapter summaries for your quick reference:

- **Chapter 1: Introduction to Enterprise Architecture with SAP**
  This chapter delved into the realm of enterprise architecture and the role of enterprise architects, positioning enterprise architecture as an indispensable cornerstone of planning modern SAP landscapes that's integral for aligning business with IT. It highlighted the role of the enterprise architect as increasingly vital within companies because it blends deep technical expertise with strategic and business acumen to steer the complexities of modern IT and business landscapes, which is especially critical during the transformation planning phase.
- **Chapter 2: The SAP Enterprise Architecture Framework**
  SAP EA Framework helps businesses gain comprehensive control over their IT landscape. It aids companies in making agile decisions and accelerates digital transformation journeys. The framework includes five primary components: methodology, reference content, tooling, practice, and services. The methodology provides a holistic view of the key components of an organization, their interrelationships, and the principles guiding their design and evolution. The reference content offers best practices, methodologies, and industry-specific insights tailored to SAP environments.
- **Chapter 3: Functional Integration of Enterprise Architecture**
  In this chapter, we went through a large variety of functions that can all play a significant role in the enterprise architecture context. But enterprise architecture should not be seen as the strategic function that orchestrates all others. Most enterprise architecture teams won't have the capacity to cover all the concerns mentioned in this chapter, so you need to establish a prioritization and an adoption plan.
- **Chapter 4: Tool Capabilities for Enterprise Architecture**
  Even using the best tools is no guarantee of ending up with good enterprise architecture. Instead, you need good methodology, good content governance, and, primarily, a mature enterprise practice. When selecting a tool, make sure it supports your current needs, but also think ahead: once your maturity improves, you might want to explore how to represent more complex concepts like representation of time or the ability to model and judge alternative architectures. At the other end of the spectrum, an easy-to-understand, less powerful tool that will be adopted more easily by a larger user base may have an overall higher impact than a niche comprehensive tool.

In Part II, you learned about a variety of enterprise architecture use cases and the architecture patterns that you can use to solve common architecture design problems that may arise in each scenario. You can also apply the patterns you use in these scenarios to additional use cases. We provide a summary of the architecture modules we use in each of the use cases in Figure 18.2.

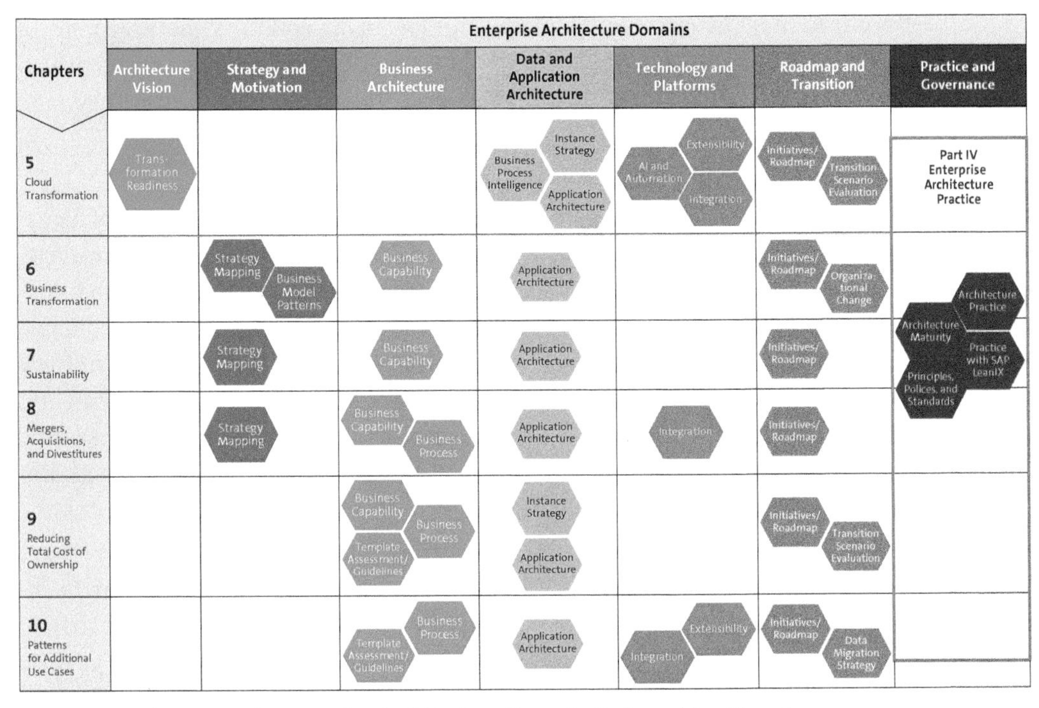

Figure 18.2 Chapter and Enterprise Architecture Domain Deliverables Mapping

Here are some chapter summaries for your quick reference:

- **Chapter 5: Cloud Transformation**
  Cloud transformation is not merely a change in infrastructure but a fundamental reimagining of how businesses operate, innovate, and compete. The heart of this transformation for many enterprises lies in their SAP systems, which form the critical backbone supporting numerous business processes spanning over several SAP and non-SAP applications. The transformation of SAP-centric landscapes into cloud architecture represents an essential opportunity to achieve agility, scalability, and digital innovation.

- **Chapter 6: Business Transformation**
  Business transformation is a comprehensive approach to evolving a company in multiple dimensions to achieve long-term success and adaptability. It encompasses strategic transformation, which redefines business models and market strategies; process transformation, which focuses on optimizing and reengineering core operations for efficiency; organizational transformation, which restructures the company to better align with strategic goals; cultural transformation, which is aimed at fostering a supportive and innovative organizational culture; and digital transformation, which integrates advanced technologies to enhance all aspects of business

operations and customer interactions. SAP EA Framework and the principles of enterprise architecture transformation can deliver a comprehensive perspective on the organization's existing situation, establish the sought-after future state, and devise a strategic roadmap to help the company reach defined objectives.

- **Chapter 7: Sustainability**
  Sustainability is the next big strategic priority for business for the purpose of digital transformation. SAP solutions are enablers of end-to-end sustainability management, and enterprise architecture and enterprise architects play a key role in bringing clarity and direction to our customers to help them become sustainable enterprises. In this chapter, we looked at how a company's sustainability strategy can be integrated into its enterprise architecture to improve operational efficiency, reduce waste and emissions, and promote social and environmental responsibility. We also learned that SAP has many new applications under SAP Cloud for Sustainable Enterprises.
- **Chapter 8: Mergers, Acquisitions, and Divestitures**
  In this chapter, we delved into the critical role of the enterprise architect in facilitating successful mergers, acquisitions, and divestitures (MAD). In the dynamic landscape of modern business, MAD has become an integral strategy for growth, consolidation, and restructuring. However, these activities pose unique challenges to enterprise architecture, and they demand a strategic approach to ensure seamless integration, optimization of resources, and preservation of business value.
- **Chapter 9: Reducing Total Cost of Ownership**
  In this chapter, we explored a case study that applied the elements of TCO and assessed how to provide recommendations for reducing TCO during transformation planning. In many organizations, IT doesn't decide on the purchase of applications; instead, the business units choose specific applications. Often, these applications are managed with a very poor governance structure, which can significantly impact costs and, consequently, the TCO for a large transformation program.
- **Chapter 10: Patterns for Additional Use Cases**
  In this chapter, we examined patterns for additional use cases from an enterprise architecture perspective. We completed our coverage of all use cases by providing real-world architecture decisions from our customers' transformation journeys. However, there are many other use cases—such as scale-out architecture, security architecture, and data architecture—that we can't cover in this book.

In Part III, you learned about SAP LeanIX concepts and the meta model, how to work with them, and the use cases for enterprise architecture transformation. Here are chapter summaries for your quick reference:

- **Chapter 11: SAP LeanIX Overview**
  This chapter introduced the main principles of SAP LeanIX: fact sheets and the relationships among them. The inventory, dashboards, reports, and diagrams are

cornerstones of the SAP LeanIX solution and will be a starting point for exploring and visualizing the fact sheet model.

- **Chapter 12: Working with SAP LeanIX**
  In this chapter, we touched on most of SAP LeanIX's main functions. Note that with agile and frequent innovations, some of these areas will evolve further quickly.
- **Chapter 13: Integration and Extensibility with SAP LeanIX**
  In this chapter, we explored different ways to extend and integrate SAP LeanIX. You can extend a meta model with new tag groups, fields, fact sheet types, and relationships, and you can do this in minutes, without any development complexities. However, we recommend that you not overconfigure the workspace by adding too many fields and tags, which may not have a clear business case or clear accountability in terms of maintenance. More advanced configurations—including the definition of custom key performance indicators (KPIs), integrations, and reports—require deeper technical understanding.
- **Chapter 14: SAP LeanIX in Practice**
  This chapter rounded out our discussion of SAP LeanIX by spotlighting two specific use cases in more detail. The first use case involved approaches to application portfolio management, and the second gave a longer example of how to model transformations and complex roadmaps.

In Part IV, you learned how to set up and improve an enterprise architecture practice, which can be anchored in the organization as a dedicated practice, exist in the form of federated groups, or be an implicit part of initiatives. Here are chapter summaries for your quick reference:

- **Chapter 15: Setting Up an Enterprise Architecture Practice**
  To establish an impactful and successful enterprise architecture practice, you need to integrate it well with different parts of the organization, especially by collaborating with other IT functions like the project management office and perhaps innovation and strategy groups. In addition to organizational accountabilities, striking a good balance between architecture development and governance and establishing clear communication with all relevant teams and stakeholders are key factors that support the success of an impactful, well-accepted architecture practice.
- **Chapter 16: Enterprise Architecture at SAP**
  This chapter explored different facets and touchpoints of enterprise architecture and its strong role in SAP. It offered new perspectives that may differ from those of the companies you have worked for or are currently working with. As discussed in Chapter 15 on setting up an enterprise architecture practice, the federated approach taken within SAP is critical to our ability to enable fluency among the different organizations at SAP. However, this federated approach can sometimes slow down the decision-making process, and it can't keep parts of the organization from diverging from the common federated approach.

Finally, in Part V, we've provided predictions from our perspective (in Chapter 17) and key takeaways (in this closing chapter).

We expect that enterprise architecture will see significant transformations over the next five to seven years, driven by advancements in technology, generative artificial intelligence (AI), and the ever-changing and dynamic business world. We foresee the following eight major trends in enterprise architecture:

1. AI-supported enterprise architecture transformation
2. The enterprise architecture network
3. Live enterprise architecture (enterprise architecture-driven innovation)
4. Clean architecture
5. Global networks on enterprise architecture using metaverse and digital twins
6. Agile and iterative architecture
7. Quantum computing

In Chapter 17, we provided our predictions on when these trends will become relevant, their potential impact, and the likelihood of our forecasts.

We conclude this final chapter with the three key takeaways from this book:

1. **You can use SAP EA Framework and SAP LeanIX to optimize your daily architecture work.**
   The book provides in-depth insights into utilizing SAP EA Framework and SAP LeanIX for effective transformation initiatives. These frameworks serve as foundational tools for managing complex IT landscapes and driving digital transformation. This book provides you with enough guidance on a day in the life of an enterprise architect to help you optimize your daily architecture work.
2. **You can learn from SAP-centric use cases that visualize enterprise architecture in action.**
   These use cases offer practical examples of business transformation, cloud integration, sustainability initiatives, and M&A activities.
3. **You can gain insights into effectively managing an integrated collaborative enterprise architecture practice.**
   The book emphasizes the importance of building an integrated and collaborative enterprise architecture practice within organizations. It highlights strategies for integrating SAP (enterprise) architecture into overall enterprise architecture approaches, making a case for enterprise architecture in the SAP context, and linking enterprise architecture to portfolio management to avoid a negative perception of policing.

In short, the book encapsulates the essence of leveraging enterprise architecture in SAP-driven environments, offering a structured pathway to help organizations navigate digital transformation. Through a synthesis of practical experience and strategic frameworks, it underscores the significance of integrating SAP architecture with broader enterprise architecture practices. The establishment of SAP Transformation Hub within SAP's Customer Services & Delivery organization exemplifies a proactive approach to delivering architecture guidance tailored to strategic customer needs, thereby influencing SAP's transformation journey and roadmap.

Appendix A
Abbreviations and Glossary

Table A.1 lists commonly used abbreviations relevant to enterprise architecture with SAP. Many are used throughout this book, and some are not used but are useful to know as an enterprise architect.

Abbreviation	Full Term
ACL	Access control list
ADM	Application Development Method (TOGAF)
AI	Artificial intelligence
ALM	Application lifecycle management
API	Application programming interface
APJ	Asia-Pacific and Japan
APoV	Architecture point of view
APQC	American Productivity & Quality Center
ATP	Available-to-promise
BPM	Business process management Business process modeling
BPMN	Business Process Model and Notation
CapEx	Capital expenditure
CDO	Chief data officer
CFO	Chief financial officer
CIO	Chief information officer
COE	Center of Excellence
CTO	Chief technology officer
CMDB	Configuration management database
COO	Chief operating officaver

Table A.1 Commonly Used Abbreviations

Abbreviation	Full Term
CRM	Customer relationship management
EA	Enterprise architecture or enterprise architect
EAM	Enterprise architecture methodology
EAF	Enterprise architecture framework
EMEA	Europe, the Middle East, and Africa
EWM	Extended warehouse management
GenAI	Generative artificial intelligence
HR	Human resources
HXM	Human experience management
IaaS	Infrastructure as a service
ISA-M	SAP Integration Solution Advisory Methodology
KPI	Key performance indicator
LLM	Large language model
M&A	Mergers and acquisitions
MAD	Mergers, acquisitions, and divestitures
NFR	Nonfunctional requirements
PaaS	Platform as a service
PMO	Project management office
ROI	Return on investment
OCM	Organizational change management
ODM	SAP One Domain Model
OpEx	Operational expenditure
R&D	Research and development
SaaS	Software as a service
SAP BTP	SAP Business Technology Platform
SAP BTS	SAP Business Transformation Services
SAP CALM	SAP Cloud ALM
SAP RBA	SAP Reference Business Architecture

Table A.1 Commonly Used Abbreviations (Cont.)

Abbreviation	Full Term
SAP RSA	SAP Reference Solution Architecture
SCM	Supply chain management
SoAW	Statement of architecture work
SSB	Solution Standardization Board
SSO	Single sign-on
TBM	Technology Business Management (the TBM Council provides a technology taxonomy)
TCI	Total cost of implementation
TCO	Total cost of ownership
TOGAF	The Open Group Architecture Framework
TSA	Transition services agreement

Table A.1 Commonly Used Abbreviations (Cont.)

Table A.2 provides a selection of key terms relevant to enterprise architecture that we have used throughout this book.

Term	Definition
American Productivity & Quality Center (APQC)	Authority in benchmarking, best practices, process and performance improvement, and knowledge management. The APQC supports decision-making and skill development for organizations worldwide.
Business capability	Business capabilities describe an organization's ability to successfully perform business activities, achieve its business objectives, and deliver value to its customers.
Business goal	Business goals are categorized into strategic, enterprise, performance, and operational levels to align objectives across an organization and link business metrics and processes for implementing value drivers.
Business model patterns	Standardized templates describe the key elements and relationships of a business model, helping organizations design and innovate their business models effectively.
Business process	A business process is the combination of value-adding business activities to fulfill a defined business goal and create a valuable result for a stakeholder.

Table A.2 Glossary

Term	Definition
Business transformation	Business transformation is a fundamental shift in how a company operates, usually aimed at coping with significant changes in the company's market or competitive environment.
Cloud transformation	Cloud transformation is a strategic process of transitioning an organization's operations, processes, data management, and IT resources from a traditional on-premise setup to a cloud-based environment.
Enterprise architecture	Enterprise architecture is a systematic approach to formally model the business strategy, organization, and processes, as well as the IT applications, data, and technology of an enterprise. It derives change opportunities and needs projects and budget, and thus closes the gap between business and IT by providing transparent midterm planning roadmaps for both business and IT.
Enterprise architecture framework	An enterprise architecture framework is a structured approach used to guide organizations through the complex process of aligning business with IT in the area of enterprise architecture.
Reference architecture	Reference architecture is a standardized blueprint that provides a common language and a set of best practices for designing and implementing business and IT architectures.
Requirement	Requirements are functional or nonfunctional business or IT needs that must be clearly identified and documented to plan, manage, and transform an architecture.
Strategic priority	A strategic priority is defined as something that creates, motivates, and fuels change in an organization.
Total cost of ownership (TCO)	Total cost of ownership (TCO) is a value used in procuring information technology that addresses all procurement, capital, and postdeployment costs. It's the complete cost of owning a product, including purchase price, maintenance, and service. For example, the TCO for a car considers the purchase price, fuel, insurance, maintenance, repairs, and depreciation.

Table A.2 Glossary (Cont.)

Term	Definition
Transition services agreement (TSA)	A transition services agreement (TSA) in the context of mergers and acquisitions (M&A) is a contract between the buyer and the seller that outlines the services and support the seller will provide to the buyer during the transitional period following the acquisition. This agreement typically includes specific details about the services or support that the seller will continue to provide to the buyer, such as IT support, human resources, accounting, and other operational functions. The TSA helps ensure a smooth transition and can facilitate the integration of the acquired business into the buyer's operations.
Value driver	A value driver is a measurable short- or midterm objective that has business relevance, as well as a target value and schedule.
Value stream	A value stream is the sequence of activities to add value to a customer from the initial request through realization of value by the customer. While a process focuses on the series of activities, a value stream focuses on its purpose to deliver value and motivates further analysis to reduce waste.

Table A.2 Glossary (Cont.)

Appendix B
The Authors

Anup Das is a chief enterprise architect and the regional delivery head for the SAP Transformation Hub in North America. He leads a team of enterprise architects and is responsible for accelerating business transformations for strategic customers using SAP EA Framework and SAP MaxAttention architecture transformation services for the Customer Services & Delivery board area.

Anup is a graduate of mechanical engineering and post-graduate of marketing management. He has more than 24 years of professional experience in SAP consulting, SAP premium engagements, people management, project management, sales and sales support, and as a trainer and coach. He also led a large team with SAP S/4HANA and supply chain skills as practice head for 7 years.

Peter Klee leads the SAP Transformation Hub within the Customer Services & Delivery board area. SAP Transformation Hub is a global group of enterprise architects supporting SAP MaxAttention customers in their transformation programs by providing architecture planning services. In addition to working directly with customers, SAP Transformation Hub constantly improves services, methodology, and tools working closely with architecture teams in development. In this capacity, Peter co-leads SAP's overall enterprise architecture methodology as part of the cross-architecture program at SAP.

Prior to joining SAP in 2011, Peter led global enterprise architecture teams at Procter & Gamble. He has more than 30 years of professional IT experience.

Peter is based at SAP's corporate headquarters in Walldorf, Germany. He holds a degree in computer science from the University of Karlsruhe and is a passionate long-distance runner in his spare time.

Johannes Reichel is a principal enterprise architect within the SAP Transformation Hub, focusing on global transformation and architecture planning for SAP clients. He is deeply involved in developing the SAP EA Framework and scaling enterprise architecture within SAP.

Prior to joining SAP, Johannes worked as a development advisor on digital transformation in Rwanda and Zambia for GIZ. He is based at SAP's corporate headquarters in Walldorf, Germany. Johannes has more than 10 years of experience in the areas of digital transformation and enterprise architecture.

Johannes holds a bachelor's degree in computational linguistics and Semitic language studies from the University of Heidelberg and a master's degree in sustainable development cooperation from the University of Kaiserslautern. He possesses certifications in TOGAF, SAFe, SAP architecture, and SAP Activate project management.

Index

C

F

G

S

T